Food Processing Waste and Utilization

Because of its high chemical oxygen demand (COD) and sheer volume, waste from food processing has significant potential to pollute land, water, and air. Both environmentally and economically, it is important to properly treat food processing wastes including the recovery of valuable products.

Food Processing Waste and Utilization: Tackling Pollution and Enhancing Product Recovery discusses possible solutions to tackle food waste generation and its further utilization. It addresses process engineering economics, microbiology of waste recycling, biochemical and nutritional aspects of food waste processing. The book includes detailed guidance and case studies about utilization/valorization of food waste.

Key Features

- Covers modern as well as conventional methods of food industry waste utilization.
- Discusses possible solutions to tackle food waste generation and its further utilization.
- Addresses socioeconomic considerations, environmental concerns and discusses regulations related to food processing waste.

Authors of this book are well-recognized researchers in their specific fields who have made important contributions to the knowledge of utilization of different food industry wastes at different levels. This book covers a wide range of breakthroughs in waste management, and is of value for students, research scholars, postdoctoral fellows, and faculties pursuing careers in fields such as Bioprocess Technology, Food Technology, Food Science and Technology, Food Biotechnology, and Fermentation and Bioengineering.

Food Processing Waste and Utilization

Tackling Pollution and Enhancing Product Recovery

Edited by
Sanju Bala Dhull, Ajay Singh, and
Pradyuman Kumar

CRC Press
Taylor & Francis Group
Boca Raton London New York

CRC Press is an imprint of the
Taylor & Francis Group, an **informa** business

First edition published 2023
by CRC Press
6000 Broken Sound Parkway NW, Suite 300, Boca Raton, FL 33487-2742

and by CRC Press
4 Park Square, Milton Park, Abingdon, Oxon, OX14 4RN

CRC Press is an imprint of Taylor & Francis Group, LLC

ISBN: 9781032062945
ISBN: 9781032075617
ISBN: 9781003207689

DOI: 10.1201/9781003207689

Typeset in Times
by Newgen Publishing UK

Contents

Preface

Food industry waste is a serious economic, environmental, and social problem. The production and processing of food commodities generate tonnes of unavoidable food waste and by-products every year. A lot of money is spent in handling this food industry waste attributable to landfill costs and other waste disposal routes. Apart from its significant quantities, the physico-chemical characteristics of the various food industry waste and by-products denote that there is immense potential for their reuse, recycle, and valorization through different processes. As a result of modern food-waste processing practices, the knowledge generated about waste utilization is huge and relevant to different research areas such as food, nutrition, pharma, nutraceuticals, textiles, and environmental studies. Therefore, with this book, the massive food processing and scientific community will have rapid access to the related advances to their field. *Food Processing Waste and Utilization: Tackling Pollution and Enhancing Product Recovery* is a reference-cum-textbook written in crisp and scientifically authentic language for teachers, scientists, researchers, students, industry managers, as well as all those who have a stake in food processing waste management and utilization. It presents the latest information on the problems of wastes generated from various food industries along with suitable utilization practices. The contents has been divided into 18 chapters, namely as follows: Utilization of food industry wastes, Recovery of valuable compounds from food industry waste using ultrafiltration, Waste management of fruits and vegetable industry, Recovery and utilization of protein from food industry waste, Fat extraction from food industry waste, Treatment of fatty effluents, Conservation of bone to edible products, Processing of coffee and tea waste, By-products from malting, brewing, and distilling, Pectate as a gelling agents in foods, Dairy waste and its valorization, Use of microbiological agents in upgrading waste for feed and food, Uses of enzymes in food industry waste utilization, The treatment of dairy industry waste, Aerobic processes for the treatment of wheat starch effluents, Anaerobic treatment of food processing wastes and agricultural effluents, Fishery by-product valorization, Food industry waste: Potential pollutants and their bioremediation strategies. Different segments of the food processing industry have been dealt with separately by specialists with respect to their waste management technology. Special emphasis has been placed on potential methods to valorize food industry waste meritoriously, to reduce biodegradable harmful compounds, recycle and recover different value-added compounds, which can be used to enhance the quality of food products as well as preparation of novel foods. The profitable utilization of food industrial wastes would not only earn extra profits for the industry but would also reduce the pollution load in the environment. The book's particular characteristic is that it covers a wide range of breakthroughs in waste management, from the earliest technology to the most recent advancements and prospective study fields. Additionally, it is of value for students, research scholars, postdoctoral fellows, and faculties who pursue academic careers in Bioprocess Technology, Food Technology, Food Science and Technology, Food Biotechnology, Fermentation and Bioengineering etc. To everyone's benefit, it is intended that the book will shed light on food waste management technologies in the fast-paced food processing business.

Authors of this book are well-recognized researchers in their specific fields. They have all made important contributions to the knowledge of utilization of different food industry waste at different levels. We profusely thank them for the time dedicated to the writing of this book, which we are sure will help readers to understand different processes and methods used in valorization of various food processing waste and their application in generation of valuable compounds. We also warmly thank our family members especially our children for the time taken from them for the preparation of this book.

Sanju Bala Dhull
Ajay Singh
Pradyuman Kumar

About the Editors

Sanju Bala Dhull, Ph.D. is an associate professor in the Department of Food Science and Technology, Chaudhary Devi Lal University, Sirsa, having more than 14 years of teaching and research experience. Her areas of interest include characterization and modification of biomolecules such as starch, gums etc., edible films, hydrogels, nanoparticles, nanoemulsions, and new product development. She has published more than 40 research papers, 3 books and 20 book chapters in books of national and international repute. She has presented more than 20 research papers at various national and international conferences. She is a life member of the Association of Food Scientists and Technologists (India) and Association of Microbiologists of India. She also serves as editorial board member and reviewer of national and international journals.

Ajay Singh, Ph.D. is an assistant professor in the Department of Food Technology, Mata Gujri College (An autonomous body, Estd. 1957), Punjab, India. He has 8 years of teaching and research experience in the field of food sciences and dairy domain. His areas of interest include the extraction and identification of bioactive compounds from different plant sources, formula refining and new product development. He has published 30 research papers and 28 book chapters in journals and books of national and international repute. Recently, he has edited five books for an international publishing house. He is a reviewer for several national and international journals.

Dr. Pradyuman Kumar is a professor in the Department of Food Engineering and Technology at Sant Longowal Institute of Engineering and Technology, Longowal under the Ministry of Education. He has 24 years of experience in teaching and research. He is a member of the academic board of Food Science and Technology in various institutions. He has published more than 200 papers in international and national journals and conferences and 14 book chapters. He has guided 9 Ph.D. and 23 MTech dissertations. Dr. Kumar has been awarded Young Scientist by AFSTI Mysore and won five best paper awards at international and national conferences. He has visited many countries for paper presentations.

Contributors

Mehwish Arshad
University of Agriculture
Faisalabad, Pakistan

Saba Anwar
University of Agriculture
Faisalabad, Pakistan

Shalini Arora
Lala Lajpat Rai University of Veterinary and
   Animal Sciences
Hisar, Haryana, India

Prajya Arya
Sant Longowal Institute of Engineering and
   Technology
Longowal, Punjab, India

Gauri Harish Athawale
MITADT University
Pune, India

Neha Rani Bhagat
Defence Institute of High Altitude Research
   (DIHAR)
Ladakh UT, India

Bindu
Lady Irwin College, University of Delhi
New Delhi, India

Om Prakash Chauhan
Defence Research and Development
   Organisation
Siddharthanagar, Mysore, India

Navnidhi Chhikara
Guru Jambheshwar University of Science and
   Technology
Hisar, Haryana, India

Rajkumar Arjun Dagadkhair
ICAR
Puttur, Karnatka, India

Mayur Suresh Dhale
ICAR
Karnal, Haryana, India

Arup Giri
Baba Mastnath University
Rohtak, Haryana, India

Simmi Goel
Mata Gujri College Fatehgarh Sahib
Punjab, India

Pankaj Kumar
Dolphin PG Institute of Medical Science and
   Research
Dehradun, India

Younis Ahmad Hajam
Sant Baba Bhag Singh University
Jalandhar, Punjab, India

Sumita S. Kadian
Department of Food Technology GJUS&T
Hissar, Haryana, India

Abhimanyu Kalne
Indira Gandhi Krishi Vishvavidyalya
Raipur, Chattisgarh, India

Gurpreet Kaur
Mata Gujri College Fatehgarh Sahib
Punjab, India

Muhammad Issa Khan
University of Agriculture
Faisalabad, Pakistan

Usman Mir Khan
University of Agriculture
Faisalabad, Pakistan

Mohd. Kashif Kidwai
Chaudhary Devilal University
Sirsa, Haryana, India

Anusha Kishore
ICAR
Karnal, Haryana, India

Rajesh Kumar
Himachal Pradesh University
Shimla, India

Maria Maqsood
University of Agriculture
Faisalabad, Pakistan

Chander Mohan
NDRI
Karnal, Punjab, India

Kiran Bala Nain
Chaudhary Devi Lal University
Sirsa, Haryana, India

Neeraj
Jharkhand Rai University
Ranchi, Jharkhan, India

Mohona Munshi
Vignan Foundation of Science, Technology
    and Research
Vadlamudi, Guntur, Andhra Pradesh, India

Shreya Panwar
Chaudhary Charan Singh Haryana Agricultural
    University
Hisar, Haryana, India

Priyanka
Mata Gujri College Fatehgarh Sahib
Punjab, India

Ubaid ur Rahman
University of Management and Technology
Lahore, Pakistan

Shweta Jeevan Raichurkar
MITADT University
Pune, India

Tilak Raj
Defence Institute of High Altitude Research
    (DIHAR)
Ladakh UT, India

Dhiraj Singh Rawat
Himachal Pradesh University
Shimla, Himachal Pradesh, India

Kritika Rawat
Mata Gujri College Fatehgarh Sahib
Punjab, India

Pawan Kumar Rose
Chaudhary Devilal University
Sirsa, Haryana, India
Siddharthanagar, Mysore, India

Amna Sahar
University of Agriculture
Faisalabad, Pakistan

Mukul Sain
National Dairy Research Institute
Karnal, Haryana

Aysha Sameen
University of Agriculture
Faisalabad, Pakistan

Pardeep Kaur Sandhu
Mata Gujri College Fatehgarh Sahib
Punjab, India

Loveleen Kaur Sarao
PAU
Ludhiana, Punjab, India

Anil Dutt Semwal
Defence Research and Development
    Organisation
Siddharthanagar, Mysore, India

Shalagha Sharma
Sobhit Deemed University
Meerut, U.P., India

Ritu Sindhu
Chaudhary Charan Singh Haryana Agricultural
    University
Hisar, Haryana, India

Richa Singh
ICAR
Karnal, Haryana, India

Muhammad Usman
University of Agriculture Faisalabad
Pakistan, India

Janifer Raj Xavier
Defence Research and Development
    Organisation

1 Utilization of Food Industry Wastes

Sumita S. Kadian,[1] *Loveleen Kaur Sarao,*[2]
and Chander Mohan[3]

[1]Department of Food Technology, GJUS&T, Hisar, Haryana, India
[2]Department of Plant Breeding and Genetics, PAU Ludhiana,
Punjab, India
[3]Dairy Technology Division, NDRI Karnal, Punjab, India

CONTENTS

DOI: 10.1201/9781003207689-1

1.1 INTRODUCTION

Globally, India is ranked fifth when it comes to production, consumption and export of processed food. It is the leading producer of some agricultural and dairy based fresh produce. This makes it possible for India to expand its food processing sector (Khedkar and Singh 2018). Huge quantities of by-products and wastes are generated by the agricultural production and agro industrial processing. The food waste results because of damage occurring at the transport, storage and processing stages. To date, there is an increasing trend of processed and frozen food products. As per the ongoing trend, processing of fresh harvests such as fruit juices, syrups, nectars, concentrates, canned products, etc., add massively to waste generation. The percentage of global food losses and wastes per year in different food industries has been depicted in Figure 1.1. The food processing wastes affects the environment in a negative way. They affect the economics and the social aspects of a nation. These wastes contribute towards greenhouse gases (Girotto et al 2015). The food recovery hierarchy has been depicted in Figure 1.2.

There are certain challenges faced by food industries related to waste utilization. These challenges have been depicted in Figure 1.3. The appropriate usage of food wastes in the form of raw materials or as food additives can assist the industry in gaining economic benefits. This will also help in tackling issues related to nutrition. Such products will have a beneficial effect on health. Ultimately, this will lead to avoidance of the mismanagement of the wastes and reduction in the negative impact on the environment. The current trend is to innovate in order to have zero waste. This is achieved by utilizing the wastes as raw material to produce new products. These efforts assist in the Millennium Development Goals, the Sustainable Development Goals, the Post 2015 Agenda and the Zero Hunger Challenge. In parts of the world, which are still under developed or developing, using food industry wastes for formulating new food products will certainly be beneficial for local communities.

1.2 MANAGING FOOD WASTES PRODUCED BY FOOD PROCESSING INDUSTRIES

The food industry processes a variety of food products and produces enormous wastes. These wastes can be utilized in several ways. Table 1.1 depicts the different types of wastes produced by various food industries and their main uses. Some of them have been enlisted below.

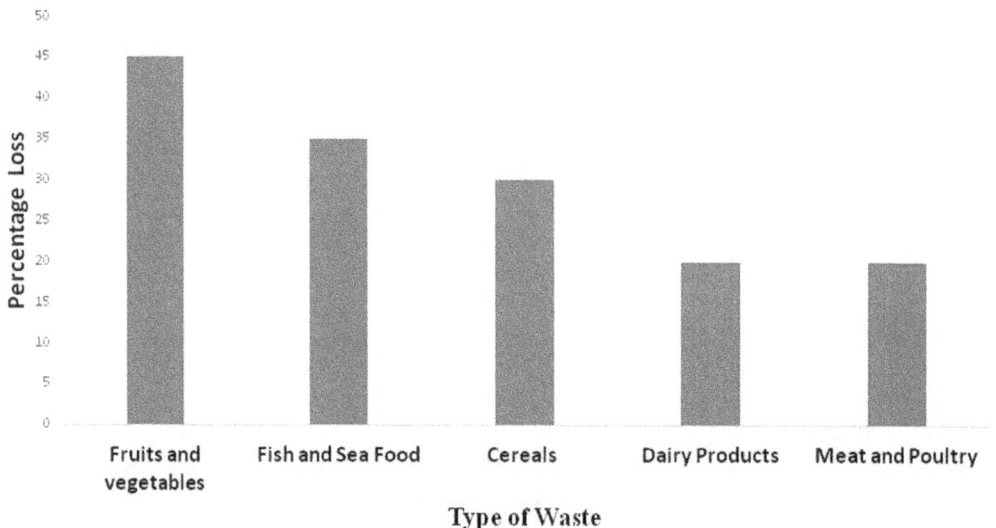

FIGURE 1.1   The percentage of global food losses and wastes per year in different food industries.

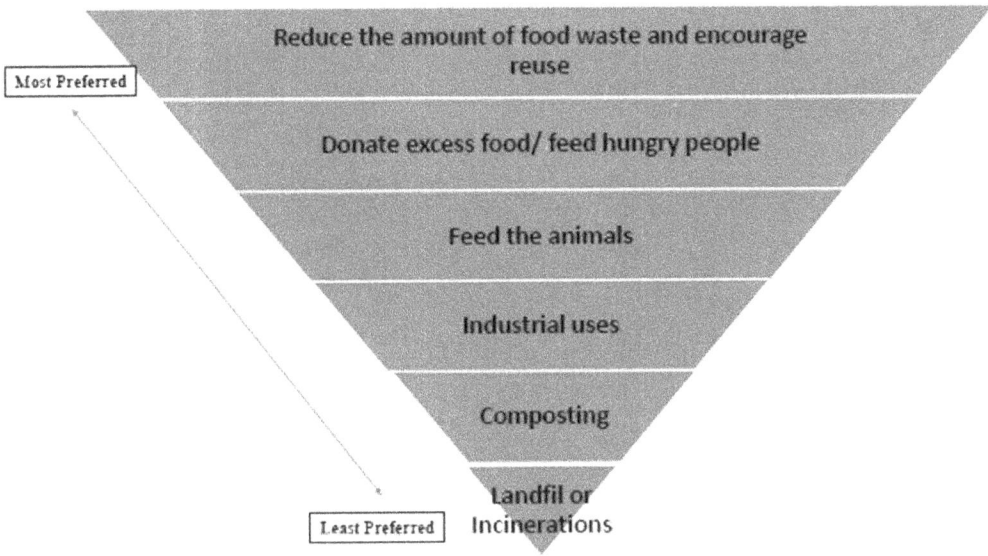

FIGURE 1.2 The food recovery hierarchy.

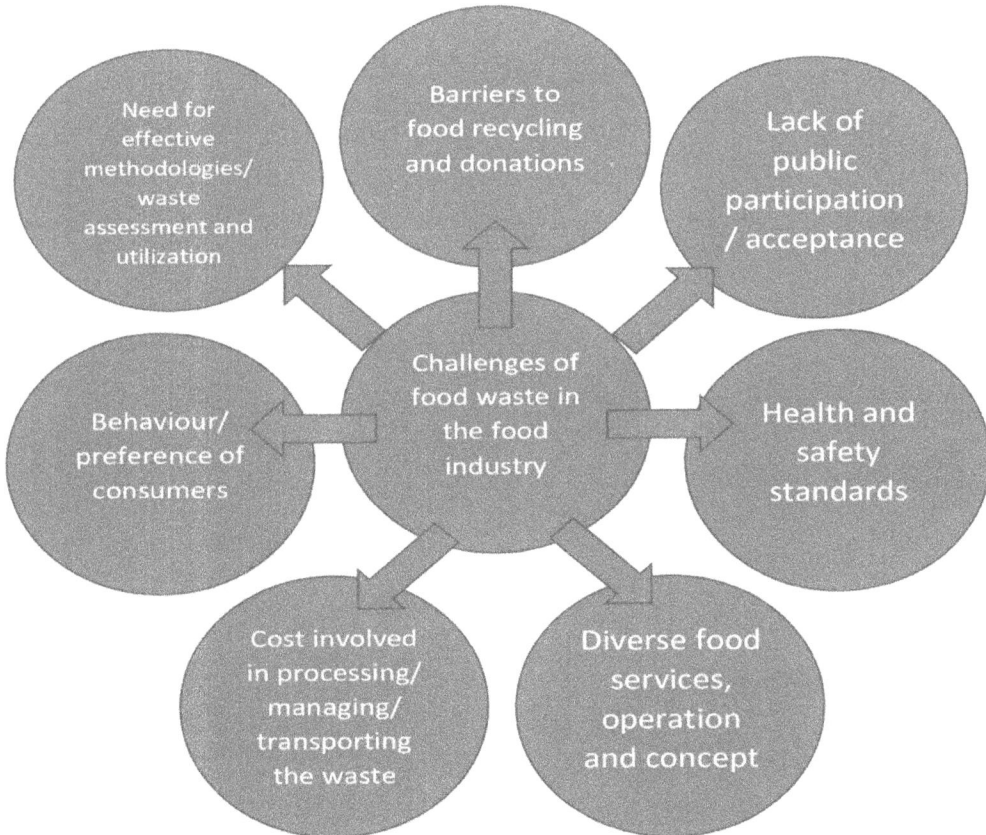

FIGURE 1.3 The challenges related to food waste in the food industry.

TABLE 1.1
Different Types of Wastes Produced by Various Food Industries and Their Main Uses

Origin of Food Waste	Type of Food Waste	Broad Usage
Cereal Processing Industry	bran, germ, endosperm, husk and the broken kernels	Refined foods, fermentation, oil production, thickening agent, fuel, abrasive packaging, production of xylitol, pullulan and polyhydroxyalkonates; biohydrogen refinery; bioethanol and electricity production
Milk Plants/Milk Processing Units/ Dairies	Whey, buttermilk, skimmed milk, curd pieces, whey permeates, effluent water, spoiled products, spilled products	Fermentation to produce ethyl alcohol, vitamins, lactic acids, Baker's yeast, SCP and xanthan gum; concentrated to produce whey protein, dried whey and lactose; Pasteurization of the whey could be done to obtain whey cream and pasteurized sweet whey; production of edible casein; production of caseinates and casein hydrolysates; production of single cell proteins
Fruits and Vegetable Processing Units	Peels, seeds, juice, pulp, pomace, empty pods, damaged fruits and vegetables, spoiled/rotten fruits and vegetables, wash water, discarded fruits after sorting	Manure; feed for livestock; production of value added products such as pectin, phenolics; production of essential oils; production of citric acid, ethanol, hydrocolloids, tartrates, dietary fiber, anthocyanins etc.
Meat Industries/Meat Shops	Bones, tendons, skin, blood, animal fats, internal organs	Used for sports equipments, leather, cosmetic products, glue and edible gelatine; manufacturing medicines; producing meat and bone meal; synthesis of vitamins; insulin extraction; biofuel production
Poultry Units	Feathers, undeveloped eggs, carcasses, offal, bones, gizzards, intestines	Feathers are used for animal feed, controlling erosion, providing thermal insulation, biodegradable composites, filtration units and fabrics; biofuel production
Coffee Industry	Used coffee grounds, husks, skin, pulp, mucilage, parchment, silverskin, spent coffee, coffee pulp, waste water	Detailed usage depicted in Table 1.2
Sugar Industry	Pressmud, bagasse, bagasse fly ash, sugar cane trash, sugar beet mud, sugar beet pulp, molasses, waste waters	Growth mixtures for earthworms; vermicomposting; fertilizers; a fuel for generating electricity; produce paper and fiber, cattle feed, potting media; value added products such as pigments, enzymes, amino acids, and drugs; ethanol production; mulch/ soil conditioner; adsorbents; removal of heavy metals; replacement for silica; organic manure; soil ameliorates; biocomposting; controlling nematodes; enhancement in the yield of root and the quality; fertigation; ruminant feed

TABLE 1.1 (Continued)
Different Types of Wastes Produced by Various Food Industries and Their Main Uses

Origin of Food Waste	Type of Food Waste	Broad Usage
Canned Food Industry	Spillage of sauces, brines and oil, condensate produced during precooking, washwater	Tuna condensate can be utilized as a cheap substrate to produce polyhydroxyalkanoate
Breweries/Wineries/ Distilleries/Malting Industries	Brewery mash, Brewers' spent grain, spent hops, and surplus yeasts, pomace, lees, wastewater, undesirable flavor compounds, distillery stillage	Soil conditioning; source of polysaccharides and volatile fatty acids natural antioxidants, including polyphenols and other bioactive compounds of interest to the pharmaceutical, cosmetic and food industries; feed and food additives; extraction of bioactive compounds; valorization; production of value added products; microalgae production, biofuel production, extraction of proteins, polyphenolic, antioxidative substances
Juice Industry/Non-alcoholic Beverages	Seeds, peels, pomace, juice, wash water, stems	Bioethanol, biogas, and biodiesel production
Baking Units/Bakeries	Dough, flour dust, sugar dust, burnt biscuits, broken biscuits, burnt loaves or rejected loaves, market returned old bakery products	Cattle feed; compost production; bioethanol production; succinic acid production
Oil Industries	Wastewater, organic solid waste (i.e., seeds and husks) and inorganic residues	Valorization; water reuse after prelimnary treatment; microalgae production; biofuel production
Confectionary/Candy Industry	Rejected products, starch, sugar, wastes of cleaning liquids, wastewaters, defective confectionary/candies	Re-using the wastewater after treatment; wastewater can be used for irrigation of agricultural land or gardens; washing of machines and equipment washing
Starch Industry	Waste starch resisues, waste wasters, effluents	Biogas production; protein feed production and extraction of pectin and dietary fiber; animal feed production; microbial processing; fermentation
Tea Industry	Tea leaves, buds and tender stems of tea plants, tea dust, tea sweepings, tea fluff, tea fiber or tea stalks	Production of instant coffee; feed for poultry, pigs and fish; used as bionutrient and biofertilizer; obtaining bioactive chemical components; auto gasification of tea waste
Cocoa Industry	Cocoa pod husks, cocoa bean shells and cocoa sweating, cocoa biomass	Extraction of useful componenets; pharmaceutical, cosmetics and agricultural industries; substitute for cocoa butter; bran powder production

1.2.1 MANAGING THE WASTES PRODUCED BY CEREAL PROCESSING INDUSTRIES

Milling of the common cereals such as the wheat, rice, barley, oat and maize is carried out keeping in mind the sensory preferences of buyers. The wastes generated by the processing of cereals are hazardous for the environment as they affect biological oxygen demand (BOD) and chemical oxygen demand (COD) levels (ElMekawy et al 2013). Some of the wastes generated by the cereal processing industry are bran, germ, endosperm, husk and broken kernels. Bran is utilized in refined foods. It is a source of dietary fibers, vitamins and minerals. Rice bran oil is good for the heart. Starch is extracted from the endosperm and is used for thickening desserts and sauces. Even the cereal husk has several uses. It finds use as a fuel, abrasive and in packagings (The Rice Association 2015). Corn is milled using either wet or dry processes. Subsequent to removal of germ and starch, protein is obtained as the major by-product. It is utilized as corn gluten.

Even the microbe-based fermentation utilizes wastes of the cereal processing industry. They are used for the production of certain organic acids such as lactic acid and citric acid. Brewer's spent grain (BSG) is utilized in *Aspergillus niger* fermentation for producing citric acid. Corn steep liquor (CSL) and BSG have been tested to produce lactic acid. Brewer's spent grain and corn steep liquor and rice bran have been used to produce xylitol, pullulan and polyhydroxyalkonates, respectively (Khedkar and Singh 2018). Wheat starch and barley malt by-products have been used in biohydrogen refineries.

The primary biomass for producing bioethanol is starch rich agro industrial by-products. These include wheat bran, rice hull, CSL and other corn milling by-products. Generation of bioelectricity is carried out by transforming lignocellulosic biomass residues into electricity via microbial fuel cell (MFC) (ElMekawy et al 2013).

1.2.2 Managing the Wastes Produced by Fruit and Vegetable Processing Industries

About 55 million tonnes of waste is generated by the processing, packaging, distribution and consumption of fruits and vegetables in India, China, Philippines and the USA (Khedkar and Singh 2018). A major part of this waste is dumped in rivers or in landfills. This leads to a negative impact on the environment. The best method to use these wastes is to utilize them as manure or as feed for livestock. They could also be used as alternative sources of energy. Even value added products can be developed from them. These are discussed below.

1.2.2.1 Utilization of Fruit and Vegetable Processing Wastes as Livestock Feed

To feed lactating animals, banana and citrus wastes can be used. The dried ripe peels of banana and citrus are mixed with wheat straw or the broiler litter to feed the lactating animals. The dried wastes of pineapple obtained after extraction of juice and the pomace of the bottle gourd juice could be utilized for replacing the roughage portion in the diet of ruminants. Fresh leaves of cabbage and cauliflower, and empty pea pods, are all a source of sugars, protein and certain micronutrients. These could be utilized as livestock feed along with cereal straws. These could be used after drying or ensiling. The reproductive system of cows could be improved by feeding them cull carrots. Even horses can be fed the flakes of carrot or dehydrated carrots. Lactating cows could be fed cull potatoes. Adult buffaloes can be fed on dried tomato pomace (Khedkar and Singh 2018).

1.2.2.2 Production of Value Added Products from Fruit and Vegetable Processing Wastes

Fruit and vegetable peels, seeds, pomace, etc., are rich in minerals, sugars, organic acids, fibers and certain bioactive compounds. Pectin is a soluble dietary fiber. It is extracted from citrus peels, guava and apple pomace. It is utilized in processed foods and beverages. Essential oils are also extracted from the peels of citrus. They are widely used in the pharmaceutical sector, cosmetic industries and beverage industries as well. Certain bioactive components are also extracted from fruit and vegetable wastes. These include phenolic compounds such as phenolic acids, lignans, lignins, tannins and simple phenolics. These compounds exhibit antibacterial, antioxidant, anticancer and antimutagenic activities. One of the richest sources of polyphenols, minerals and dietary fibers is apple pomace. The main isolated components are catechins, phloretin glycosides, hydroxycinnamates and quercetin glycosides.

An enormous amount of waste is generated when wine is produced from grapes. These wastes could be utilized to produce grape seed oil, citric acid, ethanol, hydrocolloids, tartrates, dietary fiber, anthocyanins, etc. The grape pomace consists of phenolic components. There are anthocyanins, flavonol glycosides, catechins, phenolic acids, alcohols and stilbenes as the principal phenolic constituents of grape pomace (Djilas et al 2009).

1.2.2.3 Managing the Wastes Produced from the Fruit/Vegetable Juice Industry

The wastes from the fruit/vegetable juice industry includes seeds, stems, peels, pulp, pomace, wash waters, etc. Juice industry wastes can be utilized for producing biofuels. This conversion has a lot of potential for being an alternative renewable energy source such as bioethanol, biogas and biodiesel (Sevil et al 2018).

1.2.3 Managing the Wastes Produced by Dairy Industries

One of the main food processing industries is the dairy industry; a large volume of wastewater is generated by it. This wastewater is produced in the milk processing section. The manufacturing of butter, ghee, cheese and skimmed milk powder also produces wastewater. The main by-products obtained after the production of these products are butter milk, whey and skimmed milk. The largest portion of the dairy industry by-products is comprised of whey. Whey is generated while producing cheese, chhana, paneer and casein (Gupta 2008). Owing to its high BOD, discharging whey creates environmental hazards. The high BOD is attributed to the presence of lactose. There is a lot of focus on whey protein and lactose. Whey can be used in different ways as discussed below.

  i. Whey could be used for fermentation to produce ethyl alcohol, vitamins, lactic acids, Baker's yeast, SCP and xanthan gum.
  ii. Whey protein could be concentrated to produce whey protein, dried whey and lactose
  iii. Pasteurization of whey could be done to obtain whey cream and pasteurized sweet whey

Ultrafiltration, reverse osmosis, electrodialysis are the modern industrial membrane processing techniques. These are used for the production of whey derivatives. These whey derivatives have different functional and nutritional properties. The use of whey for separating proteins from permeate consisting of lactose is done via ultrafiltration. Focus has shifted on the concentration of whey solids by using ultra filtration and reverse osmosis methods. Based on its nutritional constituents, the whey protein concentrate finds a place in international markets.

Skimmed milk is utilized for the production of edible casein. It finds use in dairy and other food products. Casein is also utilized for the production of caseinates and casein hydrolysates. It is also utilized in glue, textile, fibers, rubber, paints, sizing, etc. Single cell proteins are produced by utilizing whey. This conversion has benefits, as it is a simple process. There is considerable BOD reduction in the process. The lactose conversion into yeast biomass takes place. In several dairy units, sugars, proteins and minerals are recovered, which are present in the waste effluents. Either crystallization, evaporation or spray drying is performed to achieve this recovery (Gupta 2008).

1.2.4 Managing the Wastes Produced from the Meat, Fish and Poultry Industries

In terms of livestock health, India stands first. This industry contributes more than 12 % GDP. If animal by-products are left unused, many pathogenic health issues will arise and create major complications to the health of local inhabitants. If the wastes from livestock industries is used efficiently, it can have an impact on economics as well as the environmental status.

Meat industries produce a lot of waste at slaughter houses. The skin and hides of animals are economically valuable. They are used in the industries dealing with sports equipment, leather, cosmetic products, glue and edible gelatin. The organs and glands of animals are utilized as food. They even find use in the medicine industry. Meat and bone meal is produced by using animal organs and bones. Gelatin is also manufactured by using bones. The brain, nervous system and spinal cord are a source of cholesterol. It is a raw material for vitamin D3. B12 is obtained from the liver extracts of cattle and pigs (Jayathilakan et al 2012).

Extraction of insulin is done from pancreas. It is used for the treatment of diabetes. Proteins, minerals and fats are obtained from fish wastes. Fish oil, fish protein hydrolysate and chitosan are the products obtained from fish waste having economic value. Massive amount of by-products generated from poultry industries as part of dressing and preparation include; feathers, claws beaks and viscera. They create environmental hazards when left untreated. Feathers find uses in several areas. They are used in animal feed, controlling erosion, providing thermal insulation, biodegradable composites, filtration units and fabrics. The feathers upon hydrolysis are employed for biofuel production (Jayathilakan et al 2012). It also finds its application as biofertilizers among agro industries due to its richness in nitrogen; in addition to this, feathers also find their way into pig feed and aquaculture as fish feed to raise the biomass.

1.2.5 Managing the Wastes Produced from the Coffee Industries

One of the most important agricultural commodities around the world is coffee. Industrially, its processing steps direct the conversion of raw coffee beans to the finished aroma rich coffee powder through succession of drying, fermentation and milling. As a result of which, a large amount of waste of utmost importance is generated (Lenka et al 2017). The primary solid by-products obtained from the cultivation and preparation of coffee are spent coffee grounds, by-products of coffee fruit (coffee cherry) and processing of coffee beans (coffee husks, peel, pulp). Enormous quantities of contaminated waters are also generated in various washing steps. These have high carbon load. Hence, they affect the environment. Besides these, small quantities of defective green coffee beans and coffee tree leaves during harvest also contribute to the waste. Based on the method opted for coffee processing, various by-products may be produced. These include the following (Cruz 2014):

 i. pre-roasting coffee by-products;
 ii. dry processing (coffee cherry husks);
 iii. semi-dry and wet processing (coffee pulp);
 iv. post-roasting coffee by-products (coffee silverskin, spent coffee grounds).

Several methods have been employed for the re-utilization of coffee cherry husks. They can be used in several ways, as depicted in Table 1.2.

The coffee husk is gluten free and can be used for people who are allergic. As it has a high quantity of fermentable sugar, coffee husk is a useful substrate for mold, yeast and enzyme production, owing to its high amount of fermentable sugars (Bondesson 2015). Coffee pulp can be used in many ways (Table 1.2); viz., for food preparation, coffee pulp can be converted into a nutritious flour, used in lots of baking or cooking formulations, and can also be added to smoothies, soups or sauces (Markham 2016).

Coffee silverskin is also known as chaff. It is the primary residue of the coffee industry. It is released during roasting. This happens in the case where leaves are not polished before shipping. Documentation regarding the reuse of coffee silverskin is considerably less. Some researchers have documented that it contains antioxidants and dietary fibers. Hence, it may be considered as a new potential functional ingredient (Cruz 2014). In addition to this, coffee silverskin can be utilized as a support and nutrient source during the fructooligosaccharides and β-fructofuranosidase production by *Aspergillus japonicus* under solid-state fermentation conditions (Table 1.2). It can be used as raw material to produce fuel bioethanol (Cruz 2014). Magnetically modified coffee silverskin has been tested for the removal of xenobiotics from wastewater (methylene blue) (Zuorro et al 2013). Owing

TABLE 1.2
Utilization of Coffee Wastes

Coffee waste	Utilization	Reference
Coffee cherry husks	substrate for biogas	Ulsido et al 2016
	alcohol production	Sahu 2014
	biosorbents for cyanide	Gebresemati et al 2017
	biosorbents for the removal of heavy metals from aqueous solutions	Oliveira et al 2008
	biosorbents for the removal of dyes from aqueous solutions	Ahalya et al 2014
	biosorbent for defluoridation of water	Getachew et al 2015
	biosorbent for lead	Berhe et al 2015
	for preparing ion exchange material	Lenka et al 2017
	converted into fuel pellets	Lenka et al 2017
	extracted for bioactive substances recovery	Cruz 2014
	substrate for edible mushrooms production	Alemu 2015
	composting	Nguyen et al 2013
	smoothies, granolas and juices.	Lenka et al 2017
	energy drinks	Lenka et al 2017
	energy bars	Lenka et al 2017
Coffee husk	allergy friendly product	Bondesson 2015
	substrate for the mold, yeast and enzyme production	Bondesson 2015
Coffee pulp	in mushroom production	Martínez-Carrera et al 2000
	composting	Ulsido and LI 2016
	biogas production	Corro et al 2014
	bioethanol production	Menezes et al 2013
	converted into fuel briquettes or pellets	Cubero-Abarca and Moya 2014
	enzymes production such as pectinases or cellulases	Boccas et al 1994
	producing nutritious flour	Markham 2016
	utilized for baking or cooking	Markham 2016
	added to smoothies, soups, or sauces	Markham 2016
Coffee silverskin	potential functional ingredient	Cruz 2014
	support and nutrient source during the fructooligosaccharides and β-furctofuranosidase production	Cruz 2014
	raw material to produce fuel bioethanol	Cruz 2014
	removing xenobiotics from wastewater	Zuorro et al 2013
	used in anti-aging cosmetics and dermaceutics	Rodrigues and Oliveira 2016
Spent coffee grounds (SCG)	source for biodiesel production	Misra et al 2008
	bioethanol production	Mussatto et al 2012
	production of fuel pellets	Haile 2014
	as burning fuel in the industrial soluble industry directly	Cruz 2014
	production of reusable cups	Woolsey 2016
	production of spirit beverage	Mussatto et al 2011
	as source of sugars	Mussatto et al 2011
	as composting material	Santos et al 2017
	as a sorbent for metal ions removal	Anastopoulos et al 2017
	as an adsorbent for dye removal	Pavlovic et al 2016
	as a substrate for mushroom production	Laura 2016
	as a source of natural phenolic antioxidants	Panusa et al 2013
	as a biomaterial in the pharmaceutical industry	Chi 2016
	in the food industry and in the polymer industry	Chi 2016

to its antioxidant content, coffee silverskin aqueous extract finds use in anti-aging cosmetics and dermaceuticals (Rodrigues and Oliveira 2016).

Spent coffee grounds (SCGs) are the waste produced from brewing coffee. Generally, coffee brews are prepared using Arabica coffee or Arabica/Robusta blends. These could be obtained from a single or different geographical origins. These are available to consumers as roasted beans, whole or ground, or even as instant/soluble coffee. Hence, the term "spent coffee grounds" includes soluble coffee industry wastes as well as the ones produced after brewing at cafeterias or at home. The uses of SCG have been enlisted in Table 1.2.

The amount of coffee wastewater varies from unit to unit based on the methods being used. Coffee wastewaters are high in organic loadings. They exhibit a high acidity. When processing of washed or semi-washed coffee is done in huge quantities, untreated effluents highly exceed the self-purification capacity of natural waterways. To tackle the pollution potential of processing wastewaters, an understanding of wastewater constitution clearly is inevitable for designing a feasible treatment system. This holds true especially when expanding wet coffee processing or setting up new large-scale processing operations. The treatment of wastewaters should be considered practically (Von 2002).

1.2.6 MANAGING AND UTILIZING THE WASTES FROM TEA INDUSTRIES

The by-products of tea have little effect on the environment. The two types of tea wastes are factory tea waste (FTW) and decaffeinated tea waste (DCTW). When tea is processed in the industries, the removal of the fiber portion is performed. It is discarded as tea waste. It comprises of tea leaves and dust. This tea waste contains tannic acid. It is one of the limiting factors for using this waste as feed for poultry and pigs (Konwar and Das 1990). The waste obtained after the extraction of caffeine from factory tea waste is referred to as DCTW. This type of waste contains a very small quantity of tannic acid. Boiled water is used for the washing of decaffeinated tea. This then turns out to be a feed for pigs, poultry and fishes.

1.2.6.1 Using the Tea Waste for Instant Coffee Production

The production of instant coffee can be done by utilizing the tea wastes. Prior to manufacturing of instant coffee, the separation of caffeine from tea waste is carried out (Gao et al 2014). When instant coffee is produced from tea waste, a reduction in the excess tannins occurs (Someswararao and Sivastav 2012).

1.2.6.2 Using the Tea Waste as Feed for Poultry, Pigs and Fish

We can utilize the tea waste after separating the caffeine. It can be used for preparing feed for pigs and poultry. Factory tea waste, which has tannic acid of more than 5%, has been reported to have harmful effects on the broiler chickens' growth and performance (Konwar and Das 1990). In broiler chicken, tannic acids act as anti-nutritive factors. Hence, the presence of tannic acid in tea waste may affect animal growth (Sarker et al 2010). Such feeds enhanced the immune response in finishing pigs (Ko and Yang 2008) and also there was an increase in the egg laying capacity in hens (Uuganbaya et al 2006).

Decaffeinated tea waste can be used as a fish feed. The feed is prepared by mixing with mustard cake and water. Decaffeinated tea waste contains eight times less tannic acid content as compared to factory tea waste. Hence, decaffeinated tea waste can be used as feed for the growth and development of fish (Anurag et al 2016).

1.2.6.3 Using Tea Waste as a Bionutrient and Biofertilizer

Several tea factories use tea waste in the plantation area. The soil acidity increases due to the caffeine content. When factory tea waste is mixed with urea (5%) and cow dung and kept in soil for about 45 days, a good quality bionutrient and biofertilizer is obtained. Appreciable amounts of *n*-triacontanol (Virdi 2014) are also present in tea waste. This component supports the growth of the tea plant. Certain physiological properties are also enhanced such as the formation of leaf primordial and helps the development of primary leaves (Rao et al 2006).

1.2.6.4   Using Tea Wastes by Separating the Bioactive Chemical Components

Tea waste can be utilized as a low cost adsorbent. This could be used for the removal of Cu and Pb from wastewater (Amarasinghe and Williams 2007). A potential low cost adsorbent for the removal of ρ-nitrophenol from wastewater is the activated tea waste. Re-processing of tea fiber waste to produce valuable biological products is a useful option. This is because tea fiber waste comprises many polyphenols and cellulosic materials (Alasalvar et al 2013). Recently, the recycling of plant fibers from *Calotropis gigantea* through enzymatic hydrolysis of cellulose to produce bioethanol production has been carried out (Moeini et al 2016).

1.2.6.5   Auto Gasification of Tea Waste

A complete characterization of tea waste was done by Tamizselvan et al (2020). They reported that auto gasification of tea waste is viable. With the development of an appropriate device and its installation at tea shops, each tea shop will be able to attain self-sufficiency in energy requirement from its tea as well as other waste. This will save the environment from harm.

1.2.7   Utilization of Confectionary/Candy Industry Waste

Confectionery plants need a lot of fresh water to function. It is required for production and for washing machines and equipment. All these activities will generate huge quantities of wastewater generation. As an alternative, wastewater from confectionary industries could be treated using pressure membrane techniques (Ewa and Anna 2019) with the aid of nano filteration. The treatment is done to an extent where it is safe for it to be released into a natural reservoir. Wastewater, which is treated using pressure membrane techniques, could be responsible for irrigation of agricultural land or gardens. It also has the potential to be reutilized for cleaning machines and equipment in industries.

1.2.8   Managing and Utilizing the Wastes from the Canning Industry

Tuna condensate is an organic rich by-product, which is obtained from the tuna canning industry. This by-product was analyzed to be used as a substrate to produce polyhydroxyalkanoate (PHA). *Cupriavidus necator* TISTR 1095 was used for this. The effect of cultivation parameters on PHA accumulation was analyzed. This included substrate concentration, initial pH value, carbon to nitrogen (C/N) ratio, besides the other fermentation strategies. A high COD removal efficiency of about 70% was attained under optimal conditions. This study demonstrated that tuna condensate can be utilized as a cheap substrate to produce PHA on an industrial scale (Kanokphorn et al 2020).

1.2.9   Managing and Using the Wastes from the Sugar Industry

Several types of wastes are generated from the sugar industry. These include bagasse, pressmud, bagasse fly ash, sugar beet mud, sugar cane trash, sugar beet pulp, molasses, etc. When mixing of these wastes is done with other organic substrates, we get an ideal mixture for the growth of earthworms (Sartaj et al 2016). If such wastes are stored in open fields these lead to contamination. They pollute the environment and lead to several diseases. One of the methods to handle this waste is vermicomposting. It is a method that synergizes microbial degradation with earthworm activity. This helps in reducing, reusing and recycling the waste materials quickly. Sugar industry wastes can be changed into valuable fertilizing material. The final product obtained is the vermicompost. It is a nutrient rich organic fertilizer having nitrogen, potassium, calcium and phosphorus as the key nutrients.

1.2.9.1 Bagasse

Bagasse is the fibrous residue obtained when the juice has been extracted from the sugarcane. The waste, bagasse, is utilized as a fuel for generating electricity. The sugarcane bagasse is used to produce paper and fiber, cattle feed, potting media. It is a source to obtain value added products such as pigments, enzymes, amino acids and drugs. It is also used to obtain energy (thermoconversion and ethanol). It is also added to clay soil to suppress the desiccation cracks and the influencing factors behind the cracking behavior, i.e., volumetric shrinkage and water retention property of the clay soil. It can be used as a mulch/soil conditioner. It can be employed for mushroom production. It can also be used as cattle feed and poultry litter. Organic manure can also be made out of it. Biochar can also be prepared from it (Harish et al 2020).

1.2.9.2 Sugarcane Bagasse Ash (SBA)

Sugarcane bagasse ash (SBA) is obtained after we burn the bagasse. This is used as fuel in the sugar industry. Bagasse ash can be used as an adsorbent. It can also be used as a manure. It can be used to remove heavy metals. Even wastewater treatment employs bagasse ash (Harish et al 2020). Owing to the high amount of SiO_2 (60–81%) and other components, SBA is a potential replacement for silica. This replacement can be done in concrete and mortars, ceramics. It can also be used as a stabilizing component in compacted clay blocks and bricks. A common practice is the field application of SBA. This can be done with or without combining the ash with the sugarcane mill's filter mud (Harish et al 2020).

1.2.9.3 Pressmud

Organic wastes, such as pressmud or filter cake, is produced as a by-product of sugarcane industries. It is soft, spongy, amorphous and dark brown to brownish material. It is produced during the purification of sugar by carbonation or sulphitation. Pressmud provides an appreciable quantity of organic manure (Harish et al 2020). It can be utilized as an alternative source of plant nutrient. It can also act as a soil ameliorate. There is 50–70 % moisture content, which is favorable for soil microorganisms and earthworms. Pressmud is used as one of the substrates in biocomposting. It is used as a reclamation agent as well as a soil conditioner. The benefits of utilizing sugarcane pressmud for soil application is its low cost, presence of trace element, slower release of nutrients, high water holding capacity and mulching properties. Pressmud is used as an organic manure; a carrier for legume inoculants; in carp culture; as an animal feed; as an organic amendment and a substrate for biocomposting.

1.2.9.4 Molasses

Molasses is a dark chocolate colored viscous liquid. It contains water (15–20%), total sugars (50–55%) and the leftover proportion as non-sugars. Around 22% of the molasses is utilized by the distillery unit within the same industry. The application of molasses was done to the soil via the sprinkler irrigation system or by the overhead boom sprayer (Harish et al 2020).

A high and damaging population of reniform nematodes led to decline in the yield of fruit and its quality in papaya plantation. On the application of molasses, there was lowering of the population of soil nematode. This lead to significant improvement in the growth of trees and the fruit harvest. When molasses were applied to the Chinese cabbage, a reduction in the *Heterodera* nematode cysts was observed after harvest (Harish et al 2020).

There was an improvement in the color and yield of onion plant by the pre-plant application of molasses (Harish et al 2020). Similarly, application of molasses to the soil supplies carbohydrates and alters the C/N ratio and hence microbial ecology of the soil get affected. It will help to reduce the plant parasitic nematodes' population. A favorable effect on the plant growth is also observed.

Many more pronounced effects were observed among plants specifically in roots and yield parameters, whenever subjected to molasses' response. Competitively, soil application mode of molasses is better than that of foliar. Every year, molasses' production occurs in huge quantities and are used for animal feeding, alcohol and fertilizers. Hence, molasses are used for reclamation of soil, as manure, silage preservative, as an animal feed, insect bait and fertigation (Harish et al 2020).

1.2.9.5 Spent-wash/Distillery Effluent

For every liter of alcohol produced in the distillery industry, around 15 liters of wastewater is produced. This raw spent-wash produced is acidic in nature. It has a dark brown color, an unpleasant odor and a high COD and BOD. Spent-wash can be utilized in biomethanation. The spent-wash can be used for fertigation, cattle feed supplement, soil ameliorent and in vermicomposting (Harish et al 2020).

1.2.9.6 Sugar Beet Pulp

These are the fragments of the fine slices of sugar beets (cossettes). These are exhausted by hot water and pressed. Upon production, pressed pulp is a fresh feed product. It is non-sterilized, hot, delivered in bulk and in a relatively humid state. It is easily altered by outside elements (air, water, bacteria, molds) and is hence, perishable. Pressed pulp conservation is easy at the farm. This holds beyond the period of delivery. Sugar beet pulp is utilized as a feed for ruminants (Harish et al 2020).

1.2.10 Managing and Utilizing the Wastes Produced from the Cocoa Industry

Worldwide, cocoa is an important agricultural product. The beans represent only 30% of the cocoa fruit. This leaves behind an enormous amount of underutilized and often wasted by-products (Karen et al 2019). The waste materials of the cocoa industry include cocoa pod husk, pulp and cocoa bean shell. After separating the shell from the seed, it is generally discarded or sold as agricultural mulch. Owing to its composition, cocoa shell has a lot of potential such as availability of dietary fibers, proteins, methylxanthines and polyphenol constituents, etc. In spite of this composition, the cocoa shell cannot be utilized for food production directly. This is because it may contain components that are harmful for humans. The cocoa shell might be a carrier of mycotoxins, certain microbes, polycyclic aromatic hydrocarbons and some heavy metals as well (Veronika et al 2020).

The high voltage electrical discharge systems represents a novel non-thermal method having a great potential for decontaminating the wastes. This could also be utilized for extracting useful components from cocoa shell. Cocoa shell can also be used in pharmaceutical, cosmetics and agricultural industries. The dietary fibers of cocoa shell consist of pectin and cellulose (Redgwell et al 2003). Besides this, cocoa shell is rich in flavanols such as catechin and epicatechin. These possess antioxidant properties. Methylxanthine, such as theobromine and caffeine present in it, has an effect on the human nervous system (Grillo et al 2019). The lipid profile of cocoa shell was studied by Okiyama et al (2019). It was concluded that it is similar to cocoa butter. This could lead to its usage as a partial substitute for cocoa butter. Cocoa bean shell is one of the wastes produced by the cocoa industry. Bran powder can be produced from cocoa bean shells. This bran powder can be utilized by mixing with other flours (Lienda et al 2019).

1.2.11 Managing the Wastes from Breweries/Wineries/Malting/Distilleries

Distilleries are among the most polluting industries. This is because ethanol fermentation leads to discharge of large quantities of high strength liquid effluents. These have high concentrations of organic matter and nitrogen compounds, low pH, high temperature, dark brown color and high salinity (Wioleta and Magdalena 2020). The most common method of managing the wastewater (distillery stillage) is to use it for soil conditioning. This requires thickening the wastewater and may cause soil pollution due to its high nitrogen content. Therefore, treatment of distillery stillage is preferable.

Besides this, special attention is paid to valorization of distillery stillage. This is a valuable source of polysaccharides and volatile fatty acids (VFAs), as well as natural antioxidants, including polyphenols and other bioactive compounds of interest to the pharmaceutical, cosmetic and food industries. There is a wide variation in the composition of distillery wastewaters. This makes them extremely difficult to bioremediate. Successful biological treatments of such wastewaters has been reported (Melamane et al 2007).

Experiences in treating wine distillery wastewaters can contribute to the field of oenology. This is because many oenologists are concerned with the selection, efficiency and economy of their wastewaters. The wastewaters from the wine distillery are strongly acidic. They have a high chemical oxygen demand, high polyphenol content and are highly variable. Primary attention is focused on the sustainable biological treatment of wine distillery wastewaters, mainly by energy-efficient anaerobic digestion in different reactor configurations from bench to pilot and full scale treatment.

Proper management of these wastes may bring economical benefits and help to protect the environment from pollution caused by their excessive accumulation. The disposal of these wastes is cumbersome for the producers, however they are suitable for reuse in the food industry. Given their composition, they can serve as a low cost and highly nutritional source of feed and food additive. They also have a potential to be a cheap material for extraction of compounds valuable for the food industry and a component of media used in biotechnological processes aimed at production of compounds and enzymes relevant for the food industry (Kamila et al 2020).

Brewery industry by-products can be used as animal feed; as food ingredients or additives. Even bioactive compounds can be extracted from these wastes. As economic impact, using the BSG by-product, which has a low monetary value, as a high nutrient biomass, will enhance the economic potential of breweries and improve the dietary attributes of food formulations. The recovered bioactive compounds and functional ingredients are also of great interest for food, pharmaceutical industry (e.g., antimicrobial activity, carrier agents, controlled release, immune modulatory effects), cosmetics, agriculture and chemical industry. The social impact of these actions refers to the fact that the complex recovery of bioactive compounds and new functional ingredients is aimed to be an efficient and at the same time affordable alternative, for all social categories, to complete their diet with an appreciable number of nutrients (Anca et al 2017).

Just like other agro food productions, the wine making process produces organic waste as a series of important by-products. These are considered to be a potential resource for valorization. The properly managed winery by-products are encouraged to be reused and exploited for alternative purposes like added value products (Sergi and José 2020). Therefore, considerable wine pomace application, highlights the strength of the potential valorization of winery by-products in the different industries (Efstathia et al 2018).

The by-products generated in a winery can be treated from two points of view, as far as their valorization is concerned. The first one refers to the possibility of extracting phytochemical compounds with added value, interesting to be used in the elaboration processes of food (Mirabella et al 2014), cosmetics (Teixeira et al 2014) or even pharmaceutical products (Maia et al 2019). The second focuses on bioconversion processes based on the use of waste generated as a raw material for the cultivation of microorganisms, either as a single substrate or as an additional nutritional supplement. The valorization of the by-products obtained in the wine production process can therefore be used to grow yeasts, molds and bacteria by applying different technological solutions.

Bioactive compounds can be recovered from brewing wastes (Anca et al 2017). The recovery and reuse of the brewing industry by-products for extracting functional compounds and developing new innovative products are a research direction of great interest. This is true from the perspective of food and health relation as well as from the environment protection and waste management per-spective. Recent advances in biotechnology ensure that brewing industry by-products are no longer regarded as a waste but rather a feedstock for producing a new generation of value added products. Based on this, it is an undeniable fact that brewing residues have their own potential for sustainable reuse through biotechnological approaches. The recent findings highlighted the potential reuse of brewery by-products and led to the idea that multidisciplinary approaches should be implemented in order to develop integrated biorefineries.

Spent grains, spent hops and spent yeasts are high energy raw materials that possess a great potential for application in the branch of biotechnology and the food industry, but these by-products are commonly used as livestock feed, disposed of in the fields, or incinerated brewery by-products can be utilized for microalgae production, biofuel production, extraction of proteins, polyphenolic, antioxidative substances, etc. (Andrea et al 2020).

1.2.12 Managing and Utilizing the Wastes from the Edible Oil Industries

The refined vegetable oil and hydrogenated vegetable oil manufacturing industry produces wastewaters and solid wastes. This includes spent earth, spent catalyst, chemical and biological sludges. The wastewater streams generally come from vat house after soap splitting, floor washing, cooling tower, boiler and filter press. On the analysis of the chemical composition of the wastewater obtained from cooling tower and boiler sections and solid waste suggested that the wastes can be recycled and reused within the process after some preliminary treatment.

For achieving zero discharge for economical and ecological gains, the recycle, recovery and reuse of the wastes were adopted in the industry (Ram et al 2003). Waste from the edible oil industry has great potential for further valorization and water reuse (Welz 2019). The land appli-cation of oil seed meal as a method of disposal is not a good strategy for disposing the waste owing to high C and N. Treating the waste using allochthonous microbial biomass may offer a novel and promising strategy for developing treatment methods before the disposal of the solid waste produced in the edible oil industry. Hence, alleviating the problems of land pollution (Jeph et al 2015).

The edible oil industry produced huge amounts of wastewater. These require extensive treatments to remove the pungent smell, high phosphate, COD and metal ions before discharge (Cho et al 2016). Traditional anaerobic and aerobic digestion could mainly reduce COD of the wastewater from oil refinery factories (WEORF). In this study, a robust oleaginous microalga *Desmodesmus* sp. more than 82% of the COD and 53% of total phosphorous were removed by *Desmodesmus* sp. S1. In addition, metal ions, including ferric, aluminum, manganese and zinc were also diminished sig-nificantly in the WEORF after microalgal growth, and the pungent smell vanished as well. This study suggested that growing microalgae in WEORF can be applied for the dual roles of nutrient removal and biofuel feedstock production.

1.2.13 Managing and Utilization of Starch Industry Wastes

A large quantity of waste residue is produced in potato starch processing, which contains various organic materials such as protein, carbohydrate, starch, etc. This leads to negative impact on the environment when this waste residue is stacked as such. Biogas production, protein feed produc-tion and extraction of pectin and dietary fiber can be done from these wastes. It was concluded that biogas production might be the most brilliant prospect in this field. Further research may be established in industrialization of this biogas producing technology, which may also contribute to future mixed culture fermentation technology, an effective way for growing energy crisis (Di 2016). Utilization of potato starch processing waste can be done for producing animal feed with high lysine content (Ying et al 2015).

The food processing waste streams can be utilized as substrates for fermentation processes. This can be done to produce useful chemicals, food ingredients or pharmaceutical products. Starch wastes, either in solid or liquid form, are a more attractive proposition for microbial processing than cellulosic wastes. This is because of the easy hydrolysis of starch to glucose, owing both to the large number of microorganisms that produce amylolytic enzymes and to the cheap commercial sources of enzymes. The use of starch industry wastes could be divided into the separation of useful materials. These include proteins present in potato wastes, bioconversion to useful chemicals or biomass for food, feed or other applications and treating the liquid streams for the production of methane, ethanol and other fuels, etc. (Rakshit 1998).

1.2.14 Managing and Utilization of the Wastes of the Baking Industry

The wastes from the bakery industry include dough, flour dust, sugar dust, burnt biscuits, broken biscuit, burnt loaves or rejected loaves, market returned old bakery products. These products can be used as cattle feed. Various bakery wastes and bulking agents, such as cow dung, to produce compost have been studied by Mugilan et al (2021). Commercially, microorganisms were utilized to study its effectiveness in composting bakery waste compared to common ways of composting. The results revealed that the bakery wastes can be transformed into compost and its quality complied with standard requirements.

The possibilities of recycling wastes from baking industry in production of bioethanol were studied by Joanna et al (2013). They demonstrated that along with development of the baking industry, the amounts of bread waste, of low value for reprocessing in the food industry, is increasing. The utilization of these wastes can be done for production of ethanol fuel. This is a cheap alternative to traditional crop raw materials.

The bakery wastes such as cakes and pastries from Starbucks Hong Kong, were evaluated for the potential of succinic acid (SA) production. Results of this study demonstrated the novel use of bakery waste as the generic feedstock for the sustainable production of SA as a platform chemical in food waste biorefinery (Andrew et al 2013).

1.3 PRODUCTION OF BIOFUELS FROM THE BY-PRODUCTS OF FOOD PROCESSING INDUSTRIES

Producing biofuels using the food processing wastes is not an appreciable job in view of economical and environment sustainability concern. Several different valorization pathways might give good returns on investment by utilizing these wastes. In several scenarios, producing biofuels is a good option among wise utilization of all the waste streams. Such waste streams have inconsistent physical and chemical characteristics. These contain small quantities of impurities, which make them unfavorable for upcycling to secondary food products to be consumed by humans and animals. These could also be available in a dilute state, which makes combining them with other organic streams, such as the dairy manure, a useful option. The viability of the wastes to biofuel transformation is determined by the transportation and other logistic inputs as well. An example can be quoted here of a food processing plant, which may opt for diversion of the waste towards anaerobic digester. The digester may be present in close proximity even with payment of a tipping fee, rather than opting for developing and managing costs of converting the same material to a secondary food product. This could generate some other revenue stream. This happens when there are small regional facilities. There is a lack of up to the mark in house research and development facilities. There could be even a lack of a desire for expanding beyond the core business areas (Trabold and Rodrıguez 2020). There are three main conversion technologies, which have already been used for converting food processing wastes into biofuels:

i. *Anaerobic digestion* for producing the hydrogen or methane rich biogas.
ii. *Fermentation* for producing liquid alcohol fuels. These include ethanol and butanol and hydrogen.
iii. *Thermochemical conversion*, which includes gasification, pyrolysis and hydrothermal liquefaction for producing hydrogen rich syngas and bio oil

Transesterification is also a widely employed method for converting waste vegetable oils to obtain fatty acid methyl esters. These are commonly referred to as biodiesel. This technology typically uses a single feedstock material (Van Gerpen 2005; Knothe et al 2015).

1.4 CONCLUSION

Globally, there is a considerable focus for applying cleaner technology techniques with zero effluents for sustainable development. Although, this is not possible all of the time. So, exploring the wastes of the various processing industries can be done in a number of ways. They are a source of several functional compounds. Their application in various sectors seems promising. This calls for a joint effort of food technologists, nutritionists, food chemists and toxicologists. Optimization should be done of the food processing technology for minimizing food wastage. Focus should be on reducing, recycling and reusing the wastes. Development should focus on economical and complete use of by-products of the food processing industries at large scale. The food and allied industries should actively participate for sustainable production and management of waste. Using the food industry wastes for feedstocks composting, manuring, production of value added products, etc., is a better option as compared to dumping the wastes, which pose environmental hazards. Effective quality control systems need to be developed for exclusion of natural and anthropogenic toxins. Plant breeding needs to be employed for minimizing the production of harmful components. The organic micronutrients and other functional compounds need to be characterized and quantified by using specific analytical methods. It is important to find out the best alternatives for utilizing such residues, which otherwise pose a risk to the environment.

REFERENCES

Ahalya, N., Chandraprabha, M.N., Kanamadi, R.D., and Ramachandra, T.V. 2014. Adsorption of fast green on to coffee husk. *Journal of Chemical Engineering and Research* (2)1:201–207.
Alasalvar, C., Pelvan, E., Ozdemir, K.S., Kocadağlı, T., Mogol, B.A., Paslı, A.A., Ozcan, N., Ozcelik, B., and Gokmen, V. 2013. Compositional, nutritional, and functional characteristics of instant teas produced from low- and high-quality black teas. *Journal of Agricultural and Food Chemistry.* 7;61(31):7529–7536.
Alemu, F. 2015. Cultivation of Shiitake Mushroom (*Lentinus edodes*) on Coffee Husk at Dilla University, Ethiopia. *Journal of Food and Nutrition Sciences* 3(2): 64–70.
Amarasinghe, B.M.W.P.K.R., and Williams, A. 2007. Tea waste as a low cost adsorbent for the removal of Cu and Pb from wastewater. *Chemical Engineering Journal* 132:299–309.
Anastopoulos, I., et al. 2017. A review for coffee adsorbents. *Journal of Molecular Liquids* 229:555–565.
Anca, C.F., Sonia, A.S., Elena, M., Francisc, V.D., Dan, C.V., Maria, T., and Liana, C.S. 2017. Exploitation of brewing industry wastes to produce functional ingredients. *Intechopen*: 137–156. DOI: 10.5772/intechopen.69231
Andrea, K., Anita, J., Nevena, C., Kristina, H., Vinko, K., and Krešimir, M. 2020. By-products in the malting and brewing industries – re-usage Possibilities. *Fermentation* 6(82):1–17.
Andrew, Y.Z., Zheng, S., Cho, C.J.L., Wei, H., Kin, Y.L., Mingji, L. and Carol, S. K.L. 2013. Valorisation of bakery waste for succinic acid production. *Green Chem* 15:690–695.
Anurag, C., Satyajit, S., Akash, C., Soumik, B., Palash, M., and Monoranjan, C. 2016. Tea waste management: a case study from West Bengal, India. *Indian Journal of Science and Technology* 9(42):1–6.

Berhe, S., et al. 2015. Adsorption efficiency of coffee husk for removal of lead (II) from industrial effluents: equilibrium and kinetic study. *International Journal of Scientific and Research Publications* 5(9):1–8.

Boccas, F., et al. 1994. Production of pectinase from coffee pulp in solid-state fermentation system - selection of wild fungal isolate of high potency by a simple 3-step screening technique. *Journal of Food Science and Technology* 31(1):22–26.

Bondesson, E. 2015. A nutritional analysis on the by-product coffee husk and its potential utilization in food production. Bachelor thesis. Faculty of natural Resources and Agricultural Sciences, Uppsala.

Chi, P.A. 2016. Recycling Coffee Grounds for Science. Science and food. [Online]. Available at: https://science andfooducla.wordpress.com/2016/07/19/recycling-coffee-groundsfor-science/

Cho, C. M., Yong, F., Fu-Li., Guang-Rong, H. 2016. Bioremediation of wastewater from edible oil refinery factory using oleaginous microalga *Desmodesmus* sp. S1. *Int J Phytoremediation* 18(12):1195.

Corro, G. et al. 2014. Enhanced biogas production from coffee pulp through deligninocellulosic photocatalytic pretreatment. *Energy Science & Engineering* 2(4):177–187.

Cruz, R. 2014. Coffee by-products: Sustainable Agro-Industrial Recovery and Impact on Vegetables Quality: dissertation thesis. Universidade de Porto.Porto.

Cubero-Abarca, R., and Moya, R. 2014. Use of coffee (*Coffea arabica*) pulp for the production of briquettes and pellets for heat generation. *Ciência e Agrotecnologia* 38(5):461–470.

Di, W. 2016. Recycle technology for waste residue in potato starch processing. *Procedia Environmental Sciences* 31;108–112.

Djilas, S., Čanadanović-Brunet, J., and Cetkovit, G. 2009. By-products of fruits processing as a source of phytochemicals. *Chem Ind Chem Eng* Q 15(4):191–202.

Efstathia, K., Iliada, L., Pavlos, B., Petros, A.T., and Efstathia, S.2018. Novel application and industrial exploitation of winery by-products. *Bioresources and Bioprocessing* 5(46). https://doi.org/10.1186/s40 643-018-0232-6

ElMekawy, A., Diels, L., Wever, H.D., and Pant, D. 2013. Valorization of cereal based biorefinery by-products: reality & expectations. *Environ Sci Tech* 47(16):9014–9027.

Ewa, P., and Anna, M. 2019. Treatment of Wastewater from the Confectionery Industry Using Pressure Membrane Processes. MDPI *Proceedings*. 1–4.

Gao, X., Zhang, B., Shao, Z., Yang, Y., and Yue, P. 2014. Separation of caffeine and tea poly-phenols from instant (soluble) tea waste liquor by macro-porous resins. *Advance Journal of Food Science and Technology* 6(6):768–773.

Gebresemati, M., et al. 2017. Sorption of cyanide from aqueous medium by coffee husk: Response surface methodology. *Journal of Applied Research and Technology* 15(1):27–35.

Getachew, T., et al. 2015. Defluoridation of water by activated carbon prepared from banana (*Musa paradisiaca*) peel and coffee (*Coffea arabica*) husk. *International Journal of Environmental Science and Technology* 12(6):1857–1866.

Girotto, F., Alibardi, L., and Cossu, R. 2015. Food waste generation and industrial uses: a review. *Waste Manage*. 45:32–41.

Grillo, G., Boffa, L., Binello, A., Mantegna, S., Cravotto, G., Chemat, F., Dizhbite, T., Lauberte, L., and Telysheva, G. 2019. Cocoa bean shell waste valorization; extraction from lab to pilot-scale cavitation reactors. *Food Res. Int* 115:200–208.

Gupta, V.K. (2008) Development in the manufacture of whey protein products, Course Compendium, Technological Advances in the Utilization of Dairy By-Products, February 27–March 18, 2008, Centre of Advanced Studies, Dairy Technology Division, N.D.R.I., Karnal, India.

Haile, M. 2014. Integrated volarization of spent coffee grounds to biofuels. *Biofuel Research Journal* 2:65–69.

Harish, M.N., Nagendra, M.S., and Amaresh, P. 2020. Utilization of sugar industry wastes in agriculture. *Agriculture & food*: e- Newsletter 2(1):17–23.

Jayathilakan, K., Sultana, K., Radhakrishna, K., Bawa, A.S. 2012. Utilization of by-products and waste materials from meat, poultry and fish processing industries: a review. *J Food Sci Technol* 49(3):278–293.

Jeph, G., Mathur, S., Relekar, P.K., Nakhate, P.H., Parecha, D.K., Gautam, H.K. 2015. Management of solid waste from edible oil industry. *Journal of Environmental and Applied Bioresearch* 3(2):92–96.

Joanna, K.R., Anna, C., and Witold, P. 2013. Some aspects of baking industry wastes utilization in bioethanol production. *Zeszyty Problemowe Postępów Nauk Rolniczych* 575:71–77.

Kamila, R., Adam, W., Klaudia, G., and Magdalena, P.B. 2020. Utilization of brewery wastes in food industry. *Peer J* 8:e9427. DOI: 10.7717/peerj.9427

Kanokphorn, S., Nisa, P., Tewan, Y., Sappasith, K., and Poonsuk, P. 2020. Utilisation of tuna condensate waste from the canning industry as a novel substrate for polyhydroxyalkanoate production. *Biomass Conversion and Biorefinery* DOI: https://doi.org/10.1007/s13399-019-00581-4

Karen, H.N.F., Nancy, V.M.G., Rocio, C.V.2019. Cocoa By-products. Food Wastes and By-products: Nutraceutical and Health Potential. 373–411. DOI:10.1002/9781119534167.ch13

Khedkar, K., and Singh. 2018. New Approaches for Food industry Waste Utilization. In book: Biologix. Publisher: Dept. of Biotechnology, Mewar institute of Management Editors: Dr. Neetu Singh, Mr. Ajay Kuma. (pp.51–65).

Knothe, G., Krahl, J., Van Gerpen, J. (Eds.). 2015. *The Biodiesel Handbook*. Elsevier.

Ko, S.Y., and Yang, C.J. 2008. Effect of green tea probiotics on the growth performance, meat quality and immune response in finishing pigs. *The Asian-Australian Journal of Animal Science* 21(9):1339–1347.

Konwar, B.K., and Das, P.C. 1990. Tea waste – a new livestock and poultry feed, Technical bulletin No. 2, AICRP on Agro- industrial By-products (ICAR) 1–9.

Laura, G. 2016. Recycled coffee grounds give rise to Fremantle mushroom farm. [Online]. Available at: www.abc.net.au/news/2016-05-10/perth-mushroom-farm-beginsproduction/7399456.

Lenka, B, Maroš, S., Alica, B., Maroš, S. 2017. Review: utilization of waste from coffee production. Faculty of Materials Science and Technology in Trnava Slovak University of Technology in Bratislava 25(40):91–101.

Lienda, H., Harry, T., and Antonius, I. 2019. Cocoa bean shell waste as potential raw material for dietary fiber powder. *International Journal of Recycling of Organic Waste in Agriculture* 8(4):85–491.

Maia, M., Ferreira, A., Laureano, G., Marques, A., Torres, V., Silva, A., Matos, A., Cordeiro, C., Figueiredo, A., and Silva, M. 2019. Vitis vinifera "Pinot noir" leaves as a source of bioactive nutraceutical compounds. *Food Funct* 10:3822–3827.

Markham, D. 2016. This company converts coffee cherry pulp into a nutritious (flourless) flour. [Online]. Available at: www.treehugger.com/green-food/company-converts-coffeecherry-pulp-nutritious-flourless-flour.html.

Martínez-Carrera, D., Aguilar, A., Martínez, w., Bonilla, M., Morales, P., and Sobal, M. 2000. Commercial production and marketing of edible mushrooms cultivated on coffee pulp in Mexico. Chapter 45 In Book Coffee biotechnology and quality, T. Sera, C. Socco l, A. Pandey, and S. Roussos, Eds. Brazil: Kluwer Academic Publishers, Dordrecht, The Neterhlands: 471–488.ISBN 0-7923-6582-8.

Melamane, X.L., Strong, P.J., and Burgess, J.E. 2007. Treatment of wine distillery wastewater: a review with emphasis on anaerobic membrane *Re S Afr J Enol Vitic* 28(1): 25–36.

Menezes, E.G.T., Carmo, R.B.D., Menezes, A.G.T., Alves, J.G.L.F., Pimenta,C.J., and Queiroz, F. 2013. Use of different extracts of coffee pulp for the production of bioethanol. *Applied Biochemistry and Biotechnology* 169:673–687.

Mirabella, N., Castellani, V., Sala, S. 2014. Current options for the valorization of food manufacturing waste: a review. *J Clean Prod* 65:28–41.

Misra, M., et al. 2008. High quality biodiesel from spent coffee grounds. *Clean Technology* 39–42.

Moeini, S.S., Dadashian, F., and Vahabzadeh, F. 2016. Recycling of calotropisgigan tea waste fiber by enzymatic hydrolysis to produce bioethanol. *Indian Journal of Science and Technology* 9(9):1–6.

Mugilan, G., Kathiresan, V., Sathasivam, and Kasi M. 2021 Waste to wealth: value recovery from bakery wastes. *Sustainability* 13(2835):1–16.

Mussatto, S.I., Carneiro, L.M., Silva, J.P.A., Roberto, I.C., and Teixeira, J.A. 2011. A study on chemical constituents and sugars extraction from spent coffee grounds. *Carbohydrate Polymers* 83:368–374.

Mussatto, S.I., Machado, E.M.S., Carneiro, L.M., and Teixeira, J.A. 2012. Sugars metabolism and ethanol production by different yeast strains from coffee industry wastes hydrolysates. *Applied Energy* 92:763–768.

Nguyen, A.D., Tran Trung Dzung T.T., and Khanh, V.T.P. 2013. Evaluation of coffee husk compost for improving soil fertility and sustainable coffee production in rural central highland of Vietnam. *Resources and Environment* 3(4):77–82.

Okiyama, D.C.G., Soares, I.D., Toda, T.A., Oliveira, A.L., and Rodrigues, C.E.C. 2019. Effect of the temperature on the kinetics of cocoa bean shell fat extraction using pressurized ethanol and evaluation of the lipid fraction and defatted meal. *Ind. Crops. Prod* 130:96–103.

Oliveira, W.E., Franca, A.S., Oliveira, L.S., and Rocha S.D. 2008. Untreated coffee husks as biosorbents for the removal of heavy metals from aqueous solutions. *Journal of Hazardous Materials* 152:1073–1081.

Panusa, A., Zuorro, A., Lavecchia, R., Marrosu, G., and Petrucci, R. 2013. Recovery of natural antioxidants from spent coffee grounds. *Journal of Agricultural and Food Chemistry* 61(17):4162–4168.

Pavlovic, M., Pavlović, M.D., Nikolić, I.R., Milutinović, M.D., Dimitrijević-Branković, S.I., Šiler-Marinković, S.S., and Antonović, D.G. 2015. Plant waste materials from restaurants as the adsorbents for dyes. *Hemijska industrija* 69(6):667–677.

Rakshit, S.K. 1998. Utilization of starch industry wastes. In: Martin A.M. (eds) Bioconversion of Waste Materials to Industrial Products. Springer, Boston, MA.

Ram, A. P., Sanyal, P.B., Chattopadhyay, N., and Kaul, S.N. 2003. Treatment and reuse of wastes of a vegetable oil refinery. *Resources Conservation and Recycling* 37(2):101–117.

Rao, J.M., Natarajan, C.P., and Seshadri, R. 2006. A study on the occurrence of N-triacontanol, a plant growth regulator, in tea. *Journal of Science Food and Agriculture* 39(2):95–9.

Redgwell, R., Trovato, V., Merinat, S., Curti, D., Hediger, S., and Manez, A. 2003. Dietary fibre in cocoa shell: characterization of component polysaccharides. *Food Chem* 81:103–112.

Rodrigues, F., and Oliveira, M.B.P.P. 2016. An overview of coffee silverskin validation as a cosmetic ingredient. *CRS Newsletter* 33(2):9–11

Sahu, O. 2014. Bioethanol production by coffee husk for rural area. *Advanced Research Journal of Biochemistry and Biotechnology* 1(1):1–5.

Santos, C., et al. 2017. Effect of different rates of spent coffee grounds (SCG) on composting process, gaseous emissions and quality of end-products. *Waste Management* 59:37–47.

Sarker, M.S.K., Yim, K.J., Ko, S.Y., Uuganbayar, D., Kim, G.M., Bae, I.H., Oh, J.I., Yee, S.T., and Yang, C.J. 2010. Green tea level growth performance and meat quality in fishing pigs. *Pakistan Journal of Nutrition.* 9(1):10–14.

Sartaj, A.B., Jaswinder, S., Adarsh, P.V. 2016. Management of sugar industrial wastes through vermitechnology. *International Letters of Natural Sciences* 55:35–43.

Sergi, M., and José, J.M. 2020. Sustainability of wine production. *Sustainability* 12(55):1–10.

Sevil, Ç.E., Şeniz, Ö.A., and Erdinç, A. 2018. Biofuel potential of fruit juice industry waste. *J Hazard Toxic Radioact Waste* 22(4): 05018002.

Someswararao, C.H., and Srivastav, P.P. 2012. A novel technology for production of instant tea powder from the existing black tea manufacturing process. *Innovative Food Science and Emerging Technologies.* Volume 16, 143–147

Tamizselvan, R., Devanand, B., Selva Nanthini, S., Bhuvana Varadha, V.P., and Kirubakaran, V. 2020. Establishing auto gasification of tea waste. *Conference Proceedings.* Volume 2225, Issue 1, DOI:10.1063/5.0005573

Teixeira, A., Baenas, N., Dominguez-Perles, R., Barros, A., Rosa, E., Moreno, D., and Garcia-Viguera, C. 2014. Natural bioactive compounds from winery by-products as health promoters: a review. *Int J Mol Sci* 15: 15638–15678.

The Rice Association. 2015. By-products. www.riceassociation.org.uk/content/1/16/by-products.html.

Trabold, T.A., and Rodrıguez, A. 2020. Valorization of food processing by-products via biofuel production. *Sustainability of the Food System* Elsevier 53–69.

Ulsido, M.D., and Li, M. 2016. Effect of organic matter from coffee pulp compost on yield response of chickpeas (*Cicer arietinum* L.) in *Ethiopia. Engineering for Rural Development* 1339–1347.

Ulsido, M.D., et al. 2016. Biogas potential assessment from a coffee husk: an option for solid waste management in Gidabo watershed of Ethiopia. *Engineering for Rural Development* 1348–1354.

Uuganbayar, D., Shin, I.S., and Yang, C.J. 2006. Comparative performance of hens fed diets containing Korean, Japanese and Chinese green tea. *The Asian-Australian Journal of Animal Science* 19(8):1190–6.

Van Gerpen, J., 2005. Biodiesel processing and production. *Fuel Processing Technology* 86(10):1097e1107.

Veronika, B., Antun, J., Ivana, F., Drago, S., Jurislav, B., Borislav, M., Kristina, D., and Đurđica, A. 2020. Difficulties with use of cocoa bean shell in food production and high voltage electrical discharge as a possible solution. *Sustainability* 12: 3981.1–11.

Virdi, M.S. 2014. Manufacture of n-triacontanol: A plant growth regulator, micro enterprises in agriculture. Daya Publishing House, Delhi 3–4.

Von, E.J.C. et al. 2002. Review of coffee waste water characteristics and approaches to treatment. Coffee Research Report: Kainantu, Papua New Guinea.

Wadhwa, M., and Bakshi, M.P.S. 2013. Utilization of fruit & vegetable waste as livestock feed and as substrates for generation of other value-added products. In: Makkar (ed) FAO document.

Welz, P.J. 2019. Edible seed oil waste: status quo and future perspectives. *Water Sci Technol* 80(11):2107–2116.

Wioleta, M., and Magdalena, Z. 2020. Distillery stillage: characteristics, treatment, and valorization. *Applied Biochemistry and Biotechnology* 192:770–793.

Woolsey, B. 2016. This Company Is Turning Coffee Grounds into Coffee Cups. [Online]. Available at: https://munchies.vice.com/en_us/article/this-company-is-turning-coffee-groundsinto-coffee-cups.

Ying, Li., Bingnan, L., Jinzhu, S., Cheng, J., and Qian, Y. 2015. Utilization of potato starch processing wastes to produce animal feed with high lysine Content. *J. Microbiol. Biotechnol* 25(2):178–184.

Zuorro, A., et al. 2013. Magnetically modified coffee silverskin for the removal of xenobiotics from wastewater. *Chemical Engineering Transactions* 35:1375–1380.

2 Recovery of Valuable Compounds from Food Industry Waste Using Ultrafiltration

Kiran Bala Nain,[1*] *Sanju Bala Dhull,*[2] *and*
Navnidhi Chhikara[3]

[1,2]Department of Food Science and Technology, Chaudhary Devi
Lal University, Sirsa (Haryana) India
[2]Department of Food Technology, Guru Jambheshwar University of
Science and Technology, Hisar (Haryana) India
*Corresponding author: kiran31nain@gmail.com

CONTENTS

2.1   INTRODUCTION

Recently, waste management from food processing plants as well as recovery of useful compounds from food wastes is a big challenge for food technologists and scientists. Such a large quantity of waste is a serious issue from an economic and environmental point of view because a considerable amount of useful compounds are lost and managing them is also a very difficult task. Food wastes are generated worldwide in a large quantity at every stage, from the agricultural stage through to consumer consumption. Literature survey shows that in developed countries, 42% of food waste is produced by households, whereas 39% of losses occur in the food manufacturing industry, 14% in the food service sector and the remaining 5% in retail and distribution (Mirabella et al., 2014). Food industry wastewater contains high organic loads and nutrient content, which is the reason for raising chemical oxygen demand (COD) about tens of thousands of milligrams per liter (Chen

DOI: 10.1201/9781003207689-2

TABLE 2.1
Estimated Food Industry Waste

Sr. No.	Industry	Amount of waste (000 t)	Waste (%)
1	Production, processing, and preserving of meat and meat products	150	2.5
2	Production and preserving of fish and fish products	8	3.5
3	Production and preserving of fruits and vegetables	279	4.5
4	Manufacture of vegetable and animal oils and fats	73	1.5
5	Dairy products and ice cream industry	404	3
6	Production of grain and starch products	245	1.5
7	Manufacture of other food products	239	2
8	Drinks' industry	492	2
	Total	1890	2.6

Source: Adopted from Baiano, 2014.

et al., 2019). Data regarding estimated food wastes generated from various fields is represented in Table 2.1. According to Galanakis (2012) wastewater generated by food industries has long been the subject of minimization, treatment, and prevention, because its disposal into the environment has harmful influences. Different kinds of membranes and configurations have been employed to treat and recover valuable compounds from food industry wastes. To recover compounds from food waste, membrane technology seems a good alternative, and ultrafiltration (UF), nanofiltration (NF), microfiltration (MF), and reverse osmosis (RO) are the most useful and all of these aim to reduce organic load from waste to lower the COD.

Food industry wastewater consists of a lot of high value added compounds, such as sugars, phenolic compounds (e.g. hydroxycinnamic acids, flavonols, flavonones, o-diphenols, phenolic alcohols, anthocyanins, secoiridoids), pectins, and proteinsetc, owing to which, it acts as a potential source for their extraction (Castro-Muñoz, 2018). Food waste and by-products (peels, shells, trimmings, stems, bran, and seeds) offer a great future prospect to lower malnutrition and hunger for developing countries, which will be helpful in improving food security (Torres-leon et al., 2018). According to Chen et al. (2019) from the fruit and vegetable processing industry, including soybean by-products, fruit and beverage, starch and edible oil industry wastewater a large number of high value added nutrients (bioactive compounds) can be recovered using various membrane based techniques as well as other techniques, for their future applications in different foods i.e. acting as a nitrogen source for food supplements, antioxidants, colorants, food packaging, fat replacer, nutraceuticals, thickener, gelling agent, and emulsifier etc. Use of membranes in various fields also has some limitations i.e. application challenges, unacceptable separation efficiencies, and excessive costs etc. Lack of quantitative assessment regarding the latest opportunities and recent practices in membrane applications is the reason for their limited use. Thus detailed analysis regarding this is now needed.

2.2 APPLICATIONS OF PRESSURE-DRIVEN MEMBRANES FOR WASTE MANAGEMENT

For waste management, pressure-driven membrane based technologies such as UF, NF, and MF etc. are becoming more attractive over classic processing and specific methodologies due to high selectivity, lower energy consumption, simplicity of operation, no phase change as well as modular nature, less degradation of target compounds and no use of solvents. According to literature surveys it can be concluded that nowadays these pressure-driven membrane processes are not only focused on pollution removal, although high added value compounds can be successfully recovered (Castro-Munoz, 2018).

These membrane techniques are most commonly used for various applications like separation, treatment of natural extracts, recovery or concentration of bioactive compounds (anthocyanins, phenolic compounds, carotenoids, polysaccharides, antioxidants) from agro-food products and by-products (wastewaters), recovery of aromas from natural and processed products, production of non-alcoholic beverages etc. (Castro-Munoz et al., 2020). In addition, Galanakis et al. (2016) reported that these techniques have been successfully employed over the past three decades in the food and beverage industries by treating fruit juices, dairy products, vegetable oils, wine, potable water, and agricultural wastewater. However, the separation efficacy of these techniques mainly depends on various factors, such as flow velocity, flow type, physico-chemical characteristics of the bulk solution (i.e. weight, type, polarity, solute charge), operating conditions (i.e. temperature, feed flow rate, transmembrane pressure), and specific membrane features (i.e. configuration of membrane separation module, membrane material, pore size) (Castro-Munoz et al., 2016). Adsorption and membrane technologies are the most commonly used methods to obtain selective recovery of polyphenols from food processing wastewater, which is a resource-efficient and sustainable source of polyphenols (Hellwig and Gasser, 2020).

The recovery rate of MF technology varies from 47 up to 100% in permeate streams, forglutamine, anthocyanins, proline, betanin, isobetanin, isoproline, sugars, galacturonic acid, and some phenolic compounds, whereas recovery rates of UF membranes range between 44–99% for these compounds, however some of these compounds can start to be rejected by the membranes (commonly through tight UF), and thus partially recovered in the retentate. In NF technology water is practically passed, resulting in concentrations of the compounds in retentate from 50 up to 99% (Castro-Muñoz, 2018). Frenkel et al. (2020) reported that membrane technology can be applied for nitrogen removal from wastewater depending on operating conditions such as feed composition, membrane characteristics, solids' retention time, and hydraulic retention time.

In the membrane processes, feed is separated into two parts: permeate and retentate streams. Membrane is a porous filtration medium, which acts as a barrier to inhibit masses' movement of selected phases in applications of water and wastewater treatment (Chen et al., 2019). Basically in membrane separation technology, compounds are selectively separated through pores/or small gaps in the molecular arrangement of a continuous structure as depicted in Figure 2.1.

Galanakis et al. (2016) reported that nutraceutical compounds like phenolic acids, flavonoids, proteins, dietary fiber, and glucosinolates have been recovered from various by-products and reused in different products due to their potential to enhance the technological properties of foods and beneficially affect human health by reducing oxidative stress, retarding aging, and preventing diseases such as arteriosclerosis and cancer etc. Table 2.2 summarizes various food applications of pressure-driven membranes.

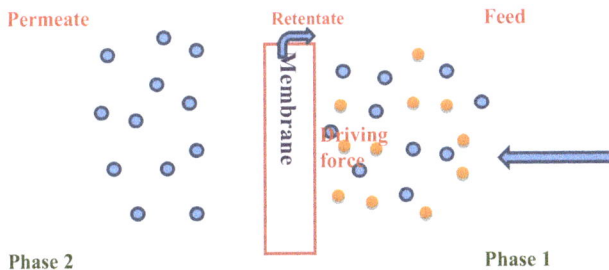

FIGURE 2.1 Represents the basic arrangement of membrane based techniques.

TABLE 2.2
Food Applications of Membrane Based Pressure-driven Techniques

S. No.	Food	Membrane type	Purpose	References
1	Cane Sugar Refinery	NF	Separate colourings and salts	Kwok, 1996
2	Fruit and Vegetable Juices	MF or UF MF and RO	Clarification Fruit juice concentrates	Todisco et al., 1998
3	Brewing Sector	MF	Mash separation, clarification of rough beer	Fillaudeau, 1999
4	Wine Making	MF	Electrodialysis Microbiological stability Tartaric stability	Daufin et al., 2001
5	Milk and Dairy Industry	NF RO MF	Whey concentration Globular milk fat fractionation, Protein extraction process Whey protein valorization	Le et al., 2014
6	Sugar Juices	MF or UF	Clarification	Herve, 1994
7	Vegetable Proteins	UF and diafiltration	High protein yield	Berot et al., 1998
8	lycopene of papaya pulp	UF	Concentration	Paes et al., 2015
9	Date palm sap syrups	UF	Retention of sucrose through tubular Membranes and reduction of syrup crystallization	Ines et al., 2016
10	Protein	UF	Separation of two proteins with a close molecular weight by stacking flat ultrafiltration with the same membranes	Feins & Sirkar, 2005
11	Cheese	UF	Separation of protein, lactose, salt and water from cheese whey and concentration of whey protein	Baldasso et al., 2011
12	Honey	UF	For improvement of honey quality	Itoh et al., 1999
13	Soy protein isolate	UF	Reduction of anti-nutritional factors found in the soy products for use in infant foods.	Yang et al., 2014
14	Fish meal effluents	UF	Recovery and concentration of proteins from fish meal effluents	Afonso et al., 2004

(UF – ultrafiltration, MF – microfiltration, NF – nanofiltration, RO – reverse osmosis)

2.3   ROLE OF ULTRAFILTRATION MEMBRANE IN WASTE MANAGEMENT

In recent years, UF has been most commonly used in the area of water treatment i.e. wastewater treatment, drinking water purification, industrial wastewater treatment, seawater desalination, and water reuse etc. around the world (Li et al., 2018). The UF technique has been widely used for fractionation, concentration, and purification of compounds in solutions on a molecular level without any phase change or addition of chemicals or solvents, and high flux rates are obtained with low pressure applications. Pore sizes of ultrafiltration membranes vary in the range of 1–100 nm and 150–500 kPa is the operating pressure, in addition, these can retain particles existing in the range of 300–1,000,000 Da molecular weight i.e. biomolecules, polymers and colloidal particles, antioxidant compounds from aqueous systems as well as emulsions and micelles (Castro-Munoz et al., 2020).

Pore size of UF membranes is very important as it decides what type of particle will pass through or not. For example, wide membranes (size ranges between 50 and 100 kDa) can pass macromolecules (such as suspended solids, carbohydrates, proteins, and pectins), whereas sizes ranging from 4 to 30 kDa are effective for concentration of high molecular weight components (such as tannins, proteins, hydrolysates, and some high molecular weight phenolicfractions), in addition, tight membranes (1 to 3 kDa) are highly used for concentration of low molecular weight compounds (such as anthocyanins, low molecular weight sugars, and peptides) (Cassano et al., 2018). Polysulfone-type materials (such as polyether sulfone, sulfonated polysulfone, polyphenyl sulfone, and so on) are mainly used for formation of these membranes (Chen et al., 2019).

UF membranes with a molecular weight cutoff of 50–100 kDa are unable to recover most compounds from food wastewater. However, β-glucan from sugars, monovalent ions, and phenols; proteins from saccharides, monovalent ions, simple sugars, and phenols; pectin from sugars, monovalent ions, anthocyanins, and phenolic classes; and polymeric from monomeric anthocyanins can be effectively separated, although membranes with molecular weight cutoff of 4–30 kDa are comparatively less effective for separations of macromolecule and micromolecule (i.e. pectin from phenolic compounds and sugars in winery sludge extracts and pigmented orange pulp wash) etc. (Galanakis et al., 2016). Castro-Muñoz (2018) also stated that on the basis of molecular weight cutoff of the UF membranes, the recovery of smaller molecules (e.g. phenolic compounds and sugars) can be attained, for which solvent and supercritical fluid extraction methodologies are most suitable. UF technology is the most common application in various water treatment technologies due to the characteristics of less land occupation, stable water quality, and high automation. In addition, it can also improve the biosafety of water by removing almost all microorganisms, viruses, colloidal substances, and suspended particles (Li et al., 2018). High value added compounds recovered from food wastes using UF membranes are listed in Table 2.3.

UF has been commonly used to treat fruit juices by concentrating high-molecular-weight macromolecules (i.e. pectin). At the same time, low-molecular-weight solutes (hydroxyl cinnamic acid derivatives, flavonoids, sucrose, acids, salts, aromas, and flavor compounds) pass through the membrane, increase the product in antioxidant compounds, and promote health effects (Galanakis et al., 2016).Nor et al. (2017) recovered the bromelain enzyme from crude pineapple waste mixture by applying integrated (two-stage) ultrafiltration process. They concluded that for obtaining appropriate flux and enzyme separation effectiveness, the process should be conducted at 2.0 bar and 0.30 m/s in UF stage 1, and 1.5 bar and 0.24 m/s in UF stage 2. Use of tight UF membranes can be a beneficial option for the environment because such narrow pore size membranes are able to provide permeate streams obtained from the fractionation of by-products, where the clear permeates basically contain low organic loads (Castro-Muñoz et al., 2018. Characteristics of commercially used UF membranes for the recovery of high-added-value compounds are represented in Table 2.4.

2.4 APPLICATION OF ULTRAFILTRATION IN VARIOUS FOOD INDUSTRIES

In developed countries most of the wastes are produced in the food industries using fruits, vegetables, dairy products, and fish etc. as raw material, whereas in developing countries mainly tropical and subtropical fruits like banana, mango, pineapple, grape, and citrus are used (Torres-leon et al., 2018).

2.4.1 FRUIT AND VEGETABLE INDUSTRY

Food processing industries generate a huge quantity of fruit and vegetable waste throughout the year. These wastes generally contain large amounts of suspended solids, and also possess high biochemical and chemical oxygen demand. This affects municipal landfills because of FW high biodegradability, leachate, and methane emissions (Misi and Forster, 2002). In comparison to other industrial wastewaters, the high contents of COD in food industry wastewater contain a

TABLE 2.3

High Value Added Compounds Recovered from Food Wastes Using UF Membranes

S. No.	Source	Membrane process	Recovered compounds
1	Orange press liquor	UF	Phenolic compounds
	Olive mill wastewater	UF	
	Fermented grape pomace	UF and Nanofiltration	
	Winery sludge from red grapes	UF	
2	Olive vegetation wastewater	UF	3,4-DHPEA, p-HPEA, 3,4-DHPEA-EDA, verbascoside, and total phenols
3	Olive mill wastewater	UF	Oleuropein
4	Olive mill wastewater	UF	Free low-MW polyphenols, hydroxytyrosol, procatechuic acid, tyrosol, oleuropein, and caffeic acid
5	Defatted milled grape seeds	UF	Proanthocyanidins
6	Olive mill wastewater	UF and Nanofiltration	Hydroxytyrosol, procatechin acid, catechol, tyrosol, caffeic acid, p-coumaric acid, and rutin
7	Soy-processing waste	UF	Isoflavones (aglycone and glucoside)
8	Brewers spent grain	UF	Proteins
	Whey from cheese production		
	Halloumi cheese whey		
	Corn cooking wastewater		
	Poultry-processing		
	Wastewater		
	Soy-processing waste		
9	Caprine whey	UF	Caseinomacropeptideimmunoglobulin G
10	Whey protein	UF	a-Lactalbumin
11	By-product of fish fillets	UF	Peptides
	Processing		
	Whitefish fillet hydrolysate		
12	Industrial fish protein hydrolysate	UF and NF	Peptides (containing proline, hydroxyproline, aspartic acid, and glutamic acid)
13	Olive mill wastewater	UF	Pectin
14	Residue of pigmented citrus	UF	Sugars
15	Oat mill waste	UF	Fiber (β-glucan)

Source: Adapted from Galanakis et al. (2016).

variety of high value bioactive substances, such as proteins, polysaccharides, flavonoids, pigments, polyphenols, and dietary fiber etc. These bioactive compounds represent an interesting opportunity for wastewater valorization, some of these might provide antioxidants, antimicrobial agents, and kinds of food additives, others might be further transformed into more sophisticated natural chemicals, macromolecules, or biofuels for various applications and even novel food might be developed with these nutrients (Chen et al., 2019).

The residue of pigmented orange pulp wash was investigated to obtain a purified sugar concentrate applying a series of resin adsorptions and membrane techniques. Final products and intermediates were characterized: anthocyanins, limonoids, flavanones, and hydroxycinnamates were absent; UF process stabilized the product by removal of enzymes and microorganisms; low acidity content was found. 80% of the water was eliminated by a RO treatment, increasing the sugar concentration

TABLE 2.4
Characteristics of Commercially Used UF Membranes

Membrane type	Supplier	Material	Configuration	pH range	Max. operating temperature (^{0}C)	Nominal MWCO (kDa)
GR40PP	Alfa Laval	Polysulfone	Spiral wound	2–10	75	100
GR51PP	Alfa Laval	Polysulfone	Spiral wound	2–10	75	50
GR60PP	Alfa Laval	Polysulfone	Spiral wound	2–10	75	25
GR70PP	Alfa Laval	Polysulfone	Spiral wound	2–10	75	20
GR81PP	Alfa Laval	Polyethersulfone	Spiral wound	2–10	75	10
GR95PP	Alfa Laval	Polyethersulfone	Spiral wound	2–10	75	–2
ETNA01PP	Alfa Laval	Fluoro polymer	Flat sheet	1–11	60	1

Source: Adapted from Galanakis et al. (2016).

by four times (Scordino et al., 2007). Performance of an integrated membrane system to treat and valorize olive mill wastewater from a traditional extraction press was evaluated and noted that a combination of UF and NF was effective in treating wastewater. In addition, residual olive fermentation brine wastewater was pretreated by ultrafiltration followed by post-UF recovery of phenolic compounds from the brine (Frenkel et al. 2020).

2.4.2 Fish and Poultry Processing

Nine commercially available ultrafiltration membranes made from polyetheersulfone and regenerated cellulose was evaluated to treat poultry processing wastewater and it was reported that up to 93% biological oxygen demand (BOD) removals, 94% chemical oxygen demand (COD) removal, and nearly 100% of TSS was obtained (Malmali et al., 2018). Using ultrafiltration, concentration, and spray-drying, cod stomach silage can give a pepsin preparation corresponding to 0.5–1 g pure pepsin per kg, while silages of both stomach and viscera provide about 100 g peptone per kg raw material (Baiano, 2014). A 5-min electrochemical coagulation pretreatment of poultry processing wastewater substantially decreased the membrane fouling in the subsequent ultrafiltration treatment stage. The reduced membrane fouling helps produce a final effluent where 85% of wastes and suspended solids were removed (Frenkel et al., 2020).

2.4.3 Beverages' Industry

Saha et al. (2019) evaluated the potential of a "coagulation-spinning basket membrane" ultrafiltration method to treat tea industry wastewater and concluded that optimizing the rotation parameters helped improve permeate clarity to nearly 95%.

2.4.4 Dairy Industry

The dairy industry containing abundance of nutrients including water, fat, minerals and protein etc. produce wastes in large quantity. It is estimated that over 75% of membrane usage is dedicated to whey processing, while 25% of UF membranes is accounted for milk processing in the dairy sector (Mohammad et al., 2012). Pereira et al. (2002) valorized bovine cheese whey and deproteinized whey, by means of thermo calcic precipitation and microfiltration followed by ultrafiltration–diafiltration

(UF/DF). Afterwards, the chemical composition of microfiltration and UF/DF retentate powders was studied and compared with that of conventional ultrafiltration powders and reported that clarification of the two cheese by-products widely improves the next filtration treatments. Another research regarding issues related to membrane fouling and wetting while using direct contact membrane distillation to treat high-salinity wastewater effluent from "hard cheese" production was conducted by Abdelkader et al. (2019) and results show that effluents treated by UF were found to have the best flux performance.

2.5 MEMBRANE FOULING IN ULTRAFILTRATION

UF technology is in line with the concept of green technology as it follows physical screening as a main retention mechanism but membrane life and membrane fouling are some problems associated with utilization of energy and resources. Due to membrane fouling, substantial flux decrease, and plant maintenance and operating costs increase. Fouling is defined as the irreversible changes in membrane properties, resulting from various interactions of feed stream components and membrane (Saxena et al., 2009). Tsagaraki and Lazarides (2012) reported that membrane is usually fouled by biofoulants such as protein and polysaccharide in food industries. Basically four general types of fouling can be identified in the food industry:

- **Organic fouling**: this type of fouling is produced after deposition of organic compounds like proteins, polysaccharides etc.
- **Inorganic (colloidal) fouling:** this fouling is created by deposition of colloidal matter such as aggregated proteins, organic colloids, colloidal silica, clay minerals, metal oxides (Fe, Al, and Mg), suspended matter and, precipitated salts etc.
- **Scaling (particulate) fouling:** scaling is generated by oversaturation of scarcely soluble salty molecules such as calcium and barium sulfates, calcium carbonate, and silica scales on the membrane surface by precipitation.
- **Biofouling:** this type of fouling is produced after the growth of microorganisms with accumulation of extracellular materials on the membrane surface (Mancinelli and Hallé, 2015).

Characteristics and activity of activated sludge biomass as well as its influence over membrane fouling when treating with different compounds were studied (Table 2.5). The ultrafiltration membrane bioreactor process was effective in removing 90% COD or higher for caffeine and triclosan and was much lower for the more recalcitrant carbamazepine (36%) (Chtourou et al., 2018). Vardanega et al. (2013) used physical strategies including turbulence generating devices, sonication, centrifuge, and use of electric and magnetic fields for reduction fouling in UF membranes. The use of membrane bioreactors for treating graywater was investigated. Membrane bioreactors were found to provide high-quality effluent and can be used in decentralized treatment if economic investment and fouling issues are addressed (Cecconet et al. (2019) and Frenkel et al. (2020)).

2.6 FOULING CONTROL

A model was developed to predict the behavior of ultrafiltration membranes with respect to membrane fouling. It was reported that a "resistance-in-series" model successfully accounted for time-dependent fouling resistance in UF membranes. It was also shown that high salt and protein levels in the influent caused a greater degree of membrane fouling (Corbatón-Báguena et al., 2018). Fouling in UF membranes used in food industries can be controlled using various techniques as shown in Table 2.6.

TABLE 2.5
Different Methods of Fouled Membrane Observation

S. No.	Method	Principle	Application
1	Direct observation of membrane (Alkhatim et al., 1998)	A microscope objective is positioned at the permeate side of a transparent membrane to observe particle deposition in real time by microscope	To directly observe particle deposition by an optical microscope
2	Optical laser sensor (Hamachi and Meitton-Peuchot, 1999)	The formation of deposit layer absorbs lights from a bypassing laser beam. The variation of the signal intensity after the laser beam traversed through the cake layer corresponds to the deposit layer thickness	To investigate the thickness of cake layer during microfiltration
3	Ultrasonic time-domain reflectometry (Mairal et al., 2000)	This technique uses sound waves to measure the location of a moving or stationary interface and can provide information on the physical characteristics of the media through which the waves travel	To investigate in situ measurement of membrane fouling. Provide information on the physical characteristics of the media
4	Electrical impedance spectroscopy (Chilcott et al., 2002; Gaedt et al., 2002)	An alternating current is injected directly into the membrane. Capacitance dispersion changes are measured to monitor in situ accumulation of particulates	To characterize membrane properties and to investigate membrane fouling
5	Scanned electron microscopy	SEM shows 3D images of cake and membrane at much higher Magnification	To investigate the membrane surface and fouling

Source: Adopted from Mohammad et al., 2012.

TABLE 2.6
Methods of Decreasing Flux Degradation

Method	Physical	Chemical
Pre-treatment	Pre-filtration	Precipitation Coagulation/flocculation Use of disinfectants Use of anti-scalants Adsorption
Design	Use of turbulence promoters Pulsed/reversed flow Rotating/vibrating membranes Additional fields (e.g. electric)	Choice of membrane material Membrane surface modification
Operation	Limit trans-membrane pressure Maintain a high cross-flow Periodic hydraulic cleaning Periodic mechanical Cleaning	Choice of cleaning chemicals Frequency of cleaning

Source: Adapted from Williams and Wakeman, 2000 and Mohammad et al., 2012.

2.7 RECENT ADVANCEMENTS IN THE FIELD OF ULTRAFILTRATION

UF membranes made up of polymers (polyvinylidiene fluoride, polyether sulfone, polysulfone, polyacrylonitrile, polyethylene, polypropylene, polyvinyl chloride) and ceramic are commonly used in the food sector. Membrane distillation and pervaporation are emerging membrane processes used for the reclamation of bioactive compounds from food (Castro-Munoz et al., 2020). Filtration technologies based on inorganic membranes such as ceramic membranes are recent advancements in the field of membrane filtration. These are better over conventional polymeric counterparts in terms of bearing harsh operating conditions of pH, temperature, pressure, and chemical stability; providing food safety, environmental friendliness, ease of cleaning, and sterilization etc. The ceramic membranes can be used for manufacturing of ultrafiltration (ranging from 0.1 μm to 10 nm of diameter), microfiltration (pore size above 0.1 μm in diameter), and nanofiltration (size below 10 nm of diameter). Although, membrane fouling and deep study of relationships between the solutes and the membrane surface at the molecular level are some issues that still have to be resolved (Mancinelli and Hallé, 2015). Ceramic membranes have been widely used to fractionate insoluble fibers like lignin and cellulose after pretreatment with alkali or ethanolic extraction (Galanakis et al., 2016).

2.8 CONCLUSIONS

Finally, it can be concluded that membrane filtration processes are gaining much more attention and focus in food industries owing to various advantages (environmental friendliness, cost saving, and product improvement) as compared with other conventional methods used for waste treatment. Currently, UF membrane is an economical and environmentally sustainable approach for recovery of valuable compounds from food industry wastes. However, membrane fouling is a major issue associated with the use of membranes. UF membranes can be easily fouled by various solutes such as protein and polysaccharide in the food industry, which increases filtration processing time, and cost. Therefore, development of fouling control and minimization by modifications and use of innovative methods is a major concern associated with recent researches and developments, which enable membrane technology to play an indispensable role in the food industry and others as well.

REFERENCES

Abdelkader, S., Gross, F., Winter, D., Went, J., Koschikowski, J., Geissen, S. U., & Bousselmi, L. (2019). Application of direct contact membrane distillation for saline dairy effluent treatment: performance and fouling analysis. *Environmental Science and Pollution Research, 26*(19), 18979–18992.

Afonso, M. D., Ferrer, J., & Bórquez, R. (2004). An economic assessment of proteins recovery from fish meal effluents by ultrafiltration. *Trends in food science & Technology, 15*(10), 506–512.

Alkhatim, H. S., Alcaina, M. I., Soriano, E., Iborra, M. I., Lora, J., & Arnal, J. (1998). Treatment of whey effluents from dairy industries by nanofiltration membranes. *Desalination, 119*(1–3), 177–183.

Baiano, A. (2014). Recovery of biomolecules from food wastes – A review. *Molecules, 19*(9), 14821–14842.

Baldasso, C., Barros, T. C., & Tessaro, I. C. (2011). Concentration and purification of whey proteins by ultrafiltration. *Desalination, 278*(1–3), 381–386.

Berot, S., Nau, F., Thapon, J.-L., Quemeneur, F., Jaouen, P., & Vandanjon, L., 1998, Vegetal and animal proteins, Membrane separations in the Processes of the Food Industry, G. DauŽn, F. Rene´, P. Aimar (Eds.) (Lavoisier Tech and Doc, Paris, France), pp. 373–417.

Cassano, A., Conidi, C., Ruby-Figueroa, R., & Castro-Muñoz, R. (2018). Nanofiltration and tight ultrafiltration membranes for the recovery of polyphenols from agro-food by-products. *International Journal of Molecular Sciences, 19*(2), 351.

Cassano, A., De Luca, G., Conidi, C., & Drioli, E. (2017). Effect of polyphenols-membrane interactions on the performance of membrane-based processes.A review. *Coordination Chemistry Reviews, 351*, 45–75.

Castro-Muñoz, R., Barragán-Huerta, B. E., Fíla, V., Denis, P. C., & Ruby-Figueroa, R. (2018). Current role of membrane technology: From the treatment of agro-industrial by-products up to the valorization of valuable compounds. *Waste and Biomass Valorization, 9*(4), 513–529.

Castro-Muñoz, R., Boczkaj, G., Gontarek, E., Cassano, A., & Fíla, V. (2020). Membrane technologies assisting plant-based and agro-food by-products processing: A comprehensive review. *Trends in Food Science & Technology*, 95, 219–232.

Castro-Muñoz, R., Yáñez-Fernández, J., & Fíla, V. (2016). Phenolic compounds recovered from agro-food by-products using membrane technologies: An overview. *Food Chemistry*, 213, 753–762.

Cecconet, D., Callegari, A., Hlavínek, P., & Capodaglio, A. G. (2019). Membrane bioreactors for sustainable, fit-for-purpose greywater treatment: A critical review. *Clean Technologies and Environmental Policy*, 21(4), 745–762.

Chen, H., Zhang, H., Tian, J., Shi, J., Linhardt, R. J., Ye, T. D. X., & Chen, S. (2019). Recovery of high value-added nutrients from fruit and vegetable industrial wastewater. *Comprehensive Reviews in Food Science and Food Safety*, 18(5), 1388–1402.

Chilcott, T. C., Chan, M., Gaedt, L., Nantawisarakul, T., Fane, A. G., & Coster, H. G. L. (2002). Electrical impedance spectroscopy characterisation of conducting membranes: I. Theory. *Journal of Membrane Science*, 195(2), 153–167.

Chtourou, M., Mallek, M., Dalmau, M., Mamo, J., Santos-Clotas, E., Salah, A. B., & Monclús, H. (2018). Triclosan, carbamazepine and caffeine removal by activated sludge system focusing on membrane bio-reactor. *Process Safety and Environmental Protection*, 118, 1–9.

Corbatón-Báguena, M. J., Álvarez-Blanco, S., & Vincent-Vela, M. C. (2018). Evaluation of fouling resistances during the ultrafiltration of whey model solutions. *Journal of Cleaner Production*, 172, 358–367.

Daufin, G., Escudier, J. P., Carrère, H., Bérot, S., Fillaudeau, L., & Decloux, M. (2001). Recent and emerging applications of membrane processes in the food and dairy industry. *Food and Bioproducts Processing*, 79(2), 89–102.

EX, L., & INNO, N. International Journal of Engineering and Advanced Technology.

Feins, M., & Sirkar, K. K. (2005). Novel internally staged ultrafiltration for protein purification. *Journal of Membrane Science*, 248(1–2), 137–148.

Fillaudeau, L. (1999). Cross-flow microfiltration in the brewing industry-an overview of uses and applications. *Brewers' Guardian*, 7, 22–30.

Frenkel, V. S., Cummings, G. A., Maillacheruvu, K. Y., & Tang, W. Z. (2020). Food-processing wastes. *Water Environment Research*, 92(10), 1726–1740.

Gaedt, L., Chilcott, T. C., Chan, M., Nantawisarakul, T., Fane, A. G., & Coster, H. G. L. (2002). Electrical impedance spectroscopy characterisation of conducting membranes: II. Experimental. *Journal of Membrane Science*, 195(2), 169–180.

Galanakis, C. M. (2012). Recovery of high added-value components from food wastes: Conventional, emerging technologies and commercialized applications. *Trends in Food Science & Technology*, 26(2), 68–87.

Galanakis, C. M., Castro-Muñoz, R., Cassano, A., & Conidi, C. (2016). Recovery of high-added-value compounds from food waste by membrane technology. In *Membrane technologies for biorefining* (pp. 189–215). Woodhead Publishing, Cambridge, England.

Hamachi, M., & Mietton-Peuchot, M. (1999). Experimental investigations of cake characteristics in crossflow microfiltration. Chemical Engineering Science, 54(18), 4023–4030.

Hellwig, V., & Gasser, J. (2020). Polyphenols from waste streams of food industry: Valorisation of blanch water from marzipan production. *Phytochemistry Reviews*, 19(6), 1539–1546.

Herve, D., 1994, Production de sucrerafŽne´ en sucrerie de canne, Industries AlimentairesetAgricoles, 111(7/8): 429–431.

Itoh, S., Yoshioka, K., Terakawa, M., Sekiguchi, Y., Kokubo, K., & Watanabe, A. (1999). The use of ultrafiltration membrane treated honey in food processing. *Journal of the Japanese Society for Food Science and Technology (Japan)*, 46(5), 293–301.

Kwok, R. J. (1996). Production of super VLC raw sugar in Hawaii.Experience with the new NAP ultrafiltration/softening process. *International Sugar Journal (Beet Sugar Edition) (United Kingdom)*, 46(8), 647–655.

Le, T. T., Cabaltica, A. D., & Bui, V. M. (2014). Membrane separations in dairy processing. *J. Food Res. Technol*, 2(1), 1–14.

Li, X., Jiang, L., & Li, H. (2018, September). Application of ultrafiltration technology in water treatment. In *IOP Conference Series: Earth and Environmental Science* (Vol. 186, No. 3, p. 012009). IOP Publishing.

Mairal, A. P., Greenberg, A. R., & Krantz, W. B. (2000). Investigation of membrane fouling and cleaning using ultrasonic time-domain reflectometry. *Desalination*, 130(1), 45–60.

Makhlouf-Gafsi, I., Baklouti, S., Mokni, A., Danthine, S., Attia, H., Blecker, C., & Masmoudi, M. (2016). Effect of ultrafiltration process on physico-chemical, rheological, microstructure and thermal properties of syrups from male and female date palm saps. *Food Chemistry*, *203*, 175–182.

Malmali, M., Askegaard, J., Sardari, K., Eswaranandam, S., Sengupta, A., & Wickramasinghe, S. R. (2018). Evaluation of ultrafiltration membranes for treating poultry processing wastewater. *Journal of water process engineering*, *22*, 218–226.

Mancinelli, D., & Hallé, C. (2015). Nano-filtration and ultra-filtration ceramic membranes for food processing: A mini review. *J. Membr. Sci. Technol.*, *5*(2), 100–140.

Meireles, M., Aimar, P., & Sanchez, V. (1991). Albumin denaturation during ultrafiltration: effects of operating conditions and consequences on membrane fouling. *Biotechnology and Bioengineering*, *38*(5), 528–534.

Mirabella, N., Castellani, V., & Sala, S. (2014). Current options for the valorization of food manufacturing waste: a review. *Journal of Cleaner Production*, *65*, 28–41.

Misi, S. N., & Forster, C. F. (2002). Semi-continuous anaerobic co-digestion of agro-wastes. *Environmental Technology*, *23*(4), 445–451.

Mohammad, A. W., Ng, C. Y., Lim, Y. P., & Ng, G. H. (2012). Ultrafiltration in food processing industry: review on application, membrane fouling, and fouling control. *Food and Bioprocess Technology*, *5*(4), 1143–1156.

Nor, M. Z. M., Ramchandran, L., Duke, M., & Vasiljevic, T. (2017). Integrated ultrafiltration process for the recovery of bromelain from pineapple waste mixture. *Journal of Food Process Engineering*, *40*(3), e12492.

Paes, J., da Cunha, C. R., & Viotto, L. A. (2015). Concentration of lycopene in the pulp of papaya (Carica papaya L.) by ultrafiltration on a pilot scale. *Food and Bioproducts Processing*, *96*, 296–305.

Pereira, C. D., Diaz, O., & Cobos, A. (2002). Valorization of by-products from ovine cheese manufacture: clarification by thermocalcic precipitation/microfiltration before ultrafiltration. *International Dairy Journal*, *12*(9), 773–783.

Saha, S., Boro, R., & Das, C. (2019). Treatment of tea industry wastewater using coagulation-spinning basket membrane ultrafiltration hybrid system. *Journal of Environmental Management*, *244*, 180–188.

Saxena, A., Tripathi, B. P., Kumar, M., & Shahi, V. K. (2009). Membrane-based techniques for the separation and purification of proteins: An overview. *Advances in Colloid and Interface Science*, *145*(1–2), 1–22.

Saxena, A., Tripathi, B. P., Kumar, M., & Shahi, V. K. (2009). Membrane-based techniques for the separation and purification of proteins: An overview. *Advances in Colloid and Interface Science*, *145*(1–2), 1–22.

Scordino, M., Di Mauro, A., Passerini, A., & Maccarone, E. (2007). Highly purified sugar concentrate from a residue of citrus pigments recovery process. *LWT-Food Science and Technology*, *40*(4), 713–721.

Todisco, S., Tallarico, P., & Drioli, E. (1998). Modelling and analysis of ultrafiltration effects on the quality of freshly squeezed orange juice. *INDUSTRIE ALIMENTARI*, 3–8.

Torres-León, C., Ramírez-Guzman, N., Londoño-Hernandez, L., Martinez-Medina, G. A., Díaz-Herrera, R., Navarro-Macias, V., & Aguilar, C. N. (2018). Food waste and by-products: An opportunity to minimize malnutrition and hunger in developing countries. *Frontiers in Sustainable Food Systems*, 2, 52.

Tsagaraki, E. V., & Lazarides, H. N. (2012). Fouling analysis and performance of tubular ultrafiltration on pretreated olive mill waste water. *Food and Bioprocess Technology*, *5*(2), 584–592.

Vardanega, R., Tres, M. V., Mazutti, M. A., Treichel, H., de Oliveira, D., Di Luccio, M., & Oliveira, J. V. (2013). Effect of magnetic field on the ultrafiltration of bovine serum albumin. *Bioprocess and Biosystems Engineering*, *36*(8), 1087–1093.

Williams, C., & Wakeman, R. (2000). Membrane fouling and alternative techniques for its alleviation. *Membrane Technology*, *2000*(124), 4–10.

Yang, J., Guo, J., Yang, X. Q., Wu, N. N., Zhang, J. B., Hou, J. J., & Xiao, W. K. (2014). A novel soy protein isolate prepared from soy protein concentrate using jet-cooking combined with enzyme-assisted ultrafiltration. *Journal of Food Engineering*, *143*, 25–32.

3 Waste Management of the Fruit and Vegetable Industry

Bindu[1] and Neeraj[2]
[1]Lady Irwin College, Department of Food and Nutrition & Food Technology, University of Delhi, New Delhi, India
[2] Department of Agriculture, Jharkhand Rai University, Ranchi, India
Corresponding author: bindu.bazaria@gmail.com

CONTENTS

3.1 INTRODUCTION

Fruit crops are those yielding fruits and berries, which generally are characterized by their sweet taste and their high content of organic acid, fiber, and pectin. Fruits are generally found in great numbers attached to the branches or stalks or trunks of the plants, in most cases singly, in other cases grouped in bunches and clusters (e.g. bananas and grapes). The commercial crops are cultivated in well-ordered orchards and compact plantations. Bananas, plantains, grapes, and dates are considered fruit crops by the Food and Agricultural Organization (FAO), while nuts, olives, and coconuts are not considered fruit crops. Fruits at times are classified into pome fruits (with seeds/pips contained in rather light endocarp, e.g. apples and pears) and stone fruits (with seeds/kernels enclosed in hard woody shells surrounded by the pulp or mesocarp, e.g. peaches and plums). Fruits are broadly classified as either sub-tropical/tropical fruits, or fruits of the temperate zones.

DOI: 10.1201/9781003207689-3

Vegetables are primarily temporary but annual crops mainly contain water, amounting to 70 to 95% of the total weight, very low dry matter, and accordingly nutrients like minerals and vitamins. Only those vegetables that are cultivated principally for human consumption belong to the major vegetable group. Vegetables grown principally for animal feed are to be excluded, as should vegetables cultivated for seeds.

Vegetables are grouped according to botanic characteristics as follows:

- Leafy or stem vegetables (e.g. cabbage)
- Fruit-bearing vegetables (e.g. melons)
- Flower vegetables (e.g. cauliflowers)
- Root, bulb, and tuberous vegetables (e.g. onion)
- Leguminous vegetables (e.g. green peas)
- Miscellaneous vegetables (e.g. green maize and mushrooms)

Fruits and vegetables have a crucial role in our diet and therefore the demand for such important food commodities has increased very significantly as a result of the growing world population and changing dietary habits (Schieber et al., 2001; Vilarino et al., 2017). Most nutritional and global recommendations include consumption of at least two servings of fruits and three servings of vegetables per day for adults (USDA, 2014).

Higher production and growth, and the lack of proper handling methods and infrastructure, have led to huge losses and waste of these important food commodities, as well as their components and by-products and residues. The "refuse", i.e. the parts of the fruit and vegetables that are discarded before consumption or processing, is quite substantial. Refuse includes tops, stems, seeds, rinds, peel, pods, damaged and withered leaves, and parts that are high in cellulose.

Around 89 million tons of food is wasted annually in the European Union (Stenmarck et al., 2016) and this value is expected to further increase by 40% in the next four years. At the same time, food production accounts for the vast majority of water used in the world today. So water wastage becomes automatically a serious concern along with the wastage of food. Losses and waste occur during all phases of the supply and handling chain, including during harvesting, transport to packing houses or markets, classification and grading, storage, marketing, processing, and at home before or after preparation. The *"losses"* occur throughout the supply chain from production throughout all post-harvest stages before consumption. It could also be due to problems in markets, institutions, and policy frameworks. (Parfitt et al., 2010). *"Waste"*, on the other hand, is food that is fit for consumption, but is not consumed and instead discarded, and this generally relates to consumer or retailer behavior (FAO, 2014). Food waste is a behavioral problem, which comes from our habits, customs, and traditions. Significant amounts of waste take place during religious holidays, wedding ceremonies, and family gatherings, and in restaurants and hotel dining.

Losses and waste can be assessed quantitatively and qualitatively (FAO, 2014) where quantitatively they refer to masses or volumes, which reduce the amount of food available for consumption and qualitatively, they represent decreases in edibility, nutrition, caloric value, consumer acceptability, economic value, which all are recognized before the food item is discarded.

According to a FAO estimate (FAO, 2011) pre-consumer phases are particularly critical in terms of fruit and vegetable waste (FVW) generations. To this regard, Segrè & Falasconi (2011) reported that, in Italy, up to 87% of fruits, vegetables, and cereals are discarded before reaching consumers.

3.2   CAUSES OF FRUIT AND VEGETABLE WASTE (FVW)

According to a FAO estimate (FAO, 2011) pre-consumer phases are particularly critical in terms of FVW generation. To this regard, Segrè & Falasconi (2011) reported that the majority of fruits and vegetables are discarded before reaching the consumer. In developing countries, wastes are mainly generated in agricultural production, post-harvest (Figure 3.2), and distribution stages (Figure 3.1),

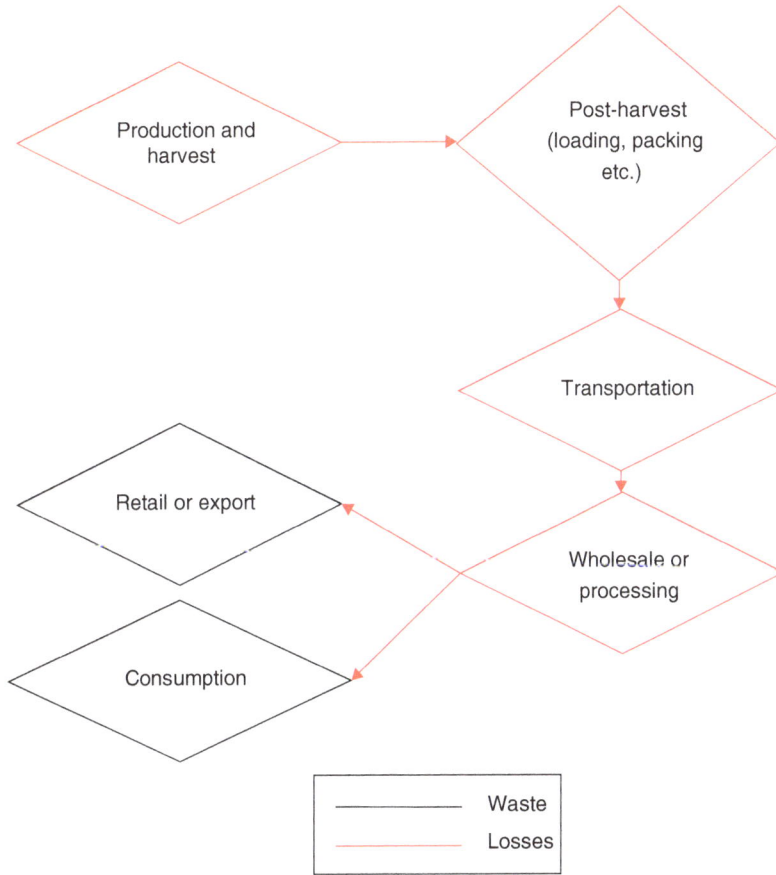

FIGURE 3.1 Post-harvest wastes and loses of fruits and vegetables in a food chain.

Source: FAO, 2019.

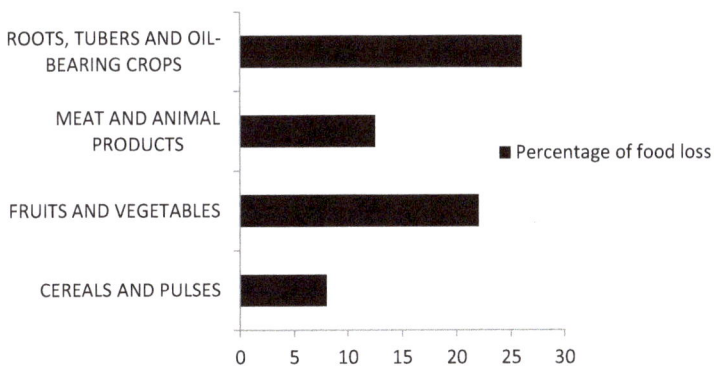

FIGURE 3.2 Food loss from post-harvest to distribution in 2016, percentages by commodity groups.

Source: FAO, 2016.

due to seasonality that leads to unsaleable excesses and to the absence of proper conservation strategies for perishable crops (Mani & Sethi, 2000). Wastes in agricultural production dominate also in industrialized countries. In this case, however, they occur mostly due to post-harvest evaluation of crops on the basis of quality standards requested by retailers and to programmed overproduction (FAO, 2011; Segrè & Falasconi, 2011).

3.3 STATUS OF FVW

Food losses in developed countries are as high as in developing countries, but in developing countries more than 40% of the food losses occur at post-harvest and processing levels (Figure 3.3), while in developed countries, more than 40% of the food losses occur at retail and consumer levels. Food waste at consumer level in industrialized countries (222 million ton) is almost as high as the total net food production in sub-Saharan Africa (230 million ton).

In the fruit and vegetable commodity group, losses in agricultural production dominate for all the developed regions, mostly due to post-harvest fruit and vegetable grading caused by quality standards set by retailers. Waste at the end of the food supply chain is also substantial in industrialized regions, with 15–30% of purchases by mass discarded by consumers (FAO, 2011). Table 3.1 demonstrates the nature and extent of FVW and losses.

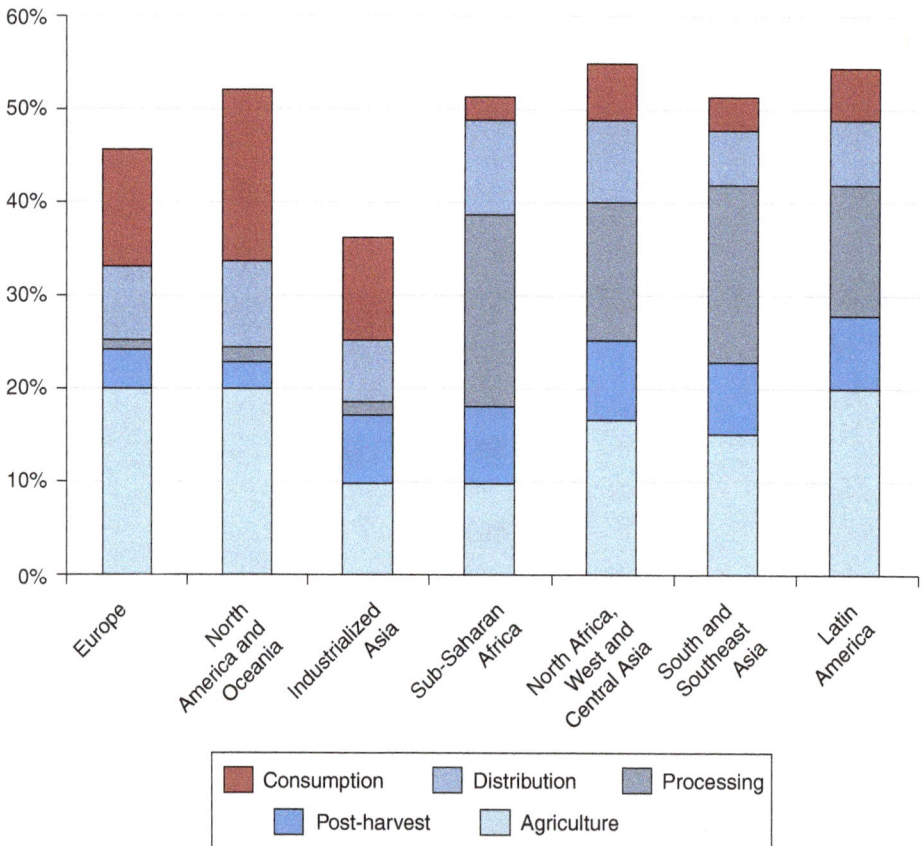

FIGURE 3.3   Food losses – fruits and vegetables in different regions.

Source: FAO, 2016.

TABLE 3.1
Nature and Extent of Fruit and Vegetable Losses, Waste

Commodity	Waste (%)	Nature of waste	Reference
Apple	20–30	Pomace, peel, and seeds	Sagar et al., 2018
Banana	20–30	Peel	Sagar et al., 2018
Citrus(Orange,Lime)	50, 60	Rag, peel, and seeds	Sagar et al., 2018
Dragon fruit	30–45	Rind and seeds	Cheok et al., 2018
Durian	60–70	Skin and seeds	Siriphanich and Yahia (2011)
Grapes	20	Steam and pomace	Gupta and Joshi (2000)
Guava	10	Core, seeds, peel	Gupta and Joshi (2000)
Jackfruit	50–70		Saxena et al., 2011
Mango	40–60	Skin and stone	Mitra et al., 2013
Papaya	10–20	Skin	Parni and Verma (2014)
Passion fruit	45–50	Rind, seeds	Almeida et al., 2015
Pineapple	35	Peel, core, trimmings, shreds	Choonut et al., 2014
Pear	-	Peel, core	Gupta and Joshi (2000)
Tomato	20–30	Core, peel, seeds	Gupta and Joshi (2000)
Potato	15	Peel, coarse solids	Sagar et al., 2018
Peas	60–65	Pod	-
Onion	-	Skin	Sagar et al., 2018

The non-edible produce of some fruits accounts between 25%–30% of the whole product (Ajila et al., 2010). The processing of fruits produces two types of waste: solid waste, e.g. peel/skin (Fernandez et al., 2004), seeds, and stones and liquid waste juice and wash water. The fruit industry typically generates large volumes of effluents and solid waste. The effluents contain high organic loads, cleansing and blanching agents, salt, and suspended solids such as fibers and soil particles. They may also contain pesticide residues from the washing of raw materials or discarded fruits.

By-products are produced during many steps of the processed fruit production chain. Such by-products may come from overstock merchandize screenings, tops, stems, pulps, pomace, skins, hulls, peels, meals, seeds, fines, green chop, pressed cake, dried fruit, and fresh fruit waste unsold merchandize because of passed "sell by" date rinse water or cooking materials below standard (size, color, and texture) merchandize trimmings' fruit harvest.

There are possible ways to use some solid fruit wastes. One major goal in using fruit wastes is to ensure a reasonable microbiological quality in them. This means that one should process waste products on the same day that they become available as the solid waste also contains moldy fruit (discarded during processing), insects, leaves, stems, soils, etc. This will contaminate any products derived from such wastes. Therefore, some preliminary separation is needed during processing, such as peel and waste pulp in one bin, moldy parts, leaves, soil, etc. in a second bin, and may be discarded stones, seeds, etc., in a third bin.

3.4 IMPORTANCE OF FVW WASTE

Nowadays, it has been a trend to find biopolymers as an alternative to synthetic substances, which are commonly used in the food industry. The various epidemiological studies have pointed out that the regular consumption of fruits imparts health benefits, e.g. reduced risk of coronary heart disease and stroke, as well as certain types of cancer. The obtained products and by-products throughout the minimal processing of the fruits have been analyzed for phytochemical content and antioxidant status. These compounds could be responsible for free radical inhibition activity. Many studies have shown that the content of phytochemical compounds is higher in peel and seeds in comparison to the

TABLE 3.2
The Nutrient Composition of the Solid Wastes from Fruits

Waste	Moisture (%)	Protein (%)	Fat (%)	Ash (%)	Fiber (%)	Carbohydrate (%)
Apple pomace	-	2.99	1.71	1.65	16.16	17.35
Mango seed kernel	8.2	8.50	8.85	3.66	-	74.49
Passion fruit peel	81.9	2.56	0.12	1.47	5.01	-
Banana peel	79.2	0.83	0.78	2.11	1.72	5.00
Jack fruit (Whole)	8.5	7.50	11.82	6.50	30.77	14.16

Source: Mani and Sethi, 2000.

edible tissue. For example, Gorinstein et al. (2001) found that the total phenolic compounds in the peels of lemons, oranges, and grapefruits were 15% higher than that of the pulp of these fruits. Other studies in parallel have reported that pomegranate peels contain 249.4 mg/g of phenolic compounds as compared to only 24.4 mg/g phenolic compounds found in the pulp of pomegranates. Apple peels were found to contain up to 3300 mg/100 g of dry weight of phenolic compounds (Wolfe and Liu, 2003). Also FVW is a potential source of flavonoids that represent a large family of low weight molecular phenolic and according to their structures involve several chemical categories such as flavones, flavonones, isoflavones, isoflavans, flavanol, and flavonols. Table 3.2 shows the nutrient composition of some of the solid wastes from fruits.

3.5 FVW MANAGEMENT

Waste management includes all measures involved in managing unwanted or discarded materials from inception to final stage of disposal. In general, it is the collection, transport, recovery, and disposal of waste, including the supervision of such operations. FVW poses disposal and environmental problems, due to its high biodegradability. In addition, it represents a loss of valuable biomass and nutrients as well as an economic loss. For these reasons, in the last years, great attention has been focused on the development of policies and methods for its management. Waste management strategies can be classified with respect to the final disposition of waste and ordered according to their priority: minimization and prevention (reduction) of waste generation, recycling and reuse, energy recovery, and landfilling. FVW can be considered a cheap, readily available feedstock for the potential recovery of energy, water, and valuable ingredients/products. To maximally exploit these potentialities, an integrated approach to FVW management should be developed by selecting, and eventually combining, the most efficacious strategies of reuse, recycle and energy recovery. Such an approach results from the application of some rational steps like waste characterization, output definition, process design, and feasibility study.

In the past, FVW was mixed into municipal waste streams and sent to landfills or incinerators (without energy recovery) for final disposal. However, this is not a good option for FVW, due to its high water content, which is, in turn, responsible for microbiological instability, formation of off-odors and leachate (Lin et al., 2011). In contrast, FVW has a great potential for reuse, recycling, and energy recovery.

Reduction has the top priority in the waste hierarchy and mostly depends on production practices (Demirbas, 2011). It has been estimated that huge amounts of fruits and vegetables are wasted because products do not fulfil quality standards set by retailers or consumers (Mena et al., 2011). This small-sized or misshaped fruit and vegetables are usually defined "substandard". Different strategies have been proposed and implemented to tackle waste of substandard fruit and vegetables. This particular practice is followed in the developed countries.

Reuse indicates the use of waste materials for other purposes without or with minor modification of their properties (Manzocco et al., 2016). Reuse strategies for FVW are nowadays limited to soil amendment and animal feed. This practice is based on the ability of organic waste to increase properties of polluted soil by immobilizing trace metals and metalloids, preventing their transfer to groundwater and living organism, and promoting the establishment of plants. The fiber content of FVW can be exploited to formulate animal feeds with increased nutritional value (San Martin et al., 2016). However, also this reuse strategy is limited by some drawbacks. The high water content, often exceeding 80%, makes these wastes prone to microbiological contamination. A partial drying is thus usually required, which add up the extra cost to the whole process. In addition, low protein content and high presence of indigestible compounds are not always suitable for animal feed (Clemente et al., 2015).

The strategies based on the recovery of waste materials after a major modification of their characteristics are defined as **recycle**. Because of its intrinsic characteristics of high content of water and fiber and low protein content, a substantial modification of FVW is usually required to maximally exploit its potentialities. Recycle of FVW offers thus more possibilities than its reuse. The recycle strategies for FVW can be divided into strategies in which the whole waste mass is recycled (composting, processing to flour, conversion into water) and strategies in which specific compounds are extracted. Processing into flour of FVW has been exploited with different purposes. The fibrous structure and the high contact surface of FVW flour has been used to adsorb pollutants such as dyes and heavy metals from water and the ground. FVW flour has also been used as an ingredient for the formulation of products rich in functional compounds such as polyphenols and fiber (Ferreira et al., 2015). Some recent studies relevant to bioactive extraction (e.g. carotenoids, essential oils, polyphenols, anthocyanins) from FVW using novel technologies include the use of ultrasounds, supercritical carbon dioxide, microwaves, and pulsed electric fields (Amiri-Rigi et al., 2016). For these reasons, extraction of specific compounds from FVW could be an affordable, sustainable, and even profitable recycle strategy for industries. However, it should be considered that novel technologies often require high initial investment and their industrial application is still limited.

Energy recovery, also called waste-to-energy, is performed in order to recover the energy contained in the waste material. Energy from waste materials can be recovered by several strategies, including thermochemical conversions, such as incineration, pyrolysis, and gasification or biochemical strategies, such as anaerobic digestion and fermentation. Anaerobic digestion (AD) has been widely used for organic waste disposal. AD is a method to decompose organic matter using several anaerobic microorganisms under oxygen-free conditions. After the treatment, the end product is represented by biogas (60% methane, 40% carbon dioxide) and digestate (Sheets et al., 2015). But, the improper application of AD digestate has led to serious environmental problems such as over-fertilization and pathogen contamination (Nkoa, 2014). Water can also be considered a valuable output of a recycle strategy.

3.6 WASTES AND BY-PRODUCTS OF FRUIT PROCESSING

Generally, the majority (60–65%) of food industry by-products become waste and a burden to the environment. In order to meet stringent environmental requirements, modern fruit processing should minimize the amount of by-products and waste, decrease energy utilization, produce high-quality foodstuffs without polluting the soil, air, and spas. Fruit and vegetable crops are processed into various preserved products, which lead to the selection and elimination of components unsuitable for human consumption, i.e. by-products and wastes. There is no sharp limit between the two categories as industries processing substances of biological origin are in close cooperation; the by-product of one industry can be a valuable secondary raw material for another. Apple processing is a good example for the aforementioned principle. Although apples (canned apples, apple juice, etc.) are the finished product, and the by-product, apple pomace, is the secondary raw material of pectin production. Fruit processing wastes differ from other wastes as they are organic and therefore, decompose. These can be utilized in different ways depending on their texture and content.

3.7   BY-PRODUCT UTILIZATION FROM APPLE PROCESSING

Apples are the most significant fruit in the temperate zone, economically. Apple production achieves approximately 10% of the world's fruit production. Pectin is the most common by-product produced from fruit pomace. The various raw materials yield different amounts of extractable pectin: pomace, 10–15%; sugar beet chips, 10–20%; sunflower-infructescence, 15–25%; and citrus peels, 20–35%. It is one of the most versatile stabilizers available with its gelling, thickening, and stabilizing properties, which make it an essential additive in the production of many food products. It produces the desired texture, limits the creation of water/juice on top of the surface as well as an even distribution of fruit in the product. Pectin is a key stabilizer and is used in many food products as fruit applications in jams, jellies, and desserts; bakery fillings and toppings in fruit preparations for dairy applications in acidified milk and protein drinks, yogurts (thickening) confectionery in fruit jellies, neutral jellies; beverages' nutritional and health products' pharmaceutical and medical applications. This wide range of applications explains the need for many different types of commercial pectin, which are sold according to their application.

After distilling the alcohol used for the precipitation of pectin and fruit extracts, such as sugar and fruit acids, the natural flavors remain in the solutions. Apple extract obtained from pomace is used as sweetening agents for the preservation of freshness and/or coloring of food. At a further processing stage, special technologies are used to remove dark natural coloring agents, mineral substances, and fruit acids from apple fruit extracts. The resulting products only contain the sugars of the respective raw material that has been processed. They are used by the food industry as sweetening agents (Khachatourians et al., 2001). A further possibility is to ferment it to form apple ethanol. After adding wine yeast to the apple pomace, remaining from fruit juice and apple pulp production, the marc is fermented at 30°C in solid phase, resulting in a liquid with a 4–5% ethyl alcohol content. After pectin extraction from apple pomace, the various residues of the original raw material are dried and pressed into pellets and have huge demand as fodder (Bennett, 2002). Apple peels were reported to have up to 3300 mg/100 g DM of phenolic content.

3.8   BY-PRODUCTS FROM GUAVA PROCESSING

Guava is one such fruit whose every part can be utilized for human consumption, either as fresh or in the processed form like many natural additives and functional food ingredients (Schieber et al., 2001). The extraction of pigment lycopene is one such possibility. Guava is one of the best natural sources of food-grade pectin, which finds uses in various food product formulations as a thickening and gelling agent. Pectin can be obtained from guava fruits by boiling with water and then precipitating with ethanol. Also the use of sodium hexametaphosphate (0.25–0.75%) or a 1:1 mixture of ammonium oxalate and oxalic acid (0.25–0.75%) increases the yield of high jelly grade pectin. Guava seeds, which are usually discarded after processing, contain 5–13% oil that is rich in essential fatty acids. Guava seed oil (GSO) is very rich in oleic (54%) and linoleic (29%) acid, making it suitable for use in salad dressing. However, linoleic acid was the major fatty acid in GSO (76.4%). Isolation of proteins having good nutritional quality from guava seeds is another possibility. The further possibility is the production of ethanol by fermentation using *Saccharomyces cerevisiae*. A yield of 5.8% (w/v) ethanol has been obtained from such waste products of guava processing plants. The results of very recent investigations indicate that peel and pulp of guava fruits could be used as a source of antioxidant dietary fiber. Fontanari et al. (2008) obtained the value of 67.00g/100g for total dietary fiber for guava seed powder and also suggested that new products based on fibers obtained from residues of this fruit can be formulated to prevent diseases, especially those related to the gastrointestinal tract and the cardiovascular system.

3.9 BY-PRODUCTS FROM PAPAYA PROCESSING

Enzyme *"Papain"* is the major by-product from dried latex derived from raw as well as ripened papaya fruit, which contains a protein hydrolyzing enzyme. This enzyme has many technological applications in the food industry like meat tenderization, beverages, and animal feeds; pharmaceutical industry; textile industry and detergents; paper and adhesives; sewage and effluent treatment. Papain in solution is easily oxidized by exposure to air, high temperature (above 70°C), or sunlight. For improving the stability of papain enzyme, a number of chemicals such as ascorbic acid, sodium ascorbate, erythorbic acid, sodium metabisulfite, 4-hexylresorcinol, TBHQ, rutin, -tocopherol and sucrose has been tried. Among the food applications, the use of papain in chill haze removal during beer clarification as well as in the tenderization of meat has shown a steady increase over the past years. Based on rat feeding studies, it has been suggested that normal consumption of ripe papaya during pregnancy may not pose any significant risk but unripe or semi-ripe papaya may be unsafe in pregnancy, as the high concentration of latex produces marked uterine contractions and can lead to the conditions of abortion (Adebowale et al., 2002).The production of papain latex is an economically important alternate product of papaya cultivation. Four to five longitudinal cuts made during the morning hours gave the highest yield of latex. Using plastic or stainless steel knives, fruits are lanced and the latex is collected in glass or porcelain containers. The latex, which hardens within 15 min, is then precipitated with alcohol, washed with acetone, and dried in a vacuum drier for obtaining good quality crude papain. The addition of potassium metabisulfite (0.05%) in small quantities to the liquid latex before drying acts as a preservative and improves the quality of crude papain powder.

The green fruits of papaya are a rich source of pectin (10% pectin on dry basis), which can be extracted for various uses in the food industry. Peel is shown to be higher in pectin content than the papaya pulp, and pectin content increases at a higher rate with fruit maturity up to a stage. The variety of the fruit, the growing conditions, and the stage of maturity of fruit are all known to influence the chemical composition of pectin.

Commercially, the three most important enzymes from fruit are papain (from papaya), bromelain (from pineapple), and ficin (from figs). Each is a protein-degrading enzyme used in such applications as meat tenderizers, and washing powders and is also used in leather tanning and beer brewing. But it is not economical to harvest these enzymes from fruit processing waste. Changes in both large-scale production with higher quality standards and use of biotechnology to produce "synthetic" enzymes also mean that small-scale producers will be unlikely to compete effectively and hence their market is therefore declining.

3.10 BY-PRODUCTS FROM CITRUS PROCESSING

Peel from citrus fruits (orange, lemon, and grapefruit) can be candied for use in, for example, baked goods and snack food. In addition, shreds of peel are used in marmalades, similar to the process of candying. Citrus peel is a potential source of certain essential oils and yields about 0.5–3 kg oil/tonne of fruit. The essential oils extracted are used for various purposes, such as pharmaceuticals, confectioneries, cosmetics, alcoholic beverages, and also for improving the shelf-life and safety of various foodstuffs.

Lemon (*Citrus limon*, family Rutaceae) peel's outer layer is called flavedo, the color of which differs from green to yellow. Flavedo is a rich source of essential oils that has been used since early times in flavoring and fragrance industries (Mohapatra et al., 2010). Orange juice by-products like peel, pulp, and seeds shows high levels of total dietary fiber and an ideal ratio between soluble and insoluble fiber. Orange fiber has proven to be a promising alternative as a fat replacer in ice cream since it led to a 70% reduction of fat without causing significant changes in product attributes, such as color, odor, and texture. The major component of lemon peel is albedo, which is a spongy and cellulosic layer under the flavedo and has high dietary fiber content. Dietary fiber from citrus fruits has higher proportion of soluble dietary fiber, when compared to other alternative sources, such as cereals (Gorinstein et al., 2001). Citrus fruits also have a better quality than other sources of dietary fiber due to the presence of associated bioactive compounds (flavonoids and vitamin C) with antioxidant properties, which may provide additional health-promoting effects. Citrus waste is a rich source of phenolic compounds because citrus peel contains a higher quantity of polyphenols in comparison with the edible part of the fruit.

3.11 BY-PRODUCTS FROM MANGO PROCESSING

Mango being seasonal fruit is used for the processing of products such as puree, nectar, leather, pickles, canned slices, and chutney. These products experience worldwide popularity and have also gained importance in national and international markets. During processing of mango, by-products such as peel and kernel are generated. Mango peels and seeds are rich in valuable bioactive compounds such as polyphenols, carotenoids, dietary fibers, enzymes phytosterols, and tocopherol; whereas and the peel extract exhibits potential antioxidant properties. Moreover the processing of mango by-products reduces the waste disposal problem, adds value to the product for food and other industrial use, and the isolated active component can be used in food fortification (Ravani and Joshi, 2013). Major by-products of mango processing are peels and stones, amounting to 35–60% of the total fruit weight. The dietary fiber contents of mango peel represent mainly cellulose fractions, which is a major part of insoluble dietary fiber. The total dietary fiber content in dry mango peel ranges from 45% to 78%. The soluble dietary fiber content in both raw and ripe mango peels are more than 35% of total dietary fiber. Insoluble dietary fiber relates to both water absorption and intestinal regulation whereas soluble dietary fiber associates with cholesterol in blood and diminishes its intestinal absorption (Palafox-Carlos et al., 2010). Mango peels are a good source of bioactive compounds such as polyphenols, carotenoids, vitamins, enzymes, and dietary fiber; its extracts have also good antioxidant properties.

3.12 BY-PRODUCT UTILIZATION FROM BANANA PROCESSING

Banana fruit is one of the most popular and highly nutritional fruit crops cultivated in more than 130 countries. The commonly available human consumable banana is a member of *Musa accuminata* species. Banana trees can bear fruit (nearly 20 fruits of banana grow to a hand (tier) and 3–20 hands can grow in a cluster) once in a lifetime, so lots of biomass waste is generated from the tree. Banana waste has been of significant importance because it has various parts that can be utilized such as banana fruit peels, trunks, pseudo-stems, leaves, and piths.

Among several wastes generated due to the banana tree and fruit, the banana peel is one of the important wastes generated in large quantities due to banana fruit consumption. Banana peel contributes about 40% of total weight of the fresh banana fruit (Anhwange et al., 2008). Banana peel contains carbon-rich organic compounds such as cellulose, hemicellulose, pectin substances, chlorophyll pigments, and some other low molecular weight compounds (Xiaomin et al., 2007). It is experimentally verified to be a good source of pectin (10–21%), lignin (6–12%), cellulose (7.6–9.6%), hemicelluloses (6.4–9.4%), and galacturonic acid (Mohapatra et al., 2010).

The pseudo-stem of the banana produces a single bunch of banana fruit before drying and being replaced by a new pseudo-stem. These banana wastes are disposed of by the cultivators into the rivers, lakes, or dump in low-lying areas; it is causing a serious threat to the biosphere due to the release of greenhouse gas (Shah et al., 2005) if its waste is not properly managed. Therefore, the selection of banana tree and fruit waste as an effective adsorbent material is a smart choice for a sustainable future. The parts of banana wastes have been extensively studied as an adsorbent against a wide range of pollutants.

Polysaccharides in banana peel can be either chemically or microbiologically transformed into poly-hydroxybutyrates (PHBs), these biopolymers are hydrophobic, and bear similar mechanical properties to polypropylene or polyethylene. Also lignocellulosic micro/nanofibers (LCMNFs) from banana leaf residue have been used to restore mechanical properties of recycled fluting paper.

3.13 BY-PRODUCT UTILIZATION FROM PINEAPPLE PROCESSING

Pineapple (*Ananas comosus L. Merr*) is an important tropical and subtropical plant widely cultivated in many places including the Philippines, Thailand, Malaysia, Mexico, South Africa, Costa Rica, Nigeria, Brazil, and China. Its waste consists of residual skin, peel, pulps, stem, and leaves, which are by-products of the pineapple processing industries. Pineapple waste is an agricultural waste, which is generally described as waste produced from farming activities; it can be from natural sources (organic) or un-natural sources (inorganic). It is mostly generated from poor handling of fresh fruit, storage, or lack of a good and reliable transportation system. Improper management of these wastes results in the deterioration of environmental quality, which can be attributed mainly to the degradation of the sugar-rich contents (Franca & Oliveira, 2009).

Bromelain is a proteolytic enzyme present in the stem of pineapple, known as stem bromelain, and also in fruit. It is widely used in pharmaceutical and food industries mostly as a tenderizer and a dietary supplement (Hebbar et al., 2008). It is a crude extract of pineapple that contains, among other components, various closely related proteinases, demonstrating, in vitro and in vivo, anti-edematous, anti-inflammatory, antithrombotic, and firinolytic activities, and has potential as an anticancer agent.

The peculiar flavor of pineapple waste has been utilized in food industries to enhance taste (citrus). Kumar et al. (2003) has reported the use of pineapple waste as a substrate to produce citric acid by *Aspergillus niger* under solid-state fermentation where variation in the methanol concentration resulted in yield increase from 37.8% to 54.2%. Pineapple waste can be further metabolized to produce ethanol. *Saccharomyces cerevisiae* and *Zymomonas mobilis* are some examples of the microorganisms used at industrial scale for this purpose. However, due to the low availability of fermentable sugars, a pre-treatment step using enzymes (cellulase, hemicellulase) is necessary. Both organisms were capable of producing about 8% ethanol from pineapple waste in 48 h after pre-treatment with enzymes.

Phytochemicals from pineapple peel and leaf also showed a high antioxidant activity with high phenolic compounds that can be harnessed and utilized in various ways the examples are myricetin, salicylic acid, tannic acid, *trans*-cinnamic acid, and *p*-coumaric acid. The leaf also has a significant amount of phytosterol content, particularly beta-sitosterol, stigmasterol, and campesterol.

Pineapple fruit residues also act as an effective bio-sorbent to remove toxic metals like Hg, Pb, Cd, Cu, Zn, and Ni. Also some researchers have reported that dietary fiber powder prepared from pineapple waste contains 70.6% total dietary fiber (TDF), which is similar to commercial dietary fiber from apple and citrus fruits. Pineapple leaves have been used to make coarse textiles and threads in some Southeast Asian countries (Tran, 2006).

3.14 BY-PRODUCT UTILIZATION FROM JACKFRUIT PROCESSING

Jackfruit is a popular gigantic fruit in the tropical parts of the world. Jackfruit (*Artocarpus heterophyllus* L., family Mulberry) tree produces around 200–500 fruits annually. Nonetheless, only 30–35% is the edible portion of the bulb, whereas the rest of the skin (55–62%) and seed (8–10%) are regarded as wastes (Saxena et al., 2011). About 59% of the fruit's outer peel is composed of fiber, which is fairly rich in calcium and pectin. In jackfruit processing industries, a huge amount of inedible parts such as peel are generated as waste, and usually they are used as animal feed. After consumption of the edible part, the fruit peels are separated and dumped into municipal landfills that causes serious pollution and disposal (solid-waste management) problems (Foo & Hameed, 2012). The utilization of jackfruit peels for food application is surprisingly low with only 16.72–17.63% of pectin extraction, while 90% is for non-food applications of biofilm, biosorbent, biohydrogen, and activated carbon. Jackfruit peel flour is a good biodegradability promoter as evidenced by tensile properties' reduction of film prepared from poly(vinyl alcohol) and jackfruit peel flour, which consequently stimulated the degradation rate (Ooi et al., 2011).

The jackfruit seed has been utilized more than the peel by 58% and 43%, respectively. As jackfruit seeds naturally possess high amounts of starch of more than 90% (Madruga et al., 2014) and high protein content, the processing of jackfruit seeds into flour has become a current trend, which covered 75% of jackfruit seed utilization. In addition, the jackfruit seed has been discovered to have antimicrobial activities against both gram positive and gram negative bacterial strains (Debnath et al., 2011).

3.15   BY-PRODUCT UTILIZATION FROM GRAPE PROCESSING

Grape is a very high value crop that is grown and produced in vineyards. According to the FAO (Food and Agriculture Organization, 2014) statistics, grape is the largest fruit crop in the world. The annual production worldwide can reach almost 70 million tons and around 80% is used for wine production, whereas 20% of processed grapes remain as pomace. Wineries constitute one of the most important agro industrial sectors world-wide (Spigno et al., 2017). The grape pomace is the solid material that remains after the pressing and the fermentation processes. It consists of two main fractions known as the seedless pomace (pulp, skin, and stem) and the grape seed itself (Beres et al., 2017). The by-products generated after grape exploitation constitute a very cheap source for the extraction and are characterized by the presence of high value ingredients including hydrocolloids, dietary fibers, lipids, proteins, and natural antioxidant flavonoids mainly in the form of phenolic, which can be used as dietary supplements, or in the production of phytochemicals, thus providing an important economic advantage. Linoleic acid that accounts for 71.965% of the grape seed oil (Da Porto et al., 2013) also exhibits anti-inflammatory and anti-aging effects when applied in facial skin care products. Extracts of grape pomace were also successfully incorporated into chitosan edible films (hydrophobic and hydrophilic), providing antioxidant properties and promising shelf-life extension (Ferreira et al., 2014). Grape pomace has an incredible potential to synthesize via fermentation hydrolytic enzymes such as xylanase, exo-polygalacturonase (exo-PG), and cellulose.

Moreover, grape pomace can be utilized as a substrate for production of microbial proteins (Matassa et al., 2016). Bioconversion of grape pomace into microbial protein could be a promising step in reducing the protein malnutrition, especially in developing countries.

3.16   BY-PRODUCT UTILIZATION FROM EXOTIC FRUIT PROCESSING

Processing of exotic fruits often generates even higher amounts of by-products than fruits from the temperate zones because in many cases, the inedible part of tropical and subtropical fruits is larger.

Dragon fruit (Hylocereus sp.) also known as "pitaya", exists in two common genotypes, which are distinguished by its flesh color of red (Hylocereus polyrhizus) and white (Hylocereus undatus). The dragon fruit comprises 22–44% of peel and 2–4% of seed discarded as wastes (Liaotrakoon et al., 2013b). These wastes are also researchers' interests to explore due to the increase in consumer demands for natural health-promoting bioactive compounds. Besides having antioxidant properties (Zhuang et al., 2012), dragon fruit peel extracts, which consist of the main components of b-amyrin, a-amyrin, and g-sitosterol, have demonstrated good cytotoxic activities against human prostate, breast, and gastric carcinoma cell lines (Luo et al., 2014). Pectin is the main substance derived from dragon fruit peel and a popular subject of recent research. Peels from white-flesh and red-flesh dragon fruits have been discovered containing significant amounts of pectic substances that were lowly methyl-esterified (Liaotrakoon et al., 2013a). Seeds of two varieties of dragon fruits, Hylocereus undatus and Hylocereus polyrhizus, have been revealed containing about 50% essential fatty acids while 48% is linoleic acid, and 1.5% is linolenic acid. Owing to the superb quality of this essential oil, dragon fruit seed oil was spray-dried microencapsulated to enhance its oxidative stability. In addition, the dragon fruit seed oil, which was observed containing a relatively high amount of tocopherols and low oxidation rate after three months of storage either at cold or room temperature, has further driven its value as a good oxidative stability essential oil (Liaotrakoon et al., 2013b).

Durian is popular as "king of fruit" in South-East Asia countries due to its rich sweet creamy delicious pulp and distinctive aroma. In recent decades, the durian fruit and its wastes have been explored as food additives for various uses as preservatives, thickening agents, antimicrobial agents, and for potential pharmaceutical applications (Ho and Bhat, 2015).

The high percentage of non-edible parts from 60 to 81% of the durian has drawn the interest of researchers in pursuance of the transformation of durian biomass into highly valuable commodity such as a biosorbent, insulator, agro-pectin derivative, polysaccharide gel, and in biotechnological development (Foo and Hameed, 2011). Obviously, the durian peel has been discovered as having more utilities than the seed, which accounted about 85% for peel and 16% for seed, respectively. Useful compounds that have been derived from the durian peel for food applications were pectin, polysaccharide gel, and fiber (Penjumras et al., 2014). The pasting properties of high dietary fiber content starch extracted from the durian seed have been characterized for its potential uses in the food, pharmaceutical, and cosmetics industries.

Passion fruit (Passiflora Edulis f. Flavicarpa L.) belongs to the family passifloraceae, has gained its popularity recently due to its pleasant taste and high nutritional values. This fruit produces 52% of residue from the juice industry (Almeida et al., 2015). The residues are mainly peel and seed, which comprised of 45–52% and 1–4%, respectively, from a whole fruit. Due to the large percentage of peels, the utilization of peel is approximately 85% while the seed is only 17%. The effectiveness of passion fruit peel extract on antihypertensive has been demonstrated in an in vivo study (Lewis et al., 2013). Besides the peel, the antifungal protein of passion fruit seeds has been described in detail. Also, the recovery of pectin from passion fruit peels is regarded as an effective way to utilize passion fruit wastes. Passion fruit peel has been revealed as possessing a high content of total dietary fiber of more than 73%. The incorporation of this passion fruit peel fiber has been discovered enhancing probiotic viability, fatty acid profile, and increased conjugated linoleic acid content in yoghurts. In non-food applications, peels of passion fruit were used as substrates for enhanced production of b-glucosidases, anenzyme that acts as catalyst for various biotechnology processes including biomass hydrolysis for bioethanol production, by Penicillium verruculosum (Almeida et al., 2015). Passion fruit peel has been recommended as an alternative low-cost material to develop biodegradable flexible films and as biosorbent to remove lead (II) from aqueous solution (Gerola et al., 2013). The passion fruit seed has been revealed to contain a rich amount of crude lipid (24.5 %) and total dietary fiber (64.8 %) for which it has been suggested to be used as a fiber source or low calorie bulk ingredient for food applications.

3.17 BY-PRODUCTS FROM VEGETABLE PROCESSING

The main constituents of by-products from vegetables such as potatoes, onions, tomatoes, and carrots are fibers, which makes up the bulk mass. As can be expected, the composition of secondary metabolites varies significantly with the source. The predominant phenolic compound of potato peels is chlorogenic acid. The peels contain also steroidal alkaloids, which are considered toxic and, therefore, need to be removed from extracts (Schieber and Aranda Saldaña, 2009).

Onion (*Allium cepa* L.) has been valued as a food and a medicinal plant since ancient times. It is widely cultivated, second only to tomato, and is a vegetable bulb crop known to most cultures and consumed worldwide (FAO, 2012). Approximately 37% of fresh onions are discarded during processing as wastes and cannot be used by customers. The onion wastes include onion skins generated during industrial peeling, two outer fleshy scales, roots, top and bottom of the bulbs, and also undersized, malformed, diseased, or the damaged bulbs. These onion waste products are a real problem for the industry because they are not suitable as a fodder to animals due to the strong characteristic pungent odor (Benitéz et al., 2013) and cannot be used as an organic fertilizer because of rapid development of phytopathogenic agents, such as, *Sclerotium cepivorum*. Onion peels may be exploited for the extraction of fiber, flavonols, fructans, and alk(en)yl cysteine sulfoxides. It has been reported that quercetin 4'-glucoside and quercetin 3, 4'-diglucoside are the main flavonols in fresh onion, while onion skins contain higher concentrations of quercetin aglycon. Quercetin has antioxidant and free radical scavenging characteristics along with the capability to protect against human cardiovascular diseases (Bonaccorsi et al., 2008). Onions also contain essential oils in very low amounts, which are mainly associated with aroma and tear-producing properties.

Vinegar made from onions is of interest as a potent new functional condiment with high mineral and polyphenol content. The concentration of total amino acids in onion vinegar has been showed to be 2100 mg/l, which is higher than those in other types of vinegar. Oleic acid (omega-9) is a monounsaturated fatty acid, one of the important fatty acids found in onion oil. Linoleic acid and linolenic acid are also found in the onion oils and are known as dietary essential fatty acids. These cannot be synthesized synthetically (Paola et al., 2004).

Onion contains toxic components that may damage red blood cells and provoke hemolytic anemia, accompanied by formation of Heinz bodies (HzB) in erythrocytes of animals, such as, cattle, buffaloes, sheep, horses, dogs, and cats.

Tomato pomace consists of the dried and crushed skins and seeds of the fruit. About 3–7% of the raw material is lost as waste during tomato juice processing. Garcia Herrera et al. (2010) has stated that tomato waste peels are rich in fiber and macronutrients (proteins, ash, total available carbohydrates, and soluble sugars). Peels are mainly composed by carbohydrates, with an average value of 80% of total dietary fiber, being insoluble fiber, and suggesting their use as a food ingredient of new functional foods. The principal carotenoid in tomato pomace is lycopene, which has attracted most attention in the valorization of this by-product, due to its putative health promoting properties. In addition, fiber, seed oil, and proteins may be recovered as valuable components (Lu et al., 2019).

Carrot yields a low level of juices and up to 50% of the material remains as pomace, which is generally disposed as feed or manure. Carrot pomace contains large amounts of valuable compounds (4–5% protein, 8–9% reducing sugar, 5–6% minerals and 37–48% total dietary fiber (on dry weight basis) such as carotenoids, dietary fiber (Nocolle et al., 2003), uronic acids and neutral sugars. Hence, by-products of carrot after juice extraction represent promising sources of compounds with bioactive properties that could be explored in the development of food ingredients and dietary supplements. In carrot pomace, considerable amounts of b-carotene and a-carotene are present because the extraction of the lipophilic carotenoids during juice production is incomplete (Surbhi et al., 2018).

Olive pomace consisting of fragments of skin, pulp, pieces of kernels, and some oil is one of the main by-products of olive food industry processing. The reported total dietary fiber content in olive pomaceis 800 g/kg dry matter with the majority being insoluble. This insoluble fraction contained primarily cellulose, xyloglucans and uronic acids while the soluble fraction was composed mainly of arabinans and uronic acids.

Pea (*Pisum sativum*) is the second most important food legume worldwide after common bean (*Phaseolus vulgaris* L.). Pea peel waste is an outer covering (pod) of pea vegetable. It is cheap and easily available lignocellulosic biomass that can be serving as potential raw material for cellulase production.

Cauliflower trimmings as cauliflower waste index are incorporated into ready-to-eat products (Stojceska et al., 2008). Waste index of cauliflower is very high and is an excellent source of protein (16.1%), cellulose (16%), and hemicellulose (8%). It is considered as a rich source of dietary fiber.

3.18 FVW AND KEY ENVIRONMENTAL RISKS

Fruit processing yields the production of large volumes of bulky perishable solid waste such as peels, stems, shells, rinds, pulps, seeds, pods, rejected raw material, etc. Then from the solid waste, liquid waste is separated by screening, sedimentation, and flotation, etc. The cost of transporting waste to approved disposal sites adds on the extra load to the environmental pollution. The microbial action in stored solid waste can produce off odors. Along with that, vermin (e.g. rats) and insects may be attracted to solid waste storage areas. The compliance with solid waste disposal regulations and sanitary regulations in importing countries also is an expensive or difficult task.

After the invention of low temperature storage and preservation of food, the fruit processing plants usually have large cold storage facilities. The refrigerants (RF) used are generally ozone-depleting chemicals, such as CFCs, the production of which is being phased out. The release of RF ammonia into atmosphere, due to leaks from cooling equipment, is a primary health and safety concern.

3.19 FVW AND FUTURE PERSPECTIVE

Several valuable substances like fibers, coloring agents, enzymes, and gelling agents can be extracted from the wastes of fruit-based products. Fruit stones constitute a significant waste disposal problem for the fruit processing industry wherein high-quality activated carbon can be produced from waste cherry stones. Fruit processing wastes including apple, cranberry, and strawberry pomace have been used as substrates for polygalacturonase production by *Lentinus edodes* through solid-state fermentation (Zheng and Shetty, 1998). Also the fruit wastes of pineapple, mixed fruit, and maosmi has been investigated as possible substrates for citric acid production by solid-state fermentation using *Aspergillus niger* (Kumar et al., 2003). The rind of watermelon has the potential to produce preserved products such as pickles, tutti-fruiti, and vadiyams. The manipulation of food processing wastes is now becoming a very serious environmental issue. The peel of *Citrus junos* fruit has been investigated to possess potent allelopathic activity and a methanol extract of the peel inhibited the growth of several weed species (Fujihara and Shimizu, 2003). Also fruit pulp can be recovered and formed into fruit pieces. Although the process is relatively simple, the demand for this product is low. The process involves preparing a concentrate by boiling the fruit pulp, followed by sterilization. A gelling agent, sodium alginate, is then combined with the cooled pulp and then mixed with a strong solution of calcium chloride. The calcium and the alginate combine to form a solid gel structure and the pulp can therefore be re-formed into fruit pieces. The most common way is to pour the mixture into fruit-shaped moulds and allow it to set. It is also possible to allow drops of the fruit/alginate mixture to fall into a bath of calcium chloride solution where they form small grains of re-formed fruit, which can be used in baked goods. Commercially, the most common product of this type is glazed cherries. Among food industry by-products of fruit or vegetable origin the products of wine industry are outstanding. The majority of grape marc goes to refuse; a small percentage is burned or used as compost but unfortunately it is also used for wine adulteration. Grape marc contains seeds that possess valuable substances such as oil, proteins, and tannins. Grape seed oil is the most valuable because of its health-protecting functional properties such as anti-inflammatory, cardioprotective, antimicrobial, and anticancer properties.

The huge amount of losses and waste occur because of a lack of adequate handling operations such as inadequate field management, harvest, classification, transportation, storage (temperature and relative humidity) and marketing in the food chain and industry infrastructure. These significant huge amounts of losses and wastes represent edible losses of food materials. The various extracted compounds from FVW can be used in food, pharmaceuticals, cosmetic, and chemical industries, and also in food research, and the development of functional foods. In summary, each of the above uses of fruit waste requires a good knowledge of the potential market for the products and the quality standards required an assessment of the economics of the production. For small-scale operations, where reducing pollution or increasing waste disposal is more important than process economics, the most likely solution is to use wastes as animal feeds.

REFERENCES

Adebowale, A., Garnesan, A.P., & Prasad, R.N.V. 2002. Papaya (*Caricapapaya*) consumption is unsafe in pregnancy: Fact or fable? Scientific evaluation of a common belief in some parts of Asia using a rat model. British Journal of Nutrition, 88:199–203.

Ajila, C.M., Aalami, M., Leelavathi, K. & Rao, U.J.S.P. 2010. Mango peels powder: a potential source of antioxidant and dietary fibre in macaroni preparations. Innovative Food Science and Emerging Technologies, 11:219–224.

Almeida, J.M., Lima, V.A., Giloni-Lima, P.C. & Knob, A. 2015. Passion fruit peel as novel substrate for enhanced β-glucosidases production by *Penicillium verruculosum*: potential of the crude extract for biomass hydrolysis. BiomassBioenergy, 72:216–226.

Amiri-Rigi, A., Abbasi, S. & Scanlon, M.G. 2016. Enhanced lycopene extraction from tomato industrial waste using microemulsion technique: Optimization of enzymatic and ultrasound pre-treatments. Innovative Food Science & Emerging Technologies, 35:160–167.

Anhwange, B.A., Ugye, T.J. & Nyiaatagher, T.D., 2008. Chemical composition of Musa sapientum (banana) peels. Electron. Journal of Environmental, Agricultural and Food Chemistry, 8:437–442.

Beres, C., Costa, G.N., Cabezudo, I., da Silva-James, N.K., Teles, A.S., Cruz, A.P. & Freitas, S.P. 2017. Towards integral utilization of grape pomace from winemaking process: a review. Waste Management, 68:581–594.

Bennett, B. 2002. Feeding Crop Waste to Livestock and the Risk of Chemical Residues, Notes Information Series, Department of Primary Industries, Victoria, Australia.

Benitéz, V., Molla, E., Martin-Cabrejas, M.A., Aguilera, Y., Lopez-Andreu, F.J., Terry, L.A. & Esteban, R.M. 2013. Food and Bioprocess Technology, 6, 1979–1989.

Bonaccorsi, P., Caristi, C., Gargiulli, C. & Leuzzi, U. 2008. Flavonol glucosides in Allium species: A comparative study by means of HPLC–DAD–ESI-MS–MS. Food Chemistry, 107(4): 1668–1673.

Cheok, C.Y., Adzahan, N.M., Rahman, R.A., Abedin, N.H.Z., Hussain, N., Sulaiman, R. & Chong, H. 2018. Current trends of tropical fruit wasteutilization. Critical Reviews in Food Science and Nutrition, 58(3):335–361.

Choonut, A., Saejong, M. & Sangkharak, K. 2014. The production of ethanol and hydrogen from pineapple peel by Saccharomyces cerevisiae and Enterobacter aerogenes. Energy Procedia, 52:242–259.

Clemente, R., Pardo, T., Madejon, P., Madej on, E. & Bernal, M.P. 2015. Food by-products as amendments in trace elements contaminated soils. Food Research International, 73:176–189.

Da Porto, C., Porretto, E. & Decorti, D. 2013. Comparison of ultrasound-assisted extraction with conventional extraction methods of oil and polyphenols from grape (*Vitis vinifera* L.) seeds. Ultrasonics Sonochemistry, 20:1076–1080.

Debnath, S., Rahman, S.M.H., Deshmukh, G., Duganath, N., Pranitha, C. & Chiranjeevi, A. 2011. Antimicrobial screening of various fruitseed extracts. Pharmacognosy Journal, 3:83–86.

Demirbas, A. 2011. Waste management, waste resource facilities and waste conversion processes. Energy Conversion and Management, 52(2):1280–1287.

FAO-Food and Agriculture Organization. 2011. Global food losses and food waste - extent, causes and prevention (Rome).

FAO 2012. World onion production. Food andAgriculture Organization of the United Nations. http://faostat.fao.org, accessed 28 November 2020.

FAO-Food and Agriculture Organization. 2014. Definitional framework of food losses and waste. Rome, Italy: FAO.

FAO-Food and Agriculture Organization of the United Nations. 2014. https://sci-hub.se/http://faostat.fao.org/site339/default.aspx. Accessed 08 December 2020.

FAO-Food and Agriculture Organization of the United Nations. 2016. The state of food and agriculture, Climate change, agriculture and foodsecurity, Rome.

FAO-Food and Agriculture Organization of the United Nations. 2019. The state of food and agriculture, Moving forward on food loss and waste reduction, Rome.

Fernandez-Lopez, J., Fernandez-Gines, J.M., Aleson-Carbonell, L., Sendra, E., Sayas-Barbera, E. & Perez-A lvarez, J.A. 2004. Application of functional citrus by-products to meat products. Trends in Food Science and Technology, 15:76–185.

Ferreira, A.S., Nunes, C. & Castro, A. 2014. Influence of grape pomace extract incorporation on chitosan films properties. Carbohydrate Polymers, 113:490–499.

Ferreira, M.S.L., Santos, M.C.P., Moro, T.M.A., Basto, G.J., Andrade, R.M.S. & Gonçalves, E.C.B.A. 2015. Formulation and characterization of functional foods based on fruit and vegetable residue flour. Journal of Food Science and Technology, 52(2): 822–830.

Fontanari, G.G., Jaco, M.C., Souza, G.R., Batistuti, J.P., Neves, V.A., Pastre, I.A. & Fertonanim, F.L. 2008. DSC studies on protein isolates of guava seeds *Psidium guajava*. Journal of Thermal Analysis and Calorimetry, 93(2): 397–402.

Foo, K.Y. & Hameed, B.H. 2011. Transformation of durian biomass into a highly valuable end commodity: Trends and opportunities. Biomass Bioenergy, 35:2470–2478.

Foo, K.Y. & Hameed, B.H. 2012. Potential of jackfruit peel as precursor for activated carbon prepared by microwave induced NaOH activation. Bioresource Technology, 112:143–150.

Franca, A.S.; Oliveira, L.S. Coffee processing solid wastes: Current uses and future perspectives (chapter 8). In Agricultural Wastes; Ashworth, G.S., Azevedo, P., Eds.; Nova Science: New York, NY, USA, 2009.

Fujihara, S. & Shimizu, T. 2003. Growth inhibitory effect of peel extract from Citrus junos. Plant Growth Regulation, 39:223–233.

Garcia Herrera, P., Sanchez-Mata, M.C. & Camara, M. 2010. Nutritional characterization of tomato über as a useful ingredient for food industry. Innovative Food Science and Emerging Technologies, 11:707–711.

Gerola, G.P., Boas, N.V., Caetano, J., Tarley, C.R.T., Goncalves, A.C. & Dragunski, D.C. 2013. Utilization of passion fruit skin by-product as lead (II) ion biosorbent. Water, Air, & Soil Pollution, 24(1446):2–11.

Gorinstein, Martin-Belloso, O., Park, Y.S., Haruenkit, R., Lojek, A., Ciz, M., Caspi, A., Libman, I. & Trakhtenberg, S. 2001. Comparisonof some biochemical characteristics of different citrusfruits. Food Chemistry, 74:309–315.

Gupta, K. & Joshi, V.K. 2000. Fermentative utilization of waste from food processing industry. In: Joshi VK, editor, Postharvest Technology of Fruits and Vegetables: Handling, Processing, Fermentation and Waste Management. New Delhi: Indus Pub Co. 1171–1193.

Hebbar, H.U., Sumana, B. & Raghavarao K.S.M.S. 2008. Use of reverse micellar systems for the extraction and purifiation of bromelain from pineapple wastes. Bioresource Technology, 99:4896–4902.

Ho, L.H. & Bhat, R. 2015. Exploring the potential nutraceutical values of durian (*Durio zibethinus* L.) an exotic tropical fruit. Food Chemistry, 168:80–89.

Khachatourians, G.G. & Arora D.K. 2001. Applied mycology and biotechnology, Vol. I. Agriculture and Food Production, Elsevier Science, The Netherlands, 353–387.

Kumar, D., Jain, V.K., Shanker, G. & Srivastava, A. 2003. Utilization of fruits waste for citric acid production by solid state fermentation, Process Biochemistry, 38 (12):1725–1729.

Lewis, B.J., Herrlinger, K.A., Craig, T.A., Mehring-Franklin, C.E., DeFreitas, Z. & Hinojosa-Laborde, C. 2013. Antihypertensive effectsof passion fruit peel extract and its major bioactive components following acute supplementation in spontaneously hypertensive rats. Journal of Nutritional Biochemistry, 24:1359–1366.

Liaotrakoon, W., Buggenhout, S.V., Christiaens, S., Houben, K., Clercq, N. & Dewettinck, K. 2013a. An explorive study on the cell wall polysaccharides in the pulp and peel of dragon fruits (*Hylocereus* spp.). European Food Research and Technology, 237:341–351.

Liaotrakoon, W., Clercq, N., Hoed, V.V. & Dewettinck, K. 2013b. Dragon fruit (*Hylocereus* spp.) seed oils: Their characterization and stability under storage conditions. Journal of the American Oil Chemists' Society, 90:207–215.

Lin, J., Zuo, J., Gan, L., Li, P., Liu, F. & Wang, K. 2011. Effects of mixture ratio on anaerobic co-digestion with fruit and vegetable waste and food waste of China. Journal of Environmental Sciences, 23(8): 1403–1408.

Lu, Z., Wang, J., Gao, R., Ye, F. & Zhao, G. 2019. Sustainable valorization of tomato pomace: a comprehensive review. Trends in Food Science and Technology, 86:172–187.

Luo, H., Cai, Y. & Peng, Z. 2014. Chemical composition and in-vitro evaluation of the cytotoxic and antioxidant activities of supercritical carbon dioxide extracts of pitaya (dragon fruit) peel. Chemistry Central Journal, 3:8(1)1.

Madruga, M.S., Albuquerque, F.S.M., Silva, I.R.A., Amaral, D.S., Magnani, M. & Neto, V.Q. 2014. Chemical, morphological and functional properties of Brazilian jackfruit (*Artocarpus heterophyllus* L.) seeds starch. Food Chemistry, 143:440–445.

Mani, S.B. & Sethi, V. 2000. Utilization of fruits and vegetables processing waste In. Postharvest Technology of Fruits and Vegetables: Handling, Processing, Fermentation and Waste Management. Verma, L.R. and Joshi, V.K. (Eds), Indus Publishing Co., New Delhi. 2:1006.

Manzocco, L., Alongi, M., Sillani, S. & Nicoli, M.C. 2016. Technological and consumer strategies to tackle food wasting. Food Engineering Reviews, 8(4):457–467.

Matassa, S., Verstraete, W., Pikaar, I. & Boon, N. 2016. Autotrophic nitrogen assimilation and carbon capture for microbial protein production by a novel enrichment of hydrogen-oxidizing bacteria. Water Research, 101:137–146.

Mena, C., Adenso-Diaz, B. & Yurt, O. 2011. The causes of food waste in the supplier-retailer interface: Evidences from the UK and Spain. Resources, Conservation and Recycling, 55(6): 648–658.

Mitra, S.K., Pathak, P.K., Devi, H.L. & Chakraborty, I. 2013. Utilization of seed and peel of mango. Acta Horticulturae, 593–596.

Mohapatra, D., Mishra, S. & Sutar, N. 2010. Banana and its by-product utilisation: an overview. Journal of Scientific and Industrial Research, 69:323–329.

Nkoa, R. 2014. Agricultural benefits and environmental risks of soil fertilization with anaerobic digestates: A review. Agronomy for Sustainable Development, 34(2): 473–492.

Nocolle, C., Cardinault, N., Aprikian, O., Busserolles, J., Grolier, P., Rock, E., Demigne, C., Mazur, A., Scalbert, A., Amouroux, P. & Remesy, C. 2003. Effect of carrot intake on cholesterol metabolism and antioxidant status in cholesterol fed rats. European Journal of Nutrition, 42:254–261.

Ooi, X.Z., Ismail, H., Abdul Aziz, N.A. & Abu Bakar, A. 2011. Preparation and properties of biodegradable polymer film based on polyvinyl alcohol and tropical fruit waste flour. Polymer-Plastics Technology and Engineering, 50:705–711.

Palafox-Carlos, H., Ayala-Zavala, F. & Gonzalez-Aguilar, G.A. 2010 The role of dietary fibre in the bioaccessibility and bioavailability of fruit and vegetable antioxidants. Journal of Food Science, 76(1): 6–15.

Parfitt, J., Barthel, M. & Macnaughton, S. 2010. Food waste within food supplychains: quantification and potential for change to 2050. Philosophical Transactions of the Royal Society B: Biological Sciences, 365:3065–3081.

Parni, B. & Verma, Y. 2014. Biochemical properties in peel, pulp and seeds of *Carica papaya*. Plant Archives, 14:565–568.

Paola, B., Gianfranco, P., Raffaella, N. & Menotti, C. 2004. Polyunsaturated fatty acids: biochemical, nutritional and epigenetic properties. Journal of the American Oil Chemists' Society, 23(4): 281–302.

Penjumras, P., Abdul Rahman, R.B., Talib, R.A. & Abdan, K. 2014. Extraction and characterization of cellulose from durian rind. Agriculture and Agricultural Science Procedia, 2:237–243.

Ravani, A. & Joshi, D.C. 2013. Mango and it's by product utilization–a review. Trends in Post-Harvest Technology, 1:55–67.

Sagar, N.A., Pareek, S., Sharma, S., Yahia, E.M. & Lobo, M.G. 2018. Fruit and vegetable waste: bioactive compounds, their extraction, and possible utilization. Comprehensive Reviews in Food Science and Food Safety, 1–20.

San Martin, D., Ramos, S. & Zufía, J. 2016. Valorisation of food waste to produce new raw materials for animal feed. Food Chemistry, 198:68–74.

Saxena, A., Bawa, A.S. & Raju, P.S. 2011. Jackfruit (*Artocarpus heterophyllus* Lam.). In: Yahia EM, editor. Postharvest Biology and Technology of Tropical and Subtropical Fruits. Cambridge: Woodhead Publishing Limited. 275–298.

Schieber, A., Stintzing, F.C. & Carle, R. 2001. By-products of plant food processing as a source of functional compounds - recent developments.Trends in Food Science and Technology, 12:401–413.

Schieber, A. & Aranda Saldaña, M.D. 2009. Potato Peels: A Source of Nutritionally and Pharmacologically Interesting Compounds – A Review. Food, 3:23–29.

Segre, A. & Falasconi, L. 2011. Il libro nero dello spreco in italia: Il cibo. Milan: Ambiente Edizioni.

Shah, M.P., Reddyb, P., Banerjee, G.V., Babu, P.R. & Kothari, I.L. 2005. Microbial degradation of banana waste under solid state bioprocessing using two lignocellulolytic fungi (*Phylosticta* spp. MPS-001 and *Aspergillus* spp. MPS-002). Process Biochemistry, 40(1): 445–451.

Sheets, J.P., Yang, L., Ge, X., Wang, Z. & Li, Y. 2015. Beyond land application: Emerging technologies for the treatment and reuse of anaerobically digested agricultural and food waste. Waste Management, 44:94–115.

Siriphanich, J. & Yahia, E.M. 2011. Durian (*Durio zibethinus* Merr.). In: Yahia EM, editor. Postharvest Biology and Technology of Tropical and Subtropical Fruits. Cambridge, GB: Woodhead Publishing. 80–114.

Spigno, G., Marinoni, L. & Garrido, G. 2017. State of the art in grape processing by-products. In: Galanakis CM, editor. Handbook of Grape Processing By-Products: Sustainable Solutions. London, UK: Academic Press, Elsevier. 1–23.

Stenmarck, Å., Jensen, C., Quested, T. & Moates, G. 2016. Estimates of European food waste 623 levels. FUSIONS EU Project–Reducing food waste through social innovation.

Stojceska, V., Ainsworth, P., Plunkett, A., Ibanoglu, E. & Ibanoglu, S. 2008. Cauliûower by-products as a new source of dietary fibre, antioxidants and proteins in cereal based ready-to-eat expanded snacks. Journal of Food Engineering, 87:554–563.

Surbhi, S., Verma, R.C., Deepak, R., Jain, H.K. & Yadav, K.K., 2018. A review: food, chemical composition and utilization of carrot (*Daucus carota* L.) pomace. International Journal of Chemical Studies, 6, 2921–2926.

Tran, A.V. 2006. Chemical analysis and pulping study of pineapple crown leaves. Industrial Crops and Products, 24:66–74.

United States Department of Agriculture, 2014. Addressing the Challenges of Conducting Effective Supplemental Nutrition Assistance Program Education (SNAP-Ed) Evaluations: A Step-by-Step Guide.

Vilariño, M.V., Franco, C. & Quarrington, C. 2017. Food loss and waste reduction as an integral part of a circular economy. Frontiers of Environmental Science, 5:1–5.

Wolfe, K.L. & Liu, R.H. 2003. Apple peels as a value-added food ingredient. Journal of Agricultural and Food Chemistry, 51:1676–83.

Xiaomin, L., Yanru, T., Zhexian, X., Yinghui, L. & Fang, L. 2007. Study on the preparation of orange peel cellulose adsorbents and biosorption of Cd^{2+} from aqueous solution. Separation and Purification Technology, 55:69–75.

Zheng, Z. & Shetty, K. 1998. Solid-state production of beneficial fungi on apple processing wastes using glucosamine as the indicator of growth. The Journal of Agricultural and Food Chemistry, 46:783–797.

Zhuang, Y., Zhang, Y. & Sun, L. 2012. Characteristics of fibre-rich powder and antioxidant activity of pitaya (*Hylocereus undatus*) peels. International Journal of Food Science & Technology, 47:1279–1285.

4 Recovery and Utilization of Protein from Food Industry Waste

Mohona Munshi,[1] Prajya Arya,[2] and Pradyuman Kumar[3,]*

[1]Assistant Professor, Department of Food Technology, Vignan Foundation of Science, Technology and Research (Deemed-to-be University), Vadlamudi, Guntur, Andhra Pradesh, India
[2]Research Scholar, Department of Food Engineering and Technology, Sant Longowal Institute of Engineering and Technology (Deemed-to-be University), Longowal, Punjab, India
[3]Professor, Department of Food Engineering and Technology, Sant Longowal Institute of Engineering and Technology (Deemed-to-be University), Longowal, Punjab, India
*Corresponding author.

CONTENTS

DOI: 10.1201/9781003207689-4

4.1 INTRODUCTION

Lowering food loss and wastage leads to the conservation of energy and resources, and ensures a sustainable global diet. Food loss and wastage indicates an overall loss of both macro- and micronutrients. Food waste is the result of negligence or a conscious decision to throw food away. This food loss or waste happens in all segments of the food cycle starting from harvesting till slaughtering and especially while processing (Kamal *et al.*, 2021). To meet the world's nutritional demand, the researcher's interest is growing towards recovery and utilization of food industry-based wastes. Food is an essential part of all living organisms, especially protein-rich food because proteins are the building blocks of life forms. They are required for critical functions of cells, tissues, organs, and systems. Among the three main chemical components of food, protein is one of the macromolecules, besides lipids, carbohydrates, that are integral to the food systems. With the predicted increase in population growth over the next two to three decades, the concern for feeding this large population nutrient-rich food is the concern of many (Nadathur *et al.*, 2017). This availability of adequate supplies of protein, which is of an appropriate quality to maintain health, has been one topic of concern and is the biggest challenge for researchers tasked with finding an alternative source to protein. The Food and Agriculture Organization of the United Nations (Bongaarts, 2007) estimates that 32% of all food produced in the world was lost or wasted in 2009. This agro-food industry produces 190 million tons of wastes in the form of plant and animal proteins. Plant proteins include soybean, wheat, pea, rice, sunflower nuts oilcakes, pomace, seeds, peel etc. These plant waste proteins also can be utilized to play an important role in daily appetite. Whereas, animal proteins include poultry carcasses, liver, skin, feathers, fish wastes like fish bones, skin, collagen, and milk waste such as whey protein etc. They have a balanced combination of all amino acids hence they are called complete proteins, which are essential for an organism's diet (Virtanen *et al.*, 2019).

So, based on protein waste availability and quantity can be valorized using various recovery techniques and its utilization. As we know, this recovery depends on the cell matrix of the product and technology used (Gençdağ *et al.*, 2021). Therefore, keeping in mind the increase in demand of the protein-rich foods and the crisis to avail it, has given rise to valorize the waste of industries rich in nutrients. This will help both to meet the protein necessity demand and will also help in zero waste produce. Industries will also benefit from this waste utilization by increasing income from the by-products (Shahid *et al.*, 2021).

To this end, this chapter suggests ways to meet protein-rich food waste sources, which can serve as the potential source of protein and reduce the waste by extracting and utilizing the protein from the waste to be used for human and animal feed.

4.2 VARIOUS SOURCES OF PROTEIN-RICH FOOD INDUSTRY WASTE

Among various types of foods, protein is one of the major bio-macromolecules that builds up muscles and entire body structure. Protein is composed of long chain amino acids and their residues. The structure of protein varies as per the arrangement and sequence of amino acids. This predicts the nucleotide sequence and results in protein folding (Burd *et al.*, 2019). In the human body, approximately, 57% growth depends on the vegetal protein source (FAO, 2016). In the universal perspective, rice, wheat, and milk are considered as a rich source of protein. The botanical protein sources are soy protein, pea protein, cereal seed protein (maize, barley rice, sorghum, oat, pearl-millet), minor seed protein (quinoa, buckwheat, lupin, locust bean gum, and bambarra groundnut), animal protein, protein from microalgae and microorganisms, and other protein sources could be oilseed cakes that are left behind after oil extraction (Hovhannisyan and Devadoss, 2020). From the literature surveyed, there is ample data available that provide food wastage evidence. There is a huge requirement to barricade the protein waste from the food industry, which could aid in providing considerable protein fulfillment in the food sector.

4.2.1 Soybean Oilseed Protein from Food Industry Waste

Soybean [*Glycine max* (L.) Merrill], which is a legume, is considered as a good source of protein that is often used to replace animal protein and has no cholesterol. The nutritional content of soybean is represented in Table 4.1. Soybean is a chief source of protein, oil, and bioactive compounds like fatty acids, sugars, and isoflavones. The soybean isoflavone content helps in controlling diseases like heart disorder, various types of cancer, and menopause. The high content of isoflavonones makes the taste of soy product become bitter (Sahin *et al.*, 2019).

The wastes from the soy milk and oil are excellent, inexpensive protein (27%) so they are commercially being used as the protein supplements. It is the only vegetable protein source that contains eight essential amino acids. Soybean is also rich in vitamin B, zinc, calcium, iron, and fiber. Soybean contains 36% protein, 18% oil, 15% soluble carbohydrate, and 15% insoluble carbohydrate. It is also used to make various unfermented products (Kumar *et al.*, 2020). Soybean comes with a variety of food products like fermented, unfermented, and second-generation food products that could be produced from soybean by-products easily. Soybean includes a range of fermented products (miso, natto, and tempeh), unfermented products (tofu, soy milk, sprouts, and soy nuts), and second-generation soy products (soy breads, soy pasta, soy cheese, and tofu sausages) (Rizzo and Baroni, 2018). After the processing of tofu, the waste released is the whey protein, which also has a high amount of protein and can be further processed to produce supplements. Among the various soy products, tofu and soy milk acquired the food market in a broad way while its other related product is still at the growing stage. After the oil has been extracted the leftover residue, which is the oil cake or also known as defatted soymeal, is very popular as it is rich in amino acids (protein). There are various studies suggesting that the consumption of soybean could aid in better heart health by lowering the cholesterol level and providing a role for β-conglycinin and its peptides in cholesterol regulation. According to the FDA, the relationship between heart diseases and soybean consumption has been studied and approved in favor of benefiting mankind and acquiring good market value in various food industries (Kumar *et al.*, 2020).

4.2.2 Peanut Oilseed Protein from Food Industry Waste

Peanut (*Arachis hypogaea*) is a legume, which belongs to the fabaceae family. It is commonly known as groundnut, monkey nut, pinder, and goober. India is the second largest producer of peanut contributing about 16% of world peanut production and a rich source of protein, oil, and fibers. The nutritional content of peanut is represented in Table 4.1. Peanut has been used in various food forms like peanut butter, extenders in meat formulations, snack and roasted peanut and soups. In recent years, peanut is considered as a complete source of nutrition in the African countries to fight as a source against malnutrition and in a diverse area like Antarctica, trekking and space peanuts and its related products could be a great dietary source of nutrition (Guimon and Guimon, 2012). Peanut protein, which is obtained as a by-product after the extraction of oil, is the third largest source of protein that is used as animal feed or in lower value-added food products. In peanut protein processing, there is utilization of low denatured prepressed leached meal, cold pressed peanut meal, and mixed defatted peanut meal, which contribute toward peanut protein extraction. In terms of typical components of peanut protein powder, it contains protein (56%), carbohydrate (32%), fat (6%), and water and ash collectively as (5%).

Peanut protein powder has white color, low sulphur content, no odor, and is cost effective. Its denatured protein and protein molecules are reoriented after processing to form a new organizational structure. The meal after protein processing is rich in fiber, which provides structural similarity with meat and could be added as a source for various food additives.

During protein processing, a huge amount of sample matrix is lost in various complicated steps like alkaline and acid hydrolysis, which leads to a greater amount of energy consumption. Generally, peanut protein is used in the form of peanut protein powder, peanut tissue protein, peanut protein isolates and concentrate, and peanut protein film (Wang, 2018).

Peanut protein is rich in oligopeptides, polyphenols, polysaccharides, and dietary fiber. Many studies reported about the protein content in peanut meal after oil extraction. Approximately, 50% protein content was left behind in the peanut meal, which could be extracted and utilized in many different ways (Zhao et al., 2011).

According to the USDA, peanut protein contains 20 various amino acids in variable proportions and studies of Protein Digestibility Corrected Amino Acid Score (PDCAAS) reported that peanut protein is equivalent to the meat and egg protein that provide nearly the same benefit to human health (USDA, 2014). Peanut protein is plant origin protein that is enriched in various bioactive compounds and has good emulsifying capacity, excellent water retention, high solubility, and foaming capacity. The recent usage of peanut protein could be as a replacement for chicken breast meat as the fibrous structure, sensory attributes, and texture has similarities with peanut protein (Zhang et al., 2019).

4.2.3 Fish Waste Protein from Food Industry Waste

The fish industry is one of the fastest growing industries with a predicted fish production of 196 million tons of fish by 2025. In fish processing, there is a huge amount of by-product generation like fins, viscera, trimmings, head, skin, and some muscles (Shahidi et al., 2019). The most underrated fish by-product includes sturgeon muscle that is either sold at low price or discarded after; caviar harvesting is a particular case of an underutilized protein-rich fish resource. These by-products contribute to the production of protein, essential amino acids, enzymes, and polyunsaturated lipids. The protein content in fish varies from 8 to 35%. Fish protein is rich in amino acids like glutamic acid, glycine, and aspartic acid as 4.28%, 3.13%, and 2.79% in major proportions while, cystine, histidine, and methionine as 0.2%, 0.55% and 0.57% respectively in minor proportions (Metwally et al., 2021). The nutritional content of fish is represented in Table 4.1. The fish residue generates a lot of waste, which is approximately 30 to 70% of its raw material by weight that depends on the processing level. This fish waste disposal is a huge issue to maintain a good environmental ecosystem that slashes the processing methods and aids in reducing global fish waste, which is around 75.24 megatons, including 4.32 megatons from India alone (FAO, 2016). Fish waste is commonly disposed of in forms of landfill and incineration that impose serious environmental effects like generation of greenhouse gases and leaching of toxic compounds directly to the soil. To reduce such a harmful impact of fish waste, there is a direct requirement for processing and converting fish waste into fish oil, silage, and fish meal. The extraction of fish protein always accounts for various methodologies that like pH shift method, solvent extraction, heat treatment, enzyme-acid hydrolysis, and combination of such methods (Shaviklo, 2013). Fish protein extraction from the pH shift method was first developed by Hultin and Kelleher (2001). In this method, fish flesh is minced, and flesh is solubilized in alkali or acidic medium, which is five to ten times the volume with water and pH is adjusted to 2.5 or 11. The mixture is further centrifuged to remove oil and other material, and due to the pH and solution isoelectric point, protein get precipitated and stored with the aid of cryoprotectants by using appropriate freezing. Fish protein is an excellent source that could aid in controlling various metabolic activities such as inflammation modulation activity, angiotensin I-converting enzyme (ACE) inhibitory activity, antioxidant activity, antimicrobial activity, antiobesity, and anticancer properties (Heffernan et al., 2021). Recent studies conducted with fish protein with respect to analyzing its health benefits, observed that fish protein aids in reducing plasma total cholesterol, enhancing the HDL cholesterol, and reducing the acyl-coenzyme A (cholesterol acyltransferase (ACAT) activity in liver. Research conducted for comparison between casein and salmon protein hydrolysate fed rats for observing the resistance against high-fat-diet. The outcome of study revealed that salmon protein hydrolysate fed rats had reduced triacylglycerol levels and postprandial plasma glucose in liver as compared with casein fed rats. It also helps in maintaining the overall body weight in the samples (Heffernan *et al.*, 2021).

4.2.4 Fruit and Vegetable Protein from Food Industry Waste

The agro-industrial residues are rich in protein, polysaccharides, fibers, carbohydrates, etc. the agriculture residues constitute the stem, leaf, etc. whereas the industrial wastes include the pomace, pulp, peels, seeds, etc. as shown in Table 4.1 obtained after food processing. So this utilization of the industrial is profitable and also reduces waste disposal and helps to produce value-added products (Gençdağ, *et al.*, 2021). Papaya has been gaining popularity as an exotic fruit; the peel and seeds of papaya are the major by-products in the papaya processing industry, which when discarded causes environment pollution. On the other hand, these peels and seeds are a good source of protein, which can be used for value addition of other food products (Parniakov *et al.*, 2015). Similarly, kiwi fruit seeds to date have been used to successfully extract oil, which is suitable for food and can be used in cosmetics. This defatted meal accounted for 150–200g/kg of seed protein, which goes underutilized. So, this seed protein can be utilized to produce protein supplements for human feed (Deng *et al.*, 2014). Pineapple skin waste has a high amount of reducing and non-reducing sugars, furthermore, it has 0.6% protein as well, which is favorable for the growth of yeast, which in turn will help to produce single cell protein (Dharumadurai *et al.*, 2011). Similarly orange-peel waste was used to produce single cell protein to utilize it as the supplement for human and animals (Zhou *et al.*, 2019). Pumpkins are gourd squashes, which are grown in tropical and sub-tropical countries. The seeds of pumpkins are rich in protein having around 24.5–36% protein, therefore, this protein extracted from the pumpkin can be used to develop edible film (Lalnunthari *et al.*, 2020). Potato starch waste is causing a serious environmental concern. This potato starch is rich in carbohydrates, which is favorable for the growth of yeast, in turn, helping to produce single cell protein (Liu *et al.*, 2014).

4.2.5 Protein-rich Whey from Dairy Industries

Whey is a waste protein extracted from cheese and casein and is the most popular waste of the dairy industry. It is released after the coagulation of the milk process. Whey is the major by-product of the dairy industry and is produced in large quantities and has high organic content, which increases the biological oxygen demand (BOD) and chemical oxygen demand (COD) content in water when released, causing environmental problems. Herein to reduce the environmental pollution and to reuse whey it needs to be valorized. Whey is a good source of lactose and soluble protein as shown in Table 4.1. So, this whey protein can be widely used to produce edible film and can be used directly as protein concentrates for the supplement food. This can reduce the environmental pollution concern (Das *et al.*, 2016).

4.3 RECOVERY TECHNIQUES OF PROTEINS FROM FOOD INDUSTRY WASTE

The global population is increasing exponentially day by day and adversely affecting natural resources (Deb *et al.*, 2020). Simultaneously, the rate of food consumption has also increased over the last couple of decades, but the ratio of food consumption between developed countries and developing countries is not the same. In that case, protein shortage is an over burning issue nowadays, mostly in the developing nations. Therefore, extraction of protein from various cheap sources, such as plant-based protein and food waste-based protein as shown in Table 4.2, is necessary. Extraction of protein from food industry-based waste is becoming an interesting topic of research among the research and industry community, as it is a known fact that proteins are present in all living organisms that include biological compounds such as enzymes, hormones, and antibodies. The concept of producing protein from waste has been applied in the extraction of crude protein from food-based

TABLE 4.1
Protein Content in the Various Waste Produced from Food Industry Waste

Food product	By-product	Protein content (%)	References
Soyabean	Defatted meal	33	Nile *et al.*, 2020
Peanut	Defatted meal	26–27	Sibt-e-Abbas *et al.*, 2015
Fish	Head, trimmings, skin, viscera, scale, bone, etc.	22	Shanthi *et al.*, 2021
Pineapple	Pomace (skin, peel, pulp)	6.4	Correia *et al.*, 2007
Watermelon	Seed	45–55	Wani *et al.*, 2015
Apple	Pomace (seed and skin)	15–20	Vendruscolo *et al.*, 2009
Pumpkin	Seeds	2.94–5.67	Bhusari *et al.*, 2008
Papaya	Seeds	25.1	Putra and Airun, 2021
Cheese	Whey	0.6–0.8*	Das *et al.*, 2016
Banana	Peel	11.32	Budhalakoti, 2021
Tomato	Pomace	19–22	Ajila *et al.*, 2012
Baby corn	Husk	11.70	Bakshi *et al.*, 2005
Peas	Empty pea pods	19.80	Wadhwa *et al.*, 2006

* (w/v)

waste (Kamal *et al.*, 2021) along with further processing and purification. The extraction methods such as enzyme-assisted extraction, cavitation-assisted extraction, ultrasound-assisted extraction, supercritical extraction process, and pulsed electric field are emerging. Researches are still continuing on the development of these techniques, whereas on the other hand the liquid biphasic flotation, Osborne fractionation, hybrid extraction (acid precipitation, alkaline extraction, aqueous extraction), and salting extraction are the conventional techniques. All of these alkali methods have resulted in higher protein extraction, therefore, alkali is the preferred method for protein extraction but due to the high alkali concentration in the protein extraction, some adverse effects on the protein extracted are observed, such as racemization of amino acids and reduction of the digestibility of protein. Therefore, more focus should be given to the green extraction techniques for the extraction of protein. To date there are still fewer researches on protein extraction techniques from food industry-based wastes. Some of the reviews have been summarized here to allow readers to understand the principles and importance of such techniques, and to gain knowledge of the techniques and their efficiency rates of extraction.

4.3.1 ENZYME-ASSISTED EXTRACTION

In recent years, uses of enzymes for the extraction of proteins are widely being used, mainly for food and nutraceuticals. Enzymes are mainly proteins from plants, micro-organisms, and animals. Enzymes basically act as a catalyst to speed up the reactions. Commercially, food grade enzymes are used for extraction purposes, where an enzyme-assisted process has been found promising in the case of dairy (cheese, yogurt), bakery (bread making), and meat processing. Commonly used enzymes in industry are carbohydrase, lipase, and mostly proteases (Raveendran *et al.*, 2018).

The enzyme-assisted extraction technique is a high bioactive yielding technology by which cell walls are broken down and desired bioactive compounds are released. It removes unnecessary components from cell walls, to improve the transparency of the system; preserves original efficacy of all the natural products including proteins to the highest. Basically, the enzymes act as an assistance for the improved protein release by degrading the walls of the cell made up of

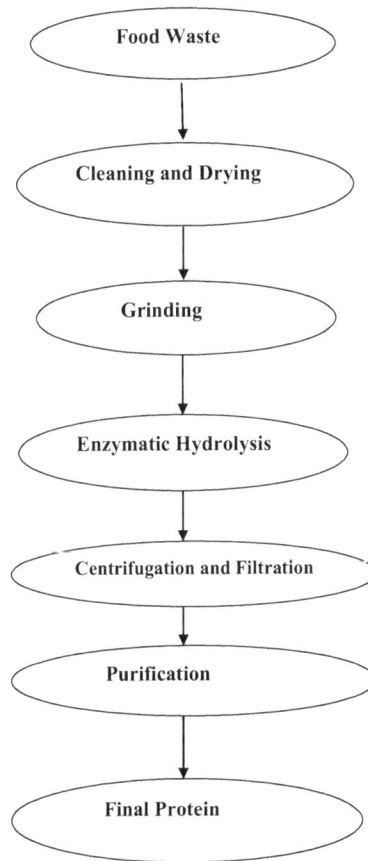

FIGURE 4.1 Enzyme-assisted extraction of protein.

Source: modified from Cheng *et al.*, 2015.

polysaccharides or increases the solubility of the protein as shown in Figure 4.1. This enzyme-assisted extraction is mostly being used as the pretreatment before ultrasound, microwave, or pulse electric field extraction of protein. Enzymatic hydrolysis is used to carry out and maintain certain operational conditions such as reaction temperature, time of extraction, pH, concentration of enzyme, and size of substrate followed by centrifugation and filtration to get the final extracted protein (Cheng *et al.*, 2015).

Research showed a positive outcome with the enzyme-assisted extraction technique, used to extract protein from plant-based waste such as seeds, roots, barks, and leaves. The enzyme-assisted extraction method used for extraction of protein from sugar beet leaves gave a protein yield of 79.01% with an increased efficiency by 43.27% (Akyüz and Ersus, 2021). Mangano *et al.*, 2021 recovered protein from fish (anchovy) wastes; the residues were converted into hydrosylate through enzymatic treatment. This hydrosylate product is rich in nutrition and has assessed beneficial effects or components of functional foods. Study also showed the possibility to produce dry powder with water activity (a_w) of 0.3–0.5 and an essential amino acids fraction of 42% over the total amino acids. Araujo *et al.* (2021) extracted protein from 1 kg of fish waste at optimal conditions and found 430 g of protein hydrolysate, 10 g of collagen, and 350 g of oil. Therefore, extraction of protein hydrolysates, collagen, and fish oil from fish waste has been done using enzyme hydrolysis for recovering various high value and value-added products.

4.3.2 Pulsed Electric Field Extraction

Extraction of protein using a pulsed electric field is an electricity-based processing technique in use for extraction of protein from various sources for the last around five decades. Pulsed electric field extraction is a promising technology mainly due to the modern developments and this technique is the most widely used technique. Compared with other techniques, preservation of nutritional value, flavor, texture, and color of products are the main advantages of the pulsed electric field technique (Buchmann *et al.*, 2019).

The mechanism of pulsed electric field is to generate short pulses of high electric fields, which is around 10 to 80 kV/cm for the time duration of microseconds to milliseconds. Food items normally transfer electricity because of the presence of several ions, which give the product a certain degree of electrical conductivity. When an electrical field is applied, electrical current flows into the product and transferred to each point in the liquid because of the charged molecules as shown in Figure 4.2 (Sharma *et al.*, 2014). The product needs to be extracted if it is placed in between a set of electrodes; the distance between electrodes is known as a pulsed electric field chamber. The high voltage is applied resulting in an electric field that inactivates microbes. After treatment, food is packed by aseptic conditions and stored in refrigerated conditions (Zhang *et al.*, 2020).

Pulsed electric field was used for the extraction of protein from papaya seed at a 13.3 kV/cm electric field for 2720 seconds with 400 pulses, and increased the yield of protein (Parniakov *et al.*, 2015). Similarly, protein extraction from sesame cake at a 13.3 kV/cm electric field showed a yield of around 30–40%. (Sarkis *et al.*, 2015). Protein from waste chicken biomass was extracted with the protein concentration of 78±8 mg mL^{-1} in the crude extract (Ghosh *et al.*, 2019. During pulsed electric field extraction, parameters such as time and temperature affects the efficiency of extraction and this is one of the major problems associated with the pulsed electric field extraction technique (Gençdağ *et al.*, 2020).

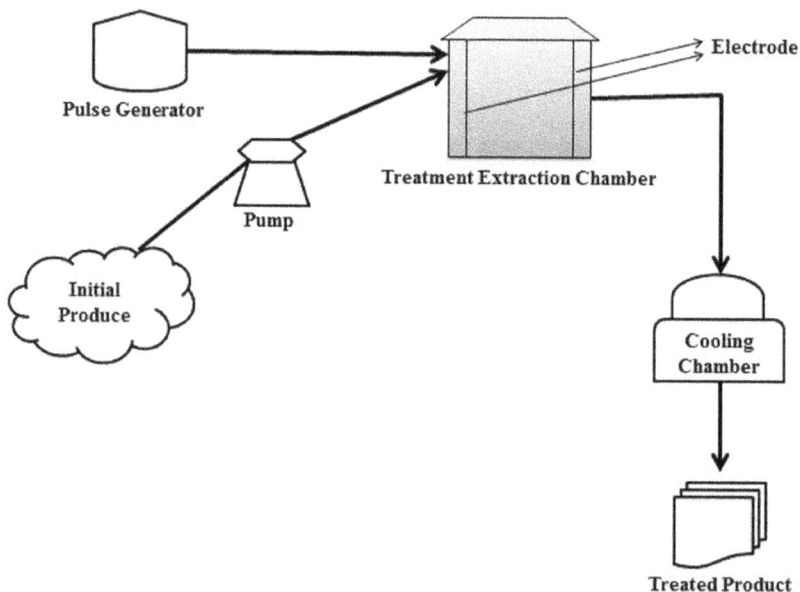

FIGURE 4.2 Pulse electric field-assisted extraction of protein.

Source: modified from Mohamed and Eissa, 2012.

4.3.3 Ultrasound-assisted Extraction

Ultrasound-assisted extraction technique is considered as a simple and more effective technique in comparison to conventional methods. Ultrasound-assisted extraction technique is one of the widely used techniques for extraction purposes for almost the last decade. Factors such as temperature, solvent characteristics, type of ultrasonic reactor (bath or probe), sonication frequency and power are the main controlling parameters of the ultrasound-assisted technique (Panda *et al.*, 2019). Ultrasound-assisted extraction creates cavitation of small bubbles in the solvent as the ultrasound waves passed, which allows more penetration within material and increase the surface area as shown in Figure 4.3 (Quintero Quiroz *et al.*, 2019). To get better efficiency during ultrasound-assisted extraction, frequency as high as 20 kHz and temperature can be used by producing cavitational bubbles to enhance the mass transfer rate from analytes to solvent. A vessel is placed in an ultrasonic bath to perform extraction and the extraction of solvent is kept in contact with the sample, where the sample is subjected to ultrasounds. As the ultrasounds are applied, it leads to formation of cavitation bubbles throughout the solvent, which is kept in contact with a sample. By collapsing, it causes pressure and temperature changes and the mass transfer rate gets enhanced from analytes to the solvent (Kamal *et al.*, 2021).

Karki *et al.* (2010) used the ultrasound-assisted extraction technique to extract protein from defatted soy flakes and the result showed a yield of 46%. During the extraction, high amplitude sonication for 120 seconds was applied, which basically increases the yield compared to non-sonicated samples. Nguyen *et al.* (2019) extracted protein from defatted peanut meal using the ultrasound-assisted extraction technique with water as a solvent and the yield of protein yield was found to be 19%. Study showed that the ultrasound-assisted technique increases the rate of protein extraction from chicken liver by 55% compared to other extraction methods, the yield of protein concentrate from rice bran using 100 W for 5 min was found to be 76.1%, wheat germ showed 70.2% yield of protein at 24 W for 120 min (Qu *et al.*, 2012), application of the ultrasound-assisted extraction technique using 630 W for 86 min showed a 61.7% yield of protein from sesame bran (Görgüç *et al.*, 2019). Cumulatively, all of these studies showed that using the ultrasound-assisted extraction technique and different operational parameters had a different effect on protein yield (Gençdağ *et al.*, 2021). Therefore, the yield from various sources solely depends on operational parameters, but

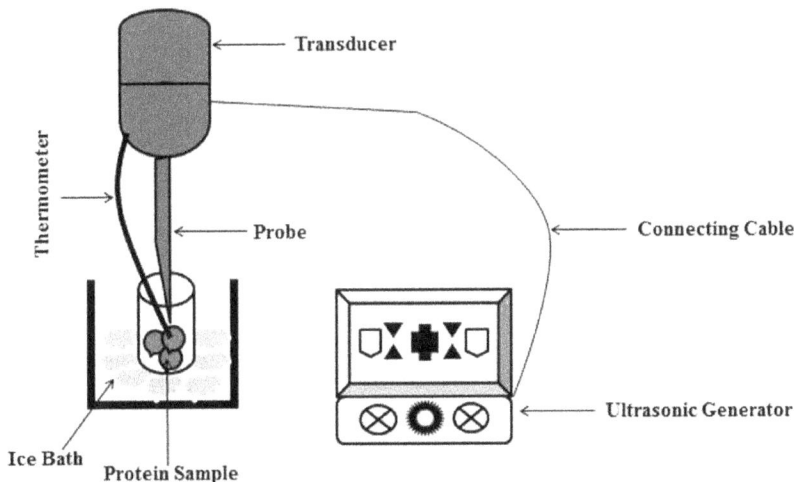

FIGURE 4.3   Ultrasound-assisted extraction of protein.

Source: modified from Moradi *et al.*, 2018.

overall, this technique has the high potentiality to extract protein with enhanced percentage of yield compare with conventional techniques from various sources.

4.3.4 MICROWAVE-ASSISTED EXTRACTION

The microwave-assisted technique started in the late 1980s and is an automated green extraction technique, which uses cost-effective extraction technology in the field of the food industry. This technique offers some of the advantages such as reduction of extraction time and rate of solvent consumption. Along with these advantages many advances have been developed in instrumentation of the microwave-assisted technique. These basically focused on pressurized and solvent-free microwave-assisted extractions. Therefore, the microwave-assisted technique became an alternative over conventional techniques for extraction of proteins (Llompart *et al.*, 2019).

Microwave-assisted technique is based on the disruption of the structure of cells when nonionizing electromagnetic waves with frequencies in the range of 300 MHz to 300 GHz are applied to the sample. As a result, transfer of energy occurs through the mechanisms of ionic conduction and simultaneously dipole rotation also occurs. As water evaporates, the intracellular pressure increases, breaking cell walls (Santos-Hernández *et al.*, 2018). During microwave-assisted extraction, the sample used is kept in the instrument and microwave (300 MHz to 300 GHz) is passed through the sample, then moisture is heated up, and evaporation of moisture takes place, where increased intra cellular pressure breaks the cell wall, as shown in Figure 4.4. In this technique, both heat and mass gradients work in the same direction towards the outside of the cells, thereby increasing the yield of extraction and reducing the process time (Gomez *et al.*, 2020). However, the efficiency of protein extraction using the microwave-assisted technique is subjective to several factors, including selection of a closed- or open-type vessel system. The extraction of protein from rice bran using the microwave-assisted extraction technique at processing conditions of 100 W for 90 seconds showed 29.4% yield of protein, which is 1.54% more in comparison to the alkaline extraction technique (Phongthai *et al.*, 2016).

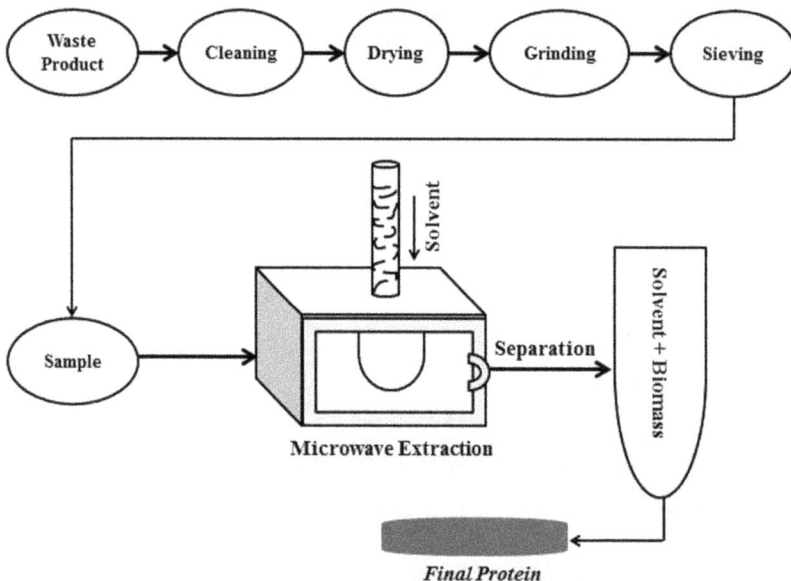

FIGURE 4.4   Microwave-assisted extraction method for protein.

Source: modified from Muley and Bolder, 2018.

4.3.5 Supercritical Fluid Extraction

Supercritical fluid extraction is a method in which solvent extraction is used to remain in the liquid phase under high temperature and pressure. Thereby high extraction efficiency is obtained. Supercritical fluids are kept in the liquid phase, and they have low surface tension and low viscosity and diffusivity that are close to gases. Mass transfer of target compounds is increased and organic solvents can be alternated using supercritical fluids for extraction purposes (Ma *et al.*, 2018). Supercritical fluid is another phase of media where it is above both its critical temperature and pressure. Supercritical fluids have high diffusion coefficients and low viscosity. It can be automated and interfaced to both chromatography and mass spectrometry. Supercritical fluid extraction technique is used for extraction of proteins from food waste with maximum efficiency, as shown in Figure 4.5. The most widely used supercritical fluid is carbon dioxide due to low cost, low toxicity, easily reachable critical parameters (31.1 °C and 7.48 MPa), ease of manipulation, good solvent strength, and compatibility with solutes (Ma *et al.*, 2018). The system contains a pump for CO_2, a pressure cell to contain the sample (sample preparation technique used to separate extract from the matrix using supercritical fluids to extract solvent), a pressure maintaining system, and collecting vessel. The liquid is pumped to a heating zone, where it is heated to supercritical conditions. Then it passes through the extraction vessel, where material to be extracted is dissolved and it rapidly diffuses into a solid matrix.

Application of supercritical fluid extraction showed protein extraction from rice bran with a 30% yield at 200 °C for 5 min; this was reported as the maximum yield (Hata *et al.*, 2008). Olive pomace undergoes supercritical fluid extraction and gave a yield of 10.8% (2.3 ml min^{-1}) at 88 °C for 30 min (Kazan *et al.*, 2015). During the experiment it was reported that temperature and flow rate have significant effects on extraction performance. Similarly, protein extraction from flaxseed meal showed a yield of 22.5% at 160 °C for 400 min. Supercritical fluid extraction technique has all the potential to extract protein from various sources with high efficiency. However, the temperature during extraction is recommended to be 130 to 160 °C for the extraction of proteins as the proteins may undergo thermal degradation (Ho *et al.*, 2019).

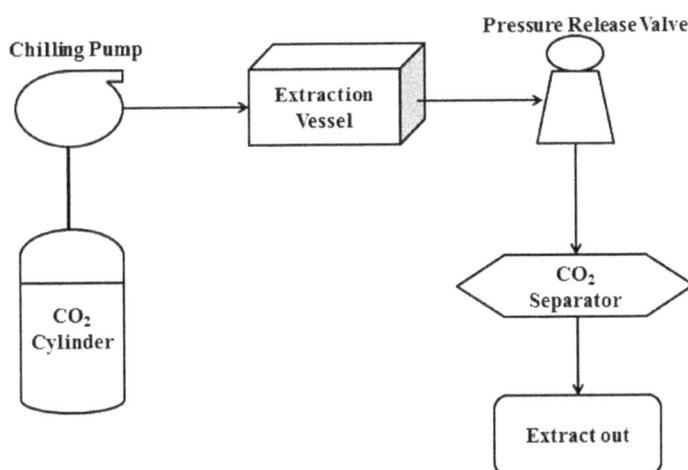

FIGURE 4.5 Supercritical fluid extraction of protein.

Source: modified from Puah *et al.*, 2005.

4.3.6 ALKALINE METHOD

Alkaline method is a hybrid extraction process where solubility is an indicator of protein extractability (Sari *et al.*, 2015). Protein extraction initiates at a specific pH (isoelectric point) by solubilizing protein, which then precipitates aiding in extraction as shown in Figure 4.6. The effect of pH on protein extraction is influenced by cell wall alterations and change in protein properties. Production of protein isolate/ hydrolysate from product is carried out by standard alkaline procedure. The resulting part is mixed with solvent and alkaline medium is created using alkaline compound such as sodium hydroxide. Due to low extraction yield, isoelectric precipitation is often additionally used to separate proteins from obtained supernatant such as protein hydrolysate is obtained after a purification step such as dialysis or ultrafiltration (Gençdağ *et al.*, 2021).

The extraction of protein using alkaline extraction technique from rice bran was carried out and the result showed a yield of 22.07% (Phongthai *et al.*, 2016). Guan *et al.* (2008) carried out extraction using the alkaline extraction technique for defatted oat bran and reported a 56.2% yield of protein. Sari *et al.* (2015) and Rommi (2016) extracted protein from rapeseed defatted meal and reported 50–80% and 53% of yield, respectively. Sedlar *et al.* (2021) extracted protein from leaves of vegetables such as cauliflower, broccoli, and beetroot with alkaline extraction at pH 10 and reported a 39.76–53.33% yield of protein. Overall, all these experiments show that the alkaline extraction technique to extract protein from various sources including food-based waste can be a promising technique with better performance and higher efficiency. However, it is advisable to apply this technique as an additional technique generally after purification to increase the rate of extraction of protein from any type of source (Gençdağ *et al.*, 2021).

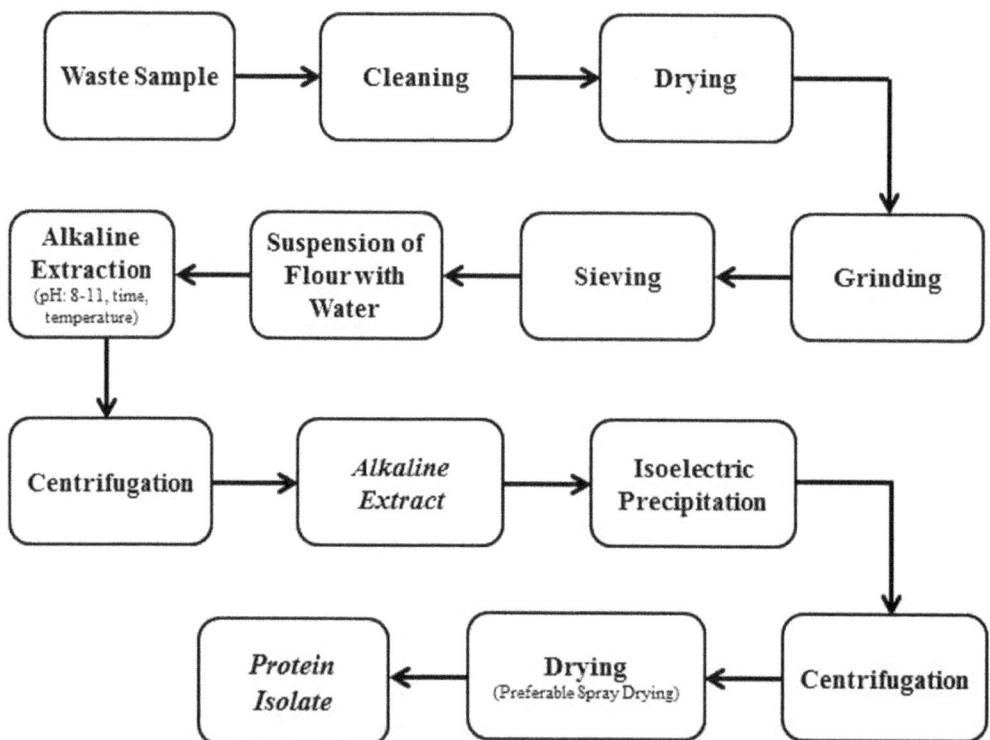

FIGURE 4.6 Alkaline extraction method for protein.

Source: modified from Barać *et al.*, 2015.

4.3.7 Osborne Fractionation Technique

It is very difficult to determine the total set of protein species that constitute cells' proteome. Protein cataloguing and the whole proteome characterization required fractionation techniques at the cellular, subcellular, protein level and therefore, there is a requirement to use complementary approaches. Protein isolates prepared by alkaline extraction have poor solubility and may cause irreversible denaturation. So, they behave poorly while used as a food ingredient. In this case the Osborne extraction method is used to fractionate the protein and better characteristics of food (Vinaya Shree *et al.*, 2021).

As per the Osborne protein extraction method the sequence of proteins according to their solubility are albumins (water soluble), globulins (dilute salt soluble), prolamins (aqueous ethanol soluble), and glutelins (dilute alkali soluble). In 1907 scientist Osborne first proposed the different classification of protein based on their solubility. Osborne protein classification was mainly proposed for wheat proteins. The Osborne method involves use of fractionation techniques at the cellular, subcellular, and protein level. It can produce proteins with better characteristics for incorporation in food. It can be combined with enzyme inactivation and downstream electrophoretic protein separation to improve yield (Romero-Rodríguez *et al.*, 2014). All the proteins that are needed to be extracted are taken into fractions. They are sequentially extracted from the material by using some of the chemicals like water, salt solutions, alkaline, and alcoholic solutions. After extraction, centrifugation is done followed by washing with water to filter the supernatants as shown in Figure 4.7 followed by calculation of protein yield normally using Kjeldal apparatus (Sardari *et al.*, 2019).

Proteins are extracted using the Osborne procedure from various sources including waste combined with alkali fraction. Vinayashree *et al.* (2021) fractionate protein from the seeds of pumpkin and found a yield of 45.82%. Tan *et al.* (2011) compared the yield of canola protein using the Osborne method with the yield of the alkaline extraction method and concluded that the Osborne method is more effective for extraction of protein and results showed higher a yield of protein. The solubility of flaxseed protein in Osborne series of solvents was compared with oilseeds and found that a major fraction (42%) was extracted and the residue of 4% was soluble in 70% ethanol. The ability of any product derived from any agricultural product crucially determined by the percentage of available protein (Žilić *et al.*, 2011). Therefore, by-products or protein contained products from different sources showed extreme complexity and various interactions with each other's, thus making them difficult to categorize and characterize. In that case Osborne protein fractionation is one of the key solutions to overcome this difficulty.

4.4   UTILIZATION OF PROTEIN FROM FOOD INDUSTRY WASTE

Food wastes are rich in carbohydrates polymers (starch, cellulose, hemicelluloses), lignin, protein, and other inorganic minerals. The food waste generated interacts with atmosphere to create a serious problem and air pollution. So, the components present in food waste can be extracted and used or reutilized in various food, pharmaceutical, cosmetics, textile industries, etc. This utilization of food industry waste, according to FAO, UNEP and stakeholder is called "Re(FED) i.e. Rethink of Food waste". Generally, as rice bran, oilseed cake, potato peel has the maximum rate of food waste, among this protein is one of the large biomolecules known for various functions in an organism. It serves as the building block of the body and is a good source of energy having 4 Kcal/g of protein (Kumar *et al.*, 2020).

Due to the increase in demand of plant protein and high consumption rate, therefore the industries are now shifting to produce proteins from the waste generated from industries, to meet this high protein demand. Examples such as waste from legumes, oilseeds, cereals are being used to produce commercial proteins (Aydemir *et al.*, 2014). Oilcake protein is now the mostly used waste all over the world to extract protein but as the years pass emerging new waste sources are also being researched for protein utilization as shown in Table 4.2. So, here we are going to discuss the processing utilization of proteins derived from agro- or food industry-based wastes.

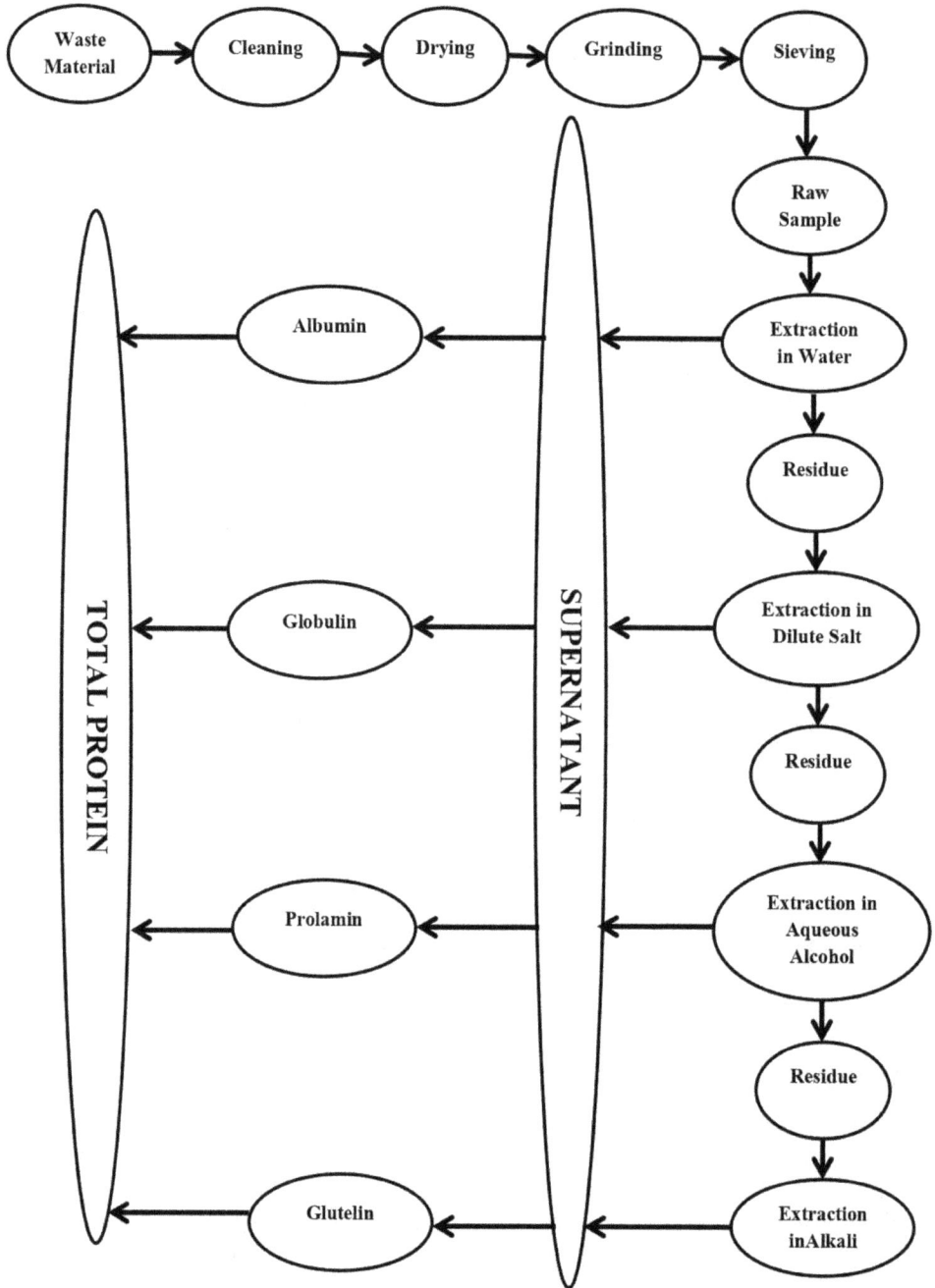

FIGURE 4.7   Osborne extraction method for protein.

4.4.1   Protein Concentrates and Isolates

Protein concentrates and isolates are the extracts of protein recovered from waste such as whey, fish waste, or some parts of fruits, vegetables, cereals, legumes having high purity, which can be used as supplements for human consumption. They undergo filtration process to get a higher quantity of protein content. The only difference between protein isolates and concentrate is that protein isolates are more pure protein than concentrate; the former contains a lesser amount of carbohydrate, fats, and calories. Here we will discuss the protein concentrates and isolates extracted from food industry-based waste for utilization as supplements for humans.

The poultry slaughterhouse produces many feathers. These feathers of the poultry contain a high amount of protein, which can be utilized only after dissolving the rigid keratin present. The feathers are processed in a reactor by the alkaline treatment and using proteinase enzyme to recover the feather protein concentrate. This feather protein concentrate was quantified, and it was found that the chemical score except lysine was remarkably high that is more than 100% in the feather protein concentrate, as mentioned by WHO experts. So, this feather protein concentrate can be used as constituents for feed (Dalev, 1994). Similarly, the protein concentrates were extracted from the tomato processing waste. The tomato paste processing unit generates almost 70–75 kg of solid waste in which it is mainly pomace. This pomace contains mainly seeds, which constitutes 22.2–33.9% of protein in dry basis. This tomato processing waste seed can be used to extract the protein isolate and concentrate, which was extracted using sugar and salt also with alkaline treatment and it was found that it had a lesser amount of cellular matter, which confirms its protein can be used for various food systems (Sogi *et al.*, 2002).

The watermelon seed oil cake has been found to be underutilized and is generally used as animal feed, but it is a rich source of protein (45–55%) and phenolic contents. As consumer demands for protein is increasing so there is a need to shift towards the by-products of fruits and vegetables. Therefore, the watermelon seed meal was prepared after oil pressing it and alkaline treatment was given to extract the protein concentrate of the seed meal, which was rich in globulin protein as characterized. This watermelon seed protein was used to produce protein-rich cookies. This protein was mixed with cookie dough and the cookies' sensory and physical analysis was done. This showed that watermelon seed protein concentrates can be used as the partial replacement for the gluten in cookies and further research can be done to produce fruit juices and other liquid products (Wani *et al.,* 2015).

Whey protein is the by-product produced from the dairy industry after the production of cheese whey. This whey contains a high amount of protein and lactose and contains a very high nutritional value and functional properties. Also, the removal of this whey (lactose and protein) can reduce the pollution that can reduce the BOD and COD content on disposal. So, the protein was extracted using ultrafiltration and later was freeze dried, which resulted in protein purity and an increase in shelf life. Therefore, this can be used as a supplement for the consumption of people for the gym and for muscle and body building (Das *et al.*, 2016).

Marine by-products have been found to be a potential source of protein. Only some of the by-products are being utilized and most are wasted. Annual discard stands at about 25 million tones. The by-products include the viscera, skin, bones and head cutoffs and viscera. These by-products contain a huge amount, which is converted into protein hydrosylates (produced by hydrolysis containing a high amount of free amino acids and peptides) for consumption. The old form is fish sauce, which is produced by the degradation of the protein fractions by the help of the various microbial and endogenous enzymes containing 8–14% digested protein. Fish silage is mostly used for the feed. Fish collagen obtained from the skin, which is purified and used for cosmetic applications, fish enzymes, etc., is being used to protein fractions for use in various applications (Rustad, 2003).

4.4.2 ANIMAL FEED PROTEIN ENRICHMENT

Fruit and vegetable residue coming from beverage companies such as residual pulp, skin, and peels need an effective way to utilize it. So, these residues are being used for bioconversion of various value-added products using enzymes or microorganisms. One of the examples of this value addition can be enriching proteins in the wastes. So that it can be used as feed and will also be economical. Therefore, the pineapple waste was taken, which was enriched by protein with the help of yeast

Saccharomyces cerevisiae by the solid-state bioprocessing along with nitrogen supplementation and not with it. It was found that the protein content, 22% of dry basis, was reached at 48 h when $(NH_4)_2SO_4$ was added to the medium and fermented pineapple waste can be a protein-rich feed by solid state bioprocessing (Correia *et al.*, 2007). Another agriculture residue such as orange, plantain, and banana wastes were used for protein yield extraction using *Candida sp.*, which can be used for both animal and human feed. Due to an alarming increase in population, food wastes are being used as a substrate to overcome nutritional problems. Most of the waste material contains an excess amount carbohydrates, which by the help of fermentation process can be converted into protein, which can be used as supplement feed for both the human and animal diet. A study was conducted to see the growth of *Candida sp.* for cell biomass production using citrus fruit wastes and it was found that it could produce the protein when a nitrogen source was provided. A low growth temperature and some supplementation will help the microorganisms help in cell biomass production. Thus, it helps in protein enrichment in feed (Adoki, 2008).

In another study protein enrichment was done in the apple pomace by solid state cultivation for using it as feed for Nile Tilapia fry. As we have already discussed, algae, fungi, and bacteria are the source of microbial single cell protein such as *S. cerevisiae* etc. So, the country producing a large quantity of apple juice produces 64% of the total of apple product waste (Bhusan *et al.*, 2008) is producing a huge amount of apple pomace (skin, seed, etc.), which is sold to farmers for use as animal feed. As this pomace has a high sugar content if left untreated, it can get fermented, resulting in intoxication of animals. So, the sugar can be utilized positively by converting this sugar as a substrate of single cell protein, which will reduce intoxication and will increase protein enrichment. As consumption of fish is increasing its production is to be also increased, so this protein enriched apple pomace can be used as feed. *G. buleri* culture therefore can be used for protein enrichment, which has the potential to increase the growth and also is economical, reduces environmental problems, and valorization to the agri-food residue (Vendruscolo *et al.*, 2009).

4.4.3 EDIBLE FILM AND COATING

Edible film and coatings are nowadays gaining popularity because of their characteristics. These are a continuous thin edible layer that is formed over the food products, especially fruits and vegetables, to increase its shelf life and protect it from microbial attacks. The edible film and coating also serve to be rich in antioxidants and flavoring agents. The only difference between edible film and coating is in their application methods that is edible coating is applied directly to the fruits simply by spraying or dipping the food and brushing with the solution. Edible film is applied over the food by extrusion, casting, molding techniques. Edible coating and film application does not change the characteristics of food.

Plant protein such as zein, gluten protein, are being used to prepare edible coatings and films. In which edible coating has been prepared using very few proteins such as zein, collagen, and gelatin protein. This zein protein powder is dissolved in an alcohol with which plasticizer is added to give mechanical strength and increase coating flexibility. Antioxidants are added to reduce lipid oxidation and this coating is formed on solvent evaporation. Zein coating is being used to coat the nuts, confectionaries, and other various types of foods (Buffo and Han, 2005). Thickness is one of the critical characteristics in edible coating. Another edible coating can be done using whey protein for strawberries to keep up its quality attributes. Rapid freezing after coating strawberries with whey protein showed successful results on keeping the anthocyanin content and helped in increasing the shelf life of the strawberries as they have very low shelf life, are prone to freezer burn and water loss, etc.; this whey protein edible coating helped to increase their shelf life along with retaining the nutritional properties of the strawberries (Soazo *et al.*, 2015). Edible films are made by adding plasticizers to the alcoholic solution such as ethanol or isopropanol having the protein extract dissolved, which is cooled and poured over a glass plate from where the alcohol vaporizes, and the film is formed, which is peeled off.

Strawberries can also be coated using apple peel polyphenol chitosan, which would increase the shelf life, reduce the post-harvest loss along with enriching it with polyphenols and natural antioxidants. Chitosan is a biopolymer generally used as edible coating to prolong the shelf life. So, this chitosan edible composite coating made from apple peel was used to coat strawberries, it decreased the total flavonoid and total anthocyanin, which delayed the deterioration and development of off- flavors and color changes (Riaz *et al.*, 2021).

Soy protein film is one of the most common and maximum used edible film. It has a unique property of being a high oxygen barrier, which makes it suitable to produce multi-layer packaging systems. When applied over meat products it protects or resists lipid oxidation and surface moisture loss. Soya protein film can also be used as a microencapsulating agent of flavors and pharmaceuticals for fruits and vegetables, and cheese. Various low sugar cakes and high moisture products are also coated using this film. Thus, this film serves as a cheap and readily available film for various food products to increase their shelf life and the nutritive value. Wheat gluten proteins are used to prepare films as they have novel functional properties for example rubber like mechanical properties and selective gas barrier properties and they are water resistant. Wheat gluten is a cohesive and elastic industrial by-product produced by wet milling. Wheat gluten is generally aimed to coat seeds, pills, foodstuffs, cosmetics, polishes, or drug capsules. It has also found its place in modified atmosphere packaging, active coating, and active packaging active coating etc. It has the potential to act as a coating material for moisture, gas and as a solute barrier, but due to the allergenic character of celiac (gluten intolerant) people, the application of this wheat gluten is limited (Buffo and Han, 2005).

Vegetables such as pumpkin seeds contain a high amount of protein, and it can be easily extracted after extracting the oil from its seeds. Depending on the quality of seeds, protein content can vary from 24.5–36.0%. This protein was extracted using the alkaline method and was analyzed for its mechanical strength and film barrier, which was found to be good except its solubility and therefore this can be used for application over cut fruits, bakery products and sweets, etc. (Lalnuthari *et al.*, 2020).

The shelf life improvement of fresh unripened curd cheese is well known in eastern European countries, by edible coating using whey protein and cinnamon bark. The cinnamon bark will act as an antimicrobial agent and the whey protein will increase the protein-rich content. This will be beneficial for both small and large scale industries as they can get the whey protein easily and soon after the extraction of this curd cheese production with minimal investment and also it was found that it reduced the moisture and microbial growth, which in turn increased or improved the shelf life of curd cheese. In future research, increasing the cinnamon concentration in the coating with the aim of better microbial resistance could be conducted (Mileriene *et al.*, 2021).

Other plant protein films, which are rarely available such as the one made from peanut protein isolate and concentrate from the peanut flour in which the water vapour barrier increased on increasing the drying temperature of the film. Cottonseed protein film was obtained from the cottonseed flour by the casting process. They are mostly used for non-food packaging systems as it has good mechanical resistance and is not soluble in water. They are used in medical packaging where biodegradability is necessary. Similarly rice protein film made from the rice bran solutions had higher water vapor permeability than synthetic films because of the hydrophilic nature. Pea protein separated from the starch and fiber after the pH method is used to make films, which show similar structure to soy protein and wheat gluten films. It has good strength and elasticity (Buffo and Han, 2005).

Apart from plant sources, animal protein, mostly fish protein, is also being used to produce protein films. The fish generally known as *Medusa* was used to extract the acid soluble collagen from the waste skin after its processing. As collagen has a unique characteristic of high tensile strength and stability by forming insoluble fibers so, this collagen protein from fish is being used for the formation of protein films. Lactic acid was found to be efficient for the extraction of collagen from the fish waste skin. This collagen film was mixed with chitosan, which showed a good result when incorporated with pomegranate peel showing good antibacterial effect against food borne pathogens. So, it can be proved that this collagen film can be used for active packaging (Bhuimbar *et al.*, 2019).

Whey protein obtained after the preparation of cottage cheese is one of the commonly used concentrates for building muscles among people. This whey protein has also proved to be one of the sources for formation of films. Whey proteins concentrate and isolate films were prepared using melanin extracted from watermelon seeds and used as the modifier to increase the UV light blocking, gas and moisture barrier to increase the mechanical strength and swelling. These modified films had high antioxidant activity but no cytotoxicity. So, whey protein isolates and concentrate film modified using melanin can serve as an active packaging system, wound dressing, and biomedical applications (Łopusiewicz *et al.*, 2020).

Other animal protein films which are limitedly available such as casein protein films, among all the casein protein fractions β-casein produces weak permeability to water vapor and caseinate films prepared mixed with whey protein isolate showed reduced puncture strength and WPC mixed casein also exhibited the same (Lacroix and Vu, 2014).

4.4.4 SINGLE CELL PROTEIN

Single cell protein (SCP) is referred to as edible protein or microbial protein, which is extracted from microbial cultures, which are dead or dried cell biomass. It is used as a protein supplement in human foods and animal feed. The fermentation of low-cost feedstock and wastes (wood, straw, cannery, and food-processing wastes, residues from alcohol production, hydrocarbons, or human and animal excreta) by the action of microorganisms like algae, fungi, yeast, and bacteria are used as a source of energy to produce biomass, protein concentrate, or amino acids. Thus, SCP is regarded as the natural source of protein concentrate and is used as an alternative to common protein sources to address the issues of increasing population and worldwide protein shortage (Mondal *et al*, 2012). SCP also contains carbohydrates, fats, nucleic acids, vitamins, and minerals. SCP can be produced from various sources of waste generated from the food industry.

The food processing sector generates lot of waste from various sectors of food industry such as dairy industry, bakery industry, fruit and vegetable processing industry, meat, fish and poultry industry, etc. These wastes generated can be used to produce SCP using different techniques. It is a beneficial method as it can reduce the amount of waste generated as well as help in producing a high protein diet. For production of single cell protein at the first collection of substrates is done and fermentation using microorganisms is the two main steps involved in the production of SCP. The substrates used for production of SCP are mostly the wastes generated from the food industry such as fishmeal, soymeal, fruit and vegetable waste, etc. The collected wastes are oven dried (45–50°C) and stored in Zip Lock pouches (Mondal *et al.*, 2012).

Microorganisms are used in fermentation for SCP production: SCP is produced by the fermentation of bacteria or yeast on the food waste. Microorganisms such as yeast and bacteria act upon the carbohydrates to convert the simple sugars for producing SCP. Extraction of SCP from fermented mass was separated by vacuum filtration from the broth and washed with sterile water. This separated portion was oven dried at 105°C followed by cooling in desiccators (Baccha *et al.*, 2011). Starch processing from potato as a raw material generates a lot of wastage, which is a big concern in the starch processing industry. Potato is rich in carbohydrates (cellulose and hemicellulose) but is very low in protein. Potato residue is converted to SCP in a two-step fermentation process, namely degradation of potato fiber using *A. niger* and fermentation using *B. lichenformis* (Liu *et al.*, 2014). SCP was also produced from date palm waste by yeast (*S. cerevisiae*) fermentation. Dates are rich in fructose and thus yeast can act upon dates for production of SCP by aerobic fermentation (Putra *et al.*, 2020). SCP was also produced from orange peel, which is waste for the fruit pulp processing industry. The orange peel has an attractive nutritional profile, which facilitates easy action for microbial community. Three strains of microbes were used to produce SCP from orange peel, namely *A. oryzae* degrades pectin present in the peel and produces proteins, *Candida tropicalis* to produce protein, and *T. koningii* degrades the cellulose and produces protein (Zhou *et al.*, 2019).

Since orange peel has low nitrogen content, an additional nitrogen source (soybean cake) was added to the substrate and then allowed for fermentation to produce SCP. SCP is also produced from waste such as fish meal, soybean meal by action of purple non-sulphur bacteria (PNSB). PNSB has been recently regarded as the group of SCP microorganisms as it can effectively control waste generated as well as aid in protein supplement production. The food waste was diluted with treated wastewater to maintain the organic loading rate (OLR) for the fermentation process, which results in the production of SCP (Liu *et al.*, 2014).

4.5 OTHER FOOD INDUSTRY WASTE PROTEIN UTILIZATION AND ITS FUTURE PERSPECTIVES

Papaya seeds known for their beneficial properties such as having numerous bioactive compounds, which can be used for both foods and pharma products. It was also used to extract proteins to valorize the waste. The pulse electric field extraction method was used with a power of 40 KV–10 kA with temperature variations. It was seen that highly acidic or basic pH had a negative impact on proteins resulting in coagulation. So, the high voltage electrical discharge (HVED)-assisted method was used, which gave good results in extraction of protein but had contaminants. So, the pulse electric method and supplementary aqueous extraction was used, which gave a good yield useful for the valorization of papaya seeds in industrial application (Paraiakov *et al.*, 2015). Modification of protein using non-thermal treatments has proved to be a boon in the food industry to overcome the lack of protein problem. The modification of protein isolates results in a change in structural and functional properties of the protein isolates making it desirable for food formulation. This is suitable for various food formulations and can be used as the protein diet for improvement in structural, rheological, and functional properties, which is done by the high intensity ultrasound treatment or other non-thermal techniques. Quinoa seeds extracted native protein structure was treated by high intensity ultrasound treatment to get desirable structural characteristics and the functional and gelling characteristic of the quinoa seed was improved, though the protein modification is still not done in waste sources. A lot more research is required to do protein modification in the waste resources for its valorization (Mir *et al.*, 2019). Residues from agri-food waste have a high amount of protein can be valorized for integration in biorefinery using various extraction techniques such as enzyme-assisted extraction and alkaline extraction (del MarContreras *et al.*, 2019). Due to an increase in population and over dependency in soya protein as a plant protein source has led to overconsumption so research for other sources is required. Kiwi fruit seeds have been found to be underutilized as solid wastes in food industries. It has been found that the seed protein has good digestibility, and it can act as an alternative to soya protein in food processing industries. Kiwi seeds have a good amount of protein and are rich in lysine and methionine. The results showed that except foaming properties kiwi seed protein was having properties better or higher than soya protein. So, it can be concluded that kiwi seed protein can be a good alternative source of plant protein to soya protein (Deng *et al.*, 2014). As we know, oilseed protein is common and good for health in the human diet. These proteins generally affected by its processing for oil so the protein extracted can be improved by the modification so that it becomes desirable to be applied are delayed deterioration removal of toxin and inhibitory compound and incorporation of nutrients and additives through covalent bonding (Moure *et al.*, 2006).

Animal protein such as shrimp processing waste protein, which is being produced from the seafood processing industries can be used for cosmetic applications. A high protein content is wasted on processing so this waste can be hydrolyzed using protease (Alcalase) and recovered as protein hydrosylate to produce chitosan for use in cosmetic applications (Gildberg and Stenberg, 2001).

More animal protein such as fish waste protein is being used as an alternative to produce wood adhesive. Generally wood adhesive is produced using formaldehyde and petroleum products, which are hazardous, therefore the protein from animals can be used to produce adhesive in an eco-friendlier way and it will be less hazardous. Recently, catfish skin has been used for wood adhesive along with

TABLE 4.2
Recovery and Utilization of Protein from Food Industry Wastes

Protein waste source	Possible recovery techniques	Possible utilization of proteins	References
Oil cake meal (soya meal, peanut meal, sesame oilcake)	Ultrasound-assisted extraction, pulse electric field extraction	Animal feed enrichment, edible film and coating, single cell protein	Karki *et al.*, 2010, Nyugen *et al.*, 2019, Sarkis *et al.*, 2015, Buffo and Tan 2005, Mondal *et al.*, 2012.
Pomace (apple, olive, pineapple)	Supercritical fluid extraction, solid state fermentation	bioethanol production (biorefinery concept), biopolymer, single cell protein	Kazan *et al.*, 2015, Vendruscolo *et al.*, 2009, Dhansekaran *et al.*, 2011
Fish waste (collagen), chicken liver	Enzyme-assisted extraction, pulse electric field extraction, ultrasound-assisted extraction	chitosan for application in cosmetics, wood adhesive	Araujo *et al.*, 2021 Arauzo *et al.*, 2019, Mangano *et al.*, 2021, Ghosh *et al.*, 2019, Zou *et al.,* 2017, Gildberg and Stenberg, 2001, Cheng *et al.*, 2021
Bran (oat, rice, sesame)	Ultrasound-assisted extraction, alkaline extraction, supercritical fluid extraction, microwave-assisted extraction	Protein isolates and concentrates for development of functional foods and supplements	Qu *et al.*, 2012, Guan *et al.*, 2008, Görgüç *et al.*, 2021 Phongthai *et al.*, 2016
Seeds (pumpkin, papaya)	Osborne method, pulse electric field, alkaline extraction	Edible film, cookies (protein enrichment)	Vinayshree *et al.*, 2021, Parniakov *et al.*, 2015, Lalnuthari *et al.*, 2020, Wani *et al.*, 2015
Peels (orange, banana)	Used as substrate precipitation method	Single cell protein, protein supplements	Zhou *et al.*, 2019, Budhalakoti, 2021
whey protein	aqueous two-phase separation technology, membrane technology	Protein concentrate, edible film	Buffo and Tan 2005, Das *et al.*, 2016, Sayed and Chase., 2011

cottonseed protein blend, which upon analysis showed good results. So, it can be suggested that cat-fish skin can be used for wood adhesive (Cheng et al., 2021).

The water drained from the canned chickpea, i.e., the water is called "aquafaba", which has 0.95–1.5% of protein and other nutrients as well. It is an alternative to milk and egg protein and is vegan friendly. It has a good emulsifying and anti-foaming agent, which will help in waste valorization and will help for the global security and agricultural sustainability. However, more thorough research is required for the aquafaba for its valorization (Mustafa and Reaney, 2020).

The recent inventions and discoveries for zero waste goals in this era is a need to have effective clean and reduced pollution free environment. Therefore, it should be kept in mind that various recoveries of protein from industry wastes need to be carefully planned. Requirements of processing units need to be developed to fulfill the direct utilization of the wastes. Still proteins are getting wasted in the form of food industry wastes. More effective research particularly in this field is necessary for the zero-waste goal, also the deficit of proteins is a major concern in the rural and semi-urban areas. Therefore, the utilization of proteins for consumable products may create change in the lifestyle of the people. More works and research on waste utilization are expected soon. Protein recovery and utilization from wastes is still under research and is an emerging development for the environment sustainability. This utilization is a way to increase the productivity of protein and increase the nutrition availability. The various green extraction technologies for the recovery of proteins, which can help to increase nutrition for consumer acceptance. Still there is the requirement of more research for the food industry-based wastes and to study the functional and structural properties of the protein extracted from the wastes, which are added to the other food products to develop improved new generation food.

4.6 CONCLUSION

Since the last decades, the recovery and utilization of food industry-based waste has become an important aspect for the environment. This utilization of the food industry-based waste will also help to give better returns of the investment made to recycle the waste for today's world, new effective research for the betterment of the environment. Researchers are attempting to valorize the waste protein from various sources using innovations. Protein availability at lower cost is essential to serve it in the rural areas. So, this chapter merges along with recovery and the utilization of protein to be processed in a cheap and eco-friendly way, which will help to supply sufficient protein to the growing global market. This chapter will help researchers to develop more innovative ideas.

REFERENCES

Adoki, A. 2008. Factors affecting yeast growth and protein yield production from orange, plantain and banana wastes processing residues using *Candida sp*. *African Journal of Biotechnology* 7(3): 290–295.

Ajila, C.M., Brar, S.K., Verma, M., Tyagi, R.D., Godbout, S. and Valéro, J.R. 2012. Bio-processing of agro-by-products to animal feed. *Critical Reviews in Biotechnology* 32(4): 382–400. https://doi.org/10.3109/07388551.2012.659172

Akyüz, A. and Ersus, S. 2021. Optimization of enzyme assisted extraction of protein from the sugar beet (*Beta vulgaris* L.) leaves for alternative plant protein concentrate production. *Food Chemistry* 335: 127673. https://doi.org/10.1016/j.foodchem.2020.127673

Araujo, J., Sica, P., Costa, C. and Márquez, M.C. 2021. Enzymatic hydrolysis of fish waste as an alternative to produce high value-added products. *Waste and Biomass Valorization* 12(2): 847–855. https://doi.org/10.1007/s12649-020-01029-x

Aydemir, L.Y., Gökbulut, A.A., Baran, Y. and Yemenicioğlu, A. 2014. Bioactive, functional and edible film-forming properties of isolated hazelnut (*Corylus avellana* L.) meal proteins. *Food Hydrocolloids* 36: 130–142. https://doi.org/10.1016/j.foodhyd.2013.09.014

Bacha, U., Nasir, M., Khalique, A., Anjum, A.A. and Jabbar, M.A. 2011. Comparative assessment of various agro-industrial wastes for *Saccharomyces cerevisiae* biomass production and its quality evaluation as single cell protein. The *Journal of Animal & Plant Sciences* 21(4): 844–849.

Bakshi, M.P.S., Kaushal, S. and Wadhwa, M. 2005. Potential of Sarson Saag Waste-a cannery waste as ruminant feed. *Asian-Australasian Journal of Animal Sciences* 18(4): 479–482. https://doi.org/10.5713/ajas.2005.479

Barać, M.B., Pešić, M.B., Stanojević, S.P., Kostić, A.Ž. and Čabrilo, S.B. 2015. Techno-functional properties of pea (Pisum sativum) protein isolates: A review. *Acta PeriodicaTechnologica* 46: 1–18.

Bhuimbar, M.V., Bhagwat, P.K. and Dandge, P.B. 2019. Extraction and characterization of acid soluble collagen from fish waste: Development of collagen-chitosan blend as food packaging film. *Journal of Environmental Chemical Engineering* 7(2): 102983. https://doi.org/10.1016/j.jece.2019.102983

Bhushan, S., Kalia, K., Sharma, M., Singh, B. and Ahuja, P.S. 2008. Processing of apple pomace for bioactive molecules. *Critical Reviews in Biotechnology* 28(4): 285–296. https://doi.org/10.1080/07388550802368895

Bongaarts, J. 2007. Food and Agriculture Organization of the United Nations: the state of food and agriculture: agricultural trade and poverty: can trade work for the poor? *Population and Development Review* 33(1): 197–198.

Buchmann, L., Brändle, I., Haberkorn, I., Hiestand, M. and Mathys, A. 2019. Pulsed electric field based cyclic protein extraction of microalgae towards closed-loop biorefinery concepts. *Bioresource Technology* 291: 121870.https://doi.org/10.1016/j.biortech.2019.121870

Budhalakoti, N. 2021. Extraction of protein from banana by-product and its characterization. *Journal of Food Measurement and Characterization* 15(3): 2202–2210. https://doi.org/10.1007/s11694-020-00803-8

Buffo, R.A. and Han, J.H. 2005. Edible films and coatings from plant origin proteins. In: *Innovations in food packaging,* Ed. J. H. Han, 277–300. Academic Press. https://doi.org/10.1016/B978-012311632-1/50049-8

Burd, N.A., McKenna, C.F., Salvador, A.F., Paulussen, K.J. and Moore, D.R. 2019. Dietary protein quantity, quality, and exercise are key to healthy living: a muscle-centric perspective across the lifespan. *Frontiers in Nutrition* 6(83): 1–12. https://doi.org/10.3389/fnut.2019.00083

Cheng, H.N., He, Z., Li, C.H., Bland, J.M. and Bechtel, P.J. 2021. Preparation and evaluation of catfish protein as a wood adhesive. *International Journal of Polymer Analysis and Characterization* 26(1): 60–67. https://doi.org/10.1080/1023666X.2020.1844361

Cheng, X., Bi, L., Zhao, Z. and Chen, Y. 2015. Advances in enzyme assisted extraction of natural products. In: *Advances in Engineering Research*, ed. P. Yarlagadda, 3:371–375. Atlantis Press.

Correia, R., Magalhaes, M. and Macêdo, G. 2007. Protein enrichment of pineapple waste with Saccharomyces cerevisiae by solid state bioprocessing. *Journal of Scientific and Industrial Research* 66: 259–262.

Dalev, P.G. 1994. Utilisation of waste feathers from poultry slaughter for production of a protein concentrate. *Bioresource Technology* 48(3): 265–267. https://doi.org/10.1016/0960-8524(94)90156-2

Das, B., Sarkar, S., Sarkar, A., Bhattacharjee, S. and Bhattacharjee, C. 2016. Recovery of whey proteins and lactose from dairy waste: A step towards green waste management. *Process Safety and Environmental Protection* 101: 27–33.https://doi.org/10.1016/j.psep.2015.05.006

Deb, S., Noori, M.T. and Rao, P.S. 2020. Application of biofloc technology for Indian major carp culture (polyculture) along with water quality management. *Aquacultural Engineering* 91: 102106. https://doi.org/10.1016/j.aquaeng.2020.102106

del Mar Contreras, M., Lama-Muñoz, A., Gutiérrez-Pérez, J.M., Espínola, F., Moya, M. and Castro, E. 2019. Protein extraction from agri-food residues for integration in biorefinery: Potential techniques and current status. *Bioresource Technology* 280: 459–477. https://doi.org/10.1016/j.biortech.2019.02.040

Deng, J., Sun, T., Cao, W., Fan, D., Cheng, N., Wang, B., Gao, H. and Yang, H. 2014. Extraction optimization and functional properties of proteins from kiwi fruit (Actinidia chinensis Planch.) seeds. *International Journal of Food Properties* 17(7): 1612–1625. https://doi.org/10.1080/10942912.2013.772197

Dharumadurai, D., Subramaniyan, L., Subhasish, S., Nooruddin, T. and Annamalai, P. 2011. Production of single cell protein from pineapple waste using yeast. *Innovative Romanian Food Biotechnology* 8: 26–32.

El-Sayed, M.M. and Chase, H.A. 2011. Trends in whey protein fractionation. *Biotechnology Letters* 33(8): 1501–1511. https://doi.org/10.1007/s10529-011-0594-8

FAO. 2016. The State of World Fisheries and Agriculture: Contributing to Food Security and Nutrition for All Food and Agricultural Organization, 200.

Gençdağ, E., Görgüç, A., Yılmaz, and F. M. 2021. Recent advances in the recovery techniques of plant-based proteins from agro-industrial by-products. *Food Reviews International* 37(4): 447–468. https://doi.org/10.1080/87559129.2019.1709203

Ghosh, S., Gillis, A., Sheviryov, J., Levkov, K. and Golberg, A. 2019. Towards waste meat biorefinery: Extraction of proteins from waste chicken meat with non-thermal pulsed electric fields and mechanical pressing. *Journal of Cleaner Production* 208: 220–231. https://doi.org/10.1016/j.jclepro.2018.10.037

Gildberg, A. and Stenberg, E. 2001. A new process for advanced utilisation of shrimp waste. *Process Biochemistry* 36(8–9): 809–812. https://doi.org/10.1016/S0032-9592(00)00278-8

Gomez, L., Tiwari, B. and Garcia-Vaquero, M. 2020. Emerging extraction techniques: microwave-assisted extraction. In: *Sustainable Seaweed Technologies*, eds. M. D. Torres, S. Kraan, and H. Dominguez 9: 207–224. Elsevier. https://doi.org/10.1016/B978-0-12-817943-7.00008-1

Görgüç, A., Özer, P. and Yılmaz, F.M. 2020. Microwave-assisted enzymatic extraction of plant protein with antioxidant compounds from the food waste sesame bran: Comparative optimization study and identification of metabolomics using LC/Q-TOF/MS. *Journal of Food Processing and Preservation*, 44(1):14304. https://doi.org/10.1111/jfpp.14304

Guan, X. and Yao, H. 2008. Optimization of Viscozyme L-assisted extraction of oat bran protein using response surface methodology. *Food Chemistry* 106(1): 345–351. https://doi.org/10.1016/j.foodchem.2007.05.041

Guimón, J. and Guimón, P. 2012. How ready-to-use therapeutic food shapes a new technological regime to treat child malnutrition. *Technological Forecasting and Social Change* 79(7): 1319–1327. https://doi.org/10.1016/j.techfore.2012.04.011

Hata, S., Wiboonsirikul, J., Maeda, A., Kimura, Y. and Adachi, S. 2008. Extraction of defatted rice bran by subcritical water treatment. *Biochemical Engineering Journal* 40(1): 44–53. https://doi.org/10.1016/j.bej.2007.11.016

Heffernan, S., Giblin, L. and O'Brien, N. 2021. Assessment of the biological activity of fish muscle protein hydrolysates using in vitro model systems. *Food Chemistry* 359: 129852. https://doi.org/10.1016/j.foodchem.2021.129852

Ho, K.S. and Chu, L.M. 2019. Characterization of food waste from different sources in Hong Kong. *Journal of the Air & Waste Management Association* 69(3): 277–288. https://doi.org/10.1080/10962247.2018.1526138

Hovhannisyan, V. and Devadoss, S. 2020. Effects of urbanization on food demand in China. *Empirical Economics* 58(2): 699–721. https://doi.org/10.1007/s00181-018-1526-4

Hultin, H.O. and Kelleher, S.D. 2001. Advanced Protein Technologies Inc., *Process for isolating a protein composition from a muscle source and protein composition*. U.S. Patent 6: 288, 216.

Kamal, H., Le, C.F., Salter, A.M. and Ali, A. 2021. Extraction of protein from food waste: An overview of current status and opportunities. *Comprehensive Reviews in Food Science and Food Safety* 20(3): 2455–2475. https://doi.org/10.1111/1541-4337.12739

Karki, B., Lamsal, B.P., Jung, S., van Leeuwen, J.H., Pometto III, A.L., Grewell, D. and Khanal, S.K. 2010. Enhancing protein and sugar release from defatted soy flakes using ultrasound technology. *Journal of Food Engineering* 96(2): 270–278. https://doi.org/10.1016/j.jfoodeng.2009.07.023

Kazan, A., Celiktas, M.S., Sargin, S. and Yesil-Celiktas, O. 2015. Bio-based fractions by hydrothermal treatment of olive pomace: process optimization and evaluation. *Energy Conversion and Management* 103: 366–373. https://doi.org/10.1016/j.enconman.2015.06.084

Kumar, S., Kushwaha, R. and Verma, M.L. 2020. Recovery and utilization of bioactives from food processing waste. In: *Biotechnological Production of Bioactive Compounds*, eds. L. V. Madan, and K. C. Anuj, 37–68. Elsevier. https://doi.org/10.1016/B978-0-444-64323-0.00002-3

Kumar, V., Rani, A., Goyal, L., Dixit, A.K., Manjaya, J.G., Dev, J. and Swamy, M. 2010. Sucrose and raffinose family oligosaccharides (RFOs) in soybean seeds as influenced by genotype and growing location. *Journal of Agricultural and Food Chemistry* 58(8): 5081–5085. https://doi.org/10.1021/jf903141s

Lacroix, M. and Vu, K.D. 2014. Edible coating and film materials: proteins. In *Innovations in food packaging*, Ed. J. H. Han, 277–304. Academic Press. https://doi.org/10.1016/B978-0-12-394601-0.00011-4

Lalnunthari, C., Devi, L.M. and Badwaik, L.S. 2020. Extraction of protein and pectin from pumpkin industry by-products and their utilization for developing edible film. *Journal of Food Science and Technology* 57(5): 1807–1816. https://doi.org/10.1007/s13197-019-04214-6

Liu, B., Li, Y., Song, J., Zhang, L., Dong, J. and Yang, Q. 2014. Production of single-cell protein with two-step fermentation for treatment of potato starch processing waste. *Cellulose* 21(5): 3637–3645. https://doi.org/10.1007/s10570-014-0400-6

Llompart, M., Celeiro, M. and Dagnac, T. 2019. Microwave-assisted extraction of pharmaceuticals, personal care products and industrial contaminants in the environment. *TrAC Trends in Analytical Chemistry* 116: 136–150. https://doi.org/10.1016/j.trac.2019.04.029

Łopusiewicz, Ł., Drozłowska, E., Trocer, P., Kostek, M., Śliwiński, M., Henriques, M.H., Bartkowiak, A. and Sobolewski, P. 2020. Whey protein concentrate/isolate biofunctional films modified with melanin from watermelon (*Citrullus lanatus*) seeds. *Materials* 13(17): 3876.https://doi.org/10.3390/ma13173876

Ma, J., Xu, R.R., Lu, Y., Ren, D.F. and Lu, J. 2018. Composition, antimicrobial and antioxidant activity of supercritical fluid extract of *Elsholtziaciliata*. *Journal of Essential OilBearing Plants* 21(2): 556–562. https://doi.org/10.1080/0972060X.2017.1409657

Mangano, V., Gervasi, T., Rotondo, A., De Pasquale, P., Dugo, G., Macrì, F. and Salvo, A. 2021. Protein hydrolysates from anchovy waste: Purification and chemical characterization. *Natural Product Research* 35(3): 399–406. https://doi.org/10.1080/14786419.2019.1634711

Metwally, R.A., Soliman, S.A., Latef, A.A.H.A. and Abdelhameed, R.E. 2021. The Individual and interactive role of arbuscular mycorrhizal fungi and Trichoderma viride on growth, protein content, amino acids fractionation, and phosphatases enzyme activities of onion plants amended with fish waste. *Ecotoxicology and Environmental Safety* 214: 112072. https://doi.org/10.1016/j.ecoenv.2021.112072

Mileriene, J., Serniene, L., Henriques, M., Gomes, D., Pereira, C., Kondrotiene, K., Kasetiene, N., Lauciene, L., Sekmokiene, D. and Malakauskas, M. 2021. Effect of liquid whey protein concentrate–based edible coating enriched with cinnamon carbon dioxide extract on the quality and shelf life of Eastern European curd cheese. *Journal of Dairy Science* 104(2): 1504–1517. https://doi.org/10.3168/jds.2020-18732

Mir, N.A., Riar, C.S. and Singh, S. 2019. Structural modification of quinoa seed protein isolates (QPIs) by variable time sonification for improving its physicochemical and functional characteristics. *Ultrasonics Sonochemistry* 58: 104700. https://doi.org/10.1016/j.ultsonch.2019.104700

Mohamed, M.E. and Eissa, A.H.A. 2012. Pulsed electric fields for food processing technology. *Structure and Function of Food Engineering* 11: 275–306.

Mondal, A.K., Sengupta, S., Bhowal, J. and Bhattacharya, D.K. 2012. Utilization of fruit wastes in producing single cell protein. *International Journal of Science, Environment and Technology*, 1(5):430–438.

Moradi, N., Rahimi, M., Moeini, A. and Parsamoghadam, M.A. 2018. Impact of ultrasound on oil yield and content of functional food ingredients at the oil extraction from sunflower. *Separation Science and Technology* 53(2): 261–276. https://doi.org/10.1080/01496395.2017.1384016

Moure, A., Sineiro, J., Domínguez, H. and Parajó, J.C. 2006. Functionality of oilseed protein products: a review. *Food Research International* 39(9): 945–963. https://doi.org/10.1016/j.foodres.2006.07.002

Muley, P. and Boldor, D. 2018. Process Intensification and Parametric Optimization in Biodiesel Synthesis Using Microwave Reactors. *Green Chemistry for Sustainable Biofuel Production* 614: 167–204.

Mustafa, R. and Reaney, M.J. 2020. Aquafaba, from food waste to a value-added product. In: *Food Wastes and By-products: Nutraceutical and Health Potential*, Eds, R. Campos-Vega, B.D. Oomah, and H.A. Vergara-Castañeda, 93–126. Wiley Online Library. https://doi.org/10.1002/9781119534167.ch4

Nadathur, S.R., Wanasundara, J.P. D. and Scanlin, L. 2017. Proteins in the diet: Challenges in feeding the global population. *Sustainable Protein Sources*. Academic Press. 1–19. https://doi.org/10.1016/B978-0-12-802778-3.00001-9

Nguyen, T.H. and Le, V.V.M. 2019. Effects of technological parameters of ultrasonic treatment on the protein extraction yield from defatted peanut meal. *International Food Research Journal* 26(3): 1079–1085.

Nile, S.H., Nile, A., Oh, J.W. and Kai, G. 2020. Soybean processing waste: Potential antioxidant, cytotoxic and enzyme inhibitory activities. *Food Bioscience* 38: 100778. https://doi.org/10.1016/j.fbio.2020.100778

Panda, D. and Manickam, S. 2019. Cavitation technology—The future of greener extraction method: A review on the extraction of natural products and process intensification mechanism and perspectives. *Applied Sciences* 9(4): 766. https://doi.org/10.3390/app9040766

Parniakov, O., Roselló-Soto, E., Barba, F.J., Grimi, N., Lebovka, N. and Vorobiev, E. 2015. New approaches for the effective valorization of papaya seeds: Extraction of proteins, phenolic compounds, carbohydrates, and isothiocyanates assisted by pulsed electric energy. *Food Research International* 77: 711–717. https://doi.org/10.1016/j.foodres.2015.03.031

Phongthai, S., Lim, S.T. and Rawdkuen, S. 2016. Optimization of microwave-assisted extraction of rice bran protein and its hydrolysates properties. *Journal of Cereal Science* 70: 146–154. https://doi.org/10.1016/j.jcs.2016.06.001

Puah, C.W., Choo, Y.M., Ma, A.N. and Chuah, C.H. 2005. Supercritical fluid extraction of palm carotenoids. *American Journal of Environmental Science* 1(4): 264–269.

Putra, M.D., Abasaeed, A.E. and Al-Zahrani, S.M. 2020. Prospective production of fructose and single cell protein from date palm waste. *Electronic Journal of Biotechnology* 48: 46–52. https://doi.org/10.1016/j.ejbt.2020.09.007

Putra, R.S. and Airun, N.H. 2021. The effect of particle size and dosage on the performance of Papaya seeds (Carica papaya) as biocoagulant on wastewater treatment of batik industry. In: *IOP Conference Series: Materials Science and Engineering* 1087(1): 012045. IOP Publishing.

Quintero Quiroz, J., Naranjo Duran, A.M., Silva Garcia, M., Ciro Gomez, G.L. and Rojas Camargo, J.J. 2019. Ultrasound-assisted extraction of bioactive compounds from annatto seeds, evaluation of their antimicrobial and antioxidant activity, and identification of main compounds by LC/ESI-MS analysis. *International Journal of Food Science*, 2019: 1–9. https://doi.org/10.1155/2019/3721828

Qu, W., Ma, H., Jia, J., He, R., Luo, L. and Pan, Z. 2012. Enzymolysis kinetics and activities of ACE inhibitory peptides from wheat germ protein prepared with SFP ultrasound-assisted processing. *Ultrasonics Sonochemistry* 19(5): 1021–1026. https://doi.org/10.1016/j.ultsonch.2012.02.006

Raveendran, S., Parameswaran, B., BeeviUmmalyma, S., Abraham, A., Kuruvilla Mathew, A., Madhavan, A., Rebello, S. and Pandey, A. 2018. Applications of microbial enzymes in food industry. *Food Technology and Biotechnology* 56(1): 16–30. https://doi.org/10.17113/ftb.56.01.18.5491

Riaz, A., Aadil, R.M., Amoussa, A.M.O., Bashari, M., Abid, M. and Hashim, M.M. 2021. Application of chitosan-based apple peel polyphenols edible coating on the preservation of strawberry (*Fragaria ananassa* cv Hongyan) fruit. *Journal of Food Processing and Preservation* 45(1): p.e15018. https://doi.org/10.1111/jfpp.15018

Rizzo, G. and Baroni, L. 2018. Soy, soy foods and their role in vegetarian diets. *Nutrients* 10(1): 43. https://doi.org/10.3390/nu10010043

Romero-Rodríguez, M.C., Maldonado-Alconada, A.M., Valledor, L. and Jorrin-Novo, J.V. 2014. Back to Osborne. Sequential protein extraction and LC-MS analysis for the characterization of the Holm oak seed proteome. *Plant Proteomics* 1072: 379–389. https://doi.org/10.1007/978-1-62703-631-3_27

Rommi, K. 2016. *Enzyme-aided recovery of protein and protein hydrolyzates from rapeseed cold-press cake* (Doctoral Dissertation, University of Helsinki).

Rustad, T. 2003. Utilisation of marine by-products. *Electronic Journal of Environmental, Agricultural and Food Chemistry* 2(4): 458–463.

Sahin, Ilyas, Birdal Bilir, Shakir Ali, Kazim Sahin, and Omer Kucuk. 2019. Soy isoflavones in integrative oncology: increased efficacy and decreased toxicity of cancer therapy. *Integrative Cancer Therapies* 18: 1534735419835310.

Santos-Hernández, A.S., Hinojosa-Reyes, L., Sáenz-Tavera, I.D.C., Hernández-Ramírez, A. and Guzmán-Mar, J.L. 2018. Atrazine and 2, 4-D determination in corn samples using microwave assisted extraction and on-line solid-phase extraction coupled to liquid chromatography. *Journal of the Mexican Chemical Society* 62(2): 282–294. https://doi.org/10.29356/jmcs.v62i2.475

Sardari, R.R., Sutiono, S., Azeem, H.A., Galbe, M., Larsson, M., Turner, C. and Nordberg Karlsson, E. 2019. Evaluation of sequential processing for the extraction of starch, lipids, and proteins from wheat bran. *Frontiers in Bioengineering and Biotechnology* 7: 413. https://doi.org/10.3389/fbioe.2019.00413

Sari, Y.W., Mulder, W.J., Sanders, J.P. and Bruins, M.E. 2015. Towards plant protein refinery: review on protein extraction using alkali and potential enzymatic assistance. *Biotechnology Journal* 10(8): 1138–1157. https://doi.org/10.1002/biot.201400569

Sarkis, J.R., Boussetta, N., Blouet, C., Tessaro, I.C., Marczak, L.D.F. and Vorobiev, E. 2015. Effect of pulsed electric fields and high voltage electrical discharges on polyphenol and protein extraction from sesame cake. *Innovative Food Science & Emerging Technologies* 29: 170–177. https://doi.org/10.1016/j.ifset.2015.02.011

Sedlar, T., Čakarević, J., Tomić, J. and Popović, L. 2021. Vegetable by-products as new sources of functional proteins. *Plant Foods for Human Nutrition* 76(1): 31–36. https://doi.org/10.1007/s11130-020-00870-8

Shahidi, F., Varatharajan, V., Peng, H. and Senadheera, R. 2019. Utilization of marine by-products for the recovery of value-added products. *Journal of Food Bioactives* 6: 10–61. https://doi.org/10.31665/JFB.2019.6184

Shahid, K., Srivastava, V. and Sillanpää, M. 2021. Protein recovery as a resource from waste specifically via membrane technology—from waste to wonder. *Environmental Science and Pollution Research* 28(8): 10262–10282. https://doi.org/10.1007/s11356-020-12290-x

Shanthi, G., Premalatha, M. and Anantharaman, N. 2021. Potential utilization of fish waste for the sustainable production of microalgae rich in renewable protein and phycocyanin-Arthrospira platensis/Spirulina. *Journal of Cleaner Production* 294: 126106.https://doi.org/10.1016/j.jclepro.2021.126106

Sharma, P., Oey, I. and Everett, D.W. 2014. Effect of pulsed electric field processing on the functional properties of bovine milk. *Trends in Food Science & Technology* 35(2): 87–101. https://doi.org/10.1016/j.tifs.2013.11.004

Shaviklo, A.R. 2015. Development of fish protein powder as an ingredient for food applications: a review. *Journal of Food Science and Technology* 52(2): 648–661. https://doi.org/10.1007/s13197-013-1042-7

Sibt-e-Abbas, M., Butt, M.S., Sultan, M.T., Sharif, M.K., Ahmad, A.N. and Batool, R. 2015. Nutritional and functional properties of protein isolates extracted from defatted peanut flour. *International Food Research Journal* 22(4): 1533–1537.

Singh, B.P., Yadav, D. and Vij, S. 2017. Soybean bioactive molecules: current trend and future prospective. In: *Bioactive Molecules in Food*, 1–29. Springer International Publishing: Berlin/Heidelberg, Germany.

Soazo, M., Pérez, L.M., Rubiolo, A.C. and Verdini, R.A. 2015. Prefreezing application of whey protein-based edible coating to maintain quality attributes of strawberries. *International Journal of Food Science & Technology* 50(3): 605–611. https://doi.org/10.1111/ijfs.12667

Sogi, D.S., Garg, S.K. and Bawa, A.S. 2002. Functional properties of seed meals and protein concentrates from tomato-processing waste. *Journal of Food Science* 67(8): 2997–3001. https://doi.org/10.1111/j.1365-2621.2002.tb08850.x

Tan, S.H., Mailer, R.J., Blanchard, C.L. and Agboola, S.O. 2011. Extraction and characterization of protein fractions from Australian canola meals. *Food Research International* 44(4): 1075–1082. https://doi.org/10.1016/j.foodres.2011.03.023

USDA. 2014. United States Department of Agriculture. Available at: www.nal.usda.gov/fnic/foodcomp/search/ (Accessed: 21 August 2014).

Vendruscolo, F., da Silva Ribeiro, C., Esposito, E. and Ninow, J.L. 2009. Protein enrichment of apple pomace and use in feed for Nile Tilapia. *Applied Biochemistry and Biotechnology* 152(1):74–87. https://doi.org/10.1007/s12010-008-8259-3

Vinayashree, S. and Prasanna V. 2021. Biochemical, nutritional and functional properties of protein isolate and fractions from pumpkin (*Cucurbita moschata* var. Kashi Harit) seeds. *Food Chemistry* 340: 128–177.

Virtanen, H.E., Voutilainen, S., Koskinen, T.T., Mursu, J., Kokko, P., Ylilauri, M., Tuomainen, T.P., Salonen, J.T. and Virtanen, J.K. 2019. Dietary proteins and protein sources and risk of death: the Kuopio Ischaemic Heart Disease Risk Factor Study. *The American Journal of Clinical Nutrition* 109(5): 1462–1471. https://doi.org/10.1093/ajcn/nqz025

Wadhwa, M., Kaushal, S. and Bakshi, M.P.S. 2006. Nutritive evaluation of vegetable wastes as complete feed for goat bucks. *Small Ruminant Research* 64(3): 279–284. https://doi.org/10.1016/j.smallrumres.2005.05.017

Wang, Q. 2018. Quality characteristics and determination methods of peanut raw materials. In: *Peanut processing characteristics and quality evaluation,* 69–125. Springer. https://doi.org/10.1007/978-981-10-6175-2_2

Wani, A.A., Sogi, D.S., Singh, P. and Khatkar, B.S. 2015. Influence of watermelon seed protein concentrates on dough handling, textural and sensory properties of cookies. *Journal of Food Science and Technology* 52(4): 2139–2147. https://doi.org/10.1007/s13197-013-1224-3

Zhang, J., Liu, L., Jiang, Y., Faisal, S., Wei, L., Cao, C., Yan, W. and Wang, Q. 2019. Converting peanut protein biomass waste into "double green" meat substitutes using a high-moisture extrusion process: A multiscale method to explore a process for forming a meat-like fibrous structure. *Journal of Agricultural and Food Chemistry* 67(38): 10713–10725. https://doi.org/10.1021/acs.jafc.9b02711

Zhang, S., Sun, L., Ju, H., Bao, Z., Zeng, X.A. and Lin, S. 2020. Research advances and application of pulsed electric field on proteins and peptides in food. *Food Research International* 139: 109914. https://doi.org/10.1016/j.foodres.2020.109914

Zhao, G., Liu, Y., Zhao, M., Ren, J. and Yang, B. 2011. Enzymatic hydrolysis and their effects on conformational and functional properties of peanut protein isolate. *Food Chemistry* 127(4): 1438–1443. https://doi.org/10.1016/j.foodchem.2011.01.046

Zhou, Y.M., Chen, Y.P., Guo, J.S., Shen, Y., Yan, P. and Yang, J.X. 2019. Recycling of orange waste for single cell protein production and the synergistic and antagonistic effects on production quality. *Journal of Cleaner Production* 213: 384–392. https://doi.org/10.1016/j.jclepro.2018.12.168

Zou, Ye, Haibo Shi, Xiao Chen, Pingping Xu, Di Jiang, Weimin Xu, and Daoying Wang. 2019. Modifying the structure, emulsifying and rheological properties of water-soluble protein from chicken liver by low-frequency ultrasound treatment. *International Journal of Biological Macromolecules* 139: 810–817.

Žilić, S., Barać, M., Pešić, M., Dodig, D. and Ignjatović-Micić, D. 2011. Characterization of proteins from grain of different bread and durum wheat genotypes. *International Journal of Molecular Sciences* 12(9): 5878–5894. https://doi.org/10.3390/ijms12095878

5 Fat Extraction from Food Industry Waste

Kiran Bala Nain

Department of Food Science and Technology, Chaudhary
Devi Lal University, Sirsa (Haryana) India
E-mail: kiran31nain@gmail.com

CONTENTS

5.1 INTRODUCTION

Globally, a huge quantity of waste is generated through food industries every year and is traditionally disposed of together with the municipal solid waste, such as landfilling or incineration. According to a report the quantity of food waste produced is about 1.51 million tons/year (Chua et al., 2019). In recent years, a large number of papers have been published regarding food industry waste management. These papers describe that food wastes cause a negative environmental impact along with significant economic losses. However, a large number of food additives can be produced using these biomaterials and there is ample potential for generating food additives, which in turn will minimize malnutrition and hunger in developing countries. In addition, useful compounds such as proteins, starch, lipids, micronutrients, bioactive compounds, and dietary fibers can also be extracted from these biomaterials (Torres-León et al., 2018). Furthermore, these compounds can be used as nutritionally and pharmacologically functional ingredients. Fruits and vegetables, cereal, dairy, seafood, meat, and poultry are the major food and fruit processing industries, which contribute contribute to waste generation. For example, the ice cream and dairy industry (21.3%), drink industry (26%), grains and starchy products (12.9%), meat processing (8%), fish and fish products (0.4%), vegetable and animal oils and fats (3.9%), preservation and production of vegetables and fruits (14.8%) (Lee et al., 2019).

Fats and oils are essentially triglycerides having straight-chain fatty acids attached, as esters, to glycerol. These can vary in chain length, may be saturated or unsaturated, and may contain an odd or even number of carbon atoms. The term "grease", as commonly used, includes fats, oils, waxes, and other related constituents found in wastewater (Wakelin and Forster, 1997). In this context, if lipid-rich wastes are not properly disposed, these can be toxic to ecosystems and form "fatbergs" that reduce sewer diameters or completely clog sewage pipes (Wallace et al., 2017). Edible-oil industries are the major producer of lipid-rich waste streams as aresult of processing, such as palm oil mill effluent, or post consumption such as waste cooking oil/frying oil. However, fatty acid distillates are a major by-product formed in the processing of vegetable oils, especially those of soybean or palm (Lad et al., 2022. According to Toldrá-Reig et al. (2020) waste animal fat is a particularly large waste stream in the livestock industry and this lipid-rich waste stream has a higher percentage of saturated fats (almost 40% total). The disposed food industry wastes have a huge amount of nutrients that may be used as a source for extraction of useful compounds, which can be further used for production of other value-added products. The extraction of different compounds will generate new options to utilize these extracts into the food industry to make better foods in the future.

5.2   SOURCES OF FAT

Food wastes are generated by various sources (Figure 5.1), including wastes generated after animal processing i.e. by-products from bred animals such as carcasses, hides, hoofs, heads, feathers, manure, offal, viscera, bones, fat and meat trimmings, blood; wastes from seafood such as skins, bones, oils, blood; wastes from the dairy processing industry such as whey, curd, and milk sludge from the separation process; vegetable-derived processing food wastes includes peelings, stems, seeds, shells, bran, trimmings residues after extraction of oil, starch, juice, and sugars (Helkar and Patil, 2016).

5.2.1   FRUIT AND VEGETABLE INDUSTRIES

Banana peel contains polyunsaturated fatty acids such as linoleic acid (omega-6) and α-linoleic acid provisional (omega-3) and 40–50% of total fatty acids are saturated fatty acids, including

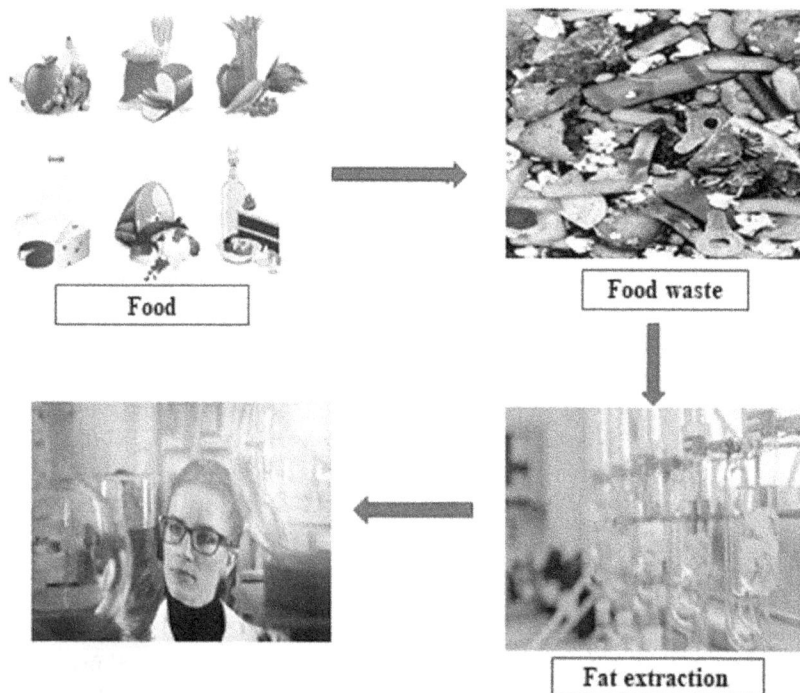

FIGURE 5.1   Fat extractions from food waste.

palmitic acid, arachidic acid, stearic acid and myristic acid. Banana peel also contains phytosterols (Torres-León et al., 2018; Venkateshwaran and Elayaperumal, 2010). Al-Wandawi et al. (1985) studied various fractions of tomato processing wastes for chemical composition and observed that seeds contain about 27% hexane-extracted lipids. Oleic acid was the most predominant fatty acid, followed by palmitic acid. Olive oil industries produce dark-colored juice having organic compounds such as lipids (1.0 to 1.5%), sugars, pectins, organic acids, poly-alcohols, colloids, and tannins (Kosseva, 2009).

5.2.2 Fish Industry

Fish waste is commonly underutilized, although it is a source of valuable compounds such as fish oils, biodiesel, enzymes, omega-3 fatty acids, and proteins. Extraction of oils is also important as oxidation of the unsaturated fatty acids present in fish oil is the major factor responsible for the offensive odor associated with fish and fish waste. Many studies have been carried out on the extraction and purification of omega-3 polyunsaturated fatty acids from fish waste (Nges et al., 2012). Fish oil from fish processing waste, marine fish, and chicken visceral wastes are rich sources of polyunsaturated fatty acid concentrates. From stomachs, viscera, and liver of fish, polyunsaturated fatty acids can be extracted and these particularly omega-3 fatty acids are useful in protecting human beings from nerve and brain disorders, cardiovascular diseases, and kidney diseases (Analava et al., 2014). Ghaly et al. (2013), the fish processing industry is a major exporter of seafood and marine products in many countries. Fish processing waste (20 to 80%) is a potential source of proteins, amino acids, and oils.

5.2.3 Meat and Poultry Processing

Meat and poultry industries contain various quantities of blood, fats, residues from the intestine, paunch grass, and manure in their waste streams (Kosseva, 2009). Khan et al. (2021) extracted fat compounds from slaughterhouse waste (suet, tongue, pancreas) of selected ruminant (cow, goat, lamb, and bull), using four selected methods including Soxhlet, acid hydrolysis, Bligh and Dyer, and Folch. The fat extraction efficiency of the methods was also compared.

5.2.4 Dairy Industry

The main wastes generated in dairy industries are whey, dairy sludges, and wastewater (processing, cleaning, and sanitary). These wastes contain various compounds, proteins, salts, fatty substances, lactose, and various kinds of cleaning chemicals (Kosseva, 2009). The dairy wastewater mainly consists of protein, fat, and sugars. Kasmi et al. (2018) reported that dairy industry wastes can be used for production of bacterial growth media. They prepared dairy curds and concluded that considerable protein and fat fractions were recovered in the dairy curds. Oleic acid, as a bacterial growth promoter, was the predominant fatty acid in the curd fat content.

5.3 EXTRACTION OF FAT FROM FOOD WASTE

Basically, there are three main facets to any practical process for extracting lipids/fats from tissue; firstly, exhaustive extraction and solubility of the lipids in organic solvent and secondly, removal of non-lipid contaminants from the extracts; thirdly, the potential toxicity of solvents to analysts. Lipids occur in tissues in a variety of physical forms. The simple lipids are easily extractable, whereas the complex lipids are found closely associated with proteins and polysaccharides (Xiao, 2010). Extraction of fat from food waste can be carried out using solid-liquid extraction, Soxhlet extraction, pressurized fluid extraction, supercritical fluid extraction, ultrasound-assisted extraction, microwave-assisted extraction, maceration and solvent extraction, green-solvent-assisted extraction, pulsed electric field extraction, and enzyme-assisted extraction etc. (Baiano, 2014). Food wastes contain a mixture of carbohydrates, protein, phosphate, and minerals. Therefore, separation of fat is required. Lipid from food waste can be extracted using solvent extraction but owing to some environmental issues, Karmee et al. (2015) isolated lipids from food waste after fungal hydrolysis and used this fat as a low-cost feedstock for biodiesel production. Lin et al. (2021) extracted lipids from food waste containing a mixture of rice, noodles, meat, and vegetables using enzymatic hydrolysis process (enzymes: glucoamylase, lipase, and protease were added at 1% (w/v) in a 10 L paddle mixer along with food waste) conducted at 45 °C for a period of 12 hour. Gas chromatography was used for lipid analysis in the mixture. According to Wakelin and Forster (1997), use of a mixed culture such as activated sludge, particularly when it has been acclimatized to the vegetable and animal fat substrate, provides a good option for the treatment of fat-rich wastewaters.

Khan et al. (2021) extracted fat from slaughterhouse waste (suet, tongue, pancreas) of ruminant (cow, goat, lamb, and bull) animals and fat extraction efficacy of Soxhlet, acid hydrolysis, Bligh and Dyer, and Folch methods was compared along with fatty acid composition. It was concluded that excellent fat and fatty acid extraction was achieved using the Folch method, as the Soxhlet method was only effective for samples with high-fat content i.e. suet, while the Bligh and Dyer method gave comparatively low lipid content.

Growth of selected microorganisms of industrial interest (*Saccharomyces cerevisiae, Kluyveromyces marxianus and kefir*) using solid state fermentation of food waste mixtures was carried out by Aggelopoulos et al. (2014) for production of single cell protein, aroma volatiles, and fat. They concluded that the substrate fermented by *K. marxianus* contained the highest sum of fat and protein concentration (59.2 % w/w dm) and therefore it could be considered for utilization of its fat content and for livestock feed enrichment. Supercritical fluid extraction was applied successfully for the extraction of oil from rice bran by Perretti et al. (2003). The extraction conducted at 10,000 psi and 80 °C with only 100 g of CO_2 gave the highest extraction yield (24.65%).

Fish oil obtained from fish processing waste can be exposed to extraction of polyunsaturated fatty acids using distillation as these have lower volatility than small-chained fatty acids. Owing to high boiling points, molecular distillation is the preferred way of separation. Although, some other methods such as low temperature crystallization, enzymatic methods, urea complexation, alkaline hydrolysis, supercritical fluid extraction, and microwave-assisted extraction, etc. can be used (Gildberg, 2004). Table 5.1 represents various extraction methods including the derivatization method, used for fat extraction of fat from food wastes.

5.4 FACTORS AFFECTING THE EXTRACTION OF FAT FROM FOOD WASTE

Various factors like temperature, pressure, pH, and substrate type, method used etc. affect the fat extraction from waste food. Lukitawesa et al. (2020) studied about factors influencing volatile fatty acids' production from food wastes via anaerobic digestion and observed that controlling the pH is most likely the most important factor determining yield. However, oxygen does not have a significant effect. Nges et al. (2012) reported that anaerobic digestion of original fish waste and the fish sludge remaining after enzymatic pre-treatment to extract fish oil and fish protein hydrolysate have potential for methane production. Table 5.2 shows various fat compounds extracted from food waste using various extraction methods.

5.5 APPLICATION OF OIL/FATS IN VALUE-ADDED PRODUCTS

Sharif et al. (2005) extracted oil from rice industrial waste and studied its effect on physico-chemical characteristics of cookies. They concluded that various treatments and storage have a very high influence on nitrogen-free extract, moisture, and fat content of cookies. Protein content

TABLE 5.1
Fat Extraction Methods from Different Food Sources

Food matrix	Food sample	Extraction method	Derivatization method
Meat	Cooked turkey breast Fresh pork loin Cooked ham Dry-cured ham Mortadella Beef burger Fresh sausage Dry-cured sausage and salami Lamb meat	Soxhlet method (PE/EE) Bligh and Dyer method Folch method Hara–Radin method Rapid microwave solvent extraction SFE	Base catalyst sodium methoxide method Acid catalyst BF3 method (BF3-MeOH method)
Nuts and seeds	Almond Apricot Cashew Hazelnut Peanut Pecan Pistachio Walnut Chia seeds oleaginous seeds	Soxhlet method (EE/Hex) UAE	BF3-MeOH method TMSH method
Butter	Animal butters Plant butter	Soxhlet method (Hex) Solvent extraction (EE/PE) Rose-Gottlieb method Bligh and Dyer method Folch method	Acid catalyst conc. H2SO4 (95%) method BF3-MeOH method
Animal feeds	Forages Cereal grains	Solvent extraction (EE/Hex) Soxhlet (diethyl ether)	BF3-MeOH method Base catalyst sodium methoxide method
Fast food	Burger French fries Pizzas Potato chips Snacks	Non-meat foods – solvent extraction (PE/Hex) Meat foods – Folch method MAE SFE	Base catalyst sodium methoxide method BF3-MeOH method
Dairy products	Milk powder Ice cream High-fat milks Low-fat milks High-fat yoghurts Low-fat yogurts High-fat cheeses Low-fat cheeses	Modified Folch method - Chloroform:methanol (2:1, v/v) Modified Bligh and Dyer Chloroform:methanol:water (1:4:0.8, v/v).	BF3 -MeOH method Base catalyst sodium methoxide/ KOH method TMAH method
Flour products and Chocolate	Cake Cream biscuits Simple biscuits Cream chocolates Simple chocolates	Solvent extraction (PE/Hex) Folch method Soxhlet extraction (Hex) MAE Bligh and Dyer	Base catalyst sodium methoxide method BF3-MeOH method TMSH method

Source: Adapted from Hewavitharana et al. (2020).

Notes: EE – ethyl ether; PE – petroleum ether; Hex – hexane; TMAH – tetramethylammonium hydroxide; TMSH – trimethylsulfoniumhydroxide; MeOH –methanol.

TABLE 5.2
Compounds Extracted from Food Waste Using Different Extraction Methods

Sr. No.	Extracted compounds	Substrate	Extraction method
1.	Polyunsaturated fatty acids	Fish wastes	Distillation, low temperature crystallization, enzymatic methods, urea complexation, alkaline hydrolysis, supercritical fluid extraction, microwave assisted extraction
2.	Collagen	Fish skin, bones and fins	Acid treatment of the by-products
3.	Gelatin	Fish skin, bones and fins	Heat denaturation of collagen
4.	Lard	Clean tissues of healthy pigs	
5.	Tallow	Fatty tissues of cattle or sheep	
6.	Tocopherols, tocotrienols, sterols, and squalene	Palm fatty acid distillate	Liquid–liquid extraction
7.	Oil	Rice bran	Supercritical fluid extraction
8.	Tocopherols and Tocotrienols	Palm fatty acid distillate	Treatment with alkyl alcohol and sodium methoxide; distillation under reduced pressure; a cooling step; passage of the filtrate through an ion-exchange column with anionic exchange resin; removal of the solvent; molecular distillation.

Adapted from Baiano (2014).

was unaffected by changing treatments and during 45 days of storage, moisture, protein, fat, and NFE content decreased significantly. In addition, the width and spread factor of cookies increased, while thickness was decreased. Lin et al. (2021) developed bio-based polyurethane materials using lipids obtained from food waste treatment by enzymatic hydrolysis and they concluded that these lipids can be alternative sources of renewable and inexpensive feedstocks and also used for bio-polyol synthesis. Polyurethanes have various applications like coatings, adhesives, sealants, foams, elastomers, furnishing, cars, wall and roofing insulation, shoes, clothing etc. Zieniuk et al. (2020) studied the possibility of using several lipid-rich food industry wastes (waste fish oil, rancid ghee, waste pork lard, and waste duck processing oil) in the culture medium on the growth of *Candida cylindracea* DSM 2031 yeast strain. They concluded that there is a potential for waste management to produce lipolytic enzymes or to produce yeast biomass. The use of waste substrates may contribute towards economy.

5.6 CONCLUSIONS

Food industries generate a huge quantity of waste throughout the year. Management and utilization of this waste for extraction of useful compounds is a challenging task these days. Concern regarding waste disposal is the main topic of research and development to food scientists and technologists today as wastes cause a negative impact on the environment. Many compounds like fats, proteins, carbohydrates, and bioactive compounds can be extracted from various food industry wastes including dairy, meat, fruit and vegetables, cereals, fish processing. Fat extraction from food waste is one of the important aspects from an economic and environmental point of view. Liquid–liquid extraction (Soxhlet extraction), microwave-assisted extraction, and ultrasonic extraction methods are widely used for the fatty acid extraction followed by catalytic derivatization. Using some advanced techniques, fat extraction can be improved and generated wastes can be utilized properly, which will reduce the negative environment influence as well as economically better management.

REFERENCES

Aggelopoulos, T., Katsieris, K., Bekatorou, A., Pandey, A., Banat, I. M., & Koutinas, A. A. (2014). Solid state fermentation of food waste mixtures for single cell protein, aroma volatiles and fat production. Food chemistry, 145, 710–716.

Al-Wandawi, H., Abdul-Rahman, M., & Al-Shaikhly, K. (1985). Tomato processing wastes as essential raw materials source. Journal of Agricultural and Food Chemistry, 33(5), 804–807.

Analava, M., Baishakhi, D., & Anindya, M. (2014). Recovery of omega-3 health boosters from fisheries and poultry wastes & their micro-delivery techniques. Int. J. Drug Deliv.Sci, 1, 1–13.

Baiano, A. (2014). Recovery of biomolecules from food wastes – A review. Molecules, 19(9), 14821–14842.

Chua, G. K., Tan, F. H. Y., Chew, F. N., & Mohd-Hairul, A. R. (2019, July). Nutrients content of food wastes from different sources and its pre-treatment. In AIP Conference Proceedings (Vol. 2124, No. 1, p. 020031). AIP Publishing LLC.

Ghaly, A. E., Ramakrishnan, V. V., Brooks, M. S., Budge, S. M., & Dave, D. (2013). Fish processing wastes as a potential source of proteins. Amino acids and oils: A critical review. J. Microb. Biochem.Technol, 5(4), 107–129.

Gildberg, A. R. (2004). Enzymes and bioactive peptides from fish waste related to fish silage, fish feed and fish sauce production. Journal of Aquatic Food Product Technology, 13(2), 3–11.

Helkar, P. B., Sahoo, A. K., & Patil, N. J. (2016). Review: Food industry by-products used as a functional food ingredients. International Journal of Waste Resources, 6(3), 1–6.

Hewavitharana, G. G., Perera, D. N., Navaratne, S. B., & Wickramasinghe, I. (2020). Extraction methods of fat from food samples and preparation of fatty acid methyl esters for gas chromatography: A review. Arabian Journal of Chemistry, 13(8), 6865–6875.

Karmee, S. K., Linardi, D., Lee, J., & Lin, C. S. K. (2015). Conversion of lipid from food waste to biodiesel. Waste Management, 41, 169–173.

Kasmi, M., Elleuch, L., Dahmeni, A., Hamdi, M., Trabelsi, I., & Snoussi, M. (2018). Novel approach for the use of dairy industry wastes for bacterial growth media production. Journal of Environmental Management, 212, 176–185.

Khan, A., Talpur, F. N., Bhanger, M. I., Musharraf, S. G., & Afridi, H. I. (2021). Extraction of fat and fatty acid composition from slaughterhouse waste by evaluating conventional analytical methods. American Journal of Analytical Chemistry, 12(5), 202–225.

Kosseva, M. R. (2009). Processing of food wastes; Chapter 3. Adv. Food Nutr. Res, 58, 57–136.

Lad, B. C., Coleman, S. M., & Alper, H. S. (2022). Microbial valorization of underutilized and nonconventional waste streams. Journal of Industrial Microbiology and Biotechnology, 49(2), kuab056.

Lee, J-K., Patel, S.K.S., Sung, B.H., Kalia, V.C. (2019). Biomolecules from Municipal and Food Industry Wastes: An Overview, Bioresource Technology, doi:10.1016/j.biotech.2019.122346.

Lin, C. S. K., Kirpluks, M., Priya, A., & Kaur, G. (2021).Conversion of food waste-derived lipid to bio-based polyurethane foam. Case Studies in Chemical and Environmental Engineering, 4, 100131.

Lukitawesa, Patinvoh, R. J., Millati, R., Sarvari-Horvath, I., & Taherzadeh, M. J. (2020). Factors influencing volatile fatty acids production from food wastes via anaerobic digestion. Bioengineered, 11(1), 39–52.

Nges, I. A., Mbatia, B., & Björnsson, L. (2012). Improved utilization of fish waste by anaerobic digestion following omega-3 fatty acids extraction. Journal of Environmental Management, 110, 159–165.

Perretti, G., Miniati, E., Montanari, L., & Fantozzi, P. (2003). Improving the value of rice by-products by SFE. The Journal of Supercritical Fluids, 26(1), 63–71.

Salas, M. E. C. (2016). Análisis de actividad metanogénica de inóculos y potencial metanogénico de biomasa residual. Master's thesis. Tesis de Maestría En Ciencias.

Sharif, K., Butt, M. S., & Huma, N. (2005). Oil extraction from rice industrial waste and its effect on physico-chemical characteristics of cookies. Nutrition & Food Science, 35(6), 416–427.

Toldrá-Reig, F., Mora, L., & Toldrá, F. (2020).Trends in biodiesel production from animal fat waste. Applied Sciences, 10(10), 3644.

Torres-León, C., Ramírez-Guzman, N., Londoño-Hernandez, L., Martinez-Medina, G. A., Díaz-Herrera, R., Navarro-Macias, V., & Aguilar, C. N. (2018). Food waste and by-products: An opportunity to minimize malnutrition and hunger in developing countries. Frontiers in Sustainable Food Systems, 2, 52.

Venkateshwaran, N., & Elayaperumal, A. (2010). Banana fiber reinforced polymer composites-a review. Journal of Reinforced Plastics and Composites, 29(15), 2387–2396.

Wakelin, N. G., & Forster, C. F. (1997). An investigation into microbial removal of fats, oils and greases. Bioresource Technology, 59(1), 37–43.

Wallace, T., Gibbons, D., O'Dwyer, M., & Curran, T. P. (2017). International evolution of fat, oil and grease (FOG) waste management–A review. Journal of Environmental Management, 187, 424–435.

Xiao, L. (2010). Evaluation of extraction methods for recovery of fatty acids from marine products (Doctoral dissertation).

Zieniuk, B., Mazurczak-Zieniuk, P., & Fabiszewska, A. (2020). Exploring the impact of lipid-rich food industry waste carbon sources on the growth of Candida cylindracea DSM 2031. Fermentation, 6(4), 122.

6 Treatment of Fatty Effluents

Muhammad Usman,[1] Aysha Sameen,[1]
Muhammad Issa Khan,[1] Ubaid ur Rahman,[2]
Usman Mir Khan,[1] Saba Anwar,[1] Mehwish Arshad,[1]
Maria Maqsood,[1] and Amna Sahar[1, 3,]*

[1]National Institute of Food Science and Technology, University of Agriculture Faisalabad, Pakistan
[2]School of Food and Agricultural Sciences, University of Management and Technology, Lahore, Pakistan
[3]Department of Food Engineering, University of Agriculture Faisalabad, Pakistan
*Corresponding author: Dr Amna Sahar (amnasahar@uaf.edu.pk)

CONTENTS

DOI: 10.1201/9781003207689-6

6.1   INTRODUCTION

Over the years, an increase in world population and industrialization have resulted in the production of waste products in high quantity. Food industries generate a huge quantity of waste as by-products during production, processing, packaging, storage, and transportation. Food and agro-allied industries discharge a significant amount of wastes that can cause considerable public and environmental issues due to inadequate waste disposal facilities. Effluents from food industries contain fats, proteins, carbohydrates, and other nutrients. These waste by-products can cause environmental pollution and serious health hazards to living organisms. All these effluents need to be treated before their disposal.

Different food industries produce different types of waste by-products. Fats, blood, bones, sludge are generated by meat processing industries and slaughterhouses. Dairy industries produce wastewater, fatty effluents, whey, and buttermilk. Fat and oil industries produce grease and fats as waste by-products. All these waste products are unfit for human consumption and are directly dumped into land or river.

6.2   MEAT PROCESSING AND SLAUGHTERHOUSES WASTE

Meat industries as well as slaughterhouses generate a large amount of wastes that results in serious environmental problems all over the world. These waste products include blood, bones, fats, sludge, and wastewater. Meat industries' effluents are rich in fats, proteins, fibers, as well as carbohydrates. All waste by-products are discharged into rivers or landfills without any treatment resulting in serious threats to the environment. These effluents can also cause serious health related issues like cholera, typhoid fever, and dysentery when microorganisms are present in effluents.

The meat industry during slaughtering is one of the major contributors to liquid waste. Carcasses, parts of animals, or products of animal origin that are unfit for human consumption are defined as animal waste. The indiscriminate release of waste from slaughterhouses causes serious environmental concerns. The cost effective treatment of solid wastes particularly wastewater effluents from the meat industry emerged as a serious concern. High fat concentrations are present in the effluents from meat industries and slaughterhouses, which can result in water toxicity, eutrophication and turbidity (Cheng et al., 2020). High levels of fats and greases result in the development of microbes and also inhibit sludge activity. These fatty effluents also accumulate on the sludge surface, which results in the reduction of oxygen level to the aerobic microbes thus causing serious problems to the environment (Cheng et al., 2020).

Meat rendering process generates liquid waste (also known as stick water) with high concentration of organic compounds including proteins, fats, and minerals. These high organic wastes need to be treated before disposal. In order to treat effluents from meat industries and slaughterhouses, different types of strategies including physical, chemical, and biological can be used. As these effluents contain a huge amount of fats and greases, it is difficult to treat them easily. So, pretreatment of fats and greases need to be done before any treatment.

The inedible parts of waste such as blood, gastrointestinal tract, tendon, bones, skin, and visceral organs are produced by the abattoir. Wastewater, gases, solid material, and volatile compounds are common waste by-products of the meat processing industry (Rahman et al., 2014). Wastewater contains an abundant amount of biodegradable organic materials and additional toxic compounds, including unionized ammonia, chromium, and tannins, which can cause death of aquatic organisms. The poultry processing wastewater is rich in total soluble solids and biochemical oxygen demand. Various suspended organic and inorganic solids increases turbidity of wastewater. Moreover, organic waste including carbohydrates, proteins, and fats consumes a large amount of dissolved oxygen and their oxidation produces various gases and biomass. Solid wastes are those meat by-products that cannot be further recycled and includes non-biodegradable materials and organic compounds. Other body parts including the head, skin, feet, rectum, udder, feathers, and lungs are termed as solid waste and must be disposed of (Barkocy-Gallagher et al., 2003; Sandvig and van Deurs, 2000).

Fat waste from the meat industry is a highly valuable by-product and can be obtained at three basic stages of processing. Fat can be recovered from the processing area and rendered to generate high grade tallow. It can be collected from primary wastewater treatment to produce low grade tallow and it can also be converted to methane by anaerobic digestion.

Regardless of indirect and direct discharge, the soluble materials especially organic materials (suspended solids including fat particles) in meat processing wastewater should remove prior to discharge in order to accomplish compliance with environmental regulations. Chemical, physical, and biological methods used in meat processing industry depending on their unique advantages. Various physiochemical methods for preliminary treatment of meat industry wastewater are increasingly used before its discharge for biological purification (Radoiu et al., 2004).

Physical treatment in the form of perforated surface or the screen is designed to separate the particulate matter while filters retain the particulate matter by media (sand or synthetic fibers). Among physical treatments, screening is regarded as a first, inexpensive, and commonly used treatment. The most common type of screen used in the poultry industry is the rotary type. Other physical methods including dissolved air flotation, sedimentation, coagulation-flocculation, membrane technology, and electrocoagulation are also used.

The meat rendering processes produce wastewater rich in minerals, proteins, and fats. The membrane distillation process recovers valuable solids and water; however, hydrophobic membranes are polluted by fats. The commercially available hydrophobic poly-tetra-fluor-ethylene membranes with hydrophilic polyurethane surface layers developed for direct membrane distillation contact on bovine stick waters, fish, and poultry (Mostafa et al., 2017). The effluent stream after elimination of coarse solids still contains suspended solids, grease, and fats. These solids have high biological oxygen demand values and cause floating scum formation, which sticks to the borders of pipes and tanks. This scum results in the blockage of pipelines thus, reduces the effectiveness of aeration and obstructs the small irrigation outlets. Hence, it is necessary to efficiently remove these materials at the start of the process. Colloidal material's turbidity and coloration are removed through extensively used flocculation or coagulation methods. A key characteristic of flocculation is removal of almost all suspended solids or organic material in wastewater. A floc-forming compound is also needed for elimination of organics and suspended solids and separated by settling, flotation, and adsorption (Al-Mutairi et al., 2004). For effective elimination of total soluble solids and organic matter, new coagulants of organic and inorganic nature are being used for wastewater treatment from slaughterhouses (Aguilar et al., 2005).

The flocculation extent and effectiveness of methods depends on temperature, composition, rate of mixing, and order of introduction of flocculants and coagulants in wastewater. Flocculants ionized or non-ionized when introduced into wastewaters while in their ionized state are termed as soluble poly-electrolytes (Radoiu et al., 2004). Settling, screening with dissolved air flotation can be used for elimination of suspended solids and wastewater fog. These results in approximately 75–80% reduction in biological oxygen demands and also have further advantage of eliminating larger amounts of phosphorus and nitrogen (Aguilar et al., 2003).

Fats, grease, and fine solids in scum have financial importance and there is significance to consume raw and processed forms of such by-products within the animal feed, soap, and cosmetics industries. The fatty components collected are classified as total fatty matter composed of non-separable and separable fractions. The approach for fats' removal depends on its quality and the amount produced. For a small low grade fatty waste, a simple fat trap can be used while for a larger volume of high grade fat waste, an additional efficient process is useful. The existing approaches are fat traps, dissolved air flotation, and chemical treatments. Fat traps work on the gravity separation principle by establishing a minimum turbulence.

Various chemical wastewater treatments are available; however, dissolved air flotation is the most popular one in the meat industry. This process includes water solid separation through introduction of air bubbles to wastewater streams. The process efficiency can be enhanced by adjusting pH and improving matter flocculation through the use of chemicals. A solid gas matrix is formed when microbubbles reach solid particulate in the wastewater. The raised matrix on the surface of water is collected and removed through mechanical skimming. A greater number of appropriate chemical flocculants are available, such as ferric sulphate, ferric chloride, aluminum sulphate, ferrous sulphate, calcium carbonate, sodium carbonate, sodium lignosulphonate, and lignin sulphonic acid.

6.2.1 Bio-treatment

Bio-treatment refers to the treatment of wastewater under a controlled environment using microorganisms. Wastewater including organic and inorganic substances and biodegradable matter are converted by microorganisms into by-products and other stable cellular masses that are removed later by physical means, viz. settling in clarifiers from remaining wastewater and gravity separation. Bio-treatments are potential cost-effective methods having greater than 90% removal efficiencies of waste products in the meat industry.

A typical biological treatment constitutes lagoons, activated sludge systems, septic tanks, and tricking filters. The effectiveness of a biological treatment method can be enhanced by using enzymatic or alkaline treatments for hydrolyzing fat in meat wastewater. The best practical application of hydrolyzing agents involves pancreatic lipase, which results in particle size reduction to 60% with associated surge in free long chain fatty acid (Grismer et al., 2002).

Anaerobic digestion treatment is an important biological waste treatment method. The efficiency of anaerobic digestion highly depends on composition and type of substance for digestion (Murto et al., 2004). Anaerobic digestion has potential to produce methane and large quantities of accumulated waste can be collected easily. Hence, anaerobic digestion can be a favorable alternative for wastewater treatment for consequently conserving energy and controlling pollution by producing methane as fuel. This method not only generates biogas but also kills pathogenic microbes and produces stabilized material for use in land application as a fertilizer (Salminen and Rintala, 2002).

Anaerobic digestion involves hydrolysis, methanogensis, and acidogenesis and can be made more efficient by optimizing the various conditions. During meat waste digestion, the ammonia toxicity inhibits the production of volatile fatty acids and reduces the biogas production. Lipids and grease can cause difficulties due to their tendency of forming floating scum. The excessive accumulation of volatile fatty acids inhibits methanogenesis and higher hydrogen concentration inhibits butyrate and propionate degrading acetogens (Magbanua et al., 2001).

Precipitation and adsorption with divalent ion cause long chain fatty acid accumulation associated with biomass and hence can be efficiently mineralized for enhancing methanogenic activity. The stable and highly loaded co-digestion of organic waste from the meat processing industry (protein rich and partly hydrolyzed organic waste) could be a better and economic alternative treatment. Aeration is an exothermic, simpler, and effective method for removal of organic pollutant and odorant. The efficient recovery and utilization of heat during aeration reduce economic cost of treatment. Moreover, aerobic treatments like activated sludge processes, aerobic lagoon, complete mixtures, trickling filters, and oxidation ditches efficiently removes the pathogens and odors.

Compositing is another extensively used treatment for organic waste prior to discharge. This aerobic biological process uses naturally present microbes for conversion of biodegradable organic matter into consumable products. It kills or retards pathogens, reduces waste volume, converts nitrogen into stable organic form from unstable ammonia and recovers the waste nature. The anaerobic digestion is a well suited treatment method for effective removal of chemical dissolved oxygen at lower cost and also generates methane rich biogas as fuel.

6.2.2 Enzymatic Treatment

Enzymes' application is one of the promising strategies in the treatment of effluents from meat industries and slaughterhouses containing a high amount of fats and greases. Enzymes act as catalysts that speed up the process, targeting specific compounds and converting into simpler compounds (complex triglycerides into free fatty acids) without affecting others. Lipases are one of the common enzymes that are used in the treatment of fatty effluents. These enzymes help in the breakdown of long chain triglycerides into simpler free fatty acids.

Research has been conducted in 2019, in which commercial lipase enzyme was used to deal with fatty effluents from poultry slaughterhouses. Results showed that long chain fatty acids were converted into simpler free fatty acids with the help of lipase enzyme (Cheng et al., 2020). Some other examples of enzymes' application in the treatment of fatty effluents coming from meat and slaughterhouses are given in Table 6.1.

6.2.3 Recovery of Fatty Effluents from the Meat Industry

A valuable and important constituent of the meat industry is fat. The two major animal fats in meat by-products are lard and tallow that are prepared by dry and wet rendering. These fats can be obtained by methods including compositing, incineration, oleo-chemical process, and anaerobic

TABLE 6.1
Enzymes' Application in Treatment of Fatty Effluents from Meat and Slaughterhouses

Industry	Enzymes	Results	Reference
Poultry slaughterhouse Wastewater	Commercial lipase enzyme	Release of long chain FAs	(Pascale et al., 2019)
Swine processing Industry	Commercial lipase enzyme	High yield of releasing FFAs	(Rigo et al., 2008)
Fish processing Industry	Lipase	COD removal up to 85.3 %	(Alexandre et al., 2011)
Animal wastewater from dairy industry	Lipase	Removal of fats Upto 80 %	(Leal et al., 2002)
Poultry industry Wastewater	Lipase	COD removal upto 96 %	(dos Santos Ferreira et al., 2020)

fermentation. These rendering processes separate tallow, lard and grease) from proteins and bones of animal tissues by use of batch cooking or continuous flow (Thimjos et al., 2014). Tallow and lard is used in margarine and shortenings. The quality of tallow/lard obtained from wet rendering is superior to dry rendering. Traditionally, lard and tallow is utilized for deep frying but due to consumer health concerns both fats' usage is declined. Rendered lard as an edible fat can be utilized without further processing and it is also used in various emulsified products and sausages. Both tallow and lard are bleached and deodorized prior to their use in food industry. Moreover, liquid tallow as an alternative has developed to prepare fast foods for low fat absorption (Ghotra et al., 2002).

6.3 DAIRY INDUSTRY

Milk is the first food fed to the newborn infants and now globally, milk and milk products are an essential food source for humanity, as they are composed of fats, minerals, protein, iron, sterols, and vitamins essential for normal body growth and maintenance. Hence, for a balanced diet and proper nutrition point of view, consumption of dairy products cannot be compromised. Milk is considered as a liquid drink but it contains 13% total solids, much more than the solid content of tomato (6%) and lettuce (5%). Approximately, 250 chemical constituents have been identified including 140 individual fatty acids. Major milk constituents are protein, fat, carbohydrates, minerals, and vitamins, which vary in quantity with environmental factors, animal species, animal health, and the production procedure of milk products (Campbell and Robert, 2016).

6.3.1 NUTRITIONAL ASPECTS OF MILK

Milk is the only food suitable for both infants and adults including all age groups. It is said to be a complete diet for the average human male or female adult. It possesses various attributes, which helps in delivery, absorption, and bioavailability of essential nutrients as necessary for a healthy immune system and regularity of functions and metabolism. Lactose, milk sugar, enhances absorption of minerals and vitamins such as phosphorus, calcium, magnesium, and vitamin D. Casein micelles coagulate inside the acidic environment of the stomach and forms gel, which provides satisfaction of drinking by reducing the digestion of milk quickly resulting in enhanced efficient digestion. Furthermore, casein protein keeps Ca-phosphate stabilized, which maximizes efficient delivery, absorption, and bioavailability of Ca-phosphate in the intestine. Dietary significance of milk can be measured by data collected in 2009 for North American diet, buffalo milk provided 8% energy, 12% dairy fats, and 16% dairy proteins. Moreover, dairy products are a source of 52–67% dietary reference intake (DRI) of calcium in diet (Chalupa-Krebzdak, 2018).

6.3.2 GLOBAL PRODUCTION OF MILK

There are two terms used for milk that are milk yield and quality. Milk yield covers the amount of milk produced in a specific time period whereas milk quality comprises two further components: (1) chemical components; (2) milk hygiene. Proteins and fats are two major constituents on which the chemical quality of milk determined as these components are key determinants of milk potential processing at an industrial level and the price of milk calculated based on yield and quality. Commonly, milk yielding cattle and annual tonnage of milk production is given in Table 6.2.

In Pakistan, milk share from domestic cattle is 62.8% (buffalo), 34.9% (cow), 2% (goat), 0.2% (camel), and 0.1% (sheep). According to data, only 3–4% of produced milk is supplied in dairy industries, which is processed in butter, butter oil, ghee, whole condensed milk, skim milk powder, skimmed condensed milk, cheese, fresh cream, condensed whey, powered whey, and yogurt (Murtaza et al., 2014; Ramirez-Rivera et al., 2019).

TABLE 6.2
Global Annual Milk Production of Bovine Cattle in 2019

Animal	Milk production (tonnes)
Cow	715922506
Buffalo	133752296
Camel	31114612
Goat	19910379
Sheep	10587020

Source: FAOSTAT, 2021.

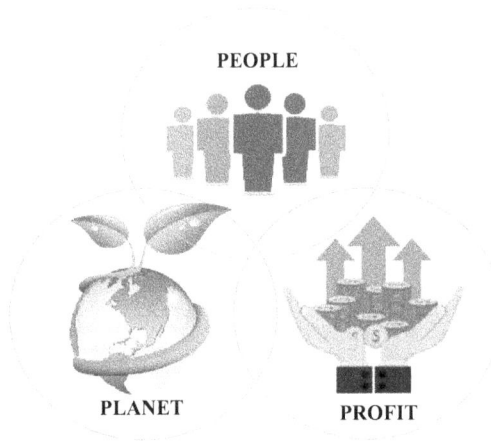

PEOPLE

PLANET   PROFIT

FIGURE 6.1 Three bottom line approach.

6.3.3 ENVIRONMENTAL CONCERNS

Increasing demands for green products by consumers, business to business relationships, and regulatory organizations has increased the interest of dairy processors on the environmental impact posed from the production line. Therefore, the pollution level increasing to a level never reached before has caught the attention of social, economic, and ecological agencies, which are also increasing pressure to replace pollution creating processes with novel technologies.

The milk industry is considered a major pollution causing industry because of the large use of water in production lines as well as during cleaning in place (CIP). Wastewater from the dairy industry easily dissolves different effluents from milk and milk products. In the development and management of industry, sustainability is a key factor to integrate with a holistic approach but there are still problems in triple bottom line aspects of sustainability as shown in Figure 6.1 (Feil et al., 2020).

6.3.4 DAIRY INDUSTRY WASTE

According to estimated data, per liter processing of milk requires approximately 1.44 L of water whereas cheese production requires even more water as 1.6–2 L per liter of milk. Almost, 80–90%

water used in dairy processing becomes wastewater that has to be purified before leaving in streams and rivers. Major dairy waste products are buttermilk and whey, although buttermilk is not very hazardous and can be used as an intermediate for the production of melted cheese. In contrast, whey is very hazardous for the environment as it contains nitrogen upto 600 mg/L, 60,000 mgO$_2$/L chemical oxygen demand (COD) and 20,000 mgO$_2$/L biological oxygen demand (BOD). Dairy waste is sometimes loaded with high organic matters reaching 15 mg/L COD (Kozlowski et al., 2019).

Dairy waste at high concentration is toxic for aquatic life including some varieties of algae and fishes. Due to high COD and BOD, bacteria and algae grows rapidly on surface of water resulting in quick consumption of oxygen and formation of suspended layer of algae and reduced penetration of light for survival of aquatic life. Moreover, CIP of the dairy production line, spoiled milk, spilled milk, and leakage from pipelines also produces an enormous amount of wastewater, which contains a considerable quantity of milk components such as protein, fats, lactose, and minerals (Birwal et al., 2017; Ahmad et al., 2019).

6.3.5 PRETREATMENT OF FATTY EFFLUENTS

Dairy wastewater is rich in fatty and nitrogenous content that are high molecular weight organic compounds with low degradability efficiency. Lipids' degradation at an industrial level has not been fully exploited even in countries with a habit of high oil food consumption, it is very difficult to discard hence fatty waste disposal in an open environment has been restricted. Some traditional methods for fatty effluents' treatment include recycling or reuse of certain waste fractions as in the case of cheese production, where whey is reused. Various pretreatment methods are applied for removal of grease, oil and fats such as the following:

(1)   Tilted plate separators.
(2)   Grease traps.
(3)   Dissolved air flotation system.

6.3.6 GREASE TRAPS

In this method, the principle of gravity separation of floatable fats and oils is applied as shown in Figure 6.2. The rectangular or circular cylinder is placed by which waste material passes using laminar flow conditions. Flow rate is set at a level that allows the fatty particles to rise and get

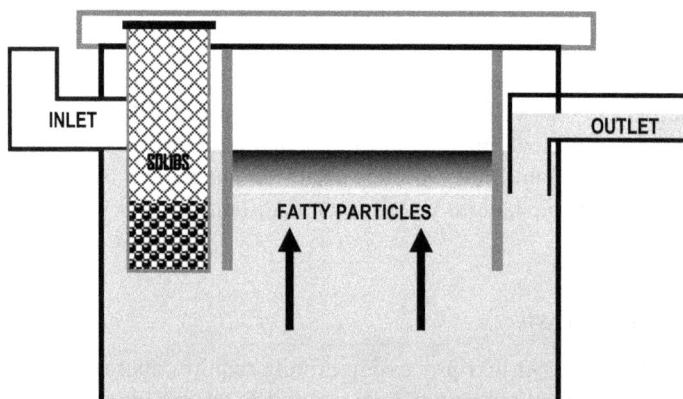

FIGURE 6.2   Grease trap.

accumulated gradually at the outlet of the cylinder under the principle of Stoke's law, which relates the theory of separation efficiency is not dependent on depth:

$$V_c = \frac{g(\rho s - \rho)d^2}{18\mu}$$

Vc is terminal velocity of fatty substances, μ is absolute velocity of waste liquid, d is diameter of fatty particles, ρ is density of waste liquid, whereas ρ_s is the density of fatty particles, and g is gravity-assisted acceleration rate. Commonly, fat traps of 1.5 meter depth with additional 0.5 m in the case of excess fat accumulation, are installed.

6.3.7 Tilted Plate Separators (TPS)

This waste removal technique was first developed in petrochemical industries where the principle of surface area was applied instead of depth. A vessel with tilted but parallel plates provides enough surface area for gravity based separation and requires approximately 10% less area as compared to traditional fat traps as shown in Figure 6.3. It is advantageous to use TPS as these come in the form of package plates, which can be removed and taken to developmental sites. However, certain problems associated with the TPS system are in consideration: (1) narrow gaps among plates susceptible to fouling if effluent is based on solid or semi-solid fats; (2) careful selection and control of waste control to avoid fluctuation and surging issues; (3) plates' removal, cleaning and reinstallation is a time consuming process.

6.3.8 Dissolved Air Floatation System

In a dissolved air floatation system (DAF), through a high pressure (6 Bar) nozzle, air is dissolved in a wastewater vessel and depressurization creates air bubbles attached with fatty particles, which enhance the rising rate of bubbles as shown in Figure 6.4. In DAF systems surface loading rate ranges from 3 to 6 m^3/m^2.h. Air solubility is decreased at elevated temperature; hence for maintaining the air solubility in a sustainable manner, temperature is a vital factor for installation of DAF systems.

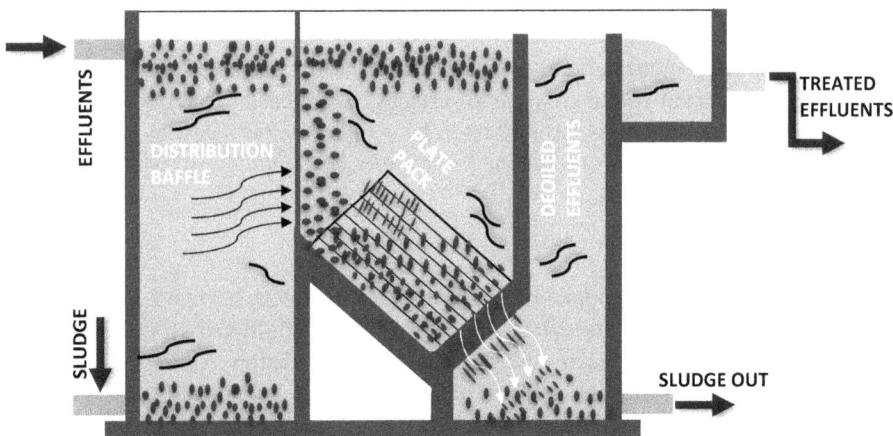

FIGURE 6.3 Tilted plate separator system.

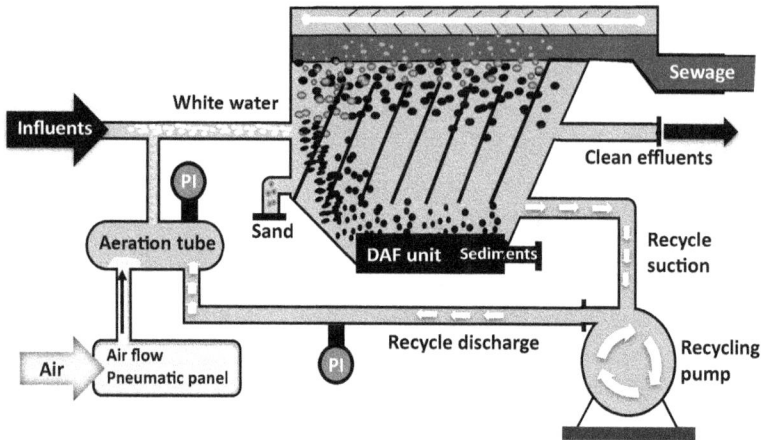

FIGURE 6.4   Dissolved air floatation system.

6.3.9   TREATMENT OF FATTY EFFLUENTS

6.3.9.1   Physical-chemical Treatment

In a physical-chemical treatment process, aluminum, sulfate, lime, or ferric chloride can be used to degrade the fatty coagulated matter and emulsions, which through sedimentation or floatation can be isolated. Floatation rate is improved by second stage floatation, which involves the addition of polyelectrolytes (0.5 to 5 mg/L) to wastewater that has been previously coagulated once. However, this process is costly because of reagents' cost, production of dissolved sludge when polyelectrolytes are used, and emulsified oil and grease removal efficiency is questionable.

6.3.9.2   Biological Treatment

The biological treatment method is highly advantageous due to thermophilic conditions, which favor the physical changes in hydrophobic compound. Various patents and research studies have been reported to use the microbial or enzymatic pool, developed in laboratories, to treat the fatty effluents with a high volume of fats and oils. A Japanese company named "Meito Sangyo Co." has developed lipase enzyme from *Candida rugosa* in the United States for efficient removal of fatty effluents from equipment in effluent treatment plants. In addition, another company named "Neozyme International Inc." also makes claims about manufacturing and selling patented bioorganic formulations that quickly degrade organic contaminants such as fats, greases, and oils.

According to a patented document, lipase producing bacteria is immobilized on non-biodegradable support medium and immersed within grease trap surfaces, where bacteria produces lipase enzyme and degrades fats from fatty traps upto a 90% hydrolysis rate. Another latest biological treatment introduced is the use of sophorolipids producing *Starmerella bombicola* yeast, which converts fatty acids into sophorolipids.

6.3.9.3   Use of Surfactants

Bio-surfactants can be easily incorporated into wastewater treatment having a high rate of fatty particles. Direct incorporation of bio-surfactants eliminates the additional processing stages and reduces the operational costs with high efficiency of fats' degradation. Although, the process is simple but the per liter bio-surfactant cost ($7–9) must be considered (Cammarota and Freire, 2006; Daverey et al., 2011; Makkar et al., 2011; Picavet, 2011; Nagappan et al., 2018).

6.3.10 UTILIZATION OF RECOVERED DAIRY FATTY WASTE

Fatty effluents from the dairy industry can be further utilized in various ways in biotechnological processes and converted into valuable products like (1) biomass; (2) biodiesel; (3) bioplastics; (4) biofuel; and (5) bio-surfactants. Microbial growth using recovered waste as a substrate can be proliferated and numerous types of products obtained from chemical reaction and biotransformation of organic waste into different products as a result of microbial activities.

*Acutodesmus dimorphus*is effective in conversion of dairy waste into biomass. *Bacillus megatarium* SRKP-3 is best known for biotransformation of dairy waste into bioplastics such as polyhydroxybutyrate (PHB), which is an easily degradable material. Fatty waste is also converted into biofuels i.e. ethanol, as it releases very minute amount of polluting agents upon ignition in contrast to fossil fuel burning processes. Many yeast types (*Kluyveromyces fragilis*, *Candida inconspicua*, and *Kluyvermyces marxianus*) are used for ethanol production in biotechnology industries with high ethanol recovery rate. Bio-surfactants are amphipathic compounds consisting of both hydrophobic and hydrophilic reaction groups hence can reduce surface tension of different liquids. These amphipathic compounds can be converted using yeast, fungus, and bacteria. *Candida bombicola* is efficient in conversion of fatty waste into sophorolipids, which can be a raw material in agriculture, cosmetics, food, and pharmaceuticals industries (Ahmad et al., 2019). A considerable amount of essential fatty acids (linoleic acid C18:2n6 and γ-linolenic acid C18:3n6) and other fatty acids (palmitic acid C16:0 and oleic acid C18:1n9) production from fatty waste of dairy industry using *Chlorella vulgaris* was reported by Rodrigues et al. (2021). The production of these fatty acids was achieved with supplementation of nitrogen to facilitate the biotransformation of waste. Further studies are needed to explore the biotransformation process with efficient use of waste, microorganisms, and the recovery of various organic products.

6.4 FAT AND OIL INDUSTRY

Lipids are a class of organic compounds that are difficult to convert into biogas. Lipids are principal components of organic matter in waste and wastewater from food processing businesses, slaughterhouses, fats and oil industries, and fat refineries, and are classified as fats, oils, and greases (Abeysinghe et al., 2002). The chemicals are glycerol and boric acid. Organic debris, suspended particles, and fats are the main contaminants in wastewater discharges from the fat industry. The raw effluent tank pumps wastewater to the fat removal tank, where it is then gravity-fed to the equalization tank, then to the UASB, aeration tank, and clarifier, and finally to the effluent tank, where it is collected and transported to the agricultural area (Babcock et al.,1992).

Carbohydrates (lactose), proteins, and fats make up the majority of fats and oil wastes. Large amounts of organic molecule, high fat, oil, and grease (FOG) contents, as well as fat solids, detergents, sanitizers, and fat wastes, characterize it. These FOGs have the potential to generate major organic loading problems in the local sewage drainage system (Bouchez et al., 2000; Abeysinghe et al., 2002). Wastewater from cleaning bulk tank, fat pipeline, fat processing units, other equipment, washing fats silo and operational malfunction are included in the fats and oil wastewater. The organic compounds in fats and oil wastewater are abundant, and the FOGs in fats and oil wastewater pose a significant environmental concern.

Fats have high molecular weight and a poor biodegradability coefficient resulting in high BOD and COD, polluting the environment and contaminating ground water. FOGs decompose slowly and present in some form of waste, forming film on the surface of water bodies that prevents adequate oxygen diffusion, resulting in death of eutrophication, aquatic flora and fauna, and ecosystem disruption (Eaton et al., 1995a: Eaton et al., 1995b).

6.4.1 Effluent Sources

The following are some of the effluent sources (Ferrari et al., 2002) (a) Fats receiving station effluent. (b) Empty cans washed in an automatic can washer automatically. (c) Pasteurization plant effluent. Wastewater from equipment washes comprising alkalies, acids, and detergents, as well as speels, floor washings, and leaks discharged from these areas. After each batch shift, it is a standard procedure to wash the entire machine. Aside from the above sources, the following parts produce a substantial amount of effluent: (VI) refrigeration and compressor section; (VII) boiler sections; (IV) fats processing section: (V) packing and filling section; (VI) boiler sections.

6.4.2 Butter and Ghee Manufacturing Wastes

Butter wastewater and floor washings, which are mostly made up of butter fats, cream, and tiny percentage of oil fats. Effluent from bottle washing plant – detergents or caustic soda in solution – is used to properly wash bottles and crates. The effluents are discharged on a regular basis, and the pasteurization of fats is done with uncontaminated cooling water. After chilling, the expended hot water can be reused.

6.4.3 Volume of the Effluent

Depending on the product manufactured, housekeeping, and available water, the volume of effluent might range from 1 to 10 lit of fats processed. The maximum discharge might be up to five times higher than the average.

6.4.4 Characteristics of the Effluent

Temperature, color, pH (6.5 to 8.0), dissolved oxygen (DO), BOD, COD and suspended particles, chlorides sulphate, and oil and grease are all properties of fats and oil effluent. Variation among these properties governs through the amount of fats processed and type of product developed. Fats and oil waste are white in color, alkaline in nature mildly and quickly becomes acidic due to high fat content.

Because they include a significant amount of fats, all of the liquors listed above have a high BOD. Because of the fine curd contained in cheese waste and suspended matter concentration of lipids' waste is high. The quick and strong oxygen demand is responsible for the polluting effect of fats and oil waste. Ghee decomposition results in massive black sludges and strong butyriform bacteria. Fats' waste pollution is characterized by the decomposition of ghee, which results in the creation of thick black sludges and pungent butyric – acid odors.

6.4.5 Effluents' Influence

FOG (fat, oil, and grease) is an ever-increasing environmental issue. Food service establishments (FSEs) and other food preparation facilities are the most common sources of FOG. Meat, sauces, gravy, dressings, deep-fried food, baked items, cheeses, and butter are among the by-products and wastes from these FSEs. When released directly into the facility's plumbing system, all of these pollutants are considered FOG and may cause FOG build-up in the sewage system. Discharges from industrial activities, such as palm oil mill effluent (POME) and automobile workshop outflow, are further sources of FOG. People's dining habits have changed recently, with more people eating outside their houses, leading in an increase in the number of food outlets, resulting in greater blockage of a city's sewer system due to FOG deposition.

In addition to detergents and sanitizers used for washings, fat and oil wastewater contains substantial amounts of fat ingredients such as fat and inorganic salts. All of these factors play a significant role in high BOD and COD, which is substantially greater than restrictions set by the Indian Standard Institute (ISI) known as the Bureau of Indian Standards (BIS) for discharge of industrial effluents. These wastes are believed to cause major environmental concerns when they are dumped into a nearby stream or land without being treated first (Goldstein et al., 1985). Fats and oil effluents breakdown quickly and quickly reduce the dissolved oxygen level in receiving streams, resulting in anaerobic conditions and the emission of strong unpleasant odors as a result of nuisance conditions. Flies and mosquitoes carrying malaria and other deadly diseases such as dengue fever, yellow fever, and chicken guniya breed in the receiving water. It has also been noted that increased fat and oil waste concentrations are hazardous to certain fish and algae (Boon et al., 2000). Fat precipitation from garbage decomposes further into foul-smelling black sludge at certain dilutions. Soluble organics, suspended particles, and trace organics are all present in fats and oil effluent.

The production of such films and greasy layers in aerobic reactors impedes the flocculation and sedimentation of biomass, interrupting pumping and, as a result, gas transfer, which was critical for biological degradation (Huban, et al., 1997). Biological treatment for fats and oil effluent appeared to be the most promising technology and aerobic treatment procedures such as trickling filters activated sludge processes, aerated lagoons, and sequencing batch reactors are commonly utilized (Keenan and Sabelnikov, 2000). Duc to floating of sludge and build-up of a fat scum layer on the reactor surface, it is expressive for anaerobic treatment (Loperena et al., 2005). Solidification of lipids at low temperatures can also cause blockage and emit an unpleasant odor. As a result, the low hydrolysis in the anaerobic reactor can be attributed to the process's poor performance. As a result, pretreatment devices are required to remove oils and fats from wastewater before biological treatment process (Loukidou et al., 2001). Pretreatment techniques that include fat hydrolysis and dissolution increase biological degradation of fatty lipids and oil wastes, speeding up the process and minimizing the amount of time it takes (Mendoza-Espinoza et al., 1996). The physicochemical pretreatment procedures cannot help in COD removal, reagents used are expensive, and processes needed strong reaction conditions, resulting in troublesome sludge production (Orhon and Artan, 1994). As a result of the rigorous environmental requirements, a clean and friendly approach of enzyme (Quan et al., 2003), mild reaction conditions, high specificity, easy availability, biochemical strategies such as utilizing particular enzymes lipases have gained consideration. Enzymes are well-known as versatile catalysts with a wide range of applications in biology. Their qualities make them an appealing waste/pollutant treatment alternative to traditional methods (Rademaker et al., 1998). In the presence of an oil/water interface, lipase break down triglyceride into free fatty acid and glycerol (Saucedo-Castaneda et al., 1994). These enzymes have shown promise in the breakdown of FOGs in wastewater from the dairy industry, slaughterhouses, and pet foods' sector (Seeley et al., 1991). Lipase Z was utilized in this investigation because it has positional selectivity against three ester linkages in triglycerides, resulting in high hydrolysis rate. Furthermore, because lipase Z catalyzes triglyceride hydrolysis at moderate temperature of roughly 30 centi-degrees Celsius, this enzymatic mechanism does not result in undesired fatty acid reactions. Furthermore, the enzymes' lower temperature reaction results in significant energy cost reductions. However, the longer time required for hydrolysis is the main worry with enzyme pretreatment. Furthermore, when the enzyme-catalyzed pretreatment was combined with ultrasonic irradiation, phenomenon of mass transfer by cavitation resulted in a significant degree of process intensification (Timmis and Pieper, 1996). Ultrasound waves generate cavitation, or the development, expansion, and implosive collapse of bubbles, which results in energy release and an increase in mass transfer rate. It is responsible for thermal and mechanical effects caused by cavitation, which improves mass transfer even more (Van Limbergen et al. 1998). As a result, it is more appealing than traditional reaction due to faster mass transfer rate and shorter reaction time. However, use of ultrasonography for lipase catalyzed fat pretreatment and oil wastewater treatment has not been investigated.

6.4.6 Conventional Stirring

In a 150 mL flat bottom baffled glass reactor with six bladed turbine glass impeller, a conventional batch experiment was carried out. In the reactor, 100 mL of prepared wastewater was added, along with 0.2% enzyme load and sodium chloride as an emulsifying agent. This assembly was placed in a water bath that was kept at 30 °C. An electric motor with a speed control mechanism was used to agitate the reaction mixture with mechanical stirrer at a speed of 200 rpm. The reactions conducted for a maximum of 24 hours. A tiny sample (2 mL) was taken from the reactor and transferred to a conical flask at regular intervals. To denature enzyme and freeze reaction 10 mL of a 50:50 mixture of acetone in ethanol was added to the sample and the reaction mixture was titrated with standard 0.02 M potassium hydroxide solution using phenolphthalein indicator to the determine acid value of fat breakage separation.

6.4.7 Ultrasound-assisted Hydrolysis

An ultrasonic bath used for ultrasound-assisted enzymatic hydrolysis. A standard experiment's flat bottom reactor was kept at a height of 2.0 cm from the base of the ultrasonic water bath. The position of the reactor in the bath was chosen based on previous research that indicated a higher cavitational intensity at that location (Versalovic et al. 1994). A transducer with an operating frequency of 25 and 40 kHz with maximum power output of roughly 200 W attached to the ultrasonic bath. Experiments were carried out in a standard batch reactor as described. Along with traditional factors, many factors such as power, duty cycle, and frequency were optimized. Figure 6.5 depicts a schematic of the reaction experiment setup.

The pretreatment of fats and oil effluent with hydrolytic enzyme lipase Z combined with an ultrasound irradiation approach is described in this work. The best enzyme loading was found to be 0.2% and a temperature of 30 °C. The varied ultrasonic power ranges from 45 to 200 W were tested. At a duty cycle of 66%, a frequency of 25 kHz and stirring speed of 200 rpm, the power at 165 W gave maximum conversion. The importance of ultrasonic irradiation for enhanced mass transfer, enhancing rate of hydrolysis, and reducing reaction time is demonstrated by a comparison of ultrasound

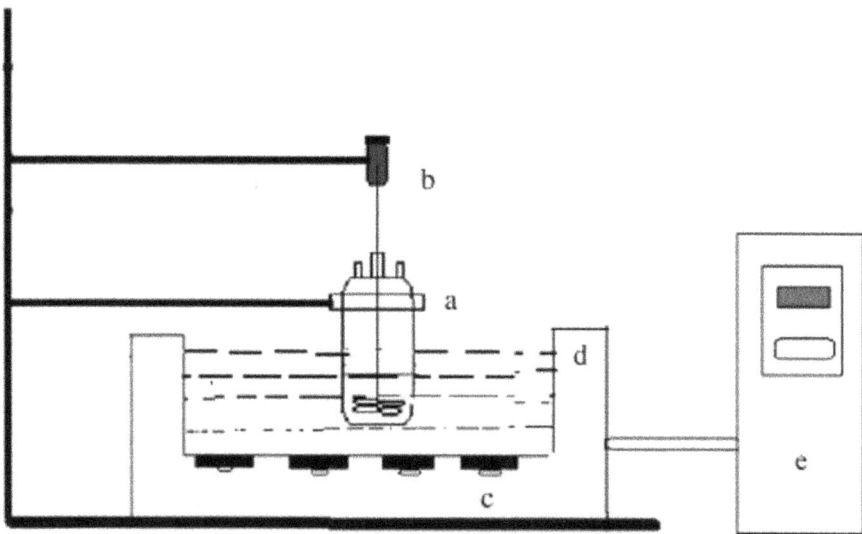

FIGURE 6.5 Schematic diagram of experimental set up of ultrasound-assisted reaction which have (a) baffled glass reactor; (b) motor; (c) transducers; (d) ultrasound water bath; € ultrasound generator.

aid methods with traditional methods. Our findings have demonstrated the utility of ultra-sonication in the realm of enzyme catalyzed reaction.

6.4.8 BIOREACTORS' DESCRIPTION AND OPERATION

Batch studies with fats and oil effluent at 4 g/L COD content were undertaken in six 0.4-L bioreactors (diameter: 75 mm, height: 140 mm). pH 7, 30 C, 200 rpm, air 0.11 L/min, and an initial cell concentration of 1.106 cfu/mL were used in the experiment. CO_2 content in exit gas stream, DO (polarographic probe), and pH were all measured during biodegradation by obtaining samples from bioreactors at regular intervals, COD, VSS, and heterotrophic number of microorganisms (cfu/mL) determined. Experiments were carried out in pairs or threes. Experiments were also carried out in a continuous mode in a 10-L bioreactor (9.5-L working volume) with a volumetric ratio of 7:3 between the aeration and settling chambers (Figure 6.6). The temperature was kept between 19 and 23 °C. After homogenization of the bioreactor material, samples were taken at regular intervals to assess the heterotrophic number of microorganisms (cfu/mL). The bioreactor's influent and effluent were also tested for soluble COD.

While the commercial and activated-sludge inoculums removed COD at similar rates in batch trials utilizing fats and oil industrial effluent as the culture medium, the commercial inoculum had a faster COD degradation rate. The activated-sludge inoculum, on the other hand, had a higher mineralization degree and produced less biomass than the commercial inoculum. In comparison to the bioreactor started with commercial inoculum, the activated-sludge inoculum trials showed more strain diversity and predominant population variability. This could be one of the reasons for the activated-sludge inoculum's greater mineralization. Despite the fact that commercial inoculum microorganisms had demonstrated the ability to grow in both fats and oil and model effluents in continuous culture experiments, and operational bioreactor conditions were determined based on their growth kinetics, other microorganisms were capable of establishing themselves and could predominate in the bioreactor (Abeysinghe et al., 2002; Babcock et al., 1992; Boon et al., 2000; Keenan and Sabelnikov, 2000). Others have recommended using different means to keep the microorganisms in the treatment tank, such as installing a biological filter at treatment tanks' outflow (Keenan and Sabelnikov, 2000) and immobilizing unique mixed culture (Quan et al., 2003). Because of the rise

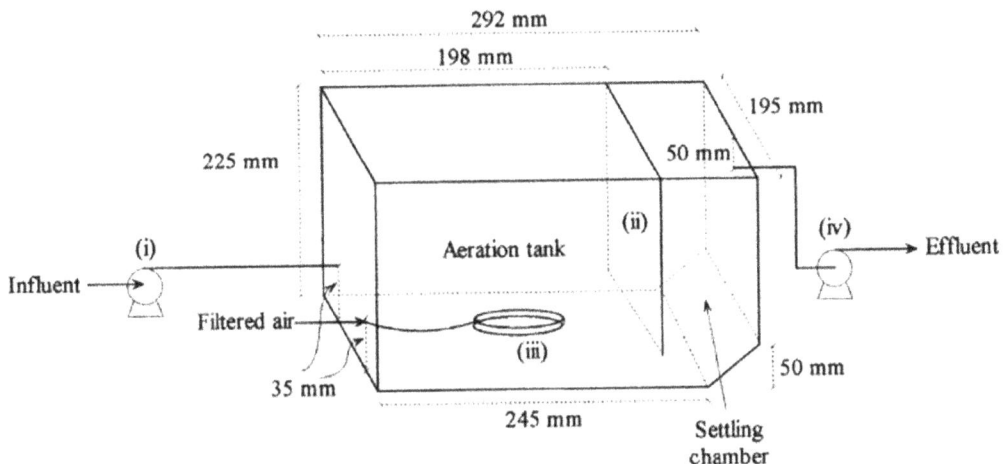

FIGURE 6.6 Schematic diagram flow of the continuous operated bioreactor with (i) inlet peristaltic pump; (ii) variable height baffle; (iii) fine bubble diffuser stone; (iv) outlet peristaltic pump.

of commercial bio-augmentation firms, all of the above mentioned factors must be considered while evaluating these products.

6.4.9 Capacity of LCFA Biodegradation and Impact of LCFA on Aceticlastic Methanogens

6.4.9.1 Sludge Source

Both sludges produced methane exclusively from biomass-associated LCFA when incubated in batch tests, and the greatest plateau attained in the cumulative methane production curves was regarded an indirect indication of the quantity of accumulated LCFA. Sludges 1 and 2 had specific LCFA concentrations of 4570 257 mg COD gVSS1 and 5200 9 mg COD gVSS1, respectively.

6.4.10 LCFA Biodegradation Assays

Parallel experiments were used to assess the LCFA biodegradation capacity of the sludge before and after degrading the biomass-associated LCFA. The sludge was incubated in 25 mL vials (12.5 mL working volume, VSS content around 5 gL1) at 37 C, 150 rpm, under stringent anaerobic conditions after being rinsed with anaerobic baseline media and centrifuged twice. The basal medium used was made up of cysteine–HCl (0.5 g L1) and sodium bicarbonate (3 g L1) and was created under strict anaerobic conditions with demineralized water. The pH was changed to 7.0–7.2 with NaOH 8N and the pH was adjusted to 7.0–7.2 with NaOH 8N. There was no calcium or trace nutrients provided.

6.4.11 Toxicity Tests

Individual toxicants oleate (for sludge 1) and palmitate (for sludge 2) were tested at doses ranging from 100 to 900 mg L1 in the toxicity experiments. To select for the aceticlastic trophic group, acetate (30 mM) was introduced. This trophic group was chosen for toxicity investigations because of its generalized increased vulnerability to unfavorable conditions (Versalovic et al., 1994) and its key metabolic role in the anaerobic digesting process (Wakelin et al., 1997). The beginning slope of the recorded methane production curve was used to calculate the specific methanogenic activity. As previously mentioned, the "toxicity" assessed by this test can also include the physical effects of transportation constraints (Yu and Mohn, 2002; Zouboulis et al., 2001).

6.4.12 Recovery of Fats and Oils from Fatty Effluents

Enzymatic prehydrolysis not only results in a process free of operational issues caused by the presence of fat, but it can also cut the process time, resulting in lower installation and operation costs in treatment facilities. Despite the stated substantial inhibition due to lipid content, some publications claim that the systems can restore themethanogenic activity and breakdown the substrate successfully due to adaptive phenomena. Long recovery durations, on the other hand, may be required for operating large-scale continuous digesters, which is undesirable.

The effects of two concurrent processes during organic matter degradation on methane generation: the influence of biomass acclimation to effluent elements and fatty effluent production, and the effect of fat accumulation in the biomass. While the first contributes to increased methane generation, the second has an impact on the activity of microbes especially methanogenic archaea, especially those that produce methane (Koster and Cramer, 1987; Lal-man and Bagley, 2002). In the studies with raw effluent, these two occurrences resulted in an increase in methane production from the first to the second batches, and then methane remained nearly stable in the third batch. Methane production reduced somewhat from the first to the second batches in the studies using previously hydrolyzed effluents, but rebounded in the third batch under all conditions. These findings imply that the biomass adapted to the medium's contents, overcoming the fat accumulation impact in the biomass.

The majority of research on LCFA's ability to block anaerobic metabolism use synthetic substrates that, in most cases, exclusively contain these lipids as a source of energy. In real-world scenarios, effluents from manufacturing plants, such as the one employed in this study, contain a variety of additional constituents that may impact the toxicity or effect of free fatty acids on biomass.

6.5   FATTY EFFLUENTS FROM CEREALS

Cereals provide nutrients and dietary energy and are considered to be a staple food in both developed and developing countries (McKevith, 2004). Cereals contributing to energy supply more than 50% in the global term, and are composed of approximately 6–15% protein, mainly starches and about 75% carbohydrates (WHO, 2003). Particularly, due to global food security, cereals gain attention on cereal production approximately 2600 million tons yearly (FAO, 2019). In the scientific literature, it has been frequently analyzed that role of cereal products is necessary for relation to nutrition value and consumption level. The best distributor regarding the dietary fiber, thiamine, niacin, folate, and iron described as cereal products such as bread product, quick bread, ready to eat cereals, bread, and rolls (Papanikolaou and Fulgoni, 2017).

6.5.1   Utilization of Cereal By-products

Cereal utilization of South Asia was more than 448 million tons in 2018, 15% of the overall global cereal production (FAOSTAT, 2019). On a global scale, in 2019 cereal utilization reached 2706 million tons whereas total consumption is expected to be 2714 million tons (FAO, 2019). In terms of calories, cereals are considered to be the most important of total food utilization. According to recent FAOSTAT data, cereals provide 50% dietary supplements to the consumer (FAOSTAT, 2019). Major cereal grains seeds from the Gramineae family such as rye, oat, millet, sorghum, maize, barley, rice, and, wheat filled the human diet with a significant proportion globally (FAO, 2019). Production ranking in the previous year was rye 15.8, oat 23.2, barley 146.3, wheat 854.9, rice paddy 949.7, corn 1253.6 tons (Papageorgiou and Skendi, 2018). During processing, a large number of by-products such as germ and bran are produced by the cereal processing industries (Anal, 2017). Generally, a by-product is produced by the cereal waste rich in phenolics, vitamins, fatty acids, lipids, dietary fibers, proteins, and mineral compounds but finally they are utilized for animal feed, bio-refinery substrate, and fuel (Verni et al., 2019).

6.5.2   Industrial Cereal Processing Waste

Cereals are composed of the bran, germ, and endosperm and cultivated for edible grains. Naturally, cereals contain the highest number of inorganic elements, carbohydrates, vitamin E and B, fats, oil, and proteins (Kumari, 2019). Globally, cereals are the most crucial food and considered an important segment of human consumption. There are two categories for cereal processing: dry milling and wet milling. Germ and bran portion is considered as the co-products or by-products, which can be removed by the dry milling method. Dry milling involves (hammer milling, impact milling, reduction, grinding, and pearling), etc. In cereal processing food industries, this abrasive technique can be utilized for the removal of the aleurone, seed coat, and sub-aleurone layers to attain polished grains such as barley, oat, and rice. While for the production of gluten and starch wet milling has been utilized, that generates the co-products such as bran, germs, and solid steep (Balandran-Quintana, 2018).

The nutritional quality of cereal-based food products is affected by certain practices, including the fractionation process and refining of the raw material and particularly in product formulation.

Highly nutritional fractions of bran and germ (protective phytochemicals, micronutrients, and fiber) were stripped from grain during the refining. Cereal-based breakfast for children is the typical example of very few micronutrients, extruded that cause high glycemic index and risk factor for development of obesity and overweight (Fardet, 2014).

6.5.3 TACKLING OF CEREAL WASTE POLLUTION

Waste of cereal food processing industries causes environmental pollution because it has an excellent amount of nutrients, solid waste, and organic load. This type of cereal industrial waste produces a huge number of algae on water and an unpleasant smell, which causes a high level of BOD that affects growth of marine animals in water (Thirugnanasambandham et al., 2013). Untreated industrial waste discharge and industrialization could contaminate the environment and have a huge impact on the ecosystem. Reduction of environmental pollution risk by cereal industrial waste is not only a concern, but it is also utilized for the production of some value-added commodities such as biofuel (bio-coal, butanol, biogas, bioethanol, and bio-hydrogen), industrial enzymes, organic acids (lactic, succinic, citric acid), biomass, proteins, polysaccharides, bio-fertilizer, and few others (Zahrim et al., 2015).

Cereal processing food industrial waste is generated in the form of non-toxic solid and liquid material (Kumar et al., 2017). Liquid waste contains corn steep liquor, bakery wastewater, parboiled rice effluents, rice milling wastewater, and wastewater from the corn tortilla food industry, while solid waste contains baking industry waste, brewers spent grain, corn grits, corn pericarp from corn food industries (Rojas-Chamorro et al., 2018; Juodeikiene et al., 2018; Haque et al., 2016). Various organic and inorganic harmful substances present in cereal industrial waste. The color of industrial liquid waste ranges from yellow to brown with higher values of COD, BOD, DO, minerals, and nutrients. Total cereal industrial waste production accounted for approximately 35% in Asia, America, and Europe while total loss and waste amount reach up to 18% in industrialized Asia (Mukherjee et al., 2016).

6.5.4 FATTY EFFLUENTS' PRODUCTION

Food processing industries particularly, cooking generate FOG, which are collected in interceptors and grease traps to prevent damage in sewage collection systems (Alqaralleh et al., 2016; Amha et al., 2017). Environmental impacts can be reduced by the management of FOG in an anaerobic digester with the implementation of conventional methods (diverting it from landfills), while that should increase the energy recovery in biogas production with substrate (Alqaralleh et al., 2016). Various studies reported that co-digesting of FOG can increase the energy recovery with other organic waste streams (Amha et al., 2017). There are some biotechnological approaches (anaerobic membrane bioreactors) that can be used to manage industrial waste and improve the biogas production and effluents' quality by combining membrane separation and anaerobic digestion. Amha et al. (2019) found that anaerobic membrane bioreactors are best for industrial waste management and approximately remove 95% COD. Furthermore, during treatment of industrial wastewater, an anaerobic membrane bioreactor removed the CH4 per gram of COD from 0.13–0.18 L (Galib et al., 2016).

Cereal industrial waste exploitation is a sustainable alternative that can also be used for production of volatile fatty acids. Volatile fatty acids contain linear short chain aliphatic monocarboxylate compounds having 2 to 6 carbon atoms. Chemical food industries utilized the volatile fatty acids extremely, while in conventional organic chemistry, the carboxylic acid is a precursor for reduction of the derivatives (alkanes, alcohols, ketones, aldehyde, and esters). Furthermore, for production of biofuels (biopolymers, hydrogen, methane), they are also well-known substrates (Domingos et al., 2017).

6.5.5 TREATMENTS

Cereal industrial waste has been treated using physical and chemical treatments as shown in Figure 6.7. These treatments are much more expensive but it takes a shorter time as compared to other biological treatments. Physical, chemical, or both in combination, are included in these treatments (Barbera and Gurnari, 2018). Cereal processing industrial waste can be processed by physical, chemical treatments for the effective removal of pollutants, which are based on advanced oxidation, ozonation, adsorption, reverse osmosis, sedimentation, precipitation, filtration, and coagulation treatments (Haq and Raj, 2020). In the last decades, several studies have been evaluated on the optimization of cereal food waste hydrolysis to enhance the volatile fatty acids. Particularly, chemical, physical, and combination treatments are preferred for improvements of volatile fatty acid content from cereal industrial waste (Bolzonella et al., 2018).

Treatments categorized as follows:

(1) Physical.
(2) Chemical.
(3) Biological.

Substrate's dimension of polymerization degree and crystallinity can be reduced by the physical treatments, and for microorganisms' reactions, they lead to a specific larger surface being available. For physical treatments, machines are adopted such as roll milling, hammer milling, compression milling, vibratory ball milling, ball milling, wet disk milling, and cutters milling (Paudel et al., 2017).

Cereal industrial food waste can be solubilized by the addition of different chemical concentrations such as alkali, acids, or organic aqueous solvent mixture (butanol, ethylene glycol, benzene, or ethanol). In this way, an alteration in lignin structure is done by increasing the surface area. Chemical treatments are not economically attractive due to their high number of reagents. It is a big drawback for chemical treatments (Passos et al., 2017).

Enzyme adaptation is involved in biological treatment, which can involve insolubilization of hemicellulose and cellulose. No chemical addition and excessive energy are needed in biological treatments; thus, they are environmentally friendly instead of chemical ones. Hydrolytic activity intensifies by the enzymatic treatment and takes part in the improvement of sugar production, which consequentially produces volatile fatty acids (Braguglia et al., 2018).

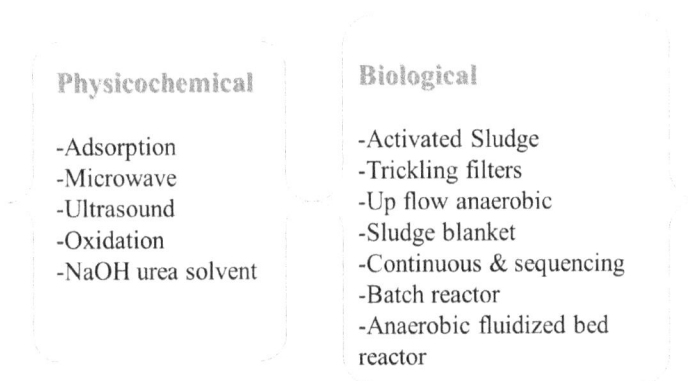

Physicochemical

-Adsorption
-Microwave
-Ultrasound
-Oxidation
-NaOH urea solvent

Biological

-Activated Sludge
-Trickling filters
-Up flow anaerobic
-Sludge blanket
-Continuous & sequencing
-Batch reactor
-Anaerobic fluidized bed reactor

FIGURE 6.7 Different treatment methods for waste by-products.

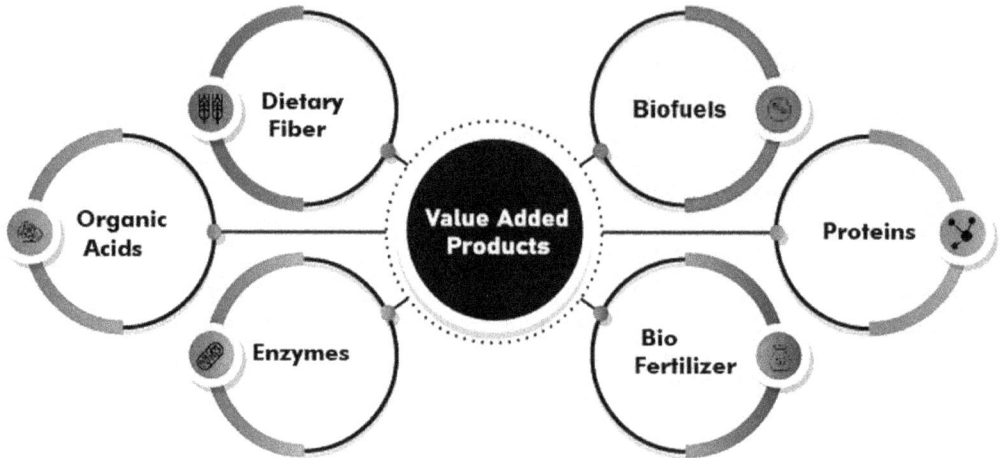

FIGURE 6.8   Different value-added products obtained from cereal waste by-products.

6.5.6   Recovery of Cereal Waste Fatty Effluents

Fatty acids obtained from the cereal industrial waste such as olive oil distillate, biodiesel fatty acid, and used form of cooking oil, can be utilized for the cheapest carbon production source. Fats from oil mills' wastewater, frying oil waste, margarine, sludge, and refinery waste are useful for polyhydroxyalkanoates (PHA) feedstock production (Rodriguez-Perez et al., 2018). Microbial growth is efficient by using fat from cereal industrial waste such as *Azotobacter vinelandii*, *Comamonas testosterone*, *Pseudomonas sp.*, and *Cupriavidus necator* and have the capacity to convert in other fatty acid-containing substrates (Cruz et al., 2015a). Fatty acid substrates (methyl esters and glycerol) are derived from different fatty wastes. The online process for monitoring is also developed for the production of fatty acid from industrial waste (Cruz et al., 2015b). On the other hand, fatty effluents utilized for the production of polysaccharides, which have functional properties and are applied in food as well as pharmaceutical industries with properties, consist of the stabilizer, detoxifiers, texturizers, and emulsifiers (Ahmad et al., 2019). Different value-added products obtained from cereal waste by-products are shown in Figure 6.8.

Fatty effluents from cereal industrial waste utilized for the recovery of commercially available enzymes. For this purpose, it needs to choose microorganisms that can be optimized, designed, and characterized for the production of the enzyme. Several types of strains such as *Monascuspurpureus sp. Bacillus sp. Streptomyces malaysiensis, Aspergillus awamori*, and *Aspergillus oryzae*, can be utilized for degradation of cereal industrial waste. All mentioned microorganisms can degrade organic compound (grease, oil, and fat) and can be used for further isolation for enzyme production (Ahmad et al., 2019). Furthermore, biofuels are important and can be produced by the fatty effluents of cereal industrial waste, it is referred to as solid, gas, and liquid fuels. Biofuels include bio-oil, biogas, bio-hydrogen bioethanol, bio-synthetic gas, coal, butanol, bio-ethanol, and biodiesel (Demirbas, 2008).

6.6   CONCLUSION

Food industries generate a huge amount of waste by-products that are unfit for human consumption. Different food industries generate different types of waste products; meat and slaughterhouses produce blood, fats, bones, wastewater, and sludge; dairy industries produce dairy effluent, buttermilk, and whey; fat and oil industries produce fats, oils, and greases; cereal industries produce corn steep liquor, bakery wastewater, rice effluents, corn grits, and wastewater. There are different physical, chemical, and biological treatments that used in different food industries for treatment of fatty effluents, ultrasound enzymatic treatment, commercial, and activated-sludge inoculums in fat and oil industries. Meat industry during slaughtering is one of the major contributors to liquid waste. Carcasses, parts of animals, and products of animal origin not intended for human consumption are defined as animal waste. The cost-effective treatment of solid wastes particularly wastewater effluents from the meat industry emerged as a serious concern. There are different methods available, which are enzymatic treatment, floatation, screening, catch basins, membrane distillation in meat, and slaughterhouses. The meat rendering processes produce wastewater rich in minerals, proteins, and fats. The membrane distillation process recovers valuable solids and water; however, hydrophobic membranes are polluted by the fats. The anaerobic digestion is a well suited treatment method for effective removal of chemical dissolved oxygen at lower cost and also generates methane rich biogas as fuel. Enzymes provide speed up of process, targeting specific compounds, and converting into simpler compounds (complex triglycerides into free fatty acids) without affecting others.

The milk industry is considered a major pollution causing industry because of the large use of water in production lines as well as during CIP. Wastewater from the dairy industry easily dissolves different effluents from milk and milk products. In the development and management of industry, sustainability is a key factor tointegrate with a holistic approach but there are still problems in triple bottom line aspects of sustainability. Various pretreatment methods are applied for removal of grease, oil and fats such as tilted plate separator, grease traps, and air flotation system. The biological treatment method is highly advantageous due to thermophilic conditions, which favor the physical changes in hydrophobic compound. Bio-surfactants can be easily incorporated into wastewater treatment having a high rate of fatty particles.

Lipids are principal components of organic matter in wastes and wastewaters from food processing businesses, slaughterhouses, fats and oil industries, and fat refineries, and are classified as fats, oils, and greases. Ghee decomposition results in massive black sludges and strong butyriform bacteria. Fats' waste pollution is characterized by the decomposition of ghee, which results in the creation of thick black sludges and pungent butyric – acid odors. Conversion of fat to trans-fatty acids and use of bioreactors and ultrasound waves results in energy release and an increase in mass transfer rate. However, use of ultrasonography for lipase catalyzed fat pretreatment and oil wastewater treatment must be investigated in the future.

Waste of cereal food processing industries cause environmental pollution because it has excellent amount of nutrients, solid waste, and organic load. This type of cereal industrial waste produces a huge number of algae on water and an unpleasant smell, which causes a high level of BOD affecting growth of marine animals in water. Cereal industrial food waste can be solubilized by precipitation, sedimentation in cereal industries. The addition of different chemical concentrations such as alkali, acids, or organic aqueous solvent mixture (butanol, ethylene glycol, benzene, or ethanol) can be used as chemical treatments. Fatty effluents from cereal industrial waste must be utilized for the production of commercially available enzymes or other by-products.

Furthermore, all types of fatty effluents can be used as biofuels that are important and which can be produced by the fatty effluents of food industrial waste, it is referred to as solid, gas, and liquid fuels. Biofuels include bio-oil, biogas, bio-hydrogen bioethanol, bio-synthetic gas, coal, butanol, bio-ethanol, and biodiesel.

REFERENCES

Abeysinghe, Dilrika H., DG Viraj De Silva, David A. Stahl, and Bruce E. Rittmann. 2002. "The effectiveness of bioaugmentation in nitrifying systems stressed by a washout condition and cold temperature." *Water Environment Research* 74, no. 2: 187–199.

Aguilar, M. I., J. Saez, M. Llorens, A. Soler, and J. F. Ortuno. 2003. "Microscopic observation of particle reduction in slaughterhouse wastewater by coagulation–flocculation using ferric sulphate as coagulant and different coagulant aids." *Water Research* 37, no. 9: 2233–2241.

Aguilar, M. I., José Sáez, Mercedes Lloréns, Antonio Soler, Juan F. Ortuño, V. Meseguer, and Ana Fuentes. 2005. "Improvement of coagulation–flocculation process using anionic polyacrylamide as coagulant aid." *Chemosphere* 58, no. 1: 47–56.

Ahmad, Talha, Rana Muhammad Aadil, Haassan Ahmed, Ubaid ur Rahman, Bruna CV Soares, Simone LQ Souza, Tatiana C. Pimentel et al. 2019. "Treatment and utilization of dairy industrial waste: A review." *Trends in Food Science & Technology* 88: 361–372.

Alexandre, V. M. F., A. M. Valente, Magali C. Cammarota, and Denise M. G. Freire. 2011. "Performance of anaerobic bioreactor treating fish-processing plant wastewater pre-hydrolyzed with a solid enzyme pool." *Renewable Energy* 36, no. 12: 3439–3444.

Al-Mutairi, N. Z., M. F. Hamoda, and I. Al-Ghusain. 2004. "Coagulant selection and sludge conditioning in a slaughterhouse wastewater treatment plant." Bioresource technology 95, no. 2: 115–119.

Alqaralleh, Rania Mona, Kevin Kennedy, Robert Delatolla, and Majid Sartaj. 2016. "Thermophilic and hyper-thermophilic co-digestion of waste activated sludge and fat, oil and grease: evaluating and modeling methane production." *Journal of Environmental Management* 183: 551–561.

Amha, Yamrot M., Michael Corbett, and Adam L. Smith. 2019. "Two-phase improves performance of anaerobic membrane bioreactor treatment of food waste at high organic loading rates." *Environmental Science & Technology* 53, no. 16: 9572–9583.

Amha, Yamrot M., Pooja Sinha, Jewls Lagman, Matt Gregori, and Adam L. Smith. 2017. "Elucidating microbial community adaptation to anaerobic co-digestion of fats, oils, and grease and food waste." *Water Research* 123: 277–289.

Anal, Anil Kumar, ed. Food processing by-products and their utilization, John Wiley & Sons, 2017, 44–67.

Babcock Jr, Roger W., Kyoung S. Ro, Chu-Chin Hsieh, and Michael K. Stenstrom. 1992. "Development of an off-line enricher-reactor process for activated sludge degradation of hazardous wastes." Water Environment Research 64, no. 6: 782–791.

Balandrán-Quintana, René R. Recovery of proteins from cereal processing by-products: In Sustainable recovery and reutilization of cereal processing by-products, Woodhead Publishing, 2018, 125–157.

Barbera, Marcella, and Giovanni Gurnari. Wastewater treatment and reuse in the food industry, Springer International Publishing, 2018, 80–89.

Barkocy-Gallagher, Genevieve A., Terrance M. Arthur, Mildred Rivera-Betancourt, Xiangwu Nou, Steven D. Shackelford, Tommy L. Wheeler, and Mohammad Koohmaraie. 2003. "Seasonal prevalence of Shiga toxin–producing Escherichia coli, including O157: H7 and non-O157 serotypes, and Salmonella in commercial beef processing plants." *Journal of Food Protection* 66, no. 11: 1978–1986.

Birwal, Preeti, G. Deshmukh, S. P. Saurabh, and S. Pragati. 2017. "Plums: a brief introduction." *Journal of Food, Nutrition and Population Health* 1, no. 1: 1–5.

Bolzonella, David, Federico Battista, Cristina Cavinato, Marco Gottardo, Federico Micolucci, Gerasimos Lyberatos, and Paolo Pavan. 2018. "Recent developments in biohythane production from household food wastes: a review." *Bioresource Technology* 257: 311–319.

Boon, Nico, Johan Goris, Paul De Vos, Willy Verstraete, and Eva M. Top. 2000. "Bioaugmentation of activated sludge by an indigenous 3-chloroaniline-degrading Comamonas testosteroni strain, I2 gfp." *Applied and Environmental Microbiology* 66, no. 7: 2906–2913.

Bouchez, T., D. Patureau, P. Dabert, S. Juretschko, J. Dore, P. Delgenes, Renatto Moletta, and M. Wagner. 2000. "Ecological study of a bioaugmentation failure." *Environmental Microbiology* 2, no. 2: 179–190.

Braguglia, Camilla M., Agata Gallipoli, Andrea Gianico, and Pamela Pagliaccia. 2017. "Anaerobic bioconversion of food waste into energy: A critical review." *Bioresource Technology* 248, no. Pt A (2017): 37–56.

Cammarota, M. C., and D. M. G. Freire. 2006. "A review on hydrolytic enzymes in the treatment of wastewater with high oil and grease content." *Bioresource Technology* 97, no. 17: 2195–2210.

Campbell, John R., and Robert T. Marshall. Dairy production and processing: The science of milk and milk products, Waveland Press, 2016, 64–76.

Chalupa-Krebzdak, Sebastian, Chloe J. Long, and Benjamin M. Bohrer. 2018. "Nutrient density and nutritional value of milk and plant-based milk alternatives." *International Dairy Journal* 87: 84–92.

Cheng, Dongle, Yi Liu, Huu Hao Ngo, Wenshan Guo, Soon Woong Chang, Dinh Duc Nguyen, Shicheng Zhang, Gang Luo, and Yiwen Liu. 2020. "A review on application of enzymatic bioprocesses in animal wastewater and manure treatment." *Bioresource Technology* 313: 123683.

Cruz, Madalena V., Filomena Freitas, Alexandre Paiva, Francisca Mano, Madalena Dionísio, Ana Maria Ramos, and Maria AM Reis. 2016. "Valorization of fatty acids-containing wastes and by-products into short-and medium-chain length polyhydroxyalkanoates." *New Biotechnology* 33, no. 1: 206–215.

Cruz, Madalena V., Mafalda Cruz Sarraguça, Filomena Freitas, João Almeida Lopes, and Maria AM Reis. 2015. "Online monitoring of P (3HB) produced from used cooking oil with near-infrared spectroscopy." *Journal of Biotechnology* 194: 1–9.

Dahiya, Shikha, Omprakash Sarkar, Y. V. Swamy, and S. Venkata Mohan. 2015. "Acidogenic fermentation of food waste for volatile fatty acid production with co-generation of biohydrogen." *Bioresource Technology* 182: 103–113.

Daverey, A., K. Pakshirajan, and S. Sumalatha. 2011. "Sophorolipids production by Candida bombicola using dairy industry wastewater." *Clean Technologies and Environmental Policy* 13, no. 3: 481–488.

Demirbas, Ayhan. 2008. "Biofuels sources, biofuel policy, biofuel economy and global biofuel projections." *Energy Conversion and Management* 49, no. 8: 2106–2116.

Dereli, Recep Kaan, Barry Heffernan, Aurelie Grelot, Frank P. van der Zee, and Jules B. van Lier. 2015. "Influence of high lipid containing wastewater on filtration performance and fouling in AnMBRs operated at different solids retention times." *Separation and Purification Technology* 139: 43–52.

Dereli, Recep Kaan, Frank P. van der Zee, Barry Heffernan, Aurelie Grelot, and Jules B. van Lier. 2014. "Effect of sludge retention time on the biological performance of anaerobic membrane bioreactors treating corn-to-ethanol thin stillage with high lipid content." *Water Research* 49: 453–464.

Domingos, Joana MB, Gonzalo A. Martinez, Alberto Scoma, Serena Fraraccio, Frederiek-Maarten Kerckhof, Nico Boon, Maria AM Reis, Fabio Fava, and Lorenzo Bertin. 2017. "Effect of operational parameters in the continuous anaerobic fermentation of cheese whey on titers, yields, productivities, and microbial community structures." *ACS Sustainable Chemistry & Engineering* 5, no. 2: 1400–1407.

dos Santos Ferreira, Janaína, Débora de Oliveira, Rafael Resende Maldonado, Eliana Setsuko Kamimura, and Agenor Furigo. 2020. "Enzymatic pretreatment and anaerobic co-digestion as a new technology to high-methane production." *Applied Microbiology and Biotechnology* 104, no. 10: 4235–4246.

Eaton, A., Clesceri, L., Greenberg, A. Standard Methods for the Examination of Water and Wastewater, 19th ed. American Public Health Association, American Water Works Association, Water Environment Federation, Washington, DC, 1995a, 2–57.

Eaton, A., Clesceri, L., Greenberg, A. Standard Methods for the Examination of Water and Wastewater, 19th ed. American Public Health Association, American Water Works Association, Water Environment Federation, Washington, DC, 1995b, 5–18.

FAO Cereal Supply and Demand Brief. World Food Situation. Cereal production and inventories to decline but overall supplies remain adequate. Food and Agriculture Organization of the United Nations. Available online: www.fao.org/worldfoodsituation/csdb/en/ (accessed on 29 January 2019).

FAO. 2019. Available online: www.fao.org/worldfoodsituation/csdb/en/ (21 November 2019, date last accessed).

FAOSTAT. 2019. Available online: http://faostat.fao.org/static/syb/syb_5000.pdf (23 January 2020, date last accessed).

FAOSTAT. 2019. Available online: Statistics division, Food and Agriculture Organization of the United Nations. Retrieved from www.fao.org/faost at/en/?#data

FAOSTAT. 2021. Available online: Food and Agriculture Organization of the United Nations.

Fardet, Anthony. 2014. "How can both the health potential and sustainability of cereal products be improved? A French perspective." *Journal of Cereal Science* 60, no. 3: 540–548.

Feil, Alexandre André, Dusan Schreiber, Claus Haetinger, Ângela Maria Haberkamp, Joice Inês Kist, Claudete Rempel, Alisson Eduardo Maehler, Mario Conill Gomes, and Gustavo Rodrigo da Silva. 2020. "Sustainability in the dairy industry: a systematic literature review." *Environmental Science and Pollution Research* 27, no. 27: 33527–33542.

Ferrari, M. D., C. Lareo, L. Loperena, D. Murro, and V. Saravia. 2002. "Uso de inóculos diseñados para la bioaumentación de sistemas de tratamiento biológico de efluentes conteniendo grasas." *Ingeniería Química* 22: 21–27.

Galib, Mohamed, Elsayed Elbeshbishy, Robertson Reid, Abid Hussain, and Hyung-Sool Lee. 2016. "Energy-positive food wastewater treatment using an anaerobic membrane bioreactor (AnMBR)." *Journal of Environmental Management* 182: 477–485.

Ghotra, Baljit S., Sandra D. Dyal, and Suresh S. Narine. 2002. "Lipid shortenings: a review." *Food Research International* 35, no. 10: 1015–1048.

Goldstein, Rebecca M., Lawrence M. Mallory and Martin Alexander. 1985. "Reasons for possible failure of inoculation to enhance biodegradation." *Applied and Environmental Microbiology* 50, no. 4: 977–983.

Grismer, Mark E., Charles C. Ross, G. Edward Valentine, Brandon M. Smith, and James L. Walsh. 2002. "Food-processing wastes." *Water Environment Research* 74, no. 4: 377–384.

Hanisah, Kamilah, Sudesh Kumar, and Aris Yang Tajul. 2014. "Characteristics of used palm olein and its bioconversion into polyhydroxybutyrate by Cupriavidus necator H16." *Malaysian Journal of Microbiology* 10, no. 2: 139–148.

Haq, Izharul, and Abhay Raj. 2020. "Pulp and paper mill wastewater: ecotoxicological effects and bioremediation approaches for environmental safety." In Bioremediation of industrial waste for environmental safety, pp. 333–356. Springer.

Haque, Md Ariful, Vasiliki Kachrimanidou, Apostolis Koutinas, and Carol Sze Ki Lin. 2016. "Valorization of bakery waste for biocolorant and enzyme production by Monascus purpureus." *Journal of Biotechnology* 231: 55–64.

Huban, Christopher M., and Robert D. Plowman. 1997. "Bioaugmentation: put microbes to work." *Chemical Engineering* 104, no. 3.

Huban, Christopher M., and Robert D. Plowman. 1997. "Bioaugmentation: put microbes to work." *Chemical Engineering* 104, no. 3.

Jeganathan, Jeganaesan, George Nakhla, and Amarjeet Bassi. 2006. "Long-term performance of high-rate anaerobic reactors for the treatment of oily wastewater." *Environmental Science & Technology* 40, no. 20: 6466–6472.

Juodeikiene, G., D. Zadeike, I. Vidziunaite, E. Bartkiene, Vadims Bartkevics, and Iveta Pugajeva. 2018. "Effect of heating method on the microbial levels and acrylamide in corn grits and subsequent use as functional ingredient for bread making." *Food and Bioproducts Processing* 112: 22–30.

Keenan, Daniel, and Alexander Sabelnikov. 2000. "Biological augmentation eliminates grease and oil in bakery wastewater." *Water Environment Research* 72, no. 2: 141–146.

Koster, I.W, Cramer, A., 1987. Inhibition of methanogenesis from acetate ingranular sludge by long-chain fatty acids. *Appl. Environ. Microb.* 53, 403–409

Kozłowski, Kamil, Maciej Pietrzykowski, Wojciech Czekała, Jacek Dach, Alina Kowalczyk-Juśko, Krzysztof Jóźwiakowski, and Michał Brzoski. 2019. "Energetic and economic analysis of biogas plant with using the dairy industry waste." *Energy* 183: 1023–1031.

Kumar, Anuj, Abhishek Roy, Rashmi Priyadarshinee, Bratin Sengupta, Alok Malaviya, Dalia Dasguptamandal, and Tamal Mandal. 2017. "Economic and sustainable management of wastes from rice industry: combating the potential threats." *Environmental Science and Pollution Research* 24, no. 34: 26279–26296.

Kumar, Upendra, and Manas Bandyopadhyay. 2006. "Sorption of cadmium from aqueous solution using pretreated rice husk." *Bioresource Technology* 97, no. 1: 104–109.

Kumari, Mamta. 2019. Cereals. Walnut Publication, 201–240.

Lalman, J., Bagley, D.M., 2002. Effects of C18 long chain fatty acids on glucose, butyrate and hydrogen degradation. Water Res. 36, 3307–3313.

Leal, M. C. M. R., M. C. Cammarota, and D. M. G. Freire. 2002. "Hydrolytic enzymes as coadjuvants in the anaerobic treatment of dairy wastewaters." *Brazilian Journal of Chemical Engineering* 19: 175–180.

Leibrock, Amy. 2017. "Are Food Waste Bans Working." Sustainable America, January, 2017, Available at: https://sustainableamerica.org/blog/are-food-waste-bans-working/

Liu, Xiaoyan, Xinfeng Wang, Jiaxing Xu, Jun Xia, Jinshun Lv, Tong Zhang, Zhen Wu, Yuanfang Deng, and Jianlong He. 2015. "Citric acid production by Yarrowia lipolytica SWJ-1b using corn steep liquor as a source of organic nitrogen and vitamins." *Industrial Crops and Products* 78: 154–160.

Long, J. Hunter, Tarek N. Aziz, L. Francis III, and Joel J. Ducoste. 2012. "Anaerobic co-digestion of fat, oil, and grease (FOG): A review of gas production and process limitations." *Process Safety and Environmental Protection* 90, no. 3: 231–245.

Loperena, Lyliam, Verónica Saravia, Daiman Murro, Mario Daniel Ferrari, and Claudia Lareo. 2006. "Kinetic properties of a commercial and a native inoculum for aerobic milk fat degradation." *Bioresource Technology* 97, no. 16: 2160–2165.

Loukidou, M. X., and A. I. Zouboulis. 2001. "Biodegradability tests of dairy and cheese whey wastewaters using enzymes." *Fresenius Environmental Bulletin* 10, no. 2: 188–192.

Magbanua Jr, Benjamin S., Thomas T. Adams, and Phillip Johnston. 2001. "Anaerobic codigestion of hog and poultry waste." *Bioresource Technology* 76, no. 2: 165–168.

Makkar, Randhir S., Swaranjit S. Cameotra, and Ibrahim M. Banat. 2011. "Advances in utilization of renewable substrates for biosurfactant production." *AMB Express* 1, no. 1: 1–19.

McKevith, Brigid. 2004. "Nutritional aspects of cereals." *Nutrition Bulletin* 29, no. 2: 111–142.

Mendoza-Espinosa, Leopoldo, and Tom Stephenson. 1996. "Grease biodegradation: is bioaugmentation more effective than natural populations for start-up?" *Water Science and Technology* 34, no. 5–6: 303–308.

Mostafa, M. G., Bo Zhu, Marlene Cran, Noel Dow, Nicholas Milne, Dilip Desai, and Mikel Duke. 2017. "Membrane distillation of meat industry effluent with hydrophilic polyurethane coated polytetrafluoro-ethylene membranes." *Membranes* 7, no. 4: 55.

Mukherjee, Chandan, Rajojit Chowdhury, Tapan Sutradhar, Momtaj Begam, Sejuti Magdalene Ghosh, Sandip Kumar Basak, and Krishna Ray. 2016. "Parboiled rice effluent: A wastewater niche for microalgae and cyanobacteria with growth coupled to comprehensive remediation and phosphorus biofertilization." *Algal Research* 19: 225–236.

Murtaza, M. A., N. Huma, A. Sameen, M. S. Murtaza, S. Mahmood, G. Mueen-ud-Din, and A. Meraj. 2014. "Texture, flavor, and sensory quality of buffalo milk Cheddar cheese as influenced by reducing sodium salt content." *Journal of Dairy Science* 97, no. 11: 6700–6707.

Murto, Marika, Lovisa Björnsson, and Bo Mattiasson. 2004. "Impact of food industrial waste on anaerobic co-digestion of sewage sludge and pig manure." *Journal of Environmental Management* 70, no. 2: 101–107.

Nagappan, Subbiah, David M. Phinney, and Dennis R. Heldman. 2018. "Management of waste streams from dairy manufacturing operations using membrane filtration and dissolved air flotation." *Applied Sciences* 8, no. 12: 2694.

Orhon, D., Artan, N. Modelling of Activated Sludge Systems. Technomic Publishing Company Inc., 1994, 45–67.

Papageorgiou, Maria, and Adriana Skendi. Introduction to cereal processing and by-product: In Sustainable Recovery and Reutilization of Cereal Processing By-Products, Woodhead Publishing, 2018, 1–25.

Papanikolaou, Yanni, and Victor L. Fulgoni. 2017. "Certain grain foods can be meaningful contributors to nutrient density in the diets of US children and adolescents: Data from the National Health and Nutrition Examination Survey, 2009–2012." *Nutrients* 9, no. 2: 160.

Pascale, N. C., J. J. Chastinet, D. M. Bila, G. L. SantAnna, S. L. Quitério, and S. M. R. Vendramel. 2019. "Enzymatic hydrolysis of floatable fatty wastes from dairy and meat food-processing industries and further anaerobic digestion." *Water Science and Technology* 79, no. 5: 985–992.

Passos, Fabiana, Valentina Ortega, and Andrés Donoso-Bravo. 2017. "Thermochemical pretreatment and anaerobic digestion of dairy cow manure: experimental and economic evaluation." *Bioresource Technology* 227: 239–246.

Paudel, Shukra Raj, Sushant Prasad Banjara, Oh Kyung Choi, Ki Young Park, Young Mo Kim, and Jae Woo Lee. 2017. "Pretreatment of agricultural biomass for anaerobic digestion: Current state and challenges." *Bioresource Technology* 245: 1194–1205.

Picavet, Merijn Amilcare. Development of the inverted anaerobic sludge blanket reactor: A novel technology for the treatment of industrial wastewater containing fat, PhD diss., Universidade do Minho (Portugal), 2011, 24–95.

Quan, Xiangchun, Hanchang Shi, Jianlong Wang, and Yi Qian. 2003. "Biodegradation of 2, 4-dichlorophenol in sequencing batch reactors augmented with immobilized mixed culture." *Chemosphere* 50, no. 8: 1069–1074.

Rademaker, J., Louws, F., de Bruijn, F.J. Characterization of the diversity of ecologically important microbes by rep-PCR genomic fingerprinting. In: Akkermans, A., van Elsas, J., De Bruijn, F. (Eds.), Molecular Microbial Ecology Manual, 3.43. Kluwer Academic Publishers, 1998, 13–47.

Radoiu, Marilena T., Diana I. Martin, Ioan Calinescu, and Horia Iovu. 2004. "Preparation of polyelectrolytes for wastewater treatment." *Journal of Hazardous Materials* 106, no. 1: 27–37.

Rahman, Ubaid ur, Amna Sahar, and Muhammad Azam Khan. 2014. "Recovery and utilization of effluents from meat processing industries." *Food Research International* 65: 322–328.

Ramírez-Rivera, E. J., J. Rodríguez-Miranda, I. R. Huerta-Mora, A. Cárdenas-Cágal, and JUAN M. Juárez-Barrientos. 2019. "Tropical milk production systems and milk quality: a review." *Tropical Animal Health and Production* 51, no. 6: 1295–1305.

Rigo, Elisandra, Roberta Eletízia Rigoni, Patrícia Lodea, Débora De Oliveira, Denise MG Freire, Helen Treichel, and Marco Di Luccio. 2008. "Comparison of two lipases in the hydrolysis of oil and grease in wastewater of the swine meat industry." *Industrial & Engineering Chemistry Research* 47, no. 6: 1760–1765.

Rodriguez-Perez, Santiago, Antonio Serrano, Alba A. Pantión, and Bernabé Alonso-Fariñas. 2018. "Challenges of scaling-up PHA production from waste streams. A review." *Journal of Environmental Management* 205: 215–230.

Rodrigues-Sousa, Ana Elisa, Ivan VO Nunes, Alex B. Muniz-Junior, João Carlos M. Carvalho, Lauris C. Mejia-da-Silva, and Marcelo C. Matsudo. 2021. "Nitrogen supplementation for the production of Chlorella vulgaris biomass in secondary effluent from dairy industry." *Biochemical Engineering Journal* 165: 107818.

Rojas-Chamorro, José A., Cristóbal Cara, Inmaculada Romero, Encarnación Ruiz, Juan M. Romero-García, Solange I. Mussatto, and Eulogio Castro. 2018. "Ethanol production from brewers' spent grain pretreated by dilute phosphoric acid." *Energy & Fuels* 32, no. 4: 5226–5233.

Salminen, E., and J. Rintala. 2002. "Anaerobic digestion of organic solid poultry slaughterhouse waste–a review." *Bioresource Technology* 83, no. 1: 13–26.

Sandvig, K., and B. Van Deurs. 2000. "Entry of ricin and Shiga toxin into cells: molecular mechanisms and medical perspectives." *The EMBO Journal* 19, no. 22: 5943–5950.

Saucedo-Castaneda, Gerardo, MdR Trejo-Hernandez, B. K. Lonsane, J. M. Navarro, Sevastianos Roussos, Dominique Dufour, and Maurice Raimbault. 1994. "On-line automated monitoring and control systems for CO2 and O2 in aerobic and anaerobic solid-state fermentations." *Process Biochemistry* 29, no. 1: 13–24.

Seeley Jr., H.W., Vandemark, P.J., Lee, J.J. Microbes in Action, fourth ed. W.H. Freeman and Company, ISB, 1991, 7167–2100.

Smith, Adam L., Lauren B. Stadler, Nancy G. Love, Steven J. Skerlos, and Lutgarde Raskin. 2012. "Perspectives on anaerobic membrane bioreactor treatment of domestic wastewater: a critical review." *Bioresource Technology* 122: 149–159.

Sousa, Diana Z., Andreia F. Salvador, Juliana Ramos, Ana P. Guedes, Sïnia Barbosa, Alfons JM Stams, M. Madalena Alves, and M. Alcina Pereira. 2013. "Activity and viability of methanogens in anaerobic digestion of unsaturated and saturated long-chain fatty acids." *Applied and Environmental Microbiology* 79, no. 14: 4239–4245.

Thimjos, N., Janne, L., Hannu, K. and Fredriksson-Ahomaa, M. Meat by-products. In P. Miguel, & L.G. Maria (Eds.), Meat Inspection and Control in the Slaughterhouse. John Wiley & Sons, Ltd, 2014, 34–57.

Thirugnanasambandham, K., V. Sivakumar, and J. Prakash Maran. 2013. "Application of chitosan as an adsorbent to treat rice mill wastewater – mechanism, modelling and optimization." *Carbohydrate Polymers* 97, no. 2: 451–457.

Timmis, K.N., Pieper, D.H. Bacteria designed for bioremediation. TIBTECH 17, 1999, 201–204.

Tisinger, J.L., Drakos, D.J. Use of respirometry to evaluate the efficacy of biological products for the biodegradation of wastewater grease and oil. In: 51st Purdue Industrial Waste Conference Proceedings, Paper 49. Ann Arbor Press, 1996, 473–481.

Van Limbergen, Heidi, E. M. Top, and Willy Verstraete. 1998. "Bioaugmentation in activated sludge: current features and future perspectives." *Applied Microbiology and Biotechnology* 50, no. 1: 16–23.

Veluchamy, C., V. Wilson Raju, and Ajay S. Kalamdhad. 2018. "Electrohydrolysis pretreatment for enhanced methane production from lignocellulose waste pulp and paper mill sludge and its kinetics." *Bioresource Technology* 252: 52–58.

Verni, Michela, Carlo Giuseppe Rizzello, and Rossana Coda. 2019. "Fermentation biotechnology applied to cereal industry by-products: nutritional and functional insights." *Frontiers in Nutrition* 6: 42.

Versalovic, J., Schneider, M., de Bruijn, F.J., Lupski, J.R., 1994. Genomic fingerprinting of bacteria using repetitive sequence based polymerase chain reaction. *Methods Mol. Cell. Biol.* 5, 25–40.

Wakelin, N. G., and C. F. Forster. 1997. "An investigation into microbial removal of fats, oils and greases." *Bioresource Technology* 59, no. 1: 37–43

Wakelin, N. G., and C. F. Forster. 1998. "The aerobic treatment of grease-containing fast food restaurant wastewaters." *Process Safety and Environmental Protection* 76, no. 1: 55–61.

Wendorff, W. L., and George FW Haenlein. Sheep Milk–Composition and Nutrition: In Handbook of milk of non-bovine mammals, 2017, 210–221.

World Health Organization. Diet, nutrition, and the prevention of chronic diseases: report of a joint WHO/FAO expert consultation. Vol. 916. World Health Organization, 2003, 34–87.

Yu, Z., Mohn, W.W., 2002. Bioaugmentation with resin acid-degrading bacterium Zoogloea resiniphila DhA-35 to counteract pH stress in an aerated lagoon treating pulp and paper mill effluent. *Water Research* 36, 2793–2801.

Zahrim, A. Y., A. Nasimah, and N. Hilal. 2015. "Coagulation/flocculation of lignin aqueous solution in single stage mixing tank system: Modeling and optimization by response surface methodology." *Journal of Environmental Chemical Engineering* 3, no. 3: 2145–2154.

Zouboulis, A.I., Loukidou, M.X., Christodoulou, K., 2001. Enzymatic treatment of sanitary landfill leachate. *Chemosphere* 44, 1103–1108.

ABBREVIATIONS

BIS	Bureau of Indian Standards
BOD	Biological Oxygen Demand
COD	Chemical Oxygen Demand
CIP	Cleaning In Place
DAF	Dissolved Air Flotation
DO	Dissolved Oxygen
DRI	Dietary Reference Intake
FAs	Fatty Acids
FAO	Food and Agriculture Organization
FAO STAT	Food and Agriculture Organization Statistic
FOG	Fat, Oil, and Grease
FSE	Food Service Establishments
ISI	Indian Standard Institute
LCFA	Long Chain Fatty Acid
OUR	Oxygen Uptake Rate
POME	Palm Oil Mill Effluent
SVI	Sludge Volume Index
TPS	Tilted Plate Separators
VSS	Volatile Suspended Solids
WHO	World Health Organization

7 Conversion of Bone to Edible Products

Pardeep Kaur Sandhu and Gurpreet Kaur*

Mata Gujri College, Fatehgarh Sahib, Punjab, India
*Corresponding author: gurpreet_adh@matagujricollege.org
pardeepkaur@matagujricollege.org

CONTENTS

DOI: 10.1201/9781003207689-7

7.1 INTRODUCTION

As per the Food and Agriculture Organization 2020 report, natural disasters like the corona pandemic, pest attack, and climate change accounts for an increase in 80–130 million people suffering from hunger and raised the alarm of food security (FAO, 2020). Healthy diet is out of reach for more than 3 billion people, making them malnourished. At a time when elimination of hunger and malnourishment is still a big challenge for humanity, it is unaffordable for reported global food loss and wastage rate exceeding 20% costing 2.5 trillion US dollars. Addressing the food waste losses along the production and supply chain is one of the agenda amongst United Nations Sustainable Development Goals (United Nations Conference on Trade and Development, 2020). The meat industry is an important part of the world's food industry with the world total meat production estimated at 337.2 million tons in 2020 with maximum contribution of poultry followed by pig, bovine and ovine (FAO, 2021). Increase in per capita consumption, population growth, and change in dietary pattern leads to a rise in total consumption of livestock products (Henchion *et al.*, 2021, Kowalski *et al.*, 2021). The average annual global consumption of meat is 122g/day of which poultry and pork is one third each, beef as one fifth and the rest from sheep, goats, and other animals (Godfray *et al.*, 2018). The overall consumption of meat is expected to rise by 70% in the period of 2000–2030 and 120% in the period of 2030–2050 (Gonzalez *et al.*, 2020). However, the current patterns of production and consumption of food are not sustainable (Steenson and Buttriss, 2020). Willett *et al.* (2019) reported that multilevel and multisector action is needed for the same and includes the need for large reduction of food loss and food waste. The meat industry is one of the most resource intensive food industries with great demand for already depleting water resources. Livestock is inefficient in converting feed to meat and 75–80% of energy is used in body maintenance or is lost in manure and by-products including bones. Moreover, the greenhouse emissions by livestock leads to a great environmental impact (Makara *et al.*, 2019). The meat industry is not only providing protein rich food but also produces a large number of by-products, more than 20 million tonnes annually only in Europe (Fu *et al.*, 2019). The under utilization of these by-products is not only the wastage of resources but also a great environmental threat (Borrajo *et al.*, 2019; Fu *et al.*, 2019). To increase the sustainability of animal source food, there is the need of adding value to existing animal by-product lines and

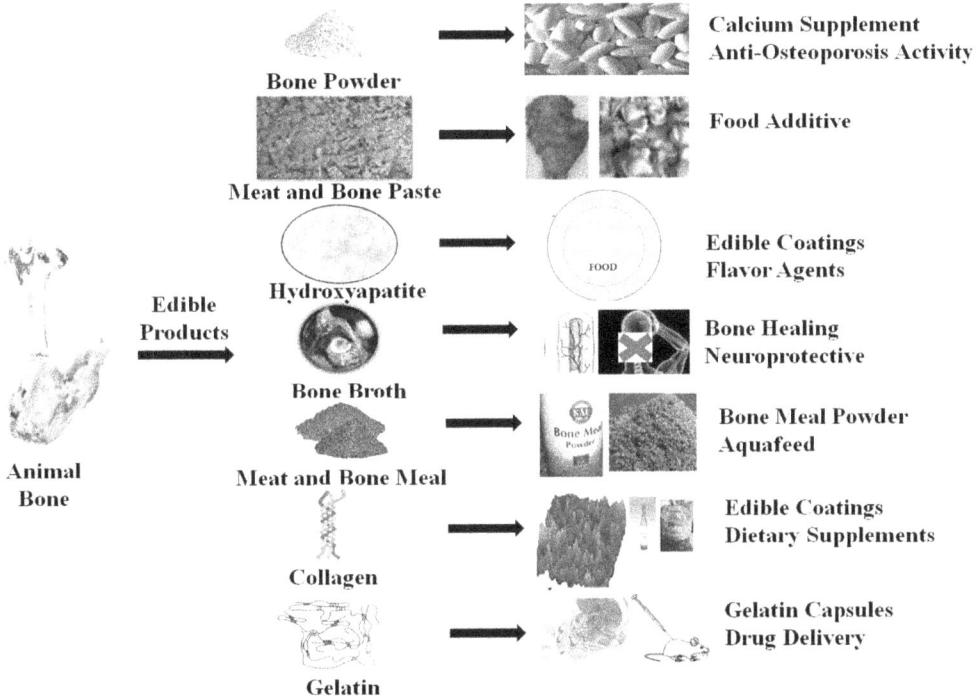

FIGURE 7.1 Edible by-products from animal bone.

also to channelize various other sources for revenue earning to bear the cost of disposal for meat industry waste.

Animal waste generated in the meat industry is divided into three categories, category 1 includes high-risk animal waste suspected of infectious diseases, category 2 includes animal by-products, which may contain contaminants, while category 3 is a low-risk waste. Bone is the category 3 waste generated in the meat industry. As per the Eurostat database, more than 50 million tonnes of bovine, poultries, and pig are slaughtered annually in the European Union. The skeletal system contributes to 20% of the carcass weight and 4 million tonnes of animal bones are produced in the EU alone (Dobbelaere, 2013; Someus, 2018). Bone is a rich source of collagen and minerals and has been used to make bone broth soups and gelatin for centuries. In recent years, the advancement in technologies led to more value-added products from bones (Figure 7.1). In addition to use in animal feed they are used to extract bioactive compounds, which have wide applications not only in enhancing sensory properties of food but also in food preservation and packaging. There is a change in people's demand for use of food not only for nutrition but also for disease prevention and treatment. This led to the development of innovative and advanced techniques of utilization of bone waste in novel edible products for use in the food and pharmaceutical industry for nourishment and alleviation of disease (Dobbelaere, 2013).

7.2 EDIBLE BY-PRODUCTS FROM BONE

Bones are considered as animal waste and are generally disposed of by the meat industry; however, it is rich source of nutrients like proteins, minerals, and fat. Various edible products like meat and bone meal, bone paste, bone powder can be prepared, which are not only used for nutrient fortification but also for improving taste, texture, and flavor of food. The bioactive compounds derived from bone waste have promising applications in the pharmaceutical and food industry.

7.2.1 Collagen and Collagen Peptide

Collagen is a huge proteinaceous constituent of extracellular matrix (ECM) in animals and is the utmost abundant protein found in several animal by-products like bone, cartilage, and skin (Toldra *et al.*, 2016; Cao *et al.*, 2020). The major morphological characteristic of the collagen proteins is a firm three-layered helical structure made by the joining of three polypeptide chains by hydrogen bonds. These polypeptide chains consist of one or more repetitive clusters of an amino acid (Gly-X-Y), where X and Y are the proline and hydroxyproline and glycine cover 1/3rd among the entire amino acids (Tan and Chang, 2018). The main reservoir of collagen is the animal tissues, particularly connective tissues and organs. More than 29 types of collagens have been screened and extracted from animal tissues. On the basis of globular and three-layered helical structure and macromolecular length, collagens are categorized into fibril (type I, II, III, V), fibrillar (type VI and IX), and network forming collagen (type IV) (Myllyharju and Kivirikko, 2004; Bay-Jensen *et al.*, 2016; Fu *et al.*,2019). Type I–III collagen is present in abundance, forming 81–91% of the total collagen of the body. Type I collagen is rich in hydroxyproline and present in skeletal tissue like in bones, skin, and tendons as "structural component". Type II collagen is found in cartilage and vitreous humour and type III collagen is found in the same tissues with type I collagen except in tendons and bones.

Collagen plays a huge role in food, pharmaceuticals, and other fields due to its bioavailability, low antigenicity, and good biological potential (Liu *et al.*, 2015). It has enhanced its market value and induces the progress of a global collagen market. Currently, consumers are more anxious about the safety of collagen obtained from terrestrial animals due to the outburst of various fatal infections. Moreover, intake of porcine collagen in food is also not favored due to some religious concerns. Thus, fish collagen generated from waste like the bones and skin has gained keen interest as a substitute for another collagen (Gómez-Guillén *et al.*, 2002; Jongjareonrak *et al.*, 2005). Terrestrial and marine animal's by-products such as bones, skin, horns, blood, hooves, feet, are the prime raw materials employed for isolating collagen and collagen peptides. Peptides are found in an inactive state in the native form of collagen. These inactive collagen peptides are converted into active form after hydrolysis of collagen through chemical or enzymatic methods. These collagen peptides consist of 2–20 amino acid residues and exhibited efficient physical, chemical, and functional properties in contrast to the collagen (Wang *et al.*, 2015). Collagen peptide possesses several biological activities like antibacterial, antioxidant, anti-inflammatory, cardioprotective, along with the beneficial effect on the skin, bones, and joints. These peptides are gaining extraordinary attention for research due to their potent effect in the food and pharmaceutical industry. It showed the existence of some active constituents, which displayed a significant part in the induction of immune response and relief from several neural and cardiovascular infections (Wang *et al.*, 2015; Halim *et al.*, 2016).

7.2.1.1 Preparation of Collagen and Collagen Peptides

The bones, feet, skin, hooves, and blood are the major animal by-products used for the extraction of collagen and its peptide.

Chemical and enzymatic hydrolysis are the two main methods employed for the extraction of collagen. Chemical hydrolysis is more commonly utilized for the isolation of collagen in comparison to enzymatic hydrolysis. However, enzymatic approaches lead to the production of a smaller amount of waste and require less time for the purification of collagen (Zavareze *et al.*, 2009). A pre-treatment with an acid or alkaline substance is required for the cleaving of covalent intra and inter-molecular cross-links. It is given to the tissues for the elimination of non-collagenous substances and to increase the extraction efficiency of chemical and enzymatic hydrolysis. Pre-treatment of tissues is done by using either an alkaline or acid solvent according to the origin and density of the tissue (Schrieber and Gareis, 2007; Ran and Wang, 2014; Yao *et al.*, 2021).

7.2.1.2 Acidic Pre-treatment

Mostly acidic solvents are employed for the partial hydrolysis of collagen to break the covalent cross-linking; however, it sustains the collagen chains unharmed (Prestes, 2013). The bone is dipped in acid till the solution creeps into the tissue. This pre-treatment is found to be superior for brittle tissues like porcine and fish tissues with minimum entangled collagen fibers (Ledward, 2000; Almeida, 2012). Inorganic acids as well as organic acids are used for the extraction of collagen, however the extraction efficiency of organic acids is found to be more in contrast to the inorganic acids (Wang et al., 2008). The acetic acid (0.5 mol/L for 24 hours) is superlative pre-treatment choice for the extraction of collagen as it takes less time for extraction and total process cost is also observed to be lower. Hydrochloric acid and EDTA (ethylene diaminetetraacetic acid) is extensively used for the removal of minerals from bone raw material (Nollet et al., 2012). It was observed that treatment of EDTA (0.25 M and 0.5 M) showed efficient decalcification in contrast to the hydrochloric acid at the similar concentration. In addition to it, other acids like citric acid and lactic acid have been used for the extraction of collagen. Sadowska et al. (2003) purified 85% of collagen protein from the *Gadus morhua* using citric acid. Moreover, lactic acid (25 mM) was observed to be a better alternative for acetic acid during the extraction of collagen (Giménez et al., 2005).

7.2.1.3 Alkaline Pre-treatment

In alkaline pre-treatment, the bone material is immersed in sodium hydroxide and calcium hydroxide, for certain weeks to remove the non-collagenous material (Prestes, 2013). Sodium hydroxide is found to be superior to calcium hydroxide as it leads to more puffiness to promote the chemical extraction by promoting the transfer proportion in a tissue matrix. Sodium hydroxide (0.05–0.1M) at temperatures of 4–20°C efficiently eliminate non-collagen proteins without any harm to the architecture of collagen. Then the washing of the tissues are performed with chilled distilled water until the neutral pH (Liu et al., 2015).

7.2.1.4 Extraction Methods

7.2.1.4.1 Chemical Hydrolysis

Inorganic acids (hydrochloric acid) and organic acids are employed for the extraction of collagen (Figure 7.2). However, it was observed that organic acids are more superior in contrast to inorganic acids (Skierka and Sadowska, 2007; Wang et al., 2008). During the extraction process, organic acids are used to dissolve non-cross-linked collagens and cleave few inter-strand cross-links in collagen, which results in more solubility of collagen (Liu et al., 2015). To obtain the acid-soluble collagen, pre-treated tissue is dipped in the acetic acid for 1-3 days with continuous mixing at 4°C. Further, filtration is done to separate the supernatant from collagen and precipitate filtrate with sodium chloride to form collagen powder. Centrifugation was done to collect the precipitate and consequently resolubilized it in acetic acid and then purify acid-soluble collagen through dialysis (Nagai and Suzuki, 2000; Nagai et al., 2015).

Nagai and Suzuki (2000) extracted the acid-solubilized bone collagen from Japanese sea bass, ayu, skipjack tuna, and horse mackerel. The yield of collagen was observed to be 53.6% for ayu, 43.5% for horse mackerel, 42.3% for skipjack tuna, 40.7% for sea bass, and 40.1% for yellow seabream.

When collagen is isolated, it can be further hydrolyzed by acids or alkaline solutions into the low molecular weight hydrophobic compounds known as "collagen peptides" or hydrolysate. Hong et al. (2018) displayed that formic acid increased the formation of collagen peptides from hens. He furthermore reported that formic acid is better for hydrolysis of collagen than other acidic solutions as it deeply enters into collagen molecules. Alkaline hydrolysis can also be conducted with

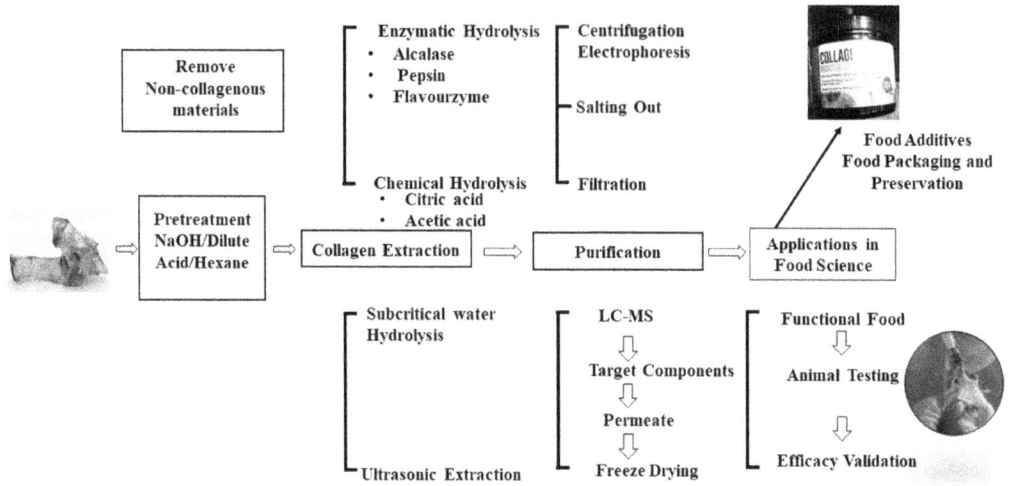

FIGURE 7.2 Extraction and applications of collagen.

vigorous chemicals like sodium hydroxide or potassium hydroxide at 130°C to 180°C. It cleaves the molecular bonding between the strands of collagen and decreases the protein fibrils of collagen into shorter peptides of about 5 kDa. However, alkaline hydrolysis is not prescribed as it deteriorates the content of amino acids (Pal and Suresh, 2016).

7.2.1.4.2 Enzymatic Hydrolysis

The enzymatic hydrolysis method reveals more promising results in contrast to chemical hydrolysis for production of nutritional and functional value products (Figure 7.2). In addition, the enzymatic method induces minimum waste and also decreases the duration of processing (Schmidt *et al.*, 2016). The bone material is immersed into an acetic acid solution consisting of enzymes like alcalase, pepsin, and flavourzyme and mixed constantly for two days at 4°C. The mixture is further filtered, and the filtrate obtained is precipitated with sodium chloride and dialyzed in a similar environment as for isolating acid-soluble collagen (Nagai and Suzuki,2000;Li *et al.*, 2009; Li *et al.*, 2013). Ogawa *et al.* (2004) reported the isolation of type I collagen from the bones and scale of black drum and seabream through chemical and enzymatic hydrolysis.

The chemical composition of collagen after the hydrolysis of pepsin showed similarity with the collagen obtained from other fishes and bovine sources. In another study, Bama *et al.* 2010 demonstrated the acid-soluble and pepsin-soluble collagen (ASC and PSC) extraction from bones of horse mackerel and croaker by using the method of Nagai and Suzuki (2000). The yield of acid-soluble bone collagen was observed to be 30.5 and 27.6 % from the bones of horse mackerel and croaker, respectively. In addition to it, the yield of pepsin-soluble bone collagen was found to be 45.1%–48.6% from horse mackerel and croaker, respectively. The yield variability of both collagen might be due to the several interchain bonds present at the telopeptide region of the collagen that leads to a reduction in acid solubility (Muyonga *et al.*, 2004). The basic and easiest technique is UV spectroscopy, which is used for the detection of collagen purity and is dependent on the highest absorption peak at 210–240 nm because of the triple helical structure of the collagen (Zhao *et al.*, 2018). The ASC isolated from the bone of mackerel and croaker revealed the greatest absorbance at 230 nm. However, a decrease in the absorbance was observed in pepsin-soluble collagen in comparison to the ASC. SDS-PAGE (sodiumdodecylsulfate- polyacrylamide gel electrophoresis) can be employed for the determination of the type of collagen and characterization of the relative molecular weight. It was observed that ASC and PSC isolated from mackerel and croaker contained

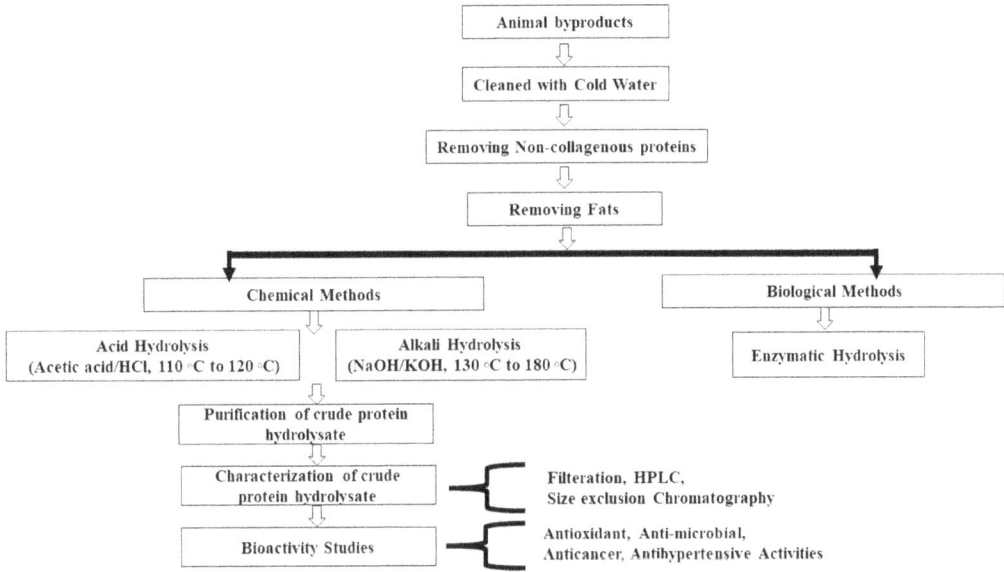

FIGURE 7.3 Extraction methods for the preparation of collagen hydrolysate.

Source: adapted from Ahmed *et al.*, 2020.

α1 chain with molecular weight of 135 KDa and α2 chain with molecular weight of 115 KDa. No remarkable variation was found in subunits of acid- and pepsin-soluble collagen, which indicates that treatment of pepsin has not affected the triple-helical architecture of collagen. It was observed that a huge constituent of isolated collagen was type I collagen (Nagai and Suzuki, 2000; Ogawa *et al.*, 2004). FT-IR spectroscopy is employed for the determination of the secondary structure of proteins and their hydrolyzed peptides (Belbachir *et al.*, 2009). The FTIR spectrum of acid- and pepsin-soluble collagen showed type I collagen, with unbound N–H stretching vibration. Nazeer *et al.* (2014) showed the identical stretching vibration of N–H in the *Donax cuneatus* collagen.

Various enzymes such as flavourzyme, papain, pepsin, alcalase, and trypsin are used for the generation of collagen peptides (Figure 7.3). It was observed that hydrolysis with pepsin, trypsin, and α-chymotrypsin produce more bioactive peptides in contrast to the hydrolysis by the single enzyme (Barbana and Boye, 2011; Himaya *et al.*, 2012). Natsir *et al.* (2021) demonstrated the formation of collagen peptides from bones of tunafish using the collagen as an enzyme. This study depicted that 10% of enzymes and a hydrolysis time of four hours are required for the maximum rate of hydrolysis. These collagen peptides showed antibacterial potential against *Escherichia coli* and *Staphylococcus aureus* and displayed inhibition zones of 11.50 mm and 12.60 mm respectively.

7.2.1.4.3 Ultrasonication Extraction

Use of ultrasonication in collagen extraction decreases the duration of the processing and enhances the yield of collagen (Kim *et al.*, 2012; Kim *et al.*, 2013; Ran and Wang, 2014; Tu *et al.*, 2015). The optimum extraction duration for ultrasonic exposure was found to be 18 h and 4°C, which revealed an extended helical structure, favorable solubility, and more thermal durability of the collagen (Ran and Wang, 2014). Similarly, Li *et al.* (2009) described the extraction of bovine collagen by using ultrasonic (40 kHz, 120 W) conjunction with pepsin, which resulted in enhanced extraction (124%), with reduction in processing time. The increment in extraction yield is due to the diffusion of pepsin with exposure to irradiation, which results in rupturing of collagen fibrils and promotes enzymatic action. Circular dichroism, atomic force microscopy, and Fourier-transform infrared spectroscopy revealed the intact three-layered structure of the collagen even after the exposure of ultrasonics.

7.2.1.4.4 Subcritical Water Hydrolysis

Subcritical water hydrolysis is an eco-friendly processing method for the production of peptides through the hydrolysis of proteins. The mode of action of subcritical water hydrolysis is the generation of ions like hydronium (H_3O+) and hydroxide ($OH-$), which serve as catalysts (Marcet *et al.*, 2016; Powell *et al.*, 2017). Numerous researchers have employed this method for the preparation of collagen from several animal waste products.

Asaduzzaman *et al.* (2020) reported the potential of subcritical water for the production of low molecular weight peptides (<1650 Da) from the pepsin-soluble bone collagen of mackerel. It was observed that these collagen hydrolysates revealed potent antioxidant activity in comparison to the pepsin-soluble collagen. It was noticed that subcritical water hydrolysis needs sophisticated instrumentation and maximal energy cost in contrast to chemical and enzymatic hydrolysis. Moreover, side chains of amino acids are modified due to the subcritical water treatment (Powell *et al.*, 2017).

7.2.1.5 Applications

Collagen hydrolysate from grass carp has high antioxidant and antihypertensive activity (Wang *et al.*, 2014; Li *et al.*, 2021). The preparation of hydrolysate from fish bone by Maillard reaction increases the proportion of umami sweet taste amino acids and causes the enhancement of volatile compounds in protein hydrolysate. Protein hydrolysate formed through the Maillard reaction showed more 2,2-diphenylpicrylhydrazyl radical scavenging and acetylcholinesterase (ACE) inhibitory activity. In contrast to this, a reduction in ACE inhibitory activity was observed in protein hydrolysate generated through the thermal degradation method as it causes the inactivation of few active proteins (Feng *et al.*, 2009; Li *et al.*, 2021).

7.2.1.5.1 Applications in the Food Industry

7.2.1.5.1.1 Meat Products   Collagen serves as a potent antioxidant, thus improving the quality of food. It is employed for the enhancement of the shelf life of food. The collagen assists in the absorption of moisture in frankfurter sausage and decreases the spoilage of meat products during thermal treatment. Various investigations indicated that the usage of collagen and gelatin coverings on fresh meat foodstuffs decreases the quantity of purge, which forms in the lowest part of meat packages during storage. Therefore, the use of edible collagen coatings can decrease the usage of absorbent pads in packaging. Subsequently collagen coatings assist in the maintenance of consistency, color and flavor of food (Mead, 2004; Antoniewski and Barringer, 2010). Collagen from chicken bone showed a marked effect on the rheological properties of sausage emulsions and it hampers the oxidation of fats without influencing their other characteristics. The collagen peptide is also used in hybrid gels which might be employed for the replacement of fats in meat products to enhance the nutritional status and safety profile of meat products (Prabhu *et al.*, 2006; Yue, 2017; Fan *et al.*, 2020).

7.2.1.5.1.2 Dairy and Beverage Products   Currently, drinks supplemented with collagen like cocoa collagen are in high demand worldwide. It is estimated that collagen drinks enhance the synthesis of collagen in the body and improve the functioning of body tissues and hence favor the reduction of wrinkles on the face (Tree, 2012). In addition to it, some collagen drinks like vitagen collagen assist in the stimulation of multiplication of growth of gut microbiota and also positively affect the skin. Collagen peptides have an efficient emulsifying ability to hinder the whey precipitation in dairy products (Najumudeen, 2012; Shori *et al.*, 2021). Many investigations suggest that collagen not only causes significant improvement in the nutritional and functional characteristics of collagen but also enhanced the number of proteins and bioavailability of juice drinks (Seda and Sibel 2015). Arely *et al.* (2020) described that collagen peptides hinder the growth of infective bacteria in the drinks. Collagen acts as a better clarifying mediator in the alcoholic beverage and also enhances sensory characteristics and storage of the beverages (Zhang *et al.*, 2018).

7.2.1.5.1.3 Role in Food Packaging and Preservation Collagen has good mechanical durability, biocompatibility, and elasticity, which make it an essential component for wrapping and storage of food. Collagen based biomaterials can be categorized into edible collagen coatings and films. Collagen edible films act as barrier membrane to shield the food from the moisture and oxygen thereby maintaining the morphological integrity and vapor penetrability to the foodstuffs (Greene, 2003; Bourtoom, 2008; Cao *et al.*, 2020), which helps in increasing the shelf life of the foodstuffs (Greene, 2003). Hou *et al.* (2020) reported the strong mechanical strength of sheep bone collagen–chitosan blend film, which can be used for packaging of food. Shao *et al.* (2020) reported that addition of collagen from bovine bone to hydroxypropylmethylcellulose (HPMC) film revealed improvement in mechanical properties, biocompatibility, and thermal stability. This study suggested that collagen serves as an active component in HPMC films and makes these films appropriate for food packaging. Edible coatings of collagen play a crucial role in the preservation of food substantiated by its ability to maintain the amount of moisture, flavor, aroma, and quality of food (Lopez-Rubio *et al.*, 2017). Wang *et al.* (2017) reported that collagen lysozyme (CLZ) coating improved the quality of fresh salmon fillets in contrast to those preserved in control coating material. CLZ coating causes the reduction in weight and total volatile basic nitrogen in salmon fillets. Moreover, these coatings act as safeguards for the protection of food from the growth of bacteria and also sustain the texture of the salmon fillets. In another study, Antoniewski and Barringer (2010) suggested the potential of collagen coatings as an active barrier on meat products for the reduction of purge, texture and aroma worsening, and put refaction and also recover the sensory characteristics.

7.2.1.5.2 Role in Functional Foods

7.2.1.5.2.1 Anti-aging Chronological aging arises with age and it affects the entire body including the face with symptoms of fine wrinkles and laxity (Helfrich *et al.*, 2008). There is great demand for anti-aging products to delay or reverse the signs of aging. The oral collagen peptide supplements are safe to use by humans and are reported to increase skin elasticity, hydration, and collagen density (Choi *et al.*, 2019). In a randomized, placebo-controlled trial on 60 healthy female volunteers, the collagen peptides with acerola extract, vitamin C and E, zinc and biotin were given as food supplements for 12 weeks. The confocal laser scanning microscopy, showed significant improvement in the collagen of the skin (Laing *et al.*, 2020). The peptide mass fingerprinting showed a high level of identity in the sequence of bovine and porcine collagen with human collagen (Buckley, 2016). Song *et al.* (2017) investigated the dose-dependent anti-aging effect of collagen peptides derived from bovine bone on a chronologically aged mouse model. The collagen peptides were prepared by alcalase and collagenase and were orally administered. The skin histology and the antioxidant indicators indicated the improvement in skin laxity; however, the moisture content was unaltered. The amount of collagen and ratio of type I to type III collagen progressively decrease with aging. Type I collagen forms bundles with more diameter than type III so a change in ratio decreases the load and tensile strength leading to aging. The collagen peptides decrease the space in dermis tissue and fibers become denser and more organized. The number of sebaceous glands increased and the ratio of type I to type III collagen was improved. The production of reactive oxygen species (ROS) increases in aging skin as it signals various signaling pathways leading to activation of transcription factor activator protein-1 (AP-1) (Callaghan *et al.*, 2008; Kammeyer and Luiten, 2015). The latter upregulate the degradation of collagen by matrix metalloprotease and downregulates the synthesis of collagen. Superoxide dismutase (SOD) and catalase (CAT) are antioxidant enzymes responsible for the inactivation of superoxide anions and hydrogen peroxide, respectively. The increase in activity of SOD and CAT on oral administration of collagen peptides to aging mice indicates the possibility of collagen peptides to decrease the ROS and hence more biosynthesis and less degradation of collagen. The collagen peptides prepared by Alcalase showed better results than those prepared by collagenase. So there is a potential for the use of bovine collagen peptides as dietary supplements against chronological aging (Zague *et al.*, 2011).

7.2.1.5.2.2 Immunomodulators

One of the strategies to control and alleviate many diseases is the modulation of immune system. A large number of *in vitro* as well as *in vivo* studies have been performed with bone collagen, which proves that it activates immune system and thus can be used among immunocompromised patients (Gao *et al.*, 2019; Si *et al.*, 2021). Gao *et al.* (2019) formulated five collagen hydrolysates of yak bone by using neutrase, trypsin, papain, pepsin, and flavoenzyme. It was observed that bone collagen after pepsin treatment produces a greater proportion of large molecular weight collagen hydrolysates (64.89% > 5 kDa). However, collagen hydrolysates produced by papain showed the largest quantity (>80.94%) of low molecular weight peptides (< 2000 Da). Out of these, hydrolysates prepared by treatment of papain showed efficient *in vitro* immunostimulatory activity as low molecular weight peptides revealed more bioaccumulation in contrast to other peptides. The proportion of the increase in splenic lymphocytes was found to be higher in hydrolysate generated by papain in contrast to the other hydrolysates. Moreover, it was noticed that hydrolysate obtained after the papain treatment showed no remarkable increment in the levels of splenic lymphocyte and reduction in nitric oxide generation upon the increment in hydrolysis time, which depicts that increase in hydrolysis time showed no significant effect on the release of immunomodulatory components. Thus, the papain hydrolysates from the bone of yak are used for the determination of immunostimulatory activity in cyclophosphamide-induced immunosuppressed mice. Treatment with cyclophosphamide markedly diminished the spleen and thymus index and caused the reduction in the serum levels of antibodies and cytokines like TNF-α, IL-1, and IL-6 (Figure 7.4). It hampered the percentage of natural killer cells and lymphocytes, however, administration of collagen peptide caused the enhancement in the concentration of cytokines, lymphocytes, and natural killer cells in the preventive model as compared to the treatment model. Similarly, Chen *et al.* (2021) showed that *Carapax trionycis* ultrafine powder

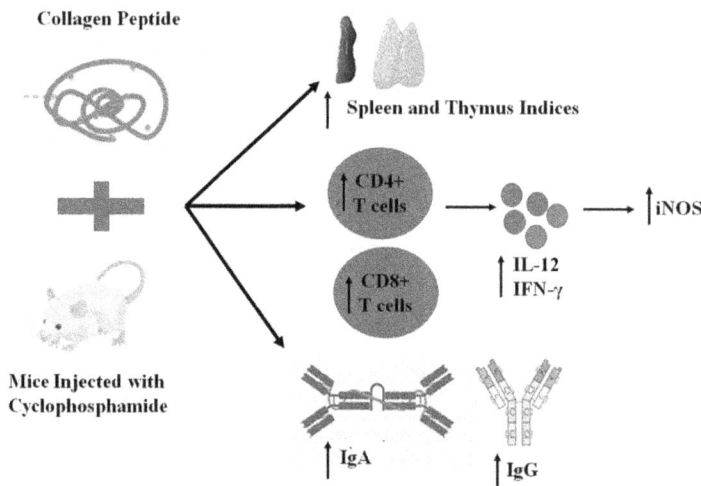

FIGURE 7.4   Immunomodulatory efficacy of collagen peptide.

administration can exhibit the immunomodulatory activity in mice immunosuppressed by cyclo-phosphamide. It was found that *Carapax trionycis* ultrafine powder treatment enhanced the humoral as well as cell-mediated immune response in mice as compared to the untreated mice. This powder form of *Carapax* causes an increment in the splenic, bone marrow, and thymus indexes and restores the architecture of the spleen. It leads to an increase in the percentage of lymphocytes and also enhances the microbicidal activity of macrophages. Moreover, immunosuppressed mice treated with *Carapax trionycis* ultrafine powder showed enhanced expression of proinflammatory cytokines in contrast to the control group. It was observed that administration of *Carapax* ultrafine powder in mice causes an increment in the antioxidant activity, which showed a protective effect on the metab-olism of calcium and restoration of bone mineral density. This study demonstrated that *Carapax trionycis* ultrafine powder can be an ideal immunomodulator in immunosuppressed mice and also might be used for the treatment of osteoporosis. Likewise, Si *et al.* (2021) also demonstrated the immunomodulatory action of bovine bone collagen peptides (300–900mg/kg body weight) in mice, which were immunosuppressed by intraperitoneal administration of dexamethasone. Mice were sacrificed on the 8th post-treatment day and spleen and thymus were taken to determine the splenic and thymus index, cytokine levels, and immunophenotyping of T cells. It was observed that levels of interleukin 2 (IL-2) and the percentage of CD4+ and CD8+ T cells were observed to be reduced in dexamethasone-treated mice. However, it was noticed that administration of collagen peptide in dexamethasone-treated mice showed higher levels of IL-12 and restored the percentage of CD4+ and CD8+ T cells and splenic and thymus index. It depicts that bovine bone collagen peptide has pro-tective potential in the recovery of steroidal immunosuppression.

7.2.1.5.2.3 Bone Health Supplements Oral administration of collagen subjected to the pro-teolytic breakdown in the intestinal region before the absorption, and collagen peptides displays direct action on target cells. Subsequently, peptides of collagen have been extensively employed for oral supplements for the enhancement of improvement in the cure of osteoporosis and have gained cumulative precise consideration. Several clinical researches depicted that collagen peptides can restore the functions of joints and recover manifestations of osteoporosis (Benitoruiz *et al.*, 2009). Ye *et al.* (2020) reported that collagen peptides from yak bone displayed potent osteogenic potential *in vitro* as well as in *in vivo* models. Collagen peptides were injected into ovariectomized rats for three months and then bioactive potential of yak bone collagen peptide was monitored in terms of microstructure and biomechanical characteristics of bone. It has been noticed that treatment with yak bone collagen peptide caused the marked change in the number of bone turnover markers in serum and hampered the destruction of microstructure and mechanical properties of bone induced due to oophorectomy. The bone turnover markers were observed to be normal in mice after treatment with yak bone collagen peptide, which depicts that yak bone collagen peptide increased the forma-tion of bone. This restorative action of yak bone collagen peptide might be due to the regulation of amino acids and lipid metabolism. It indicates that the bone collagen peptide of yak is a good replacement for health-enhancing functional foods and these peptides could be potential candidates for the cure of osteoporosis. In another study, Dressler *et al.* (2018) reported the improvement of chronic ankle instability in athletes upon intake of 5 g of collagen peptide-based supplements for six months. Similarly, Zdzieblik *et al.* (2017) also described the remarkable reduction in knee pain after the intake of bioactive collagen peptides for 12 weeks. Less pain was observed by athletes during activity after the administration of collagen peptide in contrast to control, which depicted the improvement in the structure and function of collagen within the knee. It might be due to the enhancement in the formation of type II collagen with decreased degradation of proteoglycan at the same time in comparison to the control group (McAlindon *et al.*, 2017). Zhu *et al.* (2020) described the potential of collagen peptides in enhancing the multiplication and differentiation of osteoblasts through the PI3K/Akt pathway. The levels of cell cycle and osteoblasts markers CDK-2, CDK-4, Cyclin B1, and Cyclin D1 were found to be upregulated after the administration of porcine bone collagen peptide. In addition, supplementation of collagen peptides leads to a reduction in the levels of caspase 3,Bax/Bcl-2,PTEN and thereby hamper cell death.

7.2.2 GELATIN AND GELATIN HYDROLYSATE

Gelatin originated from the term "gelatinous", which means firm, robust, and physically frozen. It is dry, condense, distasteful, edible, and transparent form. Chiefly, gelatin is a protein substance that is derived from collagen, a protein present in natural sources like bovine and porcine by-products. Over the years, demand of gelatin is increasing in the food and pharmaceutical sector. It is a widespread hydrocolloid used in bio-degradable packaging and food processing. It has common usage as a food additive in food processing industries, due to its number of functional values (Mariod and Adam, 2013). Mostly porcine and bovine skin and bones are favored for gelatin extraction, followed by several fish species and avian sources (Haug and Draget, 2011). The infections like bovine spongiform encephalopathy (BSE) have affected the rate of production of gelatin. It encourages the use of fish bones and skin to develop better alternatives to mammal-derived gelatin (Elgadir et al., 2013; Alfaro et al., 2015). It was even observed that saltwater fish gelatin showed more similarity of physical characteristics with mammalian derived gelatin (Haug and Draget, 2011). Karim and Bhat (2009) reported that gelatin extracted from fish has a low melting point and greater rate of dissolution, creating the superlative substitute for mammalian gelatin. Earlier studies indicated that a huge volume of fish waste is generated during fish processing and 30% of the waste was produced from fish skins and bone (Zakaria and Bakar, 2015;Arnesen and Gildberg, 2006).Various studies reported that the production of gelatin from the skin and bones of fishes is an efficient approach for the removal of waste and also avoids religious matters (Gudmundsson and Hafsteinsson, 1997; Choi and Regenstein, 2000).

7.2.2.1 Extraction of Gelatin

7.2.2.1.1 Conventional Method

The production process includes the preparation of tissue (scraping off the source tissues) succeeded by pre-treatment and extraction. Initially, the raw material is rinsed to scrape off contaminants. Before processing, animal bones are screened for quality and filled into chopping machines that trim the parts of bone into several little sections (approximately 12.7 cm in diameter). Bone parts are further washed under high-pressure water stream to remove any remaining debris, and immersed in warm water decrease the amount of fat by 2%. A conveyor belt shifts the de-greased bones to an industrial drier where these bones are flamed for 25–35 min at about 100°C. The de-greased bones are furthermore dipped in 4–7% hydrochloric acid for 48 h, this process is called demineralization, which leads to replacement of bone minerals like hydroxyapatite and calcium carbonate. The bone material is subjected to pre-treatment with acidic or alkaline solvent according to the collagen source (Haug and Draget, 2011; Alipal et al., 2020).

7.2.2.1.1.1 Pre-treatment This is the most widely adopted and conventional extraction method where mild to harsh acidic or alkaline treatment is used. Depending on the type of extraction processes, gelatin can be divided into type A and type B categories. The raw material is immersed in an acidic solvent in type A gelatin and in the alkaline solvent in B type gelatin. Mostly A type gelatin is extracted from porcine and fish bones and B type gelatin is isolated from beef and pig bones (Haugand Draget, 2011; Alipal et al., 2021).

7.2.2.1.1.1.1 ACIDIC PRE-TREATMENT The bone material is dipped in hydrochloric acid for 10–14 days and it can vary according to the thickness and size of the raw material. It washes off several minerals and bacteria and results in the discharge of collagen. The bone is further neutralized by rinsing in flowing water (Haug and Draget, 2011; Alipal et al., 2020). The extraction method can affect

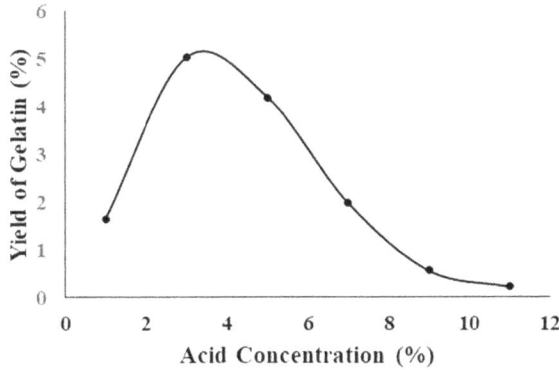

FIGURE 7.5 Yield of bone gelatin at the different concentration of hydrochloric acid.

Source: adapted from Panjaitan, 2017.

the polypeptide chains and functional characteristics of the gelatin. Optimum temperature plays an important role in gelatin production as it slows down the protein damage, which occurs at higher temperatures. Acid treatment change the triple-helical collagen fibers into single fibers; however, alkaline treatment merely produces double chains collagen fiber (Karimand and Bhat, 2009). In demineralization process, when different concentrations of hydrochloric acid {1-11% (v/v)} were used for the extraction of fish gelatin, the maximum yield of gelatin was extracted from 3% hydrochloric acid concentration with 9% moisture content (Figure 7.5), 8% ash, 0.3% fat, 80% protein content (Panjaitan, 2017). However, Arafah et al. (2008) employed 5–6% hydrochloric acid for the extraction of gelatin from the bone of snakehead fish. Other acids like phosphoric and organic acids can also be used for the extraction of gelatin. Yuliani et al. (2019) isolated gelatin from chicken bones by using 8–10% phosphoric acid. The extraction process carried out in a water bath at 55–75 °C for 4 h. On increasing phosphoric acid concentration and soaking time of chicken bone, an increase in ossein weight was depicted (Figure 7.6). An increment in ossein weight showed more interaction among the bone and phosphoric acid. It was observed that phosphoric acid after 12 h of soaking time provided 8.53%, 9.51%, and 8.62% of gelatin yield, respectively. Furthermore, 24 h exposure to phosphoric acid concentration, the gelatin was found to be 13.4%, 15.4%, and 14.6% respectively. It indicates that prolonging the soaking time of bones in phosphoric acid leads to an increment in the production of gelatin (Figure 7.7). In another study, the yield of bone gelatin was found to be 22.6% from the chicken feet after the exposure of phosphoric acid (Rahman and Jamalulail, 2012). Some investigations suggested that mild acidic solvents like citric acid or agro solvents should be used for bone pre-treatment as they are a harmless and naturally green biodegradable catalyst (Chemat et al., 2012; Mohamadpour and Feilizadeh, 2019). One of the maximum extraction products of fish-based gelatin derived from *Pangasius* catfish employed a mixture of mild acid solvent and water. Gelatin from *Pangasius* catfish bones showed identical physical properties with traditional gelatin (Mahmoodani et al., 2014a). Moreover, Maroid and Fadul (2013) reported that pre-treatment with citric acid generates gelatin of higher quality as citric acid showed negligible adverse effects up to 1200 mg/kg/d (Karlaganis, 2001). In another study, in the pre-treatment, crushed bone material was dipped into four types of citrus fruit extracts named citrus A, B, C, and D with bone: solvent ratio 1:5 (w/v). Pre-treatment was conducted at 37 °C at different times (24, 36, 48, 56 h) for demineralization. The main extraction was conducted for 5 h at various temperatures (45, 55, 65, 75 °C). This study indicates that fish gelatin was efficiently isolated by using citrus C and D in the pre-treatment phase for 48 h and main extraction using water at 55°C. The fish gelatin consists of 1.6-3.8g/100 g protein and 0.7-1.7g/100 g hydroxyproline. The yield and gel strength of gelatin were observed to be 6.2% and 451 g respectively. Moreover, the

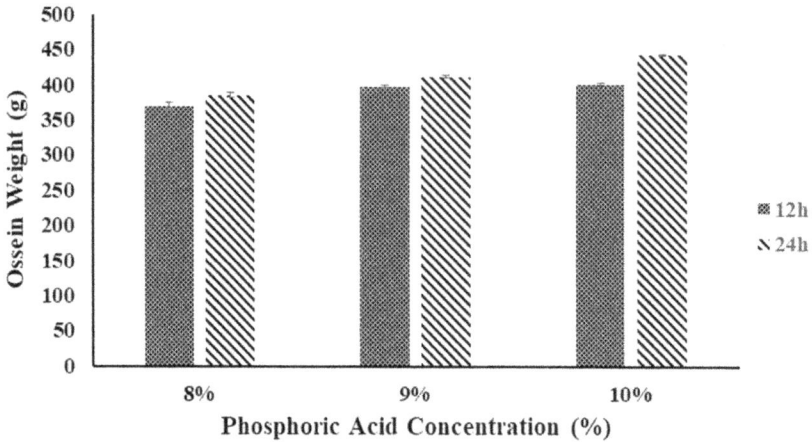

FIGURE 7.6 Effect of different concentration of phosphoric acid on Ossein weight of chicken bone.

Source: adapted from Yuliani *et al.*, 2019.

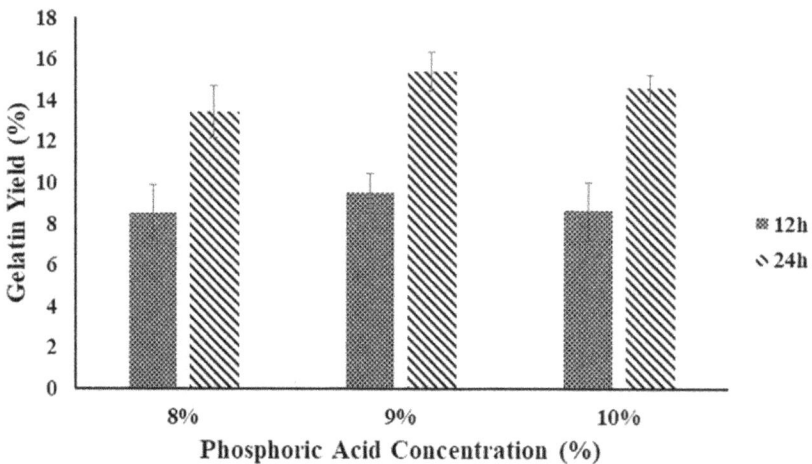

FIGURE 7.7 Effect of different concentration of phosphoric acid on the gelatin yield of chicken bone.

Source: adapted from Yuliani *et al.*, 2019.

hardness, gumminess, and chewiness of gelatin were observed to be 10.33 N, 9.81 N, and 14.32 N respectively. The proximate properties of gelatin are crude protein (58.47%), moisture (8.81%), fat (4.13%), and ash (1.12%) (Atma and Taufik, 2021). Likewise in another study, the optimal extraction environment to isolate fish bone gelatin from *Pangasius* catfish bones using pineapple liquid waste was 56 h pre-treatment and 5 h of the main extraction at a temperature 75°C. Pineapple liquid waste consists of about 0.18–0.32% citric acid. The gel electrophoresis revealed that Pangasius fish bone gelatin displayed a band at a molecular weight of 120 kDa, depicting the α-chain. Moreover, the yield and gel strength of gelatin from Pangasius fish bone isolated with pineapple waste were found to be 6.1% and 430 g respectively (Atma and Ramdhani, 2017). Likewise, Zhang *et al.* (2011) displayed the molecular weights of gelatin purified from the grass carp were observed to be 117 kDa (α1), 107 kDa (α2), and 200 kDa (β). The fish bone gelatin isolated from the king weakfish had a molecular weight ~100 kDa (α2) and ~110 kDa (α1). Gelatin with a greater amount of α-chain generally gives a higher yield in contrast to β-chain gelatin (da TrindadeAlfaro *et al.*, 2009). The fish bone gelatin isolated from the channel catfish revealed a molecular weight greater than 200 kDa, 100 kDa, and less than 97 kDa (Liu *et al.*, 2009). Lizardfish bones were observed to reveal sharp bands and have greater molecular weight portions at 100–120 kDa (Taheri *et al.*, 2009). The proximate attributes of gelatin extracted by using pineapple waste are crude protein 47.60%, moisture 8.6%, fat 7.7%, and ash 0.9% (Atma *et al.*, 2018). Thus, these green-based extraction solvents provide an ideal substitute for toxic chemicals used for gelatin extraction and also decrease fishery and agriculture waste concerning environmental sustainability and natural concerns. In another study pre-treatment of quail bone was performed by soaking into 0.1 M HCl and 0.1 M citric acid for one day at 7°C. After soaking, neutralization of the quail bone was done with tap water (Nik, 2015). The bone sample was transferred into a beaker and put in the water bath for 2 h at 75°C. Then the sample was sieved by Whatman filter paper 4 (Jamilah and Harvinder, 2002) and gelatin powder was taken to the freeze-drying process at -40°C. The yield of bone gelatin was observed to be 1.7% and 1.35% after the treatment of citric acid and HCl. In view of Tavakolipour (2011), organic acids like citric acid are found to be more efficient than HCl because strong acids stimulated denaturation to the collagen protein, which will affect the quality of gelatin. An earlier investigation by Liu *et al.* (2001) described that chicken gelatin isolated using HCl displayed the minimum yield in contrast to other acid treatments (Sarbon *et al.*, 2013). It can be verified by Schrieber and Gareis (2007) in which bone collagen dissolves in the aqueous phase during gelatin extraction and caused lower yield. Many factors like proximate composition, age of animals, collagen content, and method of extraction also can impact the proportion of gelatin (Songchotikunpan *et al.*, 2008).

7.2.2.1.1.1.2 Alkaline Pre-treatment   Several alkaline chemicals may be employed for the incubation of bone materials but saturated lime water is the most preferred and used for gelatin extraction. The sample is incubated in lime water at a temperature below 24°C with intermittent agitation for 20 days up to 6 months. Then the sample is rinsed with water to a neutral environment until the attainment of extraction pH (Haug and Draget, 2011; Alipal *et al.*, 2021). Response surface methodology of Erge and Zorba (2018) demonstrated that the optimum concentration of sodium hydroxide is 2.9–3.4%, temperature, and time ranging from 76 to 82°C and 105–183 min was required for the extraction of gelatin. Sometimes both acidic and alkaline solvents are employed to remove bacteria and minerals from the tissue. Gelatin isolation from the pangasius skin and bone was achieved by using an acid-base extraction process, where the skin and bones were washed constantly under water to scrap off extra materials and then subjected to 0.2% sodium hydroxide for 45 min. Bones were incubated in 0.2% sulfuric acid solution and then ethylenediaminetetraacetic acid (EDTA, pH 7.7) treatment was provided with intermittent shaking (Table 7.1). The bones were further incubated with acetic acid for 40 min. Samples were rinsed and extraction was completed in water at 55°C for 16 h. The filtrate was taken after filtration and evaporated for half an hour. The obtained gelatin was investigated depending on physical characteristics. The yield of bone gelatin from *Pangasius* was observed to be 5.2%. The ash content was remarkably greater in bone gelatin than moisture and fat content. It was found that the amount of bone gelatin was observed to be 73.4% in contrast to the content of commercial gelatin (92.4%) (Barman *et al.*, 2020).

  Poultry bone gelatin was isolated by using the method of Rahmanand Jamalulail (2012) with some modification. After defatting (100°C for 40 min), the poultry bones were dehydrated at 50°C for 17–18 h. Then these bones were water-logged in 4% hydrochloric acid at 27°C and submerged in sodium hydroxide for 20 days. The sample obtained after 20 days was rinsed seven times with distilled water and soaked for 24–48 h in distilled water. After pre-treatment, the extraction was conducted in distilled water at the temperature range from 40–60°C overnight without mixing. The resultant sample mixture was sieved with a Whatman filter paper in a funnel. The obtained filtrate

TABLE 7.1

Extraction of Bone Gelatin from Various Sources

Chemical/Enzymes Used	Source of Bone	Yield of Gelatin (%)	References
Papain	Buffalo	29.92	Samatra *et al.*, 2020
Pepsin +citric acid	Bovine	11.75	Cao *et al.*, 2019
Pepsin	Pig	13.5	Ma *et al.*, 2019
Papain	Alaska Pollock (*Theragrachalco gramma*) and Yellowfin Sole (*Limanda aspera*)	9.07	Mi *et al.*, 2019
NaOH	Chicken	4.2 (55°C) 2.2 (65°C) 2.23 (75°C)	Yuliani *et al.*, 2019
Alcalase	Sea bream	3.55 g/100 g	Akagündüz *et al.*, 2014
Citric acid from Citrus Fruits	*Pangasius sutchi*	6.26	Atma and Taufik, 2021
Citric acid	*Pangasius sutchi*	6.14	Pertiwi *et al.*, 2018
Pineapple liquid waste	Pangasius cat- fish	6.12	Atma and Ramdhani, 2017
Citric acid	Tuna fish	9.3	Hapsari *et al.*, 2017
HCl	Snakehead fish	3.2	Arafah *et al.*, 2008
HCl	Pangasius catfish	68.75 g/100 g	Mahmoodani *et al.*, 2014
NaOH	Grouper fish	15.00 g/100 g	Shakila *et al.*, 2012
HCl	*Clariasgariepinus*	17.52%	Sanaei *et al.*, 2013
NaOH	Red snapper	13.00 g/100 g	Shakila *et al.*, 2012
NaOH+H$_2$SO$_4$+Citric Acid	Tiger-toothed croaker	4.57%	Koli *et al.*, 2012
NaOH+H$_2$SO$_4$+Citric Acid	Pink perch	0.74 g/100 g	Koli *et al.*, 2012
NaOH	King weakfish	14.80 g/100 g	da Trindade Alfaro *et al.*, 2009
NaOH	Lizardfish	21.28 g/100 g	Taheri *et al.*, 2009

was dehydrated in a vacuum oven drier at 45°C for 18 h. The resultant dry matter was termed as "gelatin powder". The yield of gelatin was isolated from poultry bone at various temperatures (40–60°C) were 7.3, 7.49, 7.50, 7.52, and 8.03% respectively. The gelatin yield of poultry bone was found to be more at 60°C (Bichukale *et al.*, 2018) (Figure 7.8).

7.2.2.1.1.1.3 EXTRACTION OF GELATIN The bone pieces are boiled into huge aluminium extractors and the gelatin-containing solvent is collected by a tube running from the extractor. Further, sterilization of the solvent is achieved by flash-heating at 140 °C for four seconds. The solvent is passed through the filters to separate stains from the bones and tissues that are still glued together. The solid gelatin is separated from the solvent by an evaporator and the solid gelatin is pressed into leaves and moved through a shredder to a fine powder (Figure 7.9). On the basis of the use of gelatin in foods, sweeteners, flavoring agents, and colorants can be added at this stage (Alipal *et al.*, 2021). The pre-treated animal by-product is in the form of mixture between the immature liquid gelatin and insoluble indigenous collagen in the early phases of gelatin extraction. Pre-treated animal products were heated in water to temperatures above 45 ºC for the cleaving of the mixture into an optimum yield of gelatin. During pre-treatment hydrochloric acid or sodium hydroxide annihilate the non-covalent bonds in the collagen to cleave the proteinaceous architecture, resulting in appropriate swelling and solubilization. At this point, extra thermal treatment is required for the disruption of collagen and conversion into dissolvable gelatin (Gomez-Guillen *et al.*, 2011).

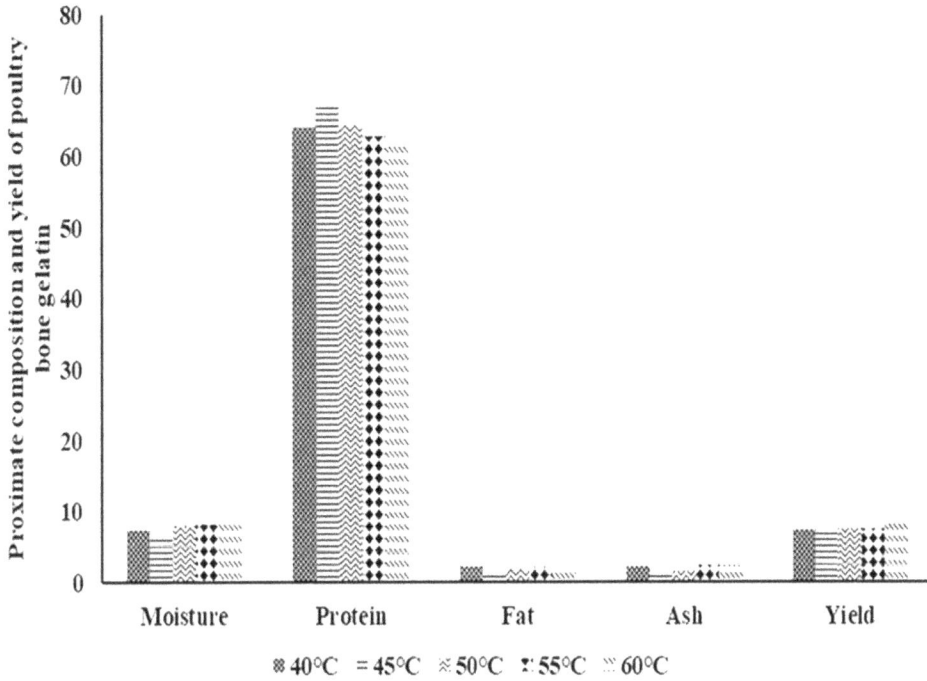

FIGURE 7.8 Effect of temperature on the proximate composition and yield of poultry bone gelatin.

Source: adapted from Bichukale *et al.*, 2018.

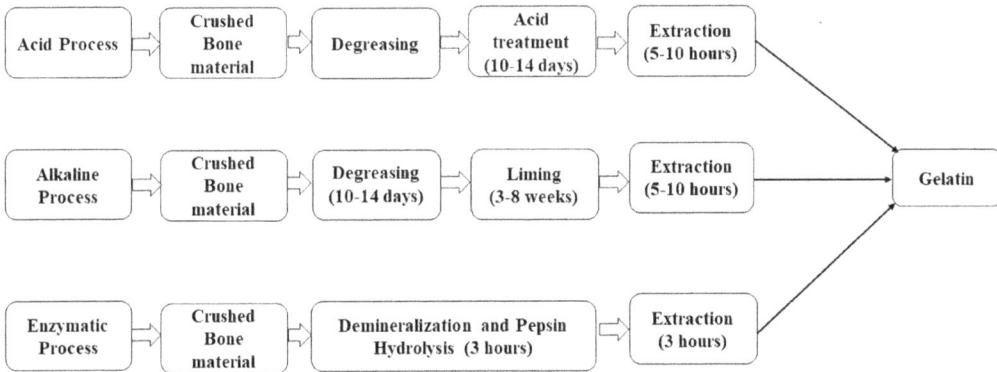

FIGURE 7.9 Various methods of extraction of gelatin.

7.2.2.1.2 Enzymatic Extraction

A lot of chemicals (acid and alkaline) are used conventionally in the extraction of gelatin from bone due to the complex structure of collagen molecules (Liu *et al.*, 2016). It produces a huge quantity of wastewater, which leads to enormous environmental anomalies. The conventional method is a prolonged process (20–60 days) for production of gelatin, thus restricting the manufacturing yield. Therefore, extraordinary attention has been given to detect the eco-friendly techniques for the isolation of gelatin from bone and decrease the effect on the environment (Dettmer *et al.*, 2013; Ma *et al.*, 2019). Bio-catalysis is gaining considerable interest to substitute the acid or alkaline method for the extraction of gelatin from animal by-products (Damrongsakkul *et al.*, 2008; Nalinanon *et al.*, 2008). Abedinia *et al.* (2017) described the significance of the enzymatic process like reduced duration for processing, more yield of gelatin, and low waste production. The enzymatic method reduces the hydrolysis time of collagen and produces a maximum yield of gelatin. Pepsin can break bonding among the peptides in the bone matrix during the demineralization, leading to greater efficacy in the pre-treatment and gelatin extraction (Table 7.1) (Ma *et al.*, 2019). The soaking of bone powder was done in HCl (1M) consisting of pepsin. Pepsin breaks down the cross-linking in the carboxy-terminal telopeptides of collagen (Nalinanon *et al.*, 2008), and will not cause the deterioration of gelatin. Pepsin averts the degeneration of collagen molecules and leads to low molecular weight gelatin chains. Then shaking of the mixture was done at 25°C–30°C for 3 h and centrifugation was performed. The obtained precipitate was washed by deionized water and purification of gelatin was done at 60°C for 3 h. The next stages of extraction were identical to those in the conventional method. The E gelatin (obtained by enzymatic method) from porcine bone reveals identical physico-chemical characteristics in contrast to type C gelatin (obtained by conventional method) except the isoelectric point (pI).The molecular weight of α1 chain in C gelatin and E gelatin was observed to be 130 and 120 KDa, respectively, and in the α2 chain was approximately 120 and 110 KDa, respectively. E gelatin showed more gel strength in contrast to the C gelatin. Type E gelatin was observed to reveal appropriate viscosity, which helps in the formulation of capsules, which depicts the significance of one-step biocatalysis in the production of bone gelatin (Zhang *et al.*, 2011; Ma *et al.*, 2019). Long alkaline treatment of raw bone material causes the deamidation of asparagine and glutamine (Ward and Courts, 1977), which results in the low pI value of type C gelatin. However, pepsin treatment of raw bone material revealed no adverse effect on glutamine and asparagine (Ma *et al.*, 2019). Zhang *et al.* (2011) reported that proteolytic enzymes at a pH of 7 and temperature of 25–40 °C for 1–12 h yield gelatin of superior quality with a gel strength of 172–219 g. Few enzymes have been employed for hydrolysis of gelatin-like flavourzyme, chymotrypsin, alcalase, trypsin, bromelain, pepsin, and papain (Himaya *et al.*, 2012; Choonpicharn *et al.*, 2015). Atma *et al.* (2018) demonstrated that the fish gelatin was superior in their biopotential when they were hydrolyzed by flavourzyme. Fish bones were soaked with a mild citric acid for 48 h in the pre-treatment. Then the ossein was partitioned from pre-treatment solvent by centrifugation at 4000 rpm for 20 min followed by the main extraction steps using warm water 75°C for 5 h. After this, the extract was filtered by using filter paper and preserved at 4°C. Then the resultant fish bone gelatin was incubated at 50°C for 10 min. Consequently, 6% flavourzyme was added and incubated for 4, 6, and 8 h. Then hydrolysis enzymatic extraction was completed through hot water at 100°C for 10 min followed by soaking in chilled water for 20 min. Finally, hydrolyzed gelatin was procured through centrifugation at 1000 rpm at the temperature of 4 °C for 10 min. The obtained supernatant is known as fish bone gelatin and preserved at -18°C. It was observed that extracted gelatin from the fish bone of Pangasius catfish contained hydroxyproline from 18.91 to 63.81mg/mL. It was observed that fish bone gelatin obtained by treating 6% flavourzyme for 4 h contained more hydroxyproline in contrast to other fish bone gelatin. Further, Cao *et al.* (2020) studied the effects of various acids and pepsin on the characteristics of gelatin isolated from bovine bone. Pre-treatment of various acids like HCl, acetic acid, and citric acid with and without pepsin was performed on a bovine bone at 70 °C for 7 h to isolate gelatin. Bovine bone incubated with pepsin revealed significantly more yield in comparison to the bone material without pepsin. SEM revealed that citric acid disarrayed the maximum of the collagen architecture in comparison to the acetic acid and HCl. Raman spectrum spectroscopy showed that all isolated gelatin had identical secondary structures. Rheological studies depicted that gelatin isolated after treatment of citric acid displayed the maximum gelling, elastic modulus, and melting temperature. The citric acid and pepsin in combination showed more yield of bone gelatin with higher integrity (Figure 7.10). This study depicts that citric acid is found to be more efficient for the isolation of bovine bone gelatin as compared to hydrochloric acid or acetic acid. Abedinia *et al.* (2017) concluded that pepsin can enhance the production of gelatin on

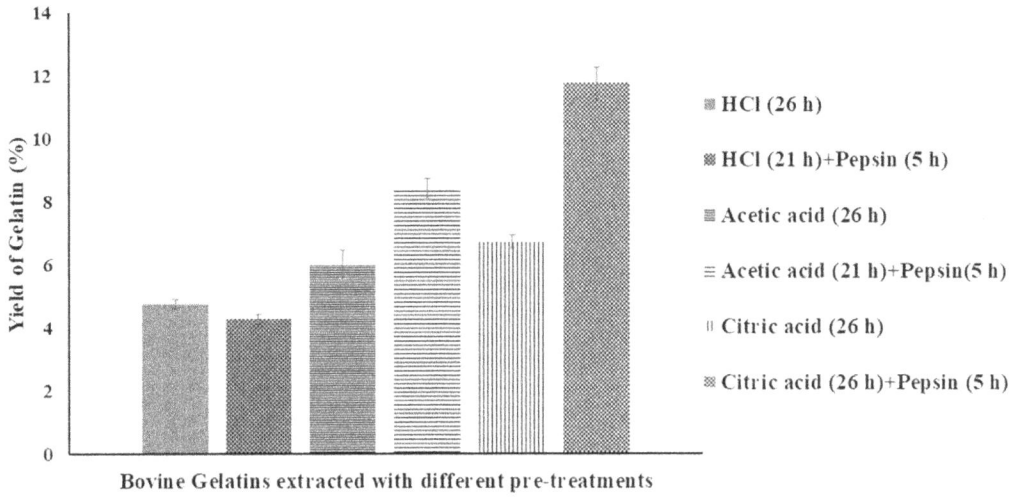

FIGURE 7.10 Effect of different pretreatment on the yield of gelatin.

Source: adapted from Cao *et al.*, 2020.

poultry sources, however, SDS-PAGE depicted more intensity of bands for α- and β-constituents. Low molecular weight peptides were isolated due to the enzymatic cleavage during gelatin extraction, which impacts the rheological characteristics of gelatin. In many studies, it is suggested that enzymes can assist in designing of bioactive gelatin peptides as an antioxidant agent (Ketnawa *et al.*, 2016). This bioactivity of gelatin is due to the existence of free amino groups in its structure, which form a firm macromolecule (Suderman and Sarbon, 2020). Gelatin cleaves into low molecular weight molecules through hydrolysis, which is named "gelatin hydrolysate". Most commonly, enzymatic proteolysis is used for the hydrolysis of gelatin. Several commercial proteases are used for production of these hydrolysates and peptides (Subara and Jaswir, 2021). The specific protease determines the concentration and composition of amino acids, which also affects the biological potential of the hydrolysates. It was found that microbial-derived proteinases are economic and safe, and these proteinases lead to more product yield (Wu *et al.*, 2003). Gelatin hydrolysates contain a greater number of amino acids and revealed potent bioactivities like microbicidal, antioxidant, and antihypertensive (Subara and Jaswir, 2021). Alcalase is a commercial microbial protease that has been employed in various researches due to its greater specificity and hydrolysis and outstanding end-product recovery and functional characteristics (Kristinsson and Rasco, 2000). The bone gelatin is sustained in water at 55 °C (pH 9) and hydrolyzed by the addition of alcalase and incubated for 6–7 h in a shaking water-bath incubator (Yang *et al.*, 2008). Then the inactivation of the enzyme was done at 90 °C for 10 min. The pH of the gelatin hydrolysate was set to neutral after cooling. Then different hydrolysates were produced by ultrafiltration with the required molecular weight. The benefit of ultrafiltration is that the molecular weight of a particular hydrolysate can be precised by embracing a suitable ultrafiltration membrane (Mahmoodani *et al.*, 2014a; Cao *et al.*, 2020).

7.2.2.2 Applications of Gelatin

Bone gelatin is a natural product that plays a significant role in the pharmaceutical and the food industry (Neffe *et al.*, 2011; Fan and Shen, 2016; Shi *et al.*, 2017).

7.2.2.2.1 Food Industry

Gelatin is a major constituent in the food industry in today's cookery due to its gelling abilities. It is used in desserts as a texturing, foaming, and stabilizing agent (Figure 7.11) (Calvarro *et al.*, 2016). In

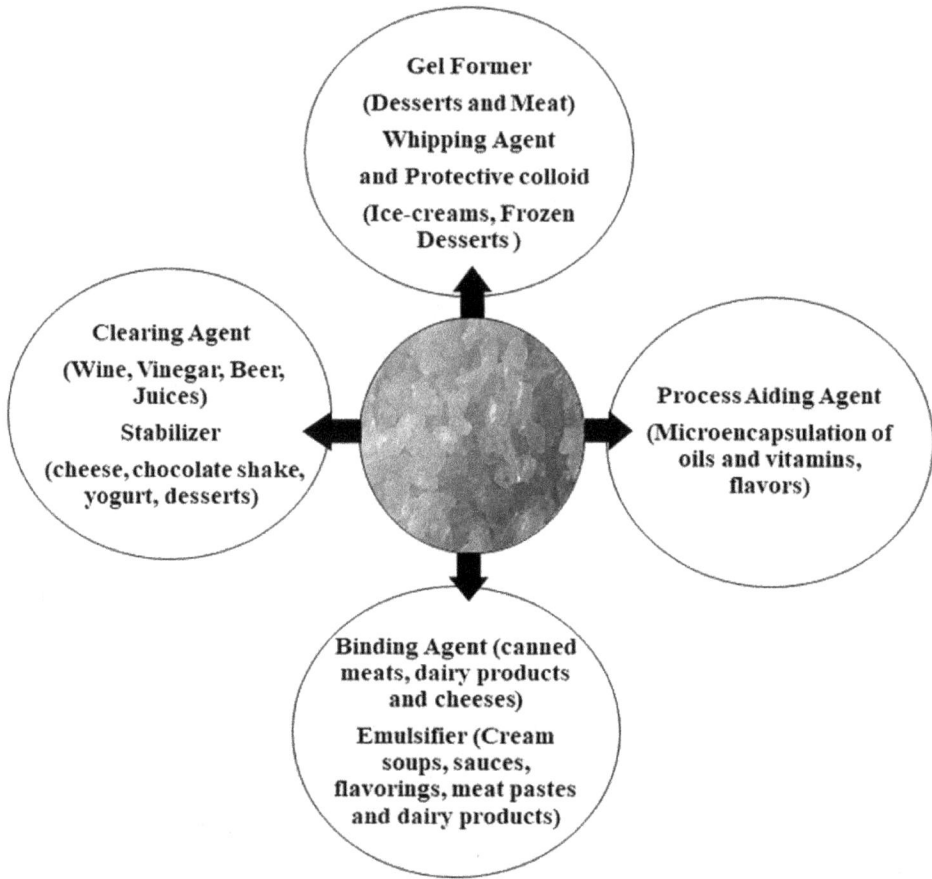

FIGURE 7.11 Functional role of gelatin in foods.

Source: adapted from Misbahul *et al.*, 2021.

canned meat products like sausages, knockwursts, gelatin withholds the loss of juices and provides an efficient heat exchange medium during cooking. Gelatin is also incorporated in preparation of ice cream, fruit salads, foam, and cottage cheese (Mariod and Adam, 2013). Gelatin possesses good film-forming ability, therefore, has the potential to be used as coating and edible films in food products (Ramos *et al.*, 2016). It is also used as a clearing agent in the wine and beer industry. Gelatin is also utilized as a food additive, that serves for the augmentation of muscle precursors of keratin and as a flavor enhancer with a low titer of fat (Siburian *et al.*, 2020). Gelatin films also act as an antimicrobial agent against microorganisms, and thus can be used for food packaging (Tymczewska *et al.*, 2021). Ataie *et al.* (2019) revealed the potential of fish bone gelatin from carp in producing low-fat mayonnaise and also recommended the applications of gelatin in developing low-fat emulsion (oil in water). Yang *et al.* (2019) reported that gelatin hydrolysate obtained from the bones of *skipjack tuna* displayed potent antioxidant properties and can be used as an active component in foods. In another study, Mahmoodani *et al.* (2014b) demonstrated the ACE (angiotensin-converting enzyme) inhibitory potential of bone and skin gelatin hydrolysate of catfish. It was observed that bioactive peptides of skin and bone gelatin with molecular weight lower than 1 kDa displayed the maximum inhibitory action and revealed more obstruction in the route of intestinal proteinases. This study concluded that gelatin hydrolysates can be employed as health-promoting constituents in foods to combat hypertension. Similarly, Cao *et al.* (2020) reported that bovine gelatin hydrolysate of bone contained 13 active peptides, which displayed potent inhibition of ACE *in vitro* as well as in animal models. It indicated the efficacy of peptides of bovine gelatin hydrolysate in the development of nutraceuticals and medicines to efficiently cure high blood pressure.

7.2.2.2.2 Pharmaceutical Industry

Gelatin is recommended for making hard and soft gelatin capsules, usually used for powders and liquids. Fish gelatin soft capsules are most commonly taken as nutritional supplements (Kathpalia *et al.*, 2014; Prasad, 2017). Mammalian and fish gelatin both have huge usage in tablet coating. Sodium acetate, sodium sulfate, sodium stearoyl lactylate, propylene glycolmonostearate, and sodium lactate along with fish gelatin are used for the coating of tablets. This coating formula causes the plane and soft finishing, which enhanced the absorption of the tablets. Moreover, it does not require any specific equipment in contrast to the other coating materials, which needed sophisticated instrumentation (Daher *et al.*, 2000).

7.2.2.2.3 Drug Delivery

Gelatin is favored for drug delivery as it is non-toxic and biodegradable (Murray *et al.*, 2015). It is used in hydrogel, nanogels, pharmaceutical additives, and implants (Gomez-Guillen *et al.*, 2011; Ma *et al.*, 2017; Kumosa *et al.*, 2018). Moreover, intake of gelatin orally can enhance bone health and joints. It plays a crucial role in medical field like hemostasis (hampering blood loss). Gelatin helps in formulation of clots at the bleeding position (Liu *et al.*, 2015). Zeng *et al* (2015) designed gelatin microcryogels that help in the migration of cells into deep layers of tissue and enhance healing of wounds. It was also observed that gelatin can hinder the comparative partition of fat and liquid in the syrup (da Trindade Alfaro *et al.*, 2015). Moreover, gelatin has an orthotropic function, therefore helps in repair of tissues, which can boost bone marrow density in osteoporosis patients (Lv *et al.*, 2019; Blakely and Johnson, 2020). Wang *et al.* (2020) demonstrated the ability of peptides of bovine gelatin hydrolysate to enhance the multiplication and differentiation of osteoblasts and consequently were found to be valuable for the secretion of extracellular matrix and remedial of fractures.

7.2.3 MEAT AND BONE MEAL

Meat and bone meal is a good source of metabolizing energy, protein, calcium, phosphorus, and minerals. The proteins present in bone contains more collagen, and hence persist low digestibility of amino acids (Garcia and Phillips, 2008). Moreover, the quality of protein and chemical composition of meat and bone meal depends upon the processing method used. Hendriks *et al.* (2002) described the variability of lysine and sulfur amino acids digestibility ranging between 45.8–89.0 and 38.2–85.5%, respectively. This variability affects the performance of the animals who depend upon the meat and bone meal for the source of amino acids. Over formulation of a diet with respect to digestible amino acid content leads to inefficient production in poultry and porcine however in the animals fed on underformulation diets leads to sub-optimum protein to energy ratio, which results in deposition of fats. The appropriate diet formulation for production animals, and the ability of the ingredients to be digestible is important. The estimation of metabolizable energy of meat and bone meal can be calculated by using ash concentration and *in vitro* disappearance of dry matter (Choi *et al.*, 2021). Protein in bone fractions consist of a greater concentration of collagen in comparison to the soft tissues, which leads to the low digestibility of amino acids. Muir *et al.* (2013) demonstrated that the pre-treatment of MBM activates the bioactive compounds and makes the MBM superior for animal growth. Pepsin is an enzyme that is secreted by the stomach and hydrolyzes the proteins to make it absorbable. Thus, for an edible meat and bone meal, the pepsin indigestible residue and crude proteins should not be more than 12% and 9% respectively. According to FDA regulation, it can be used in the preparation of feed for livestock, poultry, and aqua culture (Ockerman and Basu, 2010).

7.2.3.1 Application of Meat and Bone Meal

7.2.3.1.1 Aquafeed

Meat and bone meal are used in aquafeed as a source of protein and essential amino acids for several fish species like Nile tilapia, rainbow trout, and African catfish (Sugiura *et al.*, 2000; Goda *et al.*, 2007). The aquafeeds, which are based on plant proteins, are poor in bioavailability of phosphorus as most of it is present as phytate phosphorus and fish have low activity of phytase (Zhou *et al.*, 2004; Goda *et al.*, 2007). Thus MBM can be added to the plant-based feed to meet the requirement of phosphorus. Suloma *et al.* (2013) observed the remarkable gain in weight, specific growth rate, feed conversion ratio, whole-body phosphorus concentration, retention of protein and phosphorus in Neal Tilapia when fed on a MBM supplemented plant-based diet.

7.2.3.1.2 Human Food Fortification

Canned tuna flesh commonly consumed by people in the eastern part of Indonesia has low calcium content, therefore its fortification with tuna meat and bone meal not only enriches the nutritional value but also improves the organoleptic quality of taste and texture (Tangke *et al.*, 2021).

7.2.4 MEAT AND BONE PASTE

For the preparation of meat and bone paste, the bones are crushed to the size of 50–70 mm. Then the bone materials are pre-frozen at a temperature of {(-18) °C - (-20) °C} in freezers for 1 h. The frozen bone materials are again crushed with the grinder and cool to a temperature of -20°C. Further, again crushing of the bone material is done with a grinder (hole diameter 5 mm). The obtained meat and bone mass is mixed in cold water in the proportion of bone mass (1): water (0.5) and grinded again with a grinder (hole diameter 3 mm). Then the cooling of meat and bone is done at 0 °C. Further, meat and bone mixture is grinded on a micro-grinder with the succeeding slits among crushing rings: 0.25 mm, 0.10mm, 0.02 mm, and the obtained meat and bone paste is stored at 4 °C (Figure 7.12) (Kakimov *et al.*, 2017). The finely dispersed paste prepared from bone raw material is rich in minerals particularly calcium, magnesium, iron, and can be used as an additive in semi-finished products like pelmeni, cutlets, meatballs, and enriching their nutritional value (Kakimov *et al.*, 2017). There is a significant level of calcium in meat and bone paste from cattle and poultry in

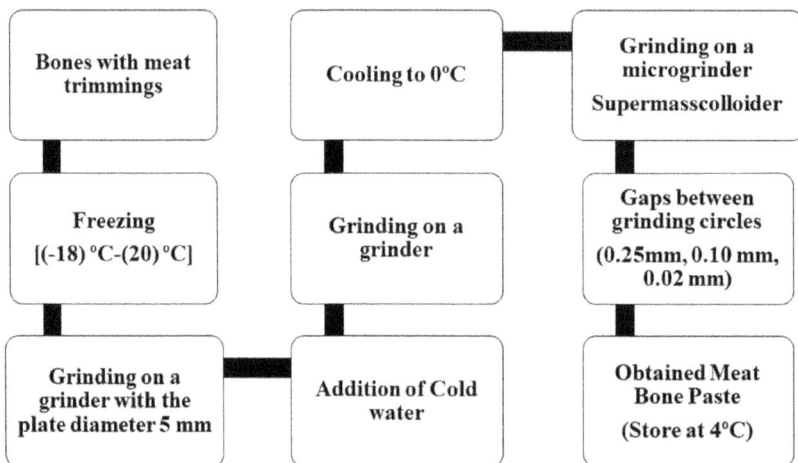

FIGURE 7.12 Preparation of meat and bone paste.

Source: adapted from Kakimov *et al.*, **2021.**

comparison to the liver (cattle liver: 5 mg/100 g; chicken liver:15 mg/100 g) excluding egg shells. As chicken bones are pneumatic, so the meat and bone paste from poultry contain 1,654.02 mg/100 g calcium while that from the cattle is significantly more i.e. 5,318.13 mg/100 g. Analysis of the mineral composition of the meat and bone paste from poultry and cattle bones indicates the considerable calcium content. In addition to it, poultry meat and bone paste contain 14.54 mg/100 g magnesium, however bovine meat and bone paste contain a notable amount of magnesium i.e. 207.62 mg/100 g (Volik et al., 2017; Kakimov et al., 2021). The fat-derived from poultry and cattle meat and bone paste has a higher content of lecithin making it a good emulsifier. The bone fat was detected to be 11.8% and 5.3% in the poultry and cattle meat bone paste respectively (Orehov, 2013).

7.2.5   Bone Powder

Bone powder has a good nutritional value and contains 20% calcium and 10% phosphorus, which make it a superlative edible resource. The bones were chopped into 1 cm pieces and then dipped in ethyl acetate for degreasing. The incubation of bone pieces was done in an incubator at 65 °C for 11–12 h till the moisture content was observed to be lower than 4%. The dried bone pieces were crudely milled (<250 μm) into powder form through a grinder and again milled bone powder was crushed into bone powder by using a planetary ball mill (Qin et al., 2020). Hemung 2013 reported the preparation of bone powder from tilapia by using autoclave processing. First, the bone was soaked in a warm alkaline solution (sodium hydroxide, bone 1: 2 solution) for 1 h. After washing with distilled water, bone samples were autoclaved for 1 h and dried in an incubator at 105 °C. The dehydrated bones were crushed into powder form and filtered through a 38 μm mesh. Then the bone powder was filled into tanks with a rotation speed of 400 rpm and pulverized for different periods with a 20 min gap every single hour to decrease the impact of heat. It was observed that the digestibility of calcium and proteins were detected to be more in bone powder prepared by autoclave processing than the only heat treatment (Nawaz et al., 2020).

7.2.5.1   Applications of Bone Powder

7.2.5.1.1   Food Industry

The bone powder is a natural reservoir of calcium source and can be used as a calcium supplement. The use of bones for the supplementation of calcium is a better approach to augment the value to the wasted animal by-products. Fish bones consist of more calcium and phosphorus in the form of hydroxyapatite along with collagen proteins and other minerals. Particularly in lactating mothers, tuna bone powder is recommended as a substitute for calcium supplement due to its improved absorbability in comparison to the calcium carbonate and its potential to thwart maternal bone damage by endorsing the formation of bone and decreasing the resorption of bone (Suntornsaratoon et al., 2018). Addition of bone powder from milk fish enhanced the quantity of calcium in the egg rolls. It was found that when 3% bone powder from milk fish is added to the eggroll, calcium-rich food standard (360 mg/100 g) possibly will be achieved (Wang, 2020).

7.2.5.1.2   Anti-osteoporosis Activity of Bone Powder

Qin et al. (2020) reported that yak bone powder significantly reduced osteoporosis in rats induced by ovariectomy. Bone markers such as Osteocalcin (OCN), C-terminated telopeptide of type I collagen (S-CTX), estradiol bone-specific (ES), and alkaline phosphatase (B-ALP) level increased after supplementation of yak bone powder, which results in the improvement of bone mineralization. It was observed that osteocalcin stimulated the differentiation in osteoblasts and higher levels of alkaline phosphatase are good inducers for the accumulation of calcium in the extracellular matrix. In addition, S-CTX assists in the formation of bone collagen. Numerous studies suggested that the decline in estradiol levels is the prime cause of the common occurrence of osteoporosis among premature and menopause women. Therefore, enhancement in the concentration of ES in osteoporotic rats depicts that bone powder has a remarkable effect on the alleviation of osteoporosis caused by post-menopause (Daniel *et al.*, 2018; He *et al.*, 2019; Qin *et al.*, 2020). Bone powder isolated from tuna showed a significant reduction in osteoporosis in mice stimulated due to glucocorticoids. Bone powder revealed encouraging anti-osteoporosis potential substantiated by its Wnt/β-catenin pathway stimulating activity, which increases bone formation and NF-κB suppressing activity that leads to the reduction in the bone resorption. In addition, supplementation of bone powder decreases the secretion of proinflammatory cytokines, restores the intestinal epithelial barrier affliction, and hampers the intensified systemic inflammation. Moreover, gene sequencing of 16S rRNA showed that supplementation of bone powder increases the richness of anti-inflammatory microorganisms and short-chain fatty acid producers. This study depicts the potential of tuna bone powder as a natural functional food to reduce osteoporosis (Li *et al.*, 2020).

7.2.6 HYDROXYAPATITE

Bones are constituted of 70% inorganic and 30% organic compounds. Hydroxyapatite (HA) is the major inorganic component existing as calcium apatite with chemical formulas $Ca_5(PO_4)_3(OH)$. Human bone is composed of HA up to 70% by weight and 50% by volume. By-products extracted from animal bones have high phosphorus (P) content 108.3 mg/g in bovine bone on dry weight basis (Beighle *et al.*, 1994). Hydroxyapatite extracted from bone waste can be used as an alternative source of various phosphorus raw materials. This not only manages the hazardous waste from the meat industry but also helps in economic benefits through material recovery from meat waste. Kowalski *et al.* (2021) reported the production of 1.0 t of crystalline HA from 3.2 tonnes of animal bone waste with a high potential economic value of 71,118 t/yHA, which amounts to 11.165 million USD.

7.2.6.1 Extraction of HA

The extraction of HA is reported from different types of animal bones including cow (Hubaidillah *et al.*, 2020), chicken, goat, pigs (Supova *et al.*, 2011; Harahap *et al.*, 2017; Ramesh *et al.*, 2018; Kumar *et al.*, 2021) and fish bones (Mondal *et al.*, 2012). A high purity HA ash with an average of 16% phosphorus can be produced by thermal treatment of animal bone, which is comparable up to 17.2% extracted from phosphate rock (Staro *et al.*, 2016). The production of HA from bone waste by thermal treatment can be a cleaner technology model of optimum utilization of waste of meat industry with energy recovery and less pollution (Kowalski and Krupauczek, 2007).When the pretreated bones are subjected to hydrolysis, protein hydrolysates and bone sludge is produced. The bone sludge produced from pig and bovine bones has an average of 14% phosphorus. Hydroxyapatite ash obtained by calcining of bone sludge in a rotary kiln with electric heating, in an air atmosphere, for 3 h at 600 °C to 950 °C with a temperature interval of 50°C. The excess air ensures the complete combustion of an organic part. HA is milled and checked for homogenous physical and chemical properties and is free from contaminants of heavy metals (Figure 7.13). The reuse of HA dust produced in calcining process ensures optimum use of bone waste for final HA production (Kowalski *et al.*, 2021).

7.2.6.2 Uses of HA

It is bioactive compound as it can form direct chemical bonds with living tissue and is biocompatible with no inflammatory and immunogenic response (Fathi *et al.*, 2008; Hu *et al.*, 2017). HA has chemical and structural similarities to bone minerals making it a good option for bone substitute. It can also be used for the production of food-grade phosphoric acid and mono and di-calcium feed phosphates (Kowalski and Krupauczek, 2007) and has the potential to be used in edible food coatings.

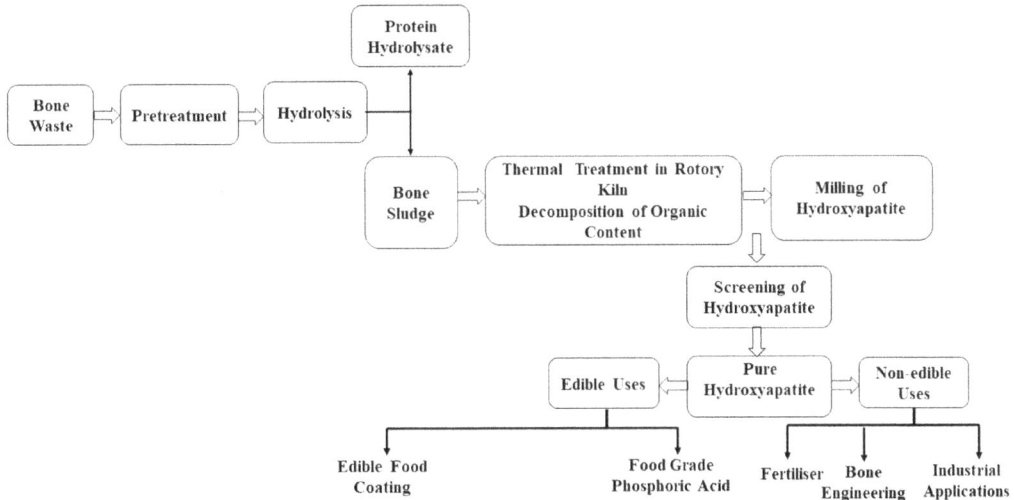

FIGURE 7.13 Extraction of hydroxyapatite from animal bone.

Source: adapted from Kowalski *et al.*, 2021.

7.2.6.2.1 Edible coating

HA nanoparticles can interact with different organic molecules, therefore can be used as a carrier for the delivery of bioactive compounds and development of edible food coatings (Fulgione *et al.*, 2019). Edible food coatings have an important role in the production of antimicrobial compounds, which prevent the deterioration of food products from contaminating microorganisms and extending the shelf life of food. *Pseudomonas spp.* are the major contaminants of fresh and processed meat products by producing proteolytic and lipolytic enzymes affecting the quality and shelf life of food. Malvano *et al.* (2021) reported in an *in-vitro* study, the fast and homogenous controlled release of antioxidant flavonoid quercetin from Quercetin-HA complex, reaching equilibrium in a day. The coating has the potential to inhibit the growth of *Pseudomonas fluorescens* leading to the increased shelf life of meat and fish products.

7.2.6.2.2 Production of Food Grade Phosphoric Acid

The thermally treated bone sludge has 16% phosphorus content and can be used as a source of food-grade phosphoric acid (PA). PA is obtained by treating HA with concentrated phosphoric acid. However, Krupazuczek *et al.* (2008) proposed a cost-effective production of phosphoric acid from incinerated bone waste in a two-step process. Firstly, the HA is treated with phosphoric acid and monocalcium phosphate, which at the second stage is treated with concentrated sulphuric acid to generate calcium sulfate and phosphoric acid. The food-grade PA is used as a preservative and flavor agent in non-alcoholic beverages (Kostik, 2014). PA increases the potential of bacteriocins like nisin in controlling the growth of pathogenic bacteria in vacuum-packed sausages (Barros, 2010). The salts of phosphoric acid increase yield and improve the sensory properties of meat products (Long *et al.*, 2011).

7.2.6.2.3 Animal Feed

The inorganic sources of phosphorus are quickly depleting and so are expensive. Moreover, exogenous phytase is used to improve the utilization of phosphorus in animal feed (Romano and Kumar, 2018). The excreted phosphorus also leads to environmental pollution. Nanoparticles of calcium phosphate, dicalcium phosphate (DCP), and hydroxyapatite are explored to increase the availability of the minerals and low environmental impact (Patra and Lalhriatpuii, 2019). Sobhi *et al.* (2020) evaluated the effect of Nano HA as an alternative foran inorganic source of calcium and phosphorus on broiler chicken and reported that dietary supplementation of 0.5% of DCP along with 0.1% Nano HA or 0.1% alone Nano HA decreased the excretion of phosphorus in urine without affecting the carcass traits and performance of the animal.

7.2.7 BONE OIL

The bone waste has a good amount of fats. Li *et al.* (2021) reported the highest fat content in porcine bone followed by bovine bone and low-fat content in chicken and rabbit bone. Bone oil extracted from rabbit bones have long chain n-6 and n-3 polyunsaturated fatty acids (PUFA) while porcine bone has high content of long chain n-6 PUFA, which is more significant than the n-6/n-3 ratio. The lipid fraction obtained through thermal hydrolysis of bone raw material can be refined into commercial fats, which according to its potential has inedible uses (Mezenova *et al.*, 2021). Various studies reported that porcine and bovine bone oil is used for the huge production of bio-diesel (Hussain *et al.*, 2021). However, deer bone oil is widely used in traditional Chinese medicines for the cure of various diseases and as a tonic for increasing testosterone and growth factors in the body by brand name pantocrine. The deer farming industry for meat is also one of the fastest growing industries estimated with the annual generation of USD 8 billion as deer venison is a lean meat and is of rising demand (Anderson *et al.*, 2017; Serrano *et al.*, 2019). *In vitro* studies by Choi *et al.* (2016) reflected the anti-inflammatory potential of methanol fraction of bone oil extracted from deer bones as it controls the production of proinflammatory cytokines and its mediators. It was observed that methanolic fraction of bone oil treatment induced the reduction in the production of Interleukin (IL)-1β, IL-12β, and COX-2 in RAW264.7 macrophages cells, which are induced by the lipopolysaccharides (LPS). The expression of iNOS was found to be downregulated in cells after treatment with methanolic fraction of bone oil in contrast to the control LPS treated group where it was found to be upregulated. Moreover, supplementation of bone oil also showed reduction in the levels of nitric oxide in lipopolysaccharides treated RAW264.7 macrophages. Further in another study, Lee *et al.* (2014) demonstrated the anti-osteoarthritic potential of deer bone oil extract against osteoporotic rats induced with monosodium iodoacetate. The proportion of IL-1β, tumor necrosis factor (TNF)-α, and IL-6 was found to be reduced by 40, 70, and 50% after treatment with deer bone oil extract in osteoporotic rats. Treatment with deer bone oil showed significant increment in the bone trabecular thickness in osteoporotic rats as compared to the monosodium iodoacetate treated control group. However micro-computed tomography (CT) analysis displayed the remarkable reduction in the trabecular separation in the impaired articular bones of monosodium iodoacetate treated rats. Further liquid chromatography quadrupole ion-trap electrospray ionization mass spectrometer revealed the presence of glycosphingolipids with mono-sialic acid in deer bone oil. This study suggested that anti-osteoporotic potential of deer bone oil might be attributed to the presence of gangliosides and bone oil can be the ideal agent against osteoporosis.

7.3 CONCLUSION

Food security as well as the availability of safe and nutritious food to the growing population through sustainable resources and this too with the least effect to the environment is a big challenge. As the meat industry is one of the important rising food industries, to make it more sustainable there is a need to reduce the waste and to recycle its by-products through value addition into inedible or edible forms. Bone is the major waste generated during slaughtering of the animal and is converted to high value edible products. These products are not only used as animal feed with advantages of contributing to animal performance but also used by humans as fortified and functional foods. These products provide adequate nutrition but also enhance taste, texture, and flavor of food. The bioactive compounds extracted from bone have strong potential in the improvement of the immune response and prevent various diseases related with bone and cardiovascular system. However, the safety concerns, technology imperfections and awareness of consumers are to be addressed for a circular economy in the meat industry.

7.4 FUTURE PERSPECTIVES

The research field addressing the comprehensive and high value utilization of edible by-products from bone is quite promising. The continued understanding and utilization of bone as a raw material in novel food products can lead to a sustainable food system. Hence, there is a need for continuous attention and development of innovative environmentally friendly sustainable technologies for efficient extraction of the bioactive compounds from bone and their utilization in functional foods and in the food processing and packaging industry. The further experimentation of the extracted bioactive compounds on animal models and standard clinical trials are needed to be in place to know the mechanism of their action, bioavailability and for addressing safety concerns.

REFERENCES

Abedinia, A., Ariffin, F., Huda, N., and A.M. Nafchi. 2017. Extraction and characterization of gelatin from the feet of Pekin duck (*Anasplatyrhynchos domestica*) as affected by acid, alkaline, and enzyme pretreatment. *International Journal of Biological Macromolecules* 98:586–594.

Akagündüz, Y., Mosquera, M., Giménez, B., Alemán, A., Montero, P. and M.C. Gómez-Guillén. 2014. Sea bream bones and scales as a source of gelatin and ACE inhibitory peptides. LWT-Food *Science and Technology* 55(2):579–585.

Alfaro A.T., Balbinot E., Weber C.I., Tonial I.B., and A. Machado-Lunkes. 2015. Fish gelatin: Characteristics, functional properties, applications and future potentials. *Food Engineering Revolution* 7(1):33–44.

Alipal, J., Pu'ad, N.M., Lee, T.C., *et al.* 2021. A review of gelatin: Properties, sources, process, applications, and commercialisation. *Materials Today: Proceedings*.

Almeida, P.F., Araújo, M.G.O. and J.C.C. Santana. 2012. Collagen extraction from chicken feet for jelly production. *Acta Scientiarum Technology* 34(3): 345–351.

Anderson, D.P., Outlaw, J.L., Earle, M. and J.W. Richardson. 2017. Economic impact of US deer breeding and hunting operations. College Station: Texas A&M University Agricultural and Food Policy Center, pp.17–24.

Antoniewski, M.N. and S.A. Barringer. 2010. Meatshelf-life and extension using collagen/gelatin coatings: a review. *Critical Reviews in Food Science and Nutrition* 50(7),644–653.

Arafah, E., Herpandi, T. Handayani. 2008. Characterization of gelatin from snakehead fish bone. In: *Proceedings of the Seminar Nasional Tahunan VHasil Penelitian Perikanandan Kelautan*, Semnaskan, 15.

Arely, L., Xochitl, A.P., Gieraldin, C., Rafael, G.C. and A. Gabriel.2020. Characterization of whey-based fermented beverages supplemented with hydrolyzed collagen: Antioxidant activity and bioavailability. *Foods* 9,1106. doi:10.3390/foods9081106.

Arnesen, J.A., and A. Gildberg, 2006. Extraction of muscle proteins and gelatine from cod head. *Process Biochemistry* 41(3):697–700.

Asaduzzaman, A.K.M., Getachew, A.T., Cho, Y.J., Park, J.S., Haq, M. and B.S. Chun. 2020. Characterization of pepsin-solubilised collagen recovered from mackerel (*Scomberjaponicus*) bone and skin using subcritical water hydrolysis. *International Journalof Biological Macromolecules*148:1290–1297.

Ataie, M.J., Shekarabi, S.P.H. and S.H. Jalili. 2019. Gelatin from bones of bighead carpas a fat replacer on physicochemical and sensory properties of low-fat mayonnaise. *Journal of Microbiology, Biotechnology and Food Sciences* 8(4):979–983.

Atma, Y., and H. Ramdhani. 2017. Gelatin extraction from the indigenous *Pangasius*catfish bone using pineapple liquid waste. *Indonesian Journal of Biotechnology* 22(2):86–91.

Atma, Y., and M. Taufik. 2021. Physico-chemical properties gelatin from bone of *Pangasiussutchi* extracted with citrus fruits. *Biointerface Research in Applied Chemistry* 11:11509–11518.

Atma, Y., Lioe, H.N., Prangdimurti, E., Seftiono, H., Taufik, M., Fitriani, D. and A. Z. Mustopa. 2018. The hydroxyproline content of fish bone gelatin from Indonesian Pangasius catfish by enzymatic hydrolysis for producing the bioactive peptide. *Asian Journal of Natural Product Biochemistry* 16(2):64–68.

Bama, P., Vijayalakshimi, M., Jayasimman, R., Kalaichelvan, P.T., Deccaraman, M. and S. Sankaranarayanan. 2010. Extraction of collagen from catfish (*Tachysurusmaculatus*) by pepsin digestion and preparation and characterization of collagen chitosan sheet. *International Journal of Pharmacy and Pharmaceutical Sciences* 2(4):133–137.

Barbana, C. and J.I. Boye. 2011. AngiotensinI-converting enzyme inhibitory propertie of lentil protein hydrolysates: Determination of the kinetics of inhibition. *Food Chemistry* 127(1):94–101.

Barman, L.C., Sikder, M.B.H., Ahmad, I., Shourove, J.H., Rashid, S.S. and A.N.M. Ramli. 2020. Gelatin extraction from the Bangladeshi Pangas Catfish (*Pangasiuspangasius*) waste and comparative study of their physicochemical properties with a commercial gelatin. *International Journal of Engineering Technology and Sciences* 7(2):13–23.

Barros, J.R.D., Kunigk, L. and C.H. Jurkiewicz. 2010. Incorporation of nisin in natural casing for the control of spoilage microorganisms in vacuum packagedsausage. *Brazilain Journal of Microbiology*, 41(4):1001–1008.

Bay-Jensen, A. C., Platt, A., Siebuhr, A. S., Christiansen, C., Byrjalsen, I., & Karsdal, M. A. (2016). Early changes in blood-based joint tissue destruction biomarkers are predictive of response to tocilizumab in the LITHE study. *Arthritus Research & Therapy*, 18(1): 1–9.

Beighle, D.E., Boyazoglu, P.A., Hemken, R.W. and Serumaga-Zake, P.A.1994. Determination of calcium, phosphorus, and magnesium values in ribbones from clinically normal cattle. *American Jounral of Veterinary Research* 55(1):85–89.

Belbachir, K., Noreen, R., Gouspillou, G. and C. Petibois. 2009. Collagen types analysisand differentiation by FTIR spectroscopy. *Analytical and Bioanalytical Chemistry* 395(3):829–837.

Benito-Ruiz, P., Camacho-Zambrano, M.M., Carrillo-Arcentales, J.N., Mestanza-Peralta, M.A., Vallejo-Flores, C.A., Vargas-López, S.V., Villacís-Tamayo, R.A.and L.A. Zurita-Gavilanes. 2009. A randomized controlled trial on the efficacy and safety of a food ingredient, collagen hydrolysate, for improving joint comfort. *International Journal of Food Sciences and Nutrition* 60: 99–113.

Bichukale, A.D., Koli, J.M., Sonavane, A.E., Vishwasrao, V.V., Pujari, K.H., and P.E. Shingare. 2018. Functional properties of gelatin extracted from poultry skin and bone waste. *International Journal of Pure & Applied Bioscience* 6(4):87–101.

Blakely, K.K. and C. Johnson. 2020. New osteoporosis treatment means new boneformation. *Nursing for Women's Health* 24(1):52–57.

Borrajo, P., Pateiro, M., Barba, F.J., Mora, L., Franco, D., Toldrá, F. and J.M.Lorenzo. 2019. Antioxidant and antimicrobial activity of peptides extracted from meat by-products: A review. *Food Analytical Methods* 12(11): 2401–2415.

Bourtoom, T. 2008. Review article. Edible films and coatings: characteristics and properties. *International Food Research Journal* 15(3):237–248.

Buckley, M. 2016. Species identification of bovine, ovine and porcine type 1 collagen; comparing peptide mass fingerprinting and LC-based proteomics methods. *International Journal of Molecular Sciences* 17(4):445.

Callaghan, T.M. and K.P. Wilhelm. 2008. A review of ageing and an examination of clinical methods in the assessment of ageing skin. Part I: Cellular and molecular perspectives of skin ageing. *International Journal of Cosmetic Science*, 30:313–322.

Calvarro, J., Perez-Palacios, T. and J. Ruiz. 2016. Modification of gelatin functionalityfor culinary applications by using transglutaminase. *International Journal of Gastronomy and Food Science* 5:27–32.

Cao, S., Wang, Y., Hao, Y., Zhang, W. and G. Zhou. 2019. Antihypertensive effects *in vitro* and *in vivo* of novel angiotensin-converting enzyme inhibitory peptides from bovine bone gelatin hydrolysate. *Journal of Agricultural and Food Chemistry* 68 (3):759–768.

Cao, S., Wang, Y., Xing, L., Zhang, W. and G. Zhou. 2020. Structure and physicalproperties of gelatin from bovine bone collagen influenced by acid pretreatment andpepsin.*Food and Bioproducts Processing*, 121:213–223.

Chemat, F., Vian, M.A., and G. Cravotto. 2012. Green extraction of natural products: Concept and principles. *International Journal of Molecular Sciences* 13(7):8615–8627.

Chen, X., Wang, S., Chen, G., Wang, Z. and J. Kan. 2021. The immunomodulatory effects of *Carapax Trionycis* ultrafine powder on cyclophosphamide-induced immunosuppression in Balb/cmice. *Journal of the Science of Food and Agiculture* 101(5):2014–2026.

Chen, Y.J., Kuo, C.Y., Kong, Z.L., Lai, C.Y., Chen, G.W., Yang, A.J., Lin, L.H. and M.F. Wang. 2021. Antifatigue effect of a dietary supplement from the fermented by-products of Taiwan Tilapia aquatic waste and *Monostromanitidum* oligosaccharide complex. *Nutrients* 13(5):1688.

Choi, F.D., Sung, C.T., Juhasz, M.L. and N.A. Mesinkovsk. 2019. Oral collagen supplementation: a systematic review of dermatological applications. *Journal of Drugs in Dermetology* 18(1):9–16.

Choi, H.S., Im, S., Park, Y., Hong, K.B. and Suh, H.J. 2016. Deer bone oil extract suppresses lipopolysaccharide-induced inflammatory responses in RAW264. 7 cells. *Biological and Pharmaceutical Bulletin*, 39 (4):593–600

Choi, H., Won, C.S. and B.G. Kim. 2021. Protein and energy concentrations of meatmeal and meat and bone meal fed to pigs based on *in vitro* assays. *Animal Nutrition* 7(1):252–257.

Choi, S.S., and J.M. Regenstein. 2000. Physicochemical and sensory characteristics of fish gelatin. *Journal of Food Science* 65(2):194–199.

Choonpicharn, S., Jaturasitha, S., Rakariyatham, N., Suree, N., and H. Niamsup. 2015. Antioxidant and antihypertensive activity of gelatin hydrolysate from *Niletilapia* skin. *Journal of Food Science and Technology* 52(5):3134–3139.

da Trindade Alfaro, A., Balbinot, E., Weber, C.I., Tonial, I.B., and A. Machado-Lunkes. 2015. Fish gelatin: characteristics, functional properties, applications and future potentials. *Food Engineering Reviews* 7(1):33–44.

da Trindade Alfaro, A., Simões da Costa, C., Graciano Fonseca, G., and C. Prentice. 2009. Effect of extraction parameters on the properties of gelatin from King weakfish (*Macrodonancylodon*) bones. *Food Science and Technology International* 15(6):553–562.

Daher, L.J., Callahan, T.P., and S.M. Lonesky. Gelatin Spray Coating. 2000. U.S. Patent 6,077,540.

Damrongsakkul, S., Ratanathammapan, K., Komolpis, K., and W. Tanthapanichakoon. 2008. Enzymatic hydrolysis of rawhide using papain and neutrase. *Journal of Industrial and Engineering Chemistry* 14 (2): 202–206

Daniel, K.N., Steffen, O., Stephan, S., Denise, Z. and G.Albert. 2018. Specific collagen peptides improve bone mineral density and bone markers in post menopausal women – a randomized controlled study. *Nutrients* 10(1):97.

Dettmer, A., Cavalli, É., Ayub, M.A., and M. Gutterres. 2013. Environmentally friendly hide unhairing: enzymatic hide processing for the replacement of sodium sulfide anddelimig. *Journal of Cleaner Production* 47:11–18.

Dobbelaere, D. 2013. Statistical overview of the animal by-products industry in the EU in 2012. *EFSA Journal*, 9:1945.

Dressler, P., Gehring, D., Zdzieblik, D., Oesser, S., Gollhofer, A. and D. König. 2018. Improvement of functional ankle properties following supplementation with specific collagen peptides in athletes with chronic ankle instability. *Journal of Sports Science & Medicine* 17(2):298.

Elgadir M.A., Mohamed E.S., Mirghani, and A. Adam. 2013. Fish gelatin and its applications in selected pharmaceutical aspects as alternative source to pork gelatin. *Journal of Food, Agriculture & Environment* 11(1):73–79.

Erge, A., and Ö. Zorba. 2018. Functional properties of gelatin and its use in food Industry. *Turkish Journal of Agriculture-Food Science and Technology* 6(7):840–849.

Fan, H., and W. Shen. 2016. Gelatin-based microporous carbon nanosheets as high performance supercapacitor electrodes. *ACS Sustainable Chemistry & Engineering* 4(3):1328–1337.

Fan, R., Zhou, D. and X. Cao. 2020. Evaluation of oat beta-glucan-marine collagenpeptide mixed gel and its application as the fat replacer in the sausage products. *PLoSOne* 15(5):e0233447. doi: 10.1371/journal.pone.0233447.

Fathi, M.H., Hanifi, A. and V. Mortazavi. 2008. Preparation and bioactivity evaluationof bone-like hydroxyapatite nanopowder. *Journal of Materials Processing Technology* 202(1–3):536–542.

Feng, X.M., Yang, X.H., Xie, W.C. *et al.* 2009. Optimization of enzymatic hydrolysis of white shrimp (*Penaeusvannamei*) head using response surface methodology. *Food Science*, 22.

FAO. 2020. Addressing the Impacts of COVID-19 in Food Crises | April–December 2020: FAO's Component of the Global COVID-19 Humanitarian Re-sponse Plan. https://doi.org/10.4060/ca8497en.

FAO. 2021. Meat Market Review: Overview of Global Meat Market Developments in 2020; FAO: Rome, Italy, 2021. www.fao.org/3/cb3700en/cb3700en.pdf.

Fu, Y., Therkildsen M., Aluko R.E., and R. Lametsch. 2019. Exploration of collagen recovered from animal by-products as a precursor of bioactive peptides: Successes and challenges. *Critical Reviews in Food Science and Nutrition* 59(13):2011–2027. doi: 10.1080/10408398.2018.1436038.

Fulgione, A., Ianniello, F., Papaianni, M., Contaldi, F., Sgamma, T., Giannini, C., Pastore, S., Velotta, R., Della Ventura, B., Roveri, N. and M. Lelli. 2019. Biomimetic hydroxyapatite nanocrystals are an active carrier for *Salmonella* bacteriophages. *International Journal of Nanomedicine* 14: 2219–2232.

Gao, S., Hong, H., Zhang, C. *et al.* 2019. Immunomodulatory effects of collagen hydrolysates from yak (*Bosgrunniens*) bone on cyclophosphamide-induced immunosuppression in BALB/cmice. *Journal of Functional Foods* 60, 103420.

Garcia, R.A. and J.G. Phillips. 2008. Physical distribution and characteristics of meat & bone meal protein. In *2008 Providence, Rhode Island, June 29–July 2, 2008* (p. 1). American Society of Agricultural and Biological Engineers.

Giménez, B., Turnay, J., Lizarbe, M.A., Montero, P. and M.C. Gómez-Guillén. 2005. Use of lactic acid for extraction of fish skin gelatin. *Food Hydrocolloids*, 19(6):941–950.

Goda, A.M., El-Haroun, E.R. and M.A. Kabir Chowdhury. 2007. Effect of totally orpartially replacing fish meal by alternative protein sources on growth of African catfish *Clariasgariepinus* (Burchell, 1822) reared in concrete tanks. *Aquaculture Research* 38(3):279–287.

Godfray, H.C.J., Aveyard, P., Garnett, T., Hall, J.W., Key,T.J., Lorimer, J., Pierrehumbert, R.T., Scarborough, P., Springmann, M. and S.A. Jebb. 2018. Meat consumption, health, and the environment. *Science*, 361:1–8.

Gómez-Guillén, M.C., Giménez, B., López-Caballero, M.A., and M.P. Montero. 2011. Functional and bioactive properties of collagen and gelatin from alternative sources: A review. *Food Hydrocolloids*, 25(8):1813–1827.

Gómez-Guillén, M.C., Turnay, J., Fernández-Dıaz, M.D., Ulmo, N., Lizarbe, M.A., and P. Montero. 2002. Structural and physical properties of gelatin extracted from different marine species: a comparative study. *Food Hydrocolloids*, 16: 25–34.

González, N., Marquès, M., Nadal, M. and J.L. Domingo. 2020. Meat consumption: Which are the current global risks? A review of recent (2010–2020) evidences. *Food Research International*, 137:109341.

Greene, D.M. 2003. Use of poultry collagen coating and antioxidants as flavor protection for cat foods made with rendered poultry fat. Virginia, United States: Virginia Polytechnic Institute and State University, MSc thesis.

Gudmundsson, M., and H. Hafsteinsson. 1997. Gelatin from cod skins as affected by chemical treatments. *Journal of Food Science* 62(1):37–39.

Halim, N.R.A., Yusof H.M., and N.M. Sarbon. 2016. Functional and bioactive properties of fish protein hydolysates and peptides: A comprehensive review. *Trends in Food Science & Technology* 51:24–33. doi: 10.1016/j.tifs.2016.02.007.

Hapsari, N., Rosida, D.F., Djajati, S., Aviskarahman, A. and R. Dewati. 2017. Physical characteristics of fish bone gelatin extracted acid. *Advanced Science Letters* 23(12):12272–12275.

Harahap, H.A.M.I.D.A.H. and Z.U.Q.N.I. Meldha. 2017. Characterization of hydroxyapatite from chicken bone via precipitation. *Key Engineering Materials*, 744.

Haug, I.J. and K.I. Draget. 2011. Gelatin. In *Handbook of food proteins*. Woodhead Publishing.

Helfrich, Y.R., Sachs, D.L. and J.J. Voorhees. 2008. Overview of skin aging and photoaging. *Dermatol. Nurs.* 20:177–183.

Henchion, M., Moloney, A.P., Hyland, J., Zimmermann, J. and S. McCarthy. 2021. Trends for meat, milk and egg consumption for the next decades and the role played by livestock systems in the global production of proteins. *Animal*, 15:1–14.

Hemung, B. 2013. Properties of tilapia bone powder and its calcium bioavailability based on transglutaminase assay. *IJBBB3*, 306–309.

Hendriks, W.H., Butts, C.A., Thomas, D.V., James, K.A.C., Morel, P.C.A. and M.W.A. Verstegen. 2002. Nutritional quality and variation of meat and bone meal. *Asian-AustralasianJournal ofAnimal Sciences* 15(10):1507–1516.

He, Y., Lu, Y., Ren, W., Shen, J., Wu, K., Xu, K., Wu, J. and Y. Hu. 2019. Glucagonlike peptide 2 has a positive impact on osteoporosis in ovariectomized rats. *Life Sciences* 226:47–56.

Himaya, S.W.A., Ngo, D.H., Ryu, B., and S.K. Kim. 2012. An active peptide purified from gastrointestinal enzyme hydrolysate of Pacific cod skin gelatin attenuates angiotensin-1 convertingenzyme (ACE) activity and cellular oxidative stress. *Food Chemistry* 132(4):1872–1882.

Hong, H., Roy, B.C., Chalamaiah, M., Bruce, H.L., and J. Wu. 2018. Pretreatment withformic acid enhances the production of small peptides from highly cross-linked collagen of spent hens. *Food Chemistry*, 258:174–180

Hou, C., Gao, L., Wang, Z., Rao, W., Du, M. and Zhang, D. 2020. Mechanical properties, thermal stability, and solubility of sheep bone collagen–chitosan films. *Journal of Food Process Engineering* 43(1):13086.

Hu, S., Jia, F., Marinescu, C., Cimpoesu, F., Qi, Y., Tao, Y., Stroppa, A. and W. Ren. 2017. Ferroelectric polarization of hydroxyapatite from density functional theory. *RSC Advances* 7(35):21375–21379.

Hubadillah, S.K., Othman, M.H.D., Tai, Z.S. *et al.* 2020. Novel hydroxyapatite-basedbio-ceramic hollow fiber membrane derived from waste cow bone for textile wastewater treatment.*Chemical Engineering Journal* 379:122396.

Hussain, F., Alshahrani, S., Abbas, M.M., Khan, H.M., Jamil, A., Yaqoob, H., Soudagar, M.E.M., Imran, M., Ahmad, M. and M. Munir. 2021. Waste animal bones as catalysts for biodiesel production; a mini review. *Catalysts*, 11(5):630.

Jamilah, B., and K.G. Harvinder. 2002. Properties of gelatins from skins of fish – black tilapia (*Oreochromismossambicus*) and red tilapia (*Oreochromisnilotica*). *Food Chemistry* 77(1):81–84.

Jongjareonrak, A., Benjakul, S., Visessanguan, W., Nagai, T., and M. Tanaka. 2005. *Food Chemistry* 93:475–484.

Kakimov, A.K., Suychinov, A.K. and ZhS. Yessimbekov. 2017. Studying the physico-chemical and microbiological characteristics, microstructure and particle-size distribution of meat-bone paste. *Bulletin of the Shakarim State University of Semey* 4(80):12–18.

Kakimov, A.K., Yessimbekov, Z.S., Kabdylzhar, B.K., Suychinov, A.K. and A.M. Baikadamova. 2021. A study on the chemical and mineral composition of the protein-mineral paste from poultry and cattle bone raw materials. *Theory and practice of meatprocessing* 6(1):39–45.

Kammeyer, A. and R.M. Luiten. 2015. Oxidation events and skin aging. *Ageing Research Reviews* 21:16–29.

Karim, A.A., and R. Bhat. 2009. Fish gelatin: properties, challenges, and prospects as an alternative to mammalian gelatins. *Food Hydrocolloids* 23(3): 563–576.

Karlaganis, G. 2001. *Citric Acid: SIDS Initial Assessment Report for 11th SIAM*. Switzerland: Unep Publication.

Kathpalia, H., Sharma, K., and G. Doshi. 2014. Recent trends in hard gelatin capsule delivery system. *Journal of Advanced Pharmacy Education & Research* 4(2).

Ketnawa, S., Martínez-Alvarez, O., Benjakul, S. and S. Rawdkuen. 2016. Gelatin hydrolysates from farmed giant catfish skin using alkaline proteases and its antioxidative function of simulated gastro-intestinal digestion. *Food chemistry* 192:34–42.

Kim, H.K., Kim, Y.H., Kim, Y.J., Park, H.J. and N.H. Lee. 2012. Effects of ultrasonic treatment on collagen extraction from skins of the sea bass *Lateolabrax japonicus*. *Fisheries Science* 78(2):485–490.

Kim, H.K., Kim, Y.H., Park, H.J. and N.H. Lee. 2013. Application of ultrasonic treatment to extraction of collagen from the skins of sea bass *Lateolabrax japonicus*. *Fisheries Science* 79(5):849–856.

Koli, J.M., Basu, S., Nayak, B.B., Patange, S.B., Pagarkar, A.U. and V. Gudipati. 2012. Functional characteristics of gelatin extracted from skin and bone of Tiger-toothed croaker (*Otolithesruber*) and Pink perch (*Nemipterus japonicus*). *Food and Bioproducts Processing* 90(3):555–562.

Kostik, V. 2014. A comprehensive study of the presence of some food additives in non-alcoholic beverages in Republic of Macedonia from the period 2008-2012. *Journal of Hygienic Engineering and Design* 7:123–131.

Kowalski, Z., and K. Krupa-Żuczek. 2007. A model of the meat waste management. *Polish Journal of Chemical Technology* 9(4):91–97.

Kowalski, Z., Kulczycka, J., Makara, A. and P. Harazin. 2021. Quantification of material recovery from meat waste incineration–An approach to an updated food waste hierarchy. *Journal of Hazardous Materials* 416:126021.

Kristinsson, H.G. and B.A. Rasco. 2000. Biochemical and functional properties of Atlantic salmon (*Salmosalar*) muscle proteins hydrolyzed with various alkaline proteases. *Journal of Agricultural and Food Chemistry* 48(3):657–666.

Krupa-Żuczek, K., Kowalski, Z. and Z.Wzorek. 2008. Manufacturing of phosphoric acid from hydroxyapatite, contained in the ashes of the incinerated meat-bone wastes. *Polish Journal of Chemical Technology* 10(3):13–20.

Kumar, K.V., Subha, T.J., Ahila, K.G. *et al.* 2021. Spectral characterization of hydroxyapatite extracted from Black Sumatra and Fighting cock bone samples: A comparative analysis. *Saudi Journal of Biological Sciences* 28(1):840–846.

Kumosa, L.S., Zetterberg, V., and J. Schouenborg. 2018. Gelatin promotes rapid restoration of the blood brain barrier after acute brain injury. *Acta biomaterialia* 65:137–149.

Laing, S., Bielfeldt, S., Ehrenberg, C. and K.P. Wilhelm. 2020. A dermonutrient containing special collagen peptides improves skin structure and function: a randomized, placebo-controlled, triple-blind trial using confocal laser scanning microscopy on the cosmetic effects and tolerance of a drinkable collagen supplement. *Journal of Medicinal Food* 23(2):147–152.

Ledward, D.A. 2000. Gelatin. In *Handbook of hydrocolloids*, ed. G.O. Philips, and P.A. Williams. 67–86. Woodhead Publishing Limited.

Lee, H., Park, Y., Ahn, C.W., Park, S.H., Jung, E.Y. and H.J. Suh. 2014. Deer bone extract suppresses articular cartilage damage induced by monosodium iodoacetate in osteoarthritic rats: An *in vivo* micro–computed tomography study. *Journal of Medicinal Food* 17(6):701–706.

Li, D., Mu, C., Cai, S. and W. Lin. 2009. Ultrasonic irradiation in the enzymatic extraction of collagen. *Ultrasonics Sonochemistry* 16(5):605–609.

Li, J., Yang, M., Lu, C. *et al.* 2020. Tuna bone powder alleviates glucocorticoid-inducedosteoporosis via coregulation of the NF-κB and Wnt/β-Catenin signaling pathways and modulation of gut microbiota composition and metabolism. *Molecular Nutrition & Food Research* 64(5). https://doi.org/ 10.1002/ mnfr.201900861.

Li, X., He, Z., Xu, J., Zhang, L., Liang, Y., Yang, S., Wang, Z., Zhang, D., Gao, F. and Li, H. 2021. Effect of nano-processing on the physicochemical properties of bovine, porcine, chicken, and rabbit bone powders. *Food Science & Nutrition*, 9:3580–3592.

Li, Y., Wang, X., Xue, Y. *et al.* 2021. The preparation and identification of characteristic flavour compounds of Maillard reaction products of protein hydrolysate from grass carp (*Ctenopharyngodonidella*) bone. *Journal of Food Quality*, 2021, 1–14.

Li, Z.R., Wang, B., Chi, C.F., *et al.* 2013. Isolation and characterization of acid solublecollagens and pepsin soluble collagens from the skin and bone of Spanish mackerel (*Scomberomorousniphonius*). *Food Hydrocolloids* 31(1):103–113.

Liu, D., Nikoo, M., Boran, G., Zhou, P. and J.M. Regenstein. 2015. Collagen andgelatin. *Annual Review of Food Science and Technology* 6:527–557.

Liu, D.C., Lin, Y.K., and M.T. Chen. 2001. Optimum condition of extracting collagen from chicken feet and its characteristics. *Asian-Australasian Journal of Animal Sciences* 14(11):1638–1644.

Liu, H. Y., Han, J., and S. D. Guo. 2009. Characteristics of the gelatin extracted from channel catfish (*Ictalurus Punctatus*) head bones. *LWT-Food Science and Technology* 42(2):540–544.

Liu, Y., Luo, D., and T. Wang. 2016. Hierarchical structures of bone and bioinspired bone tissue engineering. *Small* 12(34):4611–4632.

Liu, D., Wei, G., Li, T., Hu, J., Lu, N., Regenstein, J.M. and P. Zhou. 2015. Effects ofalkaline pretreatments and acid extraction conditions on the acid-soluble collagen fromgrasscarp (*Ctenopharyngodonidella*) skin. *Food Chemistry*, 172:836–843.

Long, N.H.B.S., Gál, R. and F. Buňka. 2011. Use of phosphates in meat products. *African Journal of Biotechnology* 10(86):19874–19882.

Lopez-Rubio, A., Fabra, M.J., Martinez-Sanz, M., Mendoza, S. and Q.V. Vuong. 2017. Biopolymer-based coatings and packaging structures for improved food quality. *Journal of Food Quality* 2017, 2351832. https://doi.org/10.1155/2017/2351832.

Lv, L.C., Huang, Q.Y., Ding, W., Xiao, X.H., Zhang, H.Y. and L.X. Xiong. 2019. Fish gelatin: The novel potential applications. *Journal of Functional Foods* 63:103581.

Ma, K., Cai, X., Zhou, Y., Wang, Y., and T. Jiang. 2017. *In vitro* and *in vivo* evaluation of tetracycline loaded chitosan-gelatin nanosphere coatings for titanium surface functionalization. *Macromolecularbioscience* 17(2):1600130.

Ma, Y., Zeng, X., Ma, X., Yang, R. and W. Zhao. 2019. A simple and eco-friendly method of gelatin production from bone: One-step biocatalysis. *Journal of Cleaner Production* 209:916–926.

Mahmoodani, F., Ardekani, V.S., See, S.F., Yusop, S.M., and A.S. Babji. 2014a. Optimization and physical properties of gelatin extracted from pangasius catfish (*Pangasiussutchi*) bone. *Journal of Food Science and Technology* 51(11):3104–3113.

Mahmoodani, F., Ghassem, M., Babji, A.S., Yusop, S.M. and R. Khosrokhavar. 2014b. ACE inhibitory activity of pangasius catfish (*Pangasius sutchi*) skin and bone gelatin hydrolysate. *Journal of Food Science and Technology* 51(9):1847–1856.

Makara, A., Kowalski, Z., Lelek, Ł. and J. Kulczycka. 2019. Comparative analyses of pig farming management systems using the life cycle assessment method. *Journal of Cleaner Production* 241:118305.

Malvano, F., Montone, A.M.I., Capparelli, R., Capuano, F. and D. Albanese. 2021. Development of a novel active edible coating containing hydroxyapatite for food shelf-life extension. *Chemical Engineering Transactions* 87:25–30.

Marcet, I., Álvarez, C., Paredes, B., and M. Díaz. 2016. The use of sub-critical water hydrolysis for the recovery of peptides and free amino acids from food processing wastes. Review of sources and main parameters. *Waste Management* 49:364–371.

Mariod, A.A., and H.F. Adam. 2013. Review: Gelatin, source, extraction and industrial applications. *Acta Scientiarum Polonorum Technologia Alimentaria* 12(2):135–147.

McAlindon, T., Bartnik, E., Ried, J.S., Teichert, L., Herrmann, M. and K. Flechsenhar. 2017. Determination of serum biomarkers in osteoarthritis patients: a previous interventional imaging study revisited. *Journal of Biomedical Research* 31(1):25.

Mead, G.M. 2004. Keeping poultry meat fresh. *Food Sci. Technol.* 19, 20–21.

Mezenova, N.Y., Agafonova, S.V., Mezenova, O.Y., Baidalinova, L.S. and T. Grimm. 2021. Obtaining and estimating the potential of protein nutraceuticals from highly mineralized collagen-containing beef raw materials. *Theory and Practice of Meat Processing*, 6(1):10–22.

Mi, H., Wang, C., Chen, J. *et al.* 2019. Characteristic and functional properties of gelatin from the bones of Alaska pollock (*Theragrachalco gramma*) and yellowfin sole (*Limandaaspera*) with papain-aided process. *Journal of Aquatic Food Product Technology* 28(3):287–297.

Misbahul, A., Azzahra, H.Y. and N. Taqiyyuddiin. 2021. Fishbone gelatin. *Asian Journal of Fisheries and Aquatic Research*, 36–44.

Mohamadpour, F., and M. Feilizadeh. 2019. Citric acid as a green and naturally biodegradable catalyst promoted convenient synthesis of polysubstituted dihydro-2-oxypyrrole derivatives via four-condensation reaction of dialkylacetylenedicarboxylate,formaldehyde andamines. *Biointerface Research in Applied Chemistry* 9:4096–4100.

Mondal, S., Mondal, B., Dey, A., and S.S. Mukhopadhyay. 2012. Studies on processing and characterization of hydroxyapatite biomaterials from different biowastes. *Journal of Minerals and Materials Characterization and Engineering* 11(1):55–67.

Muir, W.I., Lynch, G.W., Williamson, P. and A.J. Cowieson. 2013. Theoral administration of meat and bone meal-derived protein fractions improved the performance of young broiler chicks. *Animal productionscience* 53(5):369–377.

Murray, O., Hall, M., Kearney, P., and R. Green. (2015). Fast-Dispersing Dosage Forms Containing Fish Gelatin. U.S. Patent 9192580.

Muyonga, J.H., Cole, C.G.B., and K.G. Duodu. 2004. Characterisation of acid solublecollagen from skins of young and adult Nile perch (*Lates niloticus*). *Food Chemistry* 85(1):81–89.

Myllyharju, J., and K.I. Kivirikko. 2004. Collagens, modifying enzymes and their mutations in humans, flies and worms. *Trends in Genetics* 20(1):33–43.doi:10.1016/j.tig.2003.11.004.

Nagai, T., Tanoue, Y., Kai, N. and N. Suzuki. 2015. Characterization of collagen from emu (*Dromaius novaehollandiae*) skins. *Journal of Food Scienceand Technology* 52(4):2344–2351.

Nagai, T. and N. Suzuki. 2000. Isolation of collagen from fish waste material – skin, bone and fins. *Food Chemistry* 68:277–281.

Najumudeen, F. 2012. Avon celebrates 20th anniversary of skincare brand. Downloaded from http://thestar.com.my/metro/story.asp?file=/2012/7/18/central/11637420&sec=centralon 4/9/2012.

Nalinanon, S., Benjakul, S., Visessanguan, W. and H. Kishimura. 2008. Improvement ofgelatin extraction from bigeye snapper skin using pepsin-aided process in combinationwithproteaseinhibitor. *Food Hydrocolloids* 22(4):615–622.

Natsir, H., Dali, S., Sartika, L. and A.R. Arif. 2021. Enzymatic hydrolysis of collagen from yellowfin tuna bones and itsp otential as antibacterial agent. *Rasayan J Chem* 14(1):594–600.

Nawaz, A., Li, E., Irshad, S., Hammad, H.H.M., Liu, J., Shahbaz, H.M., Ahmed, W. and Regenstein, J.M. 2020. Improved effect of autoclave processing on size reduction, chemical structure, nutritional, mechanical and *in vitro* digestibility properties of fish bone powder. *Advanced Powder Technology* 31(6):2513–2520.

Nazeer, R.A., Kavitha, R., Ganesh, R.J., Naqash, S.Y., Kumar, N.S. and R. Ranjith. 2014. Detection of collagen through FTIR and HPLC from the body and foot of *Donax cuneatus* Linnaeus, 1758. *Journal of Food Science and Technology* 51(4):750–755.

Neffe, A.T., Loebus, A., Zaupa, A., Stoetzel, C., Müller, F.A., and A. Lendlein. 2011. Gelatin functionalization with tyrosine derived moieties to increase the interaction withhydroxyapatitefillers. *Acta biomaterialia* 7(4):1693–1701.

Nik, A.N.M. 2015. *Effect of various acid pretreatment on physicochemical properties and sensory profile of duck feet gelatine and application in surimi*. Dissertation, Universiti Sains Malaysia.

Nollet, L.M., Toldra, F., Benjakul, S., Paliyath, G. and Y. Hui. 2012. Food biochemistry and food processing. John Wiley & Sons.

Ockerman, H.W. and Basu, L., 2010. Edible rendering-rendered products for human use. In D. L. Meeker (Ed.). *Essential rendering: All about the animal by-products industry* (pp. 95–110). Virginia, National Renderers Association.

Ogawa, M., Moody, M.W., Portier, R.J., Bell, J., Schexnayder, M.A., and J.N. Losso. 2004. Biochemical properties of bone and scale collagens isolated from the subtropical fish black drum (*Pogoniacromis*) and sheeps head seabream (*Archosargusprobatocephalus*). *Food Chemistry* 88:495–501.

Orehov, O.G. 2013. Justification obtain natural glue from the bone residue of broiler chickens. *Proceedings of the Voronezh State University of Engineering Technologies* 3:130–134.

Pal, G.K. and P.V. Suresh. 2016. Sustainable valorisation of sea food by-products: Recovery of collagen and development of collagen-based novel functional food ingredients. *Innovative Food Science & Emerging Technologies* 37:201–215.

Panjaitan, T.F.C. 2017. Optimization of gelatin extraction from tuna fish bone (*Thunnusalbacares*). *Wiyata Journal: Science and Health Research* 3(1):11–16.

Patra, A. and M. Lalhriatpuii. 2019. Progress and prospect of essential mineral nanoparticles in poultry nutrition and feeding – a review. *Biological Trace Element Research*, 1–21.

Pertiwi, M., Atma, Y., Mustopa, A.Z. and R. Maisarah. 2018. Karakteristik fisik dan kimia gelatin dari tulang ikan patin dengan pre-treatment asam sitrat. *Jurnal Aplikasi Teknologi Pangan* 7(2).

Powell, T., Bowra, S., and H. J. Cooper. 2017. Subcritical water hydrolysis of peptides: Aminoacidside-chainmodifications. *Journal of the American Society for Mass Spectrometry* 28(9):1775–1786.

Prabhu, G.A., Doerscher, D.R., and D.H. Hull. 2006. Utilization of pork collagen protein in emulsified and whole muscle meat products. *Journal of Food Science* 69(5):388–392.

Prasad, V.D. 2017. Formulation and modifying drug release from hard and soft gelatin capsules for oral drug delivery. *Int. J. Res. Dev. Pharm. Life Sci.* 6:2663–2677.

Prestes, R.C. 2013. Colágeno e seus derivados: características eaplicações em produtos cárneos. *Revista Unopar Científica Ciências Biológicas e da Saúde* 15(1):65–74.

Qin, X., Shen, Q., Zhang, C., Zhang, H. and W. Jia. 2020. Preparation of instant yak bone powder by using instant catapults team explosion and its physicochemical properties. *Transactions of the Chinese Society of Agricultural Engineering* 36(4):307–315

Rahman, M.N.A. and S.A.S.K. Jamalulail. 2012. Extraction, physicochemical characterizations and sensory quality of chicken feet gelatin. *Borneo Science* 30:1–13.

Ramesh, S., Loo, Z.Z., Tan, C.Y. *et al.* 2018. Characterization of biogenichydroxyapatite derived from animal bones for biomedical applications. *Ceramics International* 44(9):10525–10530.

Ramos, M., Valdes, A., Beltran, A., and M.C. Garrigós. 2016. Gelatin-based films and coatings for food packaging applications. *Coatings* 6(4):41.

Ran, X.G. and L.Y. Wang. 2014. Use of ultrasonic and pepsin treatment in tandem forcollagen extraction from meat industry by-products. *Journal of the Science of Food and Agriculture* 94(3):585–590.

Romano, N. and V. Kumar. 2018. Phytase in animal feed. In *Enzymes in Human and Animal Nutrition* (pp. 73–88). Academic Press.

Sadowska, M., Kołodziejska, I., and C. Niecikowska. 2003. Isolation of collagen fromtheskins of Baltic cod (*Gadusmorhua*). *Food Chemistry*, 81(2):257–262.

Samatra, M.Y., Azmi, A., Shaarani, S.M. and U.H.M. Razali. 2020. Characterisation of gelatin extracted from buffalo (Bubalusbubalis) bone using papain pre-treatment. *Journal of Agricultural and Food Engineering* 4:0027.

Sanaei, A.V., Mahmoodani, F., See, S.F., Yusop, S.M. and A.S. Babji. 2013. Optimization of gelatin extraction and physico-chemical properties of catfish (*Clariasgariepinus*) bone gelatin. *International Food Research Journal* 20(1):423.

Sarbon, N.M., Badii, F. and N.K. Howell. 2013. Preparation and characterisation ofchicken skin gelatin as an alternative to mammalian gelatin. *Food Hydrocolloids* 30(1):143–151.

Schmidt, M.M., Dornelles, R.C.P., Mello, R.O. *et al.* 2016. Collagen extraction process. *International Food Research Journal*, 23(3).

Schrieber, R., and H. Gareis. 2007. *Gelatine handbook: theory and industrial practice.* John Wiley & Sons.

Seda, E. B. and K. B. Sibel. 2015. Fruit juice drink production containing hydrolyzed collagen. *Journal of Functional Foods* 14 562–569. doi:10.1016/j.jff.2015.02.024.

Serrano, M.P., Maggiolino, A., Pateiro, M., Landete-Castillejos, T., Domínguez, R., García, A., Franco, D., Gallego, L., DePalo, P. and J.M. Lorenzo. 2019. Carcass characteristics and meat quality of deer. In *More than Beef, Pork and Chicken–The Production, Processing, and Quality Traits of Other Sources of Meat for Human Diet* (pp. 227–268). Springer,Cham.

Shakila, R.J., Jeevithan, E., Varatharajakumar, A., Jeyasekaran, G. and Sukumar, D. 2012. Functional characterization of gelatin extracted from bones of red snapper andgrouper in comparison with mammalian gelatin. *LWT-Food Science and Technology* 48(1):30–36.

Shao, X., Sun, H., Zhou, R. *et al.* 2020. Effect of bovine bone collagen and nano-TiO2 on the properties of hydroxypropylmethylcellulose films. *International Journal of Biological Macromolecules* 158:937–944.

Shi, W., Sun, M., Hu, X. *et al.* 2017. Structurally and functionally optimized silk-fibroin–gelatin scaffold using 3D printing to repair cartilage injury *in vitro* and *in vivo*. *Advanced Materials* 29(29):1701089.

Shori, A.B., Ming, K.S. and A.S. Baba. 2021. The effects of *Lyciumbarbarum* water extract and fish collagen on milk proteolysis and *in vitro* angiotens in I-convertingenzyme inhibitory activity of yogurt. *Biotechnology and Applied Biochemistry* 68(2):221–4.

Si, S., Guo, Y., Xu, B., Qin, Y. and S. Song. 2021. Protective effects of collagen peptides on the dexamethasone-induced immunosuppression in mice. *International Journal of Peptide Research and Therapeutics* 27(2):1493–1499.

Siburian, W.Z., Rochima, E., Andriani, Y., and D. Praseptiangga. 2020. Fish gelatin (definition, manufacture, analysis of quality characteristics, and application): A review. *InternationalJournal of Fisheries and AquaticStudies* 8(4):90–95.

Skierka, E. and M. Sadowska. 2007. The influence of different acids and pepsin on the extractability of collagen from the skin of Baltic cod (*Gadusmorhua*). *Food Chemistry* 105(3):1302–1306.

Sobhi, B.M., Ismael, E.Y., Mansour, A.S., Elsabagh, M. and K.N.E.D. Fahmy.2020. Effect of nano-hydroxyapatite as an alternative to inorganic dicalcium phosphate on growth performance, carcass traits, and calcium and phosphorus metabolism of broiler chickens. *Journalof Advanced Veterinary Research* 10(4):250–256.

Someus, E. 2018. Concentrated phosphorus recovery from food grade animal bones for agronomical efficient Innovative Fertiliser Applications in the Organic and Low Input Farming Sectors. *Preprints*, 10:1–17.

Song, H., Zhang, S., Zhang, L., and B. Li. 2017. Effect of orally administered collagen peptides from bovine bone on skin aging in chronologically aged mice. *Nutrients* 9(11):1209. https://doi.org/10.3390/nu9111209.

Songchotikunpan, P., Tattiyakul, J., and P. Supaphol. 2008. Extraction and electrospinning of gelatin from fish skin. *International Journal of Biological Macromolecules* 42(3):247–255.

Staroń, P., Kowalski, Z., Staroń, A., Seidlerová, J. and M. Banach. 2016. Residues fromthe thermal conversion of waste from the meat industry as a source of valuable macro-andmicronutrients. *Waste Management* 49:337–345.

Steenson, S. and J.L. Buttriss. The challenges of defining a healthy and 'sustainable' diet. *Nutrition Bulletin* 45:206–222.

Subara, D. and I. Jaswir. 2021. The effect of processing parameters on the properties of fish gelatin hydrolysate nanoparticle. *Journal of Science and Applicative Technology* 5(1):17–24.

Sugiura, S.H., Babbitt, J.K., Dong, F.M. and R.W. Hardy. 2000. Utilization of fish and animal by-product meals in low-pollution feeds for rainbow trout *Oncorhynchusmykiss* (Walbaum). *Aquaculture Research* 31(7):585–593.

Suloma, A., Mabroke, R.S. and E.R. El-Haroun. 2013. Meat and bone meal as a potential source of phosphorus in plant-protein-based diets for Nile tilapia (*Oreochromisniloticus*). *Aquaculture International* 21(2):375–385.

Suderman, N. and N.M. Sarbon. 2020. Optimization of chicken skin gelatin film production with different glycerol concentrations by response surface methodology (RSM) approach. *Journal of Food Science and Technology* 57(2):463–472.

Suntornsaratoon, P., Charoenphandhu, N. and N. Krishnamra. 2018. Fortified tuna bone powder supplementation increases bone mineral density of lactating rats and their offspring. *Journal of theScienceof Food and Agriculture* 98(5): 2027–2034.

Supová, M., Simha Martynková, G., and Z. Sucharda. 2011. Bioapatite made from chicken femur bone. *Ceramics – Silikaty* 55(3):256–260.

Taheri, A., Abedian Kenari, A.M., Gildberg, A., and S. Behnam. 2009. Extraction and physicochemical characterization of greater lizardfish (*Sauridatumbil*) skin and bone gelatin. *Journal of Food Science* 74(3):E160-E165.

Tan, Y., and S. Chang. 2018. Isolation and characterization of collagen extracted from channel catfish (*Ictaluruspunctatus*) skin. *Food Chemistry,* 242:147–55. doi:10.1016/j.foodchem.2017.09.013.

Tangke, U., Daeng, R.A. and B. Katiandagho. 2021. Organoleptic Quality of Tuna Porridge Canned with Fortified Tuna Bone Meal. In *IOP Conference Series: Earth and Environmental Science* (Vol. 750, No. 1, p. 012047). IOP Publishing.

Tavakolipour, H. 2011. Extraction and evaluation of gelatin from silver carp waste.*World Journal Offish and Marine Sciences* 3(1):10–15.

Toldra, F., Mora, L. and M. Reig. 2016. New insights into meat by-product utilization. *Meat Science* 120:54–9. doi: 10.1016/j.meatsci.2016.04.021.

Tree, A. 2012. What is a collagen drink? Downloaded from www.wisegeek.com/what-is-a-collagen-drink.htm on 7/8/2012.

Tu, Z.C., Huang, T., Wang, H., *et al.* 2015. Physicochemical properties of gelatin from big head carp (*Hypophthalmichthysnobilis*) scales by ultrasound assisted extraction. *Journal of Food Science and Technology* 52(4):2166–2174.

Tymczewska, A., Furtado, B.U., Nowaczyk, J., Hrynkiewicz, K., and A. Szydłowska-Czerniak. 2021. Development and characterization of active gelatin films loaded with rapeseed meal extracts. *Materials* 14(11):2869.

United Nations Conference on Trade and Development, 2020. COVID-19 and Food Security in Vulnerable Countries. https://unctad.org/en/pages/newsdetails.aspx?OriginalVersionID=2331 (accessed 30 June 2020).

Volik, V.G., Ismailova, D.Y., Zinov'ev, S.V. and O.N. Erokhina. 2017. Modern technologies of processing secondary raw materials of meat and piper processing industry. *Krolikovodstvoizverovodstvo*, 11–15.

Wang, J., Zhang, B., Lu, W. *et al.* 2020. Cell proliferation stimulation ability and osteogenic activity of low molecular weight peptides derived from bovine gelatin hydrolysates. *Journal of Agricultural and Food Chemistry* 68(29):7630–7640.

Wang, L., Wang, Q., Qian, J., Liang, Q., Wang, Z., Xu, J., He, S. and H. Ma. 2015. Bioavailability and bioavailable forms of collagen after oral administration to rats. *Journal of Agricultural and Food Chemistry* 63(14):3752–6. doi:10.1021/jf5057502.

Wang, L., Yang, B., Du, X., Yang, Y. and J. Liu. 2008. Optimization of conditions for extraction of acid soluble collagen from grass carp (*Ctenopharyngodonidella*) by response surface methodology. *Innovative Food Science and Emerging Technologies* 9(4):604–607.

Wang, S., Lin, L.M., Wu, Y.N. *et al.* 2014. Angiotensin I converting enzyme (ACE) inhibitory activity and antihypertensive effects of grass carp peptides. *Food Science and Biotechnology* 23(5):1661–1666.

Wang, T. 2020. Preparation of milk fish bone powder by a novel bone-embrittlementing technique and used in manufacturing high calcium eggroll. *Journal of Food and Nutrition Research* 8(2):95–101.

Wang, Z., Hu, S., Gao, Y., Ye, C. and H. Wang. 2017. Effect of collagen-lysozyme coating on fresh-salmon fillets preservation. *LWT* 75:59–64.

Ward, A.G., and A. Courts.1977. *Science and technology of gelatin.* Academic Press.

Willett, W., Rockström, J., Loken, B., Springmann, M., Lang, T., Vermeulen, S., Garnett, T., Tilman, D., DeClerck, F., Wood, A. and M. Jonell. 2019. Food in the Anthropocene: the EAT–Lancet Commission on healthy diets from sustainable food systems. *The Lancet* 393(10170):447–492.

Wu, H.C., Chen, H.M. and C.Y. Shiau. 2003. Free amino acids and peptides as related to antioxidant properties in protein hydrolysates of mackerel (*Scomberaustriasicus*). *Food Research International* 36(9–10),949–957.

Yang, J.I., Ho, H.Y., Chu, Y.J. and C.J. Chow. 2008. Characteristic and antioxidantactivity of retorted gelatin hydrolysates from cobia (*Rachycentron canadum*) skin. *Food Chemistry* 110(1):128–136.

Yang, X.R., Zhao, Y.Q., Qiu, Y.T., Chi, C.F. and B. Wang. 2019. Preparation andcharacterization of gelatin and antioxidant peptides from gelatin hydrolysate of skipjack tuna (*Katsuwonuspelamis*) bone stimulated by *in vitro* gastrointestinal digestion. *Marine Drugs* 17(2):78.

Yao, Y., Yuan, X., Wang, M., Han, L. and X. Liu. 2021. Efficient pre-treatment of waste protein recovery from bovine bones and its underlying mechanisms. *Waste and BiomassValorization*, 1–11.

Ye, M., Zhang, C., Jia, W., Shen, Q., Qin, X., Zhang, H. and Zhu, L. 2020. Metabolomics strategy reveals the osteogenic mechanism of yak (Bos grunniens) bone collagen peptides on ovariectomy-induced osteoporosis in rats. *Food & Function* 11(2):1498–1512.

Yue, J. 2017. Study of properties and application of chicken bone protein gel. Master's thesis. Chinese Academy of Agricultural Sciences.

Yuliani, D., Awalsasi, D.R. and A. Jannah. 2019. Characterization of gelatin profile ofchicken broiler (*Gallus domestica*) bone using SDS-PAGE electrophoresis. *Alchemy* 7(1):7–12.

Zague, V., de Freitas, V., Rosa, M.D.C., de Castro, G.Á., Jaeger, R.G. and G.M. Machado-Santelli. 2011. Collagen hydrolysate intake increases skin collagen expression and suppresses matrix metalloproteinase 2 activity. *J.Med. Food* 14, 618–624.

Zakaria S., and N.H.A. Bakar. 2015. Extraction and characterization of gelatin from Black Tilapia (*Oreochromis niloticus*) scales and bones. International Conference on Advances In Science, Engineering, Technology And Natural Resources. ICASETNR-15, 27–28.

Zavareze, E. D. R., Silva, C. M., Salas-Mellado, M. and C. Prentice-Hernández. (2009). Functionality of bluewing searobin (*Prionotus punctatus*) protein hydrolysates obtained from different microbial proteases. *Química Nova* 32:1739-1743.

Zdzieblik, D., Oesser, S., Gollhofer, A. and D. König. 2017. Improvement of activity-related kneejoint discomfort following supplementation of specific collagen peptides. *Applied Physiology, Nutrition, and Metabolism* 42(6):588–595

Zeng, Y., Zhu, L., Han, Q., *et al.* 2015. Preformed gelatinmicrocryogels as injectablecellcarriers forenhancedskin woundhealing. *Actabiomaterialia* 25:291–303.

Zhang, F., Xu, S., and Z. Wang. 2011. Pre-treatment optimization and properties of gelatin from fresh water fish scales. *Food and Bioproducts Processing* 89(3):185–193.

Zhang, Q., Fu, R., Yao, K., Jia, D., He, Q. and Y. Chi. 2018. Clarification effect of collagen hydrolysate clarifieron chrysanthemum beverage. *LWT* 91:70–76. doi:10.1016/j.lwt.2018.01.041.

Zhao, W. H., Chi, C. F., Zhao, Y. Q. and B. Wang. 2018. Preparation, physicochemical and antioxidant properties of acid-and pepsin-soluble collagens from the swim bladders of miiuy croaker (*Miichthys miiuy*). *Marine drugs* 16(5):161.

Zhou, Q.C., Tan, B.P., Mai, K.S. and Y.J. Liu. 2004. Apparent digestibility of selected feed ingredients for juvenile cobia *Rachycentron canadum*. *Aquaculture* 241(1–4):441–451.

Zhu, L., Xie, Y., Wen, B., Ye, M., Liu, Y., Imam, K.M.S.U., Cai, H., Zhang, C., Wang, F. and F. Xin. 2020. Porcine bonecollagen peptidespromote osteoblast proliferation and differentiation by activating the PI3K/Akt signalling pathway. *Journal of Functional Foods* 64:103697.

8 Processing of Coffee and Tea Waste

Prajya Arya[1] and Pradyuman Kumar[2,]*
[1]Research Scholar, Department of Food Engineering and Technology,
Sant Longowal Institute of Engineering and Technology
(Deemed-to-be University), Longowal Punjab, India.
[2]Professor, Department of Food Engineering and Technology,
Sant Longowal Institute of Engineering and Technology
(Deemed-to-be University), Longowal Punjab, India.
*Corresponding author.

CONTENTS

8.1 INTRODUCTION

After water, coffee and tea are the second most squandered drinkables in the world. Coffee and tea are a consumed drink for caffeine intake by adults, although their popularity depends on the country along with regional locations (USA, NCA, 2014). Coffee and tea intake in small amounts become habitual for adults, middle, and old age people (Bhatti et al., 2013). As per reports, coffee consumption and waste generation increased simultaneously. Recent reports on the estimated international production of coffee are approximately 169 million of 60 kg bags of coffee from 2019–2020 and it is increasing year by year (International Coffee Organization, 2019). Generally, to meet the production criteria, there are some mischievous practices occurring, such as adulteration. In the case of coffee, adulteration is done by addition of color, chicory powder, and tamarind seed to fulfil the market demand. Tea adulteration is commonly done by addition of exhausted tea leaves, hawthorn and beech leaves and dried bilberry leaves, cereal starch is also added as an adulterant (Sharma et al., 2020)

Reports on the intake of coffee and tea could provide a brief idea about the health benefits, like cardiovascular effects, weight loss, and long-term life expectancy (O'Keefe et al., 2013). The health effect of coffee has been studied for an overall decrease in cardiovascular diseases when four cups of coffee are consumed per day. It might reduce mortality rates as per the European Prospective Investigation into Cancer (EPIC) (Gunter et al., 2017). While in the case of tea, suppression of cancer mortality and cardiovascular diseases' mortality was observed with cohort studies (Tang et al., 2015). The likelihood of preferring coffee or tea might depend on taste preferences, socio-economic factors, genetics, health, and lifestyle (van den Brandt, 2018).

The consumption of coffee and tea by a huge population generates a vast amount of waste acquired from the landmass. Coffee and tea consumption are heightening year by year and in its residual production. The waste generated by coffee and tea producing industries creates an issue for its appropriate utilization and disposal by leaving behind minimal waste back to the environment (Morikawa and Saigusa, 2008). Most of the waste is generated from ready to drink beverages, which produce a considerable amount of tea waste and coffee grounds. The sustainable usage of coffee and tea waste is important for left-over utilization (Morikawa and Saigusab, 2011). There is a dire requirement to investigate the matter of waste management generated from the coffee and tea industries. There are various sorts of methods adopted by different scientists for coffee and tea waste disposal. Waste utilization of coffee and tea was studied by Silva et al. (1998) and Yang et al. (2003) in the form of fuel production and broiler diet that affects cholesterol reduction in meat and body weight while keeping the efficiency unaffected. Waste generation is one of the eye-catching problems of many scientists in the world because it is increasing day by day. Its disposal is also an alarming problem, which was previously resolved by landfilling and combustion. When waste materials are landfilled, they take a huge amount of time to degrade while affecting nearby life cycles (Joseph et al., 2012). While in the case of combustion, waste material is utilized as a fuel in the incinerators of waste power plants (Eboh et al., 2019). When these wastes are not filtered properly, they may pollute air by releasing a harmful gaseous substance into the air aid, generating lung and cardiovascular diseases (Cohen et al., 2005). This could be resolved by choosing to recycle the waste, which increases the provision of new materials less harmful to the environment (Ma et al., 2019). This chapter focuses on waste utilization from coffee and tea processing and their usage in developing various natural products that benefit the environment and mankind.

8.1.1 Coffee

Coffee scientifically belongs to the Rubiaceae family of the Coffea genus. Coffee is one of the most consumed beverages in the world and approximately 103 species are recognized to date but only two species are more well known: arabica and cenophora (Naranjo et al., 2011). Out of these two, the arabica variety has established more than 60% of the coffee trade in the international market. It is also enlisted as the best coffee due to its exceptional organoleptic features (Suarez et al., 2012) due to the presence of a variety of chemical compounds that cause better sensory quality that triggers the nervous system (Gotteland et al., 2007). Coffee is generally prepared from roasted beans, where conversion from green to red indicates their ripeness and finally they are picked, processed, and dried. Dried coffee seeds are generally called beans and depend upon the degree of roasting to generate the desired flavors in the coffee. The roasted coffee beans are ground and brewed to prepare coffee as a drink as shown in Figure 8.1. Coffee is usually dark-colored, bitter, slightly acidic, and neuro stimulant due to caffeine content in coffee beans (Cappelletti et al., 2015). A general overview represents the million cubic tons of waste generated from the coffee processing industry. Coffee is enriched in phenolic compounds, phytonutrients, and melanoidins, contributing to better health (Arya et al., 2021). After coffee production, there is a huge number of residues left behind, which hinder the natural environment. In the sector of the agricultural value-chain, the generation of by-products is always alarming, because it is generally 80% of the total volume. In the case of coffee industries, coffee pulp, silver-skin, husks, peel, and spent coffee grounds are the most common by-products that create approximately 2 billion tons of agro-waste, which cause damage to the environment (Bandara and Chalamaiah, 2018).

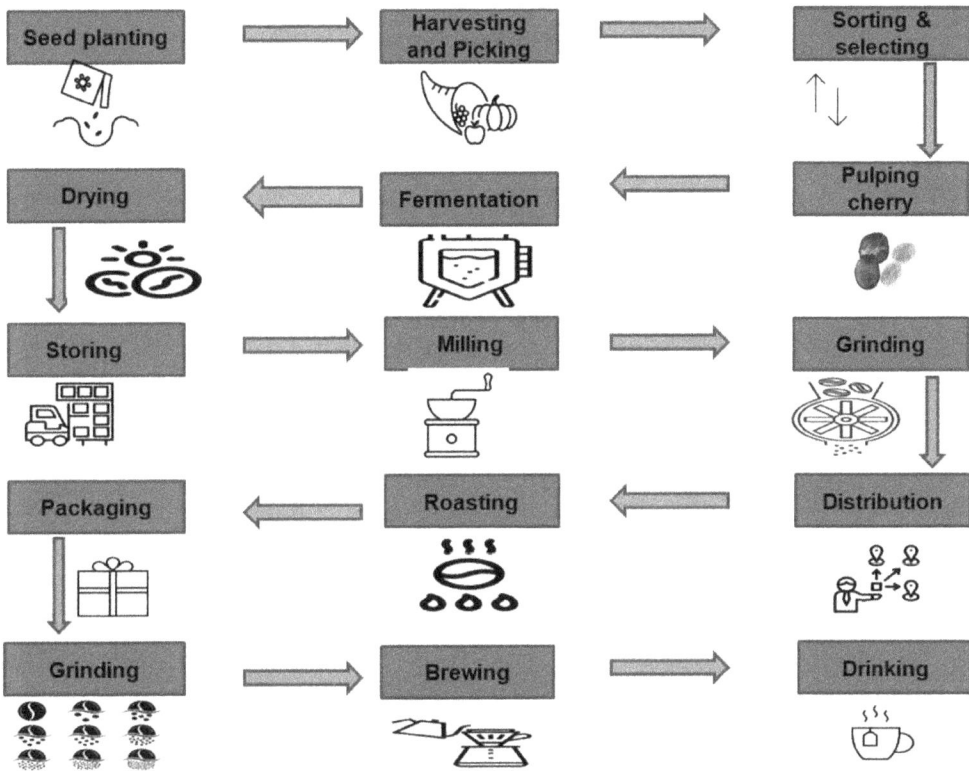

FIGURE 8.1 Overview of coffee processing from farm to cup.

8.1.2 TEA

Tea is the second most-consumed aromatic beverage made by pouring hot water over fresh or cured leaves of Camellia sinensis, which belongs to the family Theaceae. The processing of tea leaves develops various types of tea flavors like nutty, sweet, floral, astringent, and grassy (Ody, 2000). Tea consumption depends on the geographical region and country. Generally, asians prefer to consume green and oolong tea while North Americans and Europeans like to have black tea due to its less astringency and mellow flavor. Major tea producing countries are China, India, Sri Lanka, and Kenya, while China is the leading producer contributing approximately 40% of tea production in the global market. The ratio of different types of teas in China in the year 2017 is as follows: green (67.3%), dark (11.6%), oolong (10.9%), black (9.1%), white (1.0%), and yellow tea (0.2%). Out of so many varieties, black and green tea cover the major tea market that accounts for 75% and 15% globally (Zhang et al., 2019). The production processes of the most consumed tea, such as white, green, black, matcha, oolong, and Pu-erh tea, are represented in Figure 8.2. Generally, tea processing industries produce a huge amount of waste, and its disposal is mandatory keeping the environment (soil, water, and air quality) safe (Mugambi, 2006). This waste could open new ways to generate the economy by processing it in the most suitable ways. Tea waste includes buds, tea leaves, and tender stems of the tea plant. Approximately 2 kg of waste is produced by processing 100 kg of tea. On the daily basis of tea processing, waste amount sums up to a huge quantity. Tea waste could be processed by mixing urea, cow dung, and other denaturants not less than 5% as per the Tea Board of India. It is used to extract caffeine, produce poultry, and piggery feed (Konwar, 1988; Chutia et al., 1983).

FIGURE 8.2 Various types of tea produced in the market.

The most common method of coffee and tea waste disposal includes landfilling, incineration, and mixing with animal feed. Due to the high amount of caffeine and tannin content, soil quality might get affected and could induce carcinogenic properties in animal feed (Arya et al., 2021). So, a requirement of coffee and tea industry waste management is, among other important criteria, to maintain environmental aspects and to generate new ways to run the economy from by-products of these industries.

8.2   COFFEE AND TEA BOARD

India is among one of the top producers of coffee and tea in the international market. The production of coffee and tea is monitored by the Coffee Board and Tea Board, respectively. The Coffee Board is a statutory organization formulated in 1942 with administrative control of the Ministry of Commerce and Industry, Government of India. The Coffee Board is located in Bangalore (head office) with a regional coffee research station at Diphu (Assam), R.V. Nagar (Andhra Pradesh), Chettalli (Karnataka), Chundale (Kerala) and Thandigudi (Tamil Nadu). The Coffee Board is focused to look after research, development, quality, economic growth, labor welfare, and external and internal promotion of coffee. The Tea Board is an apex body that looks after the overall development of the tea industry and was set up in 1954, under the Tea act, 1953. The Tea Board is in Kolkata (head office) with two zonal offices Jorhat (Assam, northeastern region) and Coonoor (Tamil Nadu, southern region) with three overseas offices located in Dubai, London, and Moscow for export and promotion of Indian tea. The major purpose of the Tea Board is to promote production, quality, market revenue, welfare measures, and research development. Along with this, the board exerts control over producers, manufactures, brokers and auction organizers via control order notified under the Tea act (Commodity Board, India).

8.3 BOTANICAL ASPECTS OF COFFEE AND TEA

Coffee and tea botanical classifications are represented in Figure 8.3. As per historic records and documentation, coffee is native to east African highlands, which grow best with frequent rain but not with high temperatures. It is cultivated in high tropical regions and is approximately 600–1200m above sea level. The coffee plant consists of a woody shrub of 12 meters high but could be pruned to 2 meters to make harvesting easier. The coffee plant contains white flowers, which produce red-colored coffee cherry fruit, and a pair of beans is enclosed inside the coffee cherry. Coffee harvesting is mainly done by hand. The coffee beans are procured after the processing of fruit and allowed to dry and roast depending on the type of coffee to be produced. Dried coffee beans hold flavor for up to a year, while roasted coffee beans might lose flavor and should be consumed within several weeks (Dicum and Nina, 1999).

Camellia sinensis is one of the oldest species of tea primitive to the Yunnan region of China as recorded by many scientists (Huang et al., 2016). Some researchers also suggested that the Northern India region Assam and Sri Lanka also contribute to major tea production with the most appropriate climate condition. Tea is mostly grown in the subtropical region with climate ranging from warm dry temperature to wet with variation between dry and moist forest life zone. Tea plants can bear an annual temperature variation from 14 to 27°C and annual precipitation of 7 to 31 cm. Tea growth could be possibly affected by soil type as for better growth of Camellia sinensis, light, friable loam, which has porous sub-soil that allows free water percolation. Tea growth is properly maintained by constant pruning of shrubs from time to time to check the growth of leaves. As the tea plant reaches a height of 1.5m, a flat-topped shrub is the time to begin pruning again by removing the bush back to between 0.6 to 1.2 m as this would aid in the rejuvenation of the tea plant. The tea plant is mainly

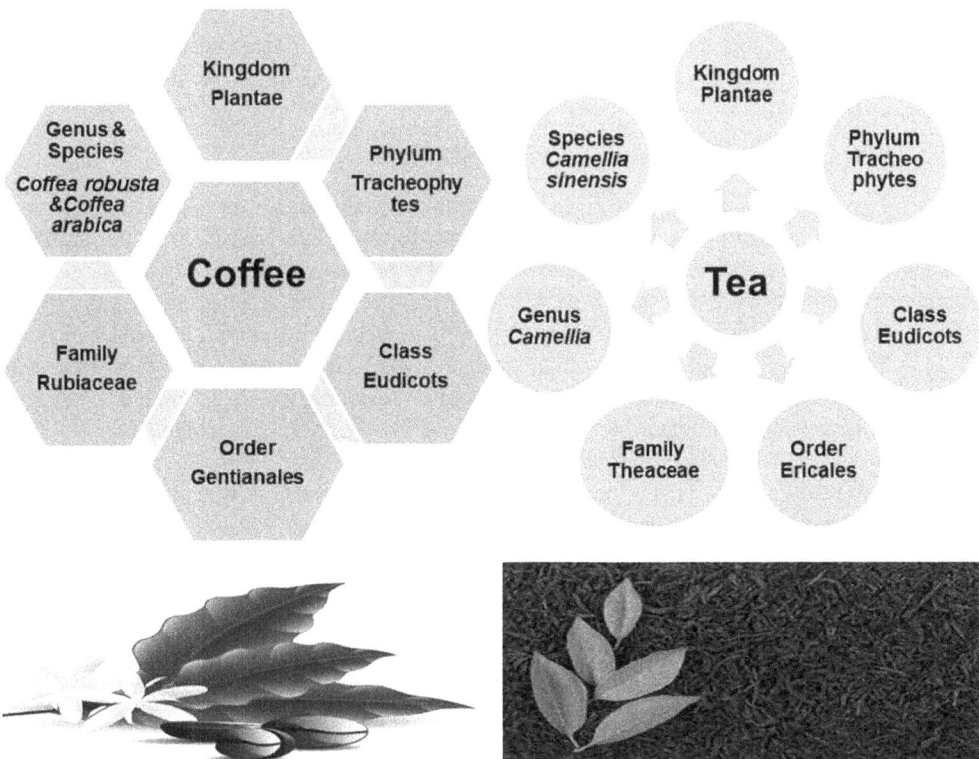

FIGURE 8.3 Biological classification of coffee (Purvot-Woehl et al., 2020) and tea (Xia et al., 2017).

used as a beverage comprised of approximately 200 species. Out of these genera, a few subgeneric into 12 sections, among them "thea" genus contains cultivated tea (Sealy, 1958). Further Camellia genus becomes complicated by cross-hybridization between species (Chuangxing, 1988). Among the so many genera of tea, Camellia has been used widely due to its commercial and taxonomical properties. Although the various types of tea are cultivated in different countries across the globe. Each type of tea has its own recognizable aromatic characteristic flavor (Wachira et al., 2013).

8.4 TOXICOLOGICAL EFFECT OF CAFFEINE CONSUMPTION

The consumption of a small amount of coffee and tea is capable of awakening our nervous system for a fixed interval of time. Coffee and tea are rich sources of caffeine (1,3,7- trimethyl xanthine), which is a psychostimulant purine like alkaloid (Heckman et al., 2010). Normally, the consumption of caffeine-containing products triggers the central nervous system (CNS) (Gracia-Lor et al., 2017). The prominent intake of caffeine is increased under the name of energy drinks due to its neurological stimulating properties. In the last few decades, it has been observed that caffeine consumption has been increased up to 80%, mainly in the USA (Verster et al., 2018). According to age group and personal choice, caffeine intake may vary but a safe limit of caffeine consumption is less than 400 mg per day and in a person approximately weighing 65 kg with 6 mg per kg body weight (Nawrot et al.2003). Generally, caffeine intake does not induce toxicity until it has been consumed within prescribed limits (Jones, 2017). In the case of excess caffeine consumption, a concentration of 15 mg/L would produce some psychological side effects like palpitations, nausea, tremor, nervousness, perspiration. When caffeine concentration is near 50 mg/L and 80 mg/L it is toxic and lethal respectively. It is generally advised that individuals suffering from cardiovascular problems should consume caffeine content less than 50 mg/L (Banerjee et al., 2014). Caffeine toxicities occur by hindering the cellular mechanism by altering the cell signaling pathway. In a caffeine affected cellular mechanism, the relevant concentration of adenosine receptors, G-protein coupled receptors and transmembrane receptors are disturbed (Kobilka, 2007). This hindered cellular signal pathway can create toxicity, which affects the neuro-transmittance of information (Wilson, 2018).

8.5 WASTE GENERATION OF COFFEE AND TEA

The waste generated from the coffee and tea industries always imposes some serious environmental threat. The requirement of policies, incentive programs, and subsidies are encouraged to utilize the coffee and tea waste more fruitfully. In the case of coffee processing, the main waste is generated during the separation of coffee beans from coffee cherries. The waste includes cherry pulp, residual water, and parchment. This waste creates surrounding that contains bad odor, pathogen growth, and pollution of nearby soil and water bodies (Beyene et al., 2012).The utilization of coffee waste could be achieved in a more suitable way by valorizing coffee residues, which generate a value added product for mankind and the environment (Murthy and Naidu, 2012). The method of landfilling in coffee and tea waste generates problems like greenhouse gas emissions (like carbon dioxide, methane) (Vaverková, 2019). This issue could be reduced to some extent by incineration with concomitant generation of thermal energy, which reduces a huge amount of solid waste volume (Lee et al., 2021). This method caused the probability for production of dioxins, particulate matter along with polycyclic aromatic hydrocarbons that impose a threat to the environment (Banu et al., 2020). This issue could be resolved by methods adopted like pyrolysis. Pyrolysis is a thermal process that includes the conversion of lignocellulosic matter into liquid (bio-oil), solid (biochar), and gaseous (pyro gas) products in absence of oxygen (Atabani et al., 2019). Pyrolysis depends on the composition of biomass as a huge quantity of oxygenated compounds in liquid products there is a requirement of catalyst (Dhyani et al., 2018). During processing, catalyst aids in the conversion of unwanted products into stable form followed by removal of oxygen (Dias et al., 2019). Some catalysts are used frequently like HZSM5, spent FCC (acidic catalyst) to $MgCO_3$, Li_2CO_3, Na_2CO_3 (basic catalyst) that aids in suppressing bio-oil acidity. So, waste from industry could be used in the production of renewable fuel (Ferreira et al., 2020).

Coffee waste comprises coffee fruit, silver-skin, leaves, and mucilage (Torres-Valenzuela et al., 2020). The removal of initial waste from coffee processing helps in achieving a quick ripening stage by hindering the fermentation process as the entire coffee bean is a rich source of water and sugar. Generally, coffee processing is done in two ways: (a) a wet process includes drying of the bean after removal of fruit covering layer; (b) a dry process involves sun-drying of coffee cherries without removal of layers. In most coffee consumption areas, wet-processing is preferred due to its better taste (Dadi et al., 2018). The main concern with coffee processing is the production of 40–45 liters of wastewater produced per kilogram of coffee processing that requires a proper disposal method (Péerez et al., 2007). This problem paved the way for the production of new methods, such as the Belcosub technique for de-pulping of fruit without water utilization. The removed fruit layer is used as an organic residue and did not generate much substantial value. This method helped in less utilization of water and averted 74% of contamination from water sources (Rodríguez et al., 2015).

Another method used to reduce coffee waste is ecomil, a method adopted for ecological mil without dumping suggested by the Federación Nacional de Cafeteros—Fedecafé, which is also recognized as the National Federation of Coffee Growers (NFCG). This technique reduced water utilization up to 0.5 L per kilogram of dry parchment. The use of stainless-steel tanks is also not required for coffee emptying. The wastewater from stainless-steel tanks is transferred to purifying tanks that contain microbes. This feed is further filtered and could directly go to water sources without creating any sort of pollution (Torres-Valenzuela et al., 2020). The overall waste generated from the coffee industry is approximately 44% of the weight of dried fruit by-products (Kovalcik et al., 2018). The suitable utilization of this product could produce valuable food products (by using coffee leftover bioactive compounds), cosmetics, and pharmaceutical industries (McNutt et al., 2019; Torres-Valenzuela et al., 2019).

Coffee processing waste includes coffee husk, which is an outermost layer with a hard structure found after coffee silver-skin separation. The coffee husk contains bioactive compounds like polyphenols, pectin, and caffeine. It contains some amount of carbohydrates especially monosaccharides and disaccharides (Janissen and Huynh, 2018). Various studies have been conducted for observing the potential of coffee husk in degradation of oxidative stress, anti-allergic properties, and antimicrobial properties.

Coffee husk helps regulate blood glucose levels and stimulate insulin regulation. As the name suggests, coffee husk is a rich source of dietary fiber and probiotic effect, which can be easily employed in different food formulations that aid in regulating satiety feelings (Gomez-Gonzalez et al., 2016). The use of coffee husk in various food products is under development of emerging nutraceutical food products. A study conducted by Arya et al (2021) equated *Aspergillus oryzae* based fructooligosaccharide formation through solid-state fermentation based on coffee husk, sugarcane bagasse, banana peel, pineapple peel, and prickly pear peel. This study depicted the best fermentation yield after 12 h was with sugarcane bagasse approximately 7.64 g/L. The coffee husk and prickly pear did not respond for fructooligosaccharide formation that attributed toward no microbial activity due to its physicochemical properties.

Other studies revealed more about the chemical properties of coffee husk, which are easily extracted by solid-state fermentation with the least operational cost. It is generally used as a substrate in various bioprocesses. Coffee husk has been used as a substrate for enzymes like cellulases, polyphenol oxidase, polygalacturonase, and xylanases. Coffee husk has also been a substrate to produce a few acids, such as gibberellic acid, citric acid, polyhydroxyalkanoates (PHA), and gallic acid. After using coffee husk in enzyme and acid production, mass production of microorganisms is another reward of coffee husk as a substrate. The various and vast usage of coffee husk makes it an attractive source of an economy base of waste recycling from the coffee industry (Kumar et al., 2018).

Apart from the coffee husk, silver-skin, spent coffee grounds (SCG), there is one more waste produced during coffee processing known as coffee flour. Coffee flour is prepared from coffee cherries by drying and milling. Coffee flour is gluten-free enriched with antioxidant properties and their extraction can be achieved using low-cost methods. It has 42% more fiber and 84% less fat compared to coconut flour. It imparts a floral, roasted fruity note, which does not taste like coffee with a lesser amount of caffeine in it. It has been used in preparation of brownies and granola with some more application in food products like pasta, beverage, and sauces. Coffee flour is also rich in fiber content and iron thrice as much as spinach. It has non-heme iron with low bioavailability due to the presence of caffeine, calcium, zinc, magnesium, and dietary fibers recognized with less absorption of nutrients. During production of coffee flour, a degradation of amino acid, vitamin A, and vitamin C can occur (Gemechu, 2020).

The waste generated from tea processing industries is abundant in hemicellulose and lignin. These wastes are a good reservoir of phenolic hydroxyl, oxyl, hydroxyl, carboxyl, oxygen, and heteroatoms containing groups. Just like coffee waste, tea waste could also be processed with the pyrolysis technique. Waste tea leaves could produce carbonaceous materials that could be used in adsorptive usage due to large surface area, much functional group, and porous structure (Hussain et al., 2018). Research conducted on the pruned tea leaves depicted the high content of nitrogen (caffeine and amino acid) that can be intituled into biochar, which is further used as a nitrogen-rich biomass precursor for developing contaminant adsorbents (Li et al., 2020). Along with this usage, biotechnological methods were also adopted to utilize tea waste. It includes the entire employment of tea waste like ensiling and composting by extraction of bioactive compounds (polyphenols, lignin, cellulose, water-soluble protein). These bioactive compounds aid in controlling blood glucose, fat, pressure by hindering oxidation resistance, and scavenging free radicals. Tea waste usage as for developing biochar, bio-adsorbent for wastewater, development of animal feed, and agricultural compost (Guo et al., 2021).

The utilization of various waste from coffee industries in the form of food products is enlisted in Table 8.1. Most of the food product developed from coffee waste are still at pilot level production and they need further scientific approval.

TABLE 8.1
Utilization of Coffee Industries' Waste in Development of Various Food Products

S. No	Waste product	Usage in Food form	Reference
1.	Flowers and leaves	Beverage as tea prepared from coffee flowers and leaves	Nguyen et al., 2019, Campa et al., 2018, Yuwono et al., 2019.
2.	Coffee pulp	Jam, jelly, pulp flour bread production	Madahava et al., 2004, Janissen and Tien, 2018
3.	Husk dried cherries	Beverage Spirits Dietary fibre source	Heeger et al., 2017 Velissariou et al., 2010 Carlsen, 2020
4.	Green unroasted beans	Dietary supplement White coffee	Marcason, 2013 Macheiner et al., 2019
5.	Silver skin	Smoke flavour additive Dietary fiber source, Beverage as tea	Costa et al., 2018 Jiménez-Zamora et al., 2015 Ateş et al., 2019
6.	Spent coffee ground	Coffee adulteration Bakery product	Sampaio et al., 2013 Vázquez-Sánchez et al., 2018
7.	Parchment	Food preservative, Antioxidant	Torres-Valenzuela et al., 2020 Mirón-Mérida et al., 2019
8.	Coffee skin flour	Potential dietary fibre source	Borrelli et al., 2004

8.6 UTILIZATION OF COFFEE AND TEA WASTE

8.6.1 Coffee Waste Use in Various Forms

Coffee waste includes spent coffee ground (solid residue produced during the processing of coffee beans). SCG also known as coffee drags are produced during coffee brewing. This waste is dispersed in nature and its collection depends on the location of consumption (coffee shop, cafeteria, home, catering services). SCG waste disposal includes the landfilling method as it contains a high amount of water and the incineration method is ruled out. SCG usage could be done in animal feed, fermentation, and composting. Silver-skin and SCG are rich in organic compounds, meaning they need a greater amount of oxygen to degrade as represented in Figure 8.4. They are also enriched in caffeine, tannin, and polyphenols, which can air toxicity in a near environment (Arya et al., 2021).

SCG utilization in production of valuable compounds was performed in many ways. The usage of SCG was studied for enhancing the nutritional value of lettuce by suppressing the pH of soil (Cervera-Mata et al., 2019). Further SCG is employed in the production of biochar for CO_2 reduction by actuating coffee grounds at 400°C (Liu and Huang, 2018). The combination of SCG with polypropylene can enhance mechanical performance of biocomposite (Essabir et al., 2018). Furthermore, use of coffee waste was studied for hydrogen holding capacity of carbon-based material using micro-pores and the wide space area of coffee beans. These studies are based on the porous nature of coffee waste (Akasaka et al., 2011). Coffee silver-skin (CS) is a by-product obtained after roasting coffee and is cutis of coffee beans. CS is rich in dietary fiber (54–62%) and polyphenol content, cellulose, hemicellulose, melanoidins compound produced by Maillard reaction during roasting and results in antioxidant-rich coffee silver-skin (Borrelli et al., 2004). CS is the best utilization in the form

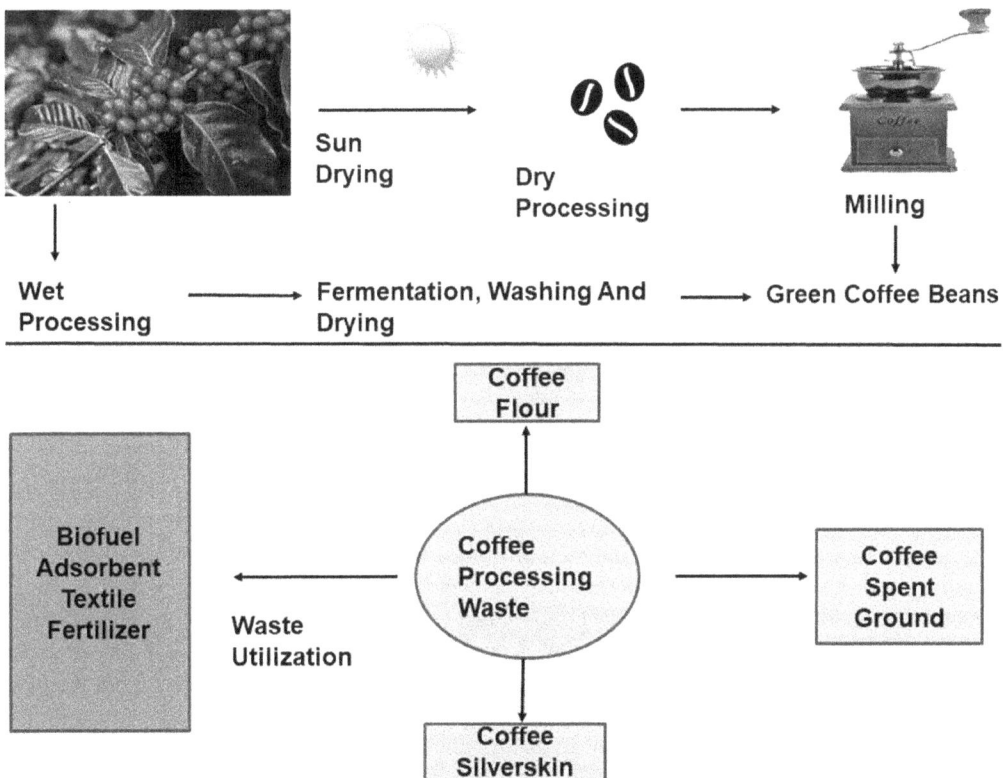

FIGURE 8.4 A systematic representation of coffee processing and coffee waste utilization.

of solid-state fermentation as CS is rich in glucose. Fermentation leads to the generation of phenolic compounds as studied with selected microbial genus like Aspergillus, Mucor, and Penicillium (Machado et al., 2012).

8.6.2 Tea Waste Usage in Various Forms

Tea waste generates an alarming situation and there is a pressing requirement to investigate the matter. Tea waste contains an elite number of polyphenols, which proves to be beneficial for human health. Tea polyphenols from waste was 47.70 mg catechin/g of spent black tea (Abdeltaif et al., 2018). These polyphenols could be recovered from tea waste by the steam explosion activation method (Sui et al., 2019). Tea polyphenol content is well recognized for improving human health by controlling blood pressure cancer cell growth due to their antioxidant potential. These bioactive compounds of tea waste are highest in black tea waste as compared with green and oolong tea. Black tea waste is enriched with antihypertensive and anticancer effects due to flavonol compounds like patuletin, quercetin, and kaempferol (Xue et al., 2018).

Along with polyphenols, tea waste is used in preparation of metal-based nanoparticles in the field of adsorbents. Tea catechins work as capping and reducing agents for production and stabilization of metal-based nanoparticles. Polyphenol structures enable delocalization of electrons, which cause high reactivity towards free radical quenching. The nanoparticles prepared from tea waste ranged from 20 to 60 nm with shapes varied as circular, triangular, and hexagonal (Rajput et al., 2020). The new ways for creating nanoparticles and their usage in the betterment of the environment is always an eye-catching area of research (Debnath et al., 2020). Tea waste prepared nanoparticles are sufficiently good for replacement of harmful dye from textile industry wastewater, which further pollutes nearby water bodies. The production of metal nanoparticles by using tea waste can be an efficient, environmentally friendly, and cheap way. This method does not include any sort of harmful chemical or toxic by-product formation (Gautam et al., 2018). Furthermore, the usage of tea waste is observed in the generation of bio-adsorbents and biochar. The pyrolysis of biomass in an oxygen deprived environment leads to the production of biochar (Tripathi et al., 2016). Biochar had been used for adsorbing heavy metals, organic dyes from wastewater. The efficiency of biochar enhanced due to its porous structure and presence of various surface functional groups led to a great affinity for metal ions (Wu et al., 2021). The preparation of cellulose nanocrystals from waste tea stalk with acid hydrolysis exhibited better thermal stability. So, tea stalk waste could also be used in the production of crystalline cellulosic derivates for analyzing the strength and potential of cellulose nanocrystal (Debnath et al., 2021).

Among the usage of coffee and tea, there is various other users as they are enlisted in Table 8.2. This usage could also impart a new way to utilize waste in a much better way.

8.7 WASTE AND ECONOMY

The concept of processing coffee and tea produces a huge amount of waste per year, and it took many steps to deliver fresh roasted coffee and tea at respective shops. The waste creates an awareness in the consumer for its appropriate disposal. But unfortunately, it is limited to bins only from the end consumer side. The question of coffee and tea waste could provide new ways to run an economy well or not. The answer lies in a circular economy that focuses on reusing, recycling, and reshaping by-products of coffee and tea waste industries. A circular economy helps various companies to prepare their economic structure well with the usage of waste products as shown in Figure 8.5. This way of economic generation could help in the reduction of a much greater amount of waste at a ground-to-ground level. It includes the most awareness related to waste usage like coffee grounds in the form of compost, skin exfoliator, and hair dye. This could be achieved by providing these

TABLE 8.2
Different Types of Waste Produced from Coffee and Tea Industries and Their Potential Usage

S. No	By-product	Utilization	References
1.	Spent coffee waste	Metal biochar formaldehyde removal	Ahn et al., 2021
2.	Coffee waste	Sound absorption in café	Yun et al., 2020
3.	Spent coffee ground	Bio-oil production by pyrolysis	Rijo et al., 2021
4.	Coffee waste	Nanogenerator self-powered sensor	Li et al., 2021
5.	Spent coffee grounds	Inhibition of cytokine production	Ho et al., 2020
6.	Dried spent coffee grounds	Nutraceutical biscuit formulation	Ali et al., 2018
7.	Waste coffee grounds	Fluorescent carbon dots	Hong et al., 2021
8.	Coffee waste	Bio-oil production by catalytic pyrolysis	Ly et al., 2022
9.	Coffee pulp waste	Rapid recovery of pectin and polyphenols	Manasa et al., 2021
10.	Spent coffee ground waste	Solid biofuel production	Lee et al., 2021
11.	Waste coffee	Biodiesel production	Uddin et al., 2019
12.	Solid coffee waste	Antioxidant and carotenoid production	Moreira et al., 2018
13.	Black tea waste	Tea waste powder	Khayum et al., 2018
14.	Spent tea waste	Fresh cow manure	Indira et al., 2018
15.	Green tea waste	Iron oxide nanoparticle adsorbate for heavy metals	Nie et al., 2021
16.	Tea waste	Biochar activator for antibiotic adsorption	He et al., 2021
17.	Tea processing waste hydrolysate	Ethanol production	Germec and Turhan, 2018
18.	Tea powder waste	Methane production co-substrate	Thanarasu et al., 2018
19.	Tea leaves waste	Mercury adsorption by magnetic tea biochar	Altaf et al., 2021
20.	Tea waste	Erichrome Black-T (EBT) dye adsoprtion	Bansal et al., 2020
21.	Black tea waste	Caffeine and catechin production	Sayar et al., 2019
22.	Tea waste compost	Better soil quality for plant health	De et al., 2020
23.	Tea waste	Chromium adsorption from wastewater	Kabir et al., 2021
24.	Black tea waste	Surplus extraction of antioxidant and phenolic compounds	Abdeltaif et al., 2018
25.	Spent Tea waste	Semi-industrial gasifier	Augustine et al., 2021
26.	Solid tea waste	Corrosion inhibition in boiler quality steel	Pal and Das 2020
27.	Tea waste	Biomass fuel pellets	Pua et al., 2020
28.	Tea waste	Cobalt nanoparticle from activate carbon	Akbayrak et al., 2020
29.	Tea leaf residue	Distillery sludge vermicomposting	Mahaly et al., 2018
30.	Tea waste	Oyster mushroom cultivation	Yang et al., 2016

leftovers to the consumers by café owners. The economy runs well when new attractive products continuously run into the market. The criteria of producing new products should be waste utilization to reduce environmental impact and create a new generation more attentive towards nature. This is based on the natural concept, as nature wastes nothing, in a similar way, a circular economy generates almost zero waste. In the past, coffee and tea waste has been utilized in manufacturing various new products like paper, charcoal, textiles, and many more. The bio-bean is another name given to biofuel, which is second generation fuel that focuses on non-food origin. Hence utilizing the waste of coffee and tea industries, researchers and scientists could help in bring forth emerging ways to develop economy in a more prominent way (Pike, 2018).

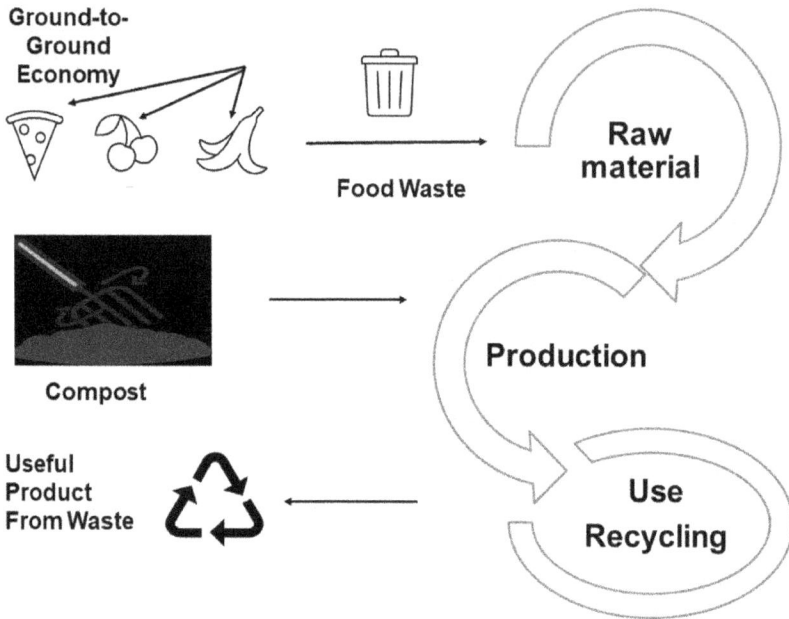

FIGURE 8.5   Representation of utilization and recycling of waste from the food industry.

8.8   CONCLUSION

The emerging concept of using waste has produced new ways for the technologist to create more useful products from waste. The waste produced from the coffee and tea industries has surpluses and researchers across the world need to investigate the matter further. This issue became more alarming when the conventional method of disposal created environmental problems: the nearby polluting environment from waste by food industries created hazardous substances. This problem could occur in the face of various serious metabolic and cellular diseases. The waste generated from coffee and tea processing plants has produced by-products that aid in the production of fertilizers, biofuel, cosmetics, adsorbents, and textile products. Along with environment efficient products, this waste can be used as tea, for preparation of bread, sprits, antioxidants, and a rich source of polyphenols. These products aid in the better economic growth of the nation and produces almost zero waste by adopting the ground-to-ground economy concept. Hence, this chapter imposed various ways to use coffee and tea processing waste in an eco-friendlier manner also by creating new ways for empowerment.

REFERENCES

Abdeltaif, Samar A., Khitma A. SirElkhatim, and Amro B. Hassan. "Estimation of phenolic and flavonoid compounds and antioxidant activity of spent coffee and black tea (processing) waste for potential recovery and reuse in Sudan." *Recycling* 3, no. 2 (2018): 27. doi:10.3390/recycling3020027.

Ahn, Yongtae, Dong-Wan Cho, Waleed Ahmad, Jungman Jo, JongsooJurng, Mayur B. Kurade, Byong-Hun Jeon, and Jaeyoung Choi. "Efficient removal of formaldehyde using metal-biochar derived from acid mine drainage sludge and spent coffee waste." *Journal of Environmental Management* 298 (2021): 113468. doi:10.1016/j.jenvman.2021.113468

Akasaka, Hiroki, TomokazuTakahata, Ikumi Toda, Hiroki Ono, Shigeo Ohshio, SyujiHimeno, ToshinoriKokubu, and Hidetoshi Saitoh. "Hydrogen storage ability of porous carbon material fabricated from coffee bean wastes." *International Journal of Hydrogen Energy* 36, no. 1 (2011): 580–585. doi:10.1016/j. ijhydene.2010.09.102.

Akbayrak, Serdar, Zehra Özçifçi, and Ahmet Tabak. "Activated carbon derived from tea waste: A promising supporting material for metal nanoparticles used as catalysts in hydrolysis of ammonia borane." *Biomass and Bioenergy* 138 (2020): 105589. doi:10.1016/j.biombioe.2020.105589

Ali, Hatem S., A. Farouk Mansour, M. M. Kamil, and A. M. S. Hussein. "Formulation of nutraceutical biscuits based on dried spent coffee grounds." *International Journal of Pharmacology* 14, no. 4 (2018): 584–594. doi:10.3923/ijp.2018.584.

Altaf, Adnan Raza, Haipeng Teng, Maoshen Zheng, Imtiaz Ashraf, Muhammad Arsalan, Ata Ur Rehman, Liu Gang, Wang Pengjie, Ren Yongqiang, and Li Xiaoyu. "One-step synthesis of renewable magnetic tea-biochar derived from waste tea leaves for the removal of Hg0 from coal-syngas." *Journal of Environmental Chemical Engineering* 9, no. 4 (2021): 105313. doi:10.1016/j.jece.2021.105313

Arya, Shalini S., Rahul Venkatram, Pavankumar R. More, and Poornima Vijayan. "The wastes of coffee bean processing for utilization in food: a review." *Journal of Food Science and Technology* (2021): 1–16. doi:10.1007/s13197-021-05032-5 2021

Atabani, A. E., H. Ala'a, Gopalakrishnan Kumar, Ganesh Dattatraya Saratale, Muhammad Aslam, Hassnain Abbas Khan, Zafar Said, and Eyas Mahmoud. "Valorization of spent coffee grounds into biofuels and value-added products: Pathway towards integrated bio-refinery." *Fuel* 254 (2019): 115640. doi:10.1016/j.fuel.2019.115640.

Ateş, Gizem, and YeşimElmacı. "Physical, chemical and sensory characteristics of fiber-enriched cakes prepared with coffee silverskin as wheat flour substitution." *Journal of Food Measurement and Characterization* 13, no. 1 (2019): 755–763. doi:10.1007/s11694-018-9988-9

Augustine, Anjan. "Spent tea waste as a biomass for co-gasification enhances the performance of semi-industrial gasifier working on groundnut shell." *Biomass and Bioenergy* 145 (2021): 105964. doi:10.1016/j.biombioe.2021.105964

Bandara, N., and M. Chalamaiah. "Bioactives from agricultural processing by-products." (2019): 472–480.

Banerjee, Priya, Zabiullah Ali, Barry Levine, and David R. Fowler. "Fatal caffeine intoxication: a series of eight cases from 1999 to 2009." *Journal of forensic sciences* 59, no. 3 (2014): 865–868.

Bansal, Megha, Prem Kishore Patnala, and Tom Dugmore. "Adsorption of Eriochrome Black-T (EBT) using tea waste as a low cost adsorbent by batch studies: A green approach for dye effluent treatments." *Current Research in Green and Sustainable Chemistry* 3 (2020): 100036. doi:10.1016/j.crgsc.2020.100036

Banu, J. Rajesh, S. Kavitha, R. YukeshKannah, M. Dinesh Kumar, A. E. Atabani, and Gopalakrishnan Kumar. "Biorefinery of spent coffee grounds waste: Viable pathway towards circular bioeconomy." *Bioresource technology* 302 (2020): 122821. doi:10.1016/j.biortech.2020.122821.

Beyene, Abebe, Yared Kassahun, Taffere Addis, Fassil Assefa, Aklilu Amsalu, Worku Legesse, Helmut Kloos, and Ludwig Triest. "The impact of traditional coffee processing on river water quality in Ethiopia and the urgency of adopting sound environmental practices." *Environmental monitoring and assessment* 184, no. 11 (2012): 7053–7063.

Bhatti, Salman K., James H. O'Keefe, and Carl J. Lavie. "Coffee and tea: perks for health and longevity?" *Current Opinion in Clinical Nutrition & Metabolic Care* 16, no. 6 (2013): 688–697. doi:10.1097/MCO.0b013e328365b9a

Borrelli, Rosa Cinzia, Fabrizio Esposito, Aurora Napolitano, Alberto Ritieni, and Vincenzo Fogliano. "Characterization of a new potential functional ingredient: coffee silverskin." *Journal of agricultural and food chemistry* 52, no. 5 (2004): 1338–1343.

Campa, Claudine, France UMR IPME, and Wize Arnaud Petitvallet. "Beneficial compounds from coffee leaves Monkey, Canada." In *Achieving sustainable cultivation of coffee*, pp. 255–276. Burleigh Dodds Science Publishing, 2018.

Cappelletti, Simone, Piacentino Daria, Gabriele Sani, and MariarosariaAromatario. "Caffeine: cognitive and physical performance enhancer or psychoactive drug?" *Current neuropharmacology* 13, no. 1 (2015): 71–88. doi:10.2174/1570159X13666141210215655

Carlsen, Z. Magic in the Moonshine: Cascara Booze Is Here. Available online: https://sprudge.com/magic-inthe-moonshine-cascara-booze-is-here-115811.html (accessed on 16 March 2020).

Cervera-Mata, Ana, Miguel Navarro-Alarcón, Gabriel Delgado, Silvia Pastoriza, Javier Montilla-Gómez, Juan Llopis, Cristina Sánchez-González, and José ÁngelRufián-Henares. "Spent coffee grounds improve the nutritional value in elements of lettuce (Lactuca sativa L.) and are an ecological alternative to inorganic fertilizers." *Food chemistry* 282 (2019): 1–8. doi:10.1016/j.foodchem.2018.12.101.

Chuangxing, Ye. "The subdivisions of genus Camellia with a discussion on their phylogenetic relationships." *Acta Botanica Yunnanica* 10 (1988): 61–67.

Chutia, S., A. Saikia, B. K. Konwar, and K. K. Baruah. "Water treated factory tea waste and pig production." In *Proceedings of the National Symposium on feeding systems for maximising livestock production, HAU, Hissar*, vol. 49, pp. 1–9. 1983.

Cohen, Aaron J., H. Ross Anderson, Bart Ostro, Kiran Dev Pandey, Michal Krzyzanowski, Nino Künzli, Kersten Gutschmidt et al. "The global burden of disease due to outdoor air pollution." *Journal of Toxicology and Environmental Health, Part A* 68, no. 13–14 (2005): 1301–1307. doi:10.1080/15287390590936166.

Commodity Board, India https://commerce.gov.in/useful-links/commodity-boards/

Costa, Anabela SG, Rita C. Alves, Ana F. Vinha, Elísio Costa, Catarina SG Costa, M. Antónia Nunes, Agostinho A. Almeida, Alice Santos-Silva, and M. Beatriz PP Oliveira. "Nutritional, chemical and antioxidant/pro-oxidant profiles of silverskin, a coffee roasting by-product." *Food Chemistry* 267 (2018): 28–35. doi:10.1016/j.foodchem.2017.03.106

Dadi, Dessalegn, EmbialleMengistie, GudinaTerefe, Tadesse Getahun, Alemayehu Haddis, Wondwossen Birke, Abebe Beyene, Patricia Luis, and Bart Van der Bruggen. "Assessment of the effluent quality of wet coffee processing wastewater and its influence on downstream water quality." *Ecohydrology & Hydrobiology* 18, no. 2 (2018): 201–211.

De Corato, Ugo. "Agricultural waste recycling in horticultural intensive farming systems by on-farm composting and compost-based tea application improves soil quality and plant health: A review under the perspective of a circular economy." *Science of the Total Environment* 738 (2020): 139840. doi:10.1016/j.scitotenv.2020.139840

Debnath, Banhishikha, Moumita Majumdar, Mahashweta Bhowmik, Kartick Lal Bhowmik, Animesh Debnath, and Dijendra Nath Roy. "The effective adsorption of tetracycline onto zirconia nanoparticles synthesized by novel microbial green technology." *Journal of Environmental Management* 261 (2020): 110235. doi:10.1016/j.jenvman.2020.110235

Debnath, Banhisikha, DibyajyotiHaldar, and Mihir Kumar Purkait. "Potential and sustainable utilization of tea waste: A review on present status and future trends." *Journal of Environmental Chemical Engineering* (2021): 106179. doi:10.1016/j.jece.2021.106179

Dhyani, Vaibhav, and Thallada Bhaskar. "A comprehensive review on the pyrolysis of lignocellulosic biomass." *Renewable Energy* 129 (2018): 695–716. doi:10.1016/j.renene.2017.04.035.

Dias, Ana Paula Soares, Filipe Rego, Frederico Fonseca, Miguel Casquilho, Fátima Rosa, and Abel Rodrigues. "Catalyzed pyrolysis of SRC poplar biomass. Alkaline carbonates and zeolites catalysts." *Energy* 183 (2019): 1114–1122. doi:10.1016/j.energy.2019.07.009.

Dicum, G., and N. Luttinger. "The Coffee Book: Anatomy from Crop to the Last Drop." (1999).

Duncan Pike What Goes Around: How Coffee Waste Is Fueling a Circular Economy (2018) https://dailycoffee ews.com/2018/10/09/what-goes-around-how-coffee-waste-is-fueling-a-circular-economy/

Eboh, Francis Chinweuba, Bengt-Åke Andersson, and Tobias Richards. "Economic evaluation of improvements in a waste-to-energy combined heat and power plant." *Waste Management* 100 (2019): 75–83. doi:10.1016/j.wasman.2019.09.008.

Essabir, Hamid, Marya Raji, Sana Ait Laaziz, Denis Rodrique, and Rachid Bouhfid. "Thermo-mechanical performances of polypropylene biocomposites based on untreated, treated and compatibilized spent coffee grounds." *Composites Part B: Engineering* 149 (2018): 1–11. doi:10.1016/j.compositesb.2018.05.020

Ferreira, Ana Filipa, and Ana Paula Soares Dias. "Pyrolysis of microalgae biomass over carbonate catalysts." *Journal of Chemical Technology & Biotechnology* 95, no. 12 (2020): 3270–3279. doi:10.1002/jctb.6506.

Gautam, Anamika, Shalu Rawat, Lata Verma, Jiwan Singh, Samiksha Sikarwar, B. C. Yadav, and Ajay S. Kalamdhad. "Green synthesis of iron nanoparticle from extract of waste tea: An application for phenol red removal from aqueous solution." *Environmental Nanotechnology, Monitoring & Management* 10 (2018): 377–387. doi:10.1016/j. enmm.2018.08.003.

Gemechu, FeyeraGobena. "Embracing nutritional qualities, biological activities and technological properties of coffee by-products in functional food formulation." *Trends in Food Science & Technology* (2020). doi:10.1016/j.tifs.2020.08.005

Germec, Mustafa, and Irfan Turhan. "Ethanol production from acid-pretreated and detoxified tea processing waste and its modeling." *Fuel* 231 (2018): 101–109. doi:10.1016/j.fuel.2018.05.070

Gomez-Gonzalez, Ricardo, Felipe J. Cerino-Córdova, Azucena M. Garcia-León, Eduardo Soto-Regalado, Nancy E. Davila-Guzman, and Jacob J. Salazar-Rabago. "Lead biosorption onto coffee grounds: Comparative analysis of several optimization techniques using equilibrium adsorption models and ANN." *Journal of the Taiwan Institute of Chemical Engineers* 68 (2016): 201–210. doi:10.1016/j.jtice.2016.08.038

Gotteland, Martín, and Saturnino De Pablo. "Algunasverdadessobreel café." *Revistachilena de nutrición* 34, no. 2 (2007): 105–115.

Gracia-Lor, Emma, Nikolaos I. Rousis, Ettore Zuccato, Richard Bade, Jose Antonio Baz-Lomba, Erika Castrignanò, Ana Causanilles et al. "Estimation of caffeine intake from analysis of caffeine metabolites in wastewater." *Science of the Total Environment* 609 (2017): 1582–1588.

Gunter, Marc J., Neil Murphy, Amanda J. Cross, Laure Dossus, Laureen Dartois, Guy Fagherazzi, Rudolf Kaaks et al. "Coffee drinking and mortality in 10 European countries: a multinational cohort study." *Annals of internal medicine* 167, no. 4 (2017): 236–247. doi:10.7326/ M16-2945.

Guo, Shasha, Mukesh Kumar Awasthi, Yuefei Wang, and Ping Xu. "Current understanding in conversion and application of tea waste biomass: A review." *Bioresource Technology* (2021): 125530. doi:10.1016/ j.biortech.2021.125530

He, Xianglei, Jialiang Li, Qingmei Meng, Ziyu Guo, Hao Zhang, and Yurong Liu. "Enhanced adsorption capacity of sulfadiazine on tea waste biochar from aqueous solutions by the two-step sintering method without corrosive activator." *Journal of Environmental Chemical Engineering* 9, no. 1 (2021): 104898.

Heeger, Andrea, Agnieszka Kosińska-Cagnazzo, Ennio Cantergiani, and Wilfried Andlauer. "Bioactives of coffee cherry pulp and its utilisation for production of Cascara beverage." *Food Chemistry* 221 (2017): 969–975. doi:10.1016/j.foodchem.2016.11.067

Heckman, Melanie A., Jorge Weil, and Elvira Gonzalez De Mejia. "Caffeine (1, 3, 7-trimethylxanthine) in foods: a comprehensive review on consumption, functionality, safety, and regulatory matters." *Journal of Food Science* 75, no. 3 (2010): R77–R87.

Ho, Khanh-Van, Kathy L. Schreiber, Jihyun Park, Phuc H. Vo, Zhentian Lei, Lloyd W. Sumner, Charles R. Brown, and Chung-Ho Lin. "Identification and quantification of bioactive molecules inhibiting pro-inflammatory cytokine production in spent coffee grounds using metabolomics analyses." *Frontiers in Pharmacology* 11 (2020): 229. doi: 10.3389/fphar.2020.00229

Hong, Woo Tae, Jin Young Park, Jong Won Chung, Hyun Kyoung Yang, and Jae-Yong Je. "Anti-counterfeiting application of fluorescent carbon dots derived from wasted coffee grounds." *Optik* (2021): 166449. doi:10.1016/j.ijleo.2021.166449

Huang, Lu-Liang, Jian-Hua Jin, Cheng Quan, and Alexei A. Oskolski. "Camellia nanningensis sp. nov.: the earliest fossil wood record of the genus Camellia (Theaceae) from East Asia." *Journal of Plant Research* 129, no. 5 (2016): 823–831. doi:10.1007/s10265-016-0846-8

Hussain, Siam, K. P. Anjali, Saima Towhida Hassan, and Priy Brat Dwivedi. "Waste tea as a novel adsorbent: a review." *Applied Water Science* 8, no. 6 (2018): 1–16.

Indira, Dash, Bhaskar Das, Harshita Bhawsar, Sahoo Moumita, EldinMaliyakkal Johnson, P. Balasubramanian, and R. Jayabalan. "Investigation on the production of bioethanol from black tea waste biomass in the seawater-based system." *Bioresource Technology Reports* 4 (2018): 209–213. doi:10.1016/ j.biteb.2018.11.003.

International Coffee Organization. "World coffee consumption." (2019).

Janissen, Brendan, and Tien Huynh. "Chemical composition and value-adding applications of coffee industry by-products: A review." *Resources, Conservation and Recycling* 128 (2018): 110–117. doi:10.1016/ j.resconrec.2017.10.001

Jiménez-Zamora, Ana, Silvia Pastoriza, and José A. Rufián-Henares. "Revalorization of coffee by-products. Prebiotic, antimicrobial and antioxidant properties." *LWT-Food Science and Technology* 61, no. 1 (2015): 12–18. doi:10.1016/j.lwt.2014.11.031

Jones, Alan Wayne. "Review of caffeine-related fatalities along with postmortem blood concentrations in 51 poisoning deaths." *Journal of Analytical Toxicology* 41, no. 3 (2017): 167–172.

Joseph, Kurian, S. Rajendiran, R. Senthilnathan, and M. Rakesh. "Integrated approach to solid waste management in Chennai: an Indian metro city." *Journal of Material Cycles and Waste Management* 14, no. 2 (2012): 75–84.

Kabir, Mohammad Mahbub, Snigdha Setu Paul Mouna, Samia Akter, Shahjalal Khandaker, Md Didar-ul-Alam, Newaz Mohammed Bahadur, Mohammad Mohinuzzaman, Md Aminul Islam, and M. A. Shenashen.

"Tea waste based natural adsorbent for toxic pollutant removal from waste samples." *Journal of Molecular Liquids* 322 (2021): 115012. doi:10.1016/j.molliq.2020.115012

Khayum, Naseem, S. Anbarasu, and S. Murugan. "Biogas potential from spent tea waste: A laboratory scale investigation of co-digestion with cow manure." *Energy* 165 (2018): 760–768. doi:10.1016/j.energy.2018.09.163.

Kobilka, Brian K. "G protein coupled receptor structure and activation." *Biochimica et Biophysica Acta (BBA)-Biomembranes* 1768, no. 4 (2007): 794–807.

Konwar, B. K. "Potentiality and viability of agro-industrial by-products in NE Region." *Annual Workshop, Sponsored by National Institute of Rural Development, Guwahati.* 1988

Kovalcik, Adriana, Stanislav Obruca, and Ivana Marova. "Valorization of spent coffee grounds: A review." *Food and Bioproducts Processing* 110 (2018): 104–119.

Kumar, Swaroop S., T. S. Swapna, and Abdulhameed Sabu. "Coffee husk: a potential agro-industrial residue for bioprocess." In *Waste to Wealth*, pp. 97–109. Springer, Singapore, 2018.

Lee, Nahyeon, Soosan Kim, and Jechan Lee. "Valorization of waste tea bags via CO2-assisted pyrolysis." *Journal of CO2 Utilization* 44 (2021): 101414. doi:10.1016/j.jcou.2020.101414.

Lee, Xin Jiat, HwaiChyuan Ong, Wei Gao, Yong Sik Ok, Wei-Hsin Chen, Brandon Han Hoe Goh, and Cheng Tung Chong. "Solid biofuel production from spent coffee ground wastes: Process optimisation, characterisation and kinetic studies." *Fuel* 292 (2021): 120309. doi:10.1016/j.fuel.2021.120309

Li, Mengjiao, Wei-Yuan Cheng, Yi-Chiun Li, Hsing-Mei Wu, Yan-Cheng Wu, Hong-Wei Lu, Shueh-Lian Cheng et al. "Deformable, resilient, and mechanically-durable triboelectric nanogenerator based on recycled coffee waste for wearable power and self-powered smart sensors." *Nano Energy* 79 (2021): 105405. doi:10.1016/j.nanoen.2020.105405

Liu, Shou-Heng, and Yi-Yang Huang. "Valorization of coffee grounds to biochar-derived adsorbents for CO2 adsorption." *Journal of Cleaner Production* 175 (2018): 354–360. doi:10.1016/j.jclepro.2017.12.076.

Ly, Hoang Vu, Boreum Lee, Jae Wook Sim, Quoc Khanh Tran, Seung-Soo Kim, Jinsoo Kim, Boris Brigljević, Hyun Tae Hwang, and Hankwon Lim. "Catalytic pyrolysis of spent coffee waste for upgrading sustainable bio-oil in a bubbling fluidized-bed reactor: Experimental and techno-economic analysis." *Chemical Engineering Journal* 427 (2022): 130956. doi:10.1016/j.cej.2021.130956

Ma, Baolong, Xiaofei Li, Zhongjun Jiang, and Jiefan Jiang. "Recycle more, waste more? When recycling efforts increase resource consumption." *Journal of Cleaner Production* 206 (2019): 870–877. doi:10.1016/j.jclepro.2018.09.063.

Machado, Ercília MS, Rosa M. Rodriguez-Jasso, José A. Teixeira, and Solange I. Mussatto. "Growth of fungal strains on coffee industry residues with removal of polyphenolic compounds." *Biochemical Engineering Journal* 60 (2012): 87–90. doi:10.1016/j.bej.2011.10.007.

Macheiner, L., Schmidt, A., Schreiner, M. and Mayer, H.K., 2019. Green coffee infusion as a source of caffeine and chlorogenic acid. *Journal of Food Composition and Analysis*, 84, p.103307.

Madahava Naidu, M., P. Vijayanada, A. Usha Devi, M. R. Vijayalakshmi, and K. Ramalakshmi. "Utilization of coffee by-products in food industry, preparation of jam using coffee pulp as raw material." *Plantation Crops Research and Development in the New Millennium, Placrosym* 14 (2004): 201–203.

Mahaly, Moorthi, Abbiramy K. Senthil Kumar, Senthil Kumar Arumugam, Chitra Priya Kaliyaperumal, and Nagarajan Karupannan. "Vermicomposting of distillery sludge waste with tea leaf residues." *Sustainable Environment Research* 28, no. 5 (2018): 223–227. doi:10.1016/j.serj.2018.02.002

Manasa, Vallamkondu, Aparna Padmanabhan, and KA Anu Appaiah. "Utilization of coffee pulp waste for rapid recovery of pectin and polyphenols for sustainable material recycle." *Waste Management* 120 (2021): 762–771. doi:10.1016/j.wasman.2020.10.045

Marcason, Wendy. "What is green coffee extract?" *Journal of the Academy of Nutrition and Dietetics* 113, no. 2 (2013): 364. doi:10.1016/j.jand.2012.12.004

McNutt, Josiah. "Spent coffee grounds: A review on current utilization." *Journal of Industrial and Engineering Chemistry* 71 (2019): 78–88.

Mirón-Mérida, Vicente A., Jorge Yáñez-Fernández, Brenda Montañez-Barragán, and Blanca E. Barragán Huerta. "Valorization of coffee parchment waste (Coffea arabica) as a source of caffeine and phenolic compounds in antifungal gellan gum films." *Lwt* 101 (2019): 167–174. doi:10.1016/j.lwt.2018.11.013

Moreira, Mariana Dias, Marcela Magalhães Melo, Jéssica Marques Coimbra, Kelly Cristina Dos Reis, Rosane Freitas Schwan, and Cristina Ferreira Silva. "Solid coffee waste as alternative to produce carotenoids with antioxidant and antimicrobial activities." *Waste Management* 82 (2018): 93–99. doi:10.1016/j.wasman.2018.10.017

Morikawa, C. K., and M. Saigusa. "Recycling coffee and tea wastes to increase plant available Fe in alkaline soils." *Plant and Soil* 304, no. 1 (2008): 249–255. doi:10.1007/s11104-008-9544-1

Morikawa, Claudio K., and M. Saigusa. "Recycling coffee grounds and tea leaf wastes to improve the yield and mineral content of grains of paddy rice." *Journal of the Science of Food and Agriculture* 91, no. 11 (2011): 2108–2111. doi:10.1002/jsfa.4444

Mugambi, M. J. "A study of waste approaches in Tea processing factories: Case study of small holder tea factories in Kenya." *Digital Repository, University of Nairobi* (2006).

Murthy, Pushpa S., and M. Madhava Naidu. "Sustainable management of coffee industry by-products and value addition—A review." *Resources, Conservation and recycling* 66 (2012): 45–58.

Naranjo, Mauricio, Luz T. Vélez, and Benjamín A. Rojano. "Actividadantioxidante de café colombiano de diferentescalidades." *RevistaCubana de PlantasMedicinales* 16, no. 2 (2011): 164–173.

National Coffee Association. "National coffee drinking trends." *New York, NY: National Coffee Association* (2014).

Nawrot, Peter, S. Jordan, J. Eastwood, J. Rotstein, A. Hugenholtz, and M. Feeley. "Effects of caffeine on human health." *Food Additives & Contaminants* 20, no. 1 (2003): 1–30.

Nguyen, Thi Minh Thu, EunJin Cho, Younho Song, Chi Hoon Oh, Ryo Funada, and Hyeun-Jong Bae. "Use of coffee flower as a novel resource for the production of bioactive compounds, melanoidins, and bio-sugars." *Food Chemistry* 299 (2019): 125120. doi:10.1016/j.foodchem.2019.125120

Nie, Lei, Pengbo Chang, Shuang Liang, Kehui Hu, Dangling Hua, Shiliang Liu, Jinfang Sun et al. "Polyphenol rich green tea waste hydrogel for removal of copper and chromium ions from aqueous solution." *Cleaner Engineering and Technology* (2021): 100167. doi:10.1016/j.clet.2021.100167

Ody, P. "The complete guide medical herbal." (2000): 75.

O'Keefe, James H., Salman K. Bhatti, Harshal R. Patil, James J. DiNicolantonio, Sean C. Lucan, and Carl J. Lavie. "Effects of habitual coffee consumption on cardiometabolic disease, cardiovascular health, and all-cause mortality." *Journal of the American College of Cardiology* 62, no. 12 (2013): 1043–1051.

Pal, Abhradip, and Chandan Das. "A novel use of solid waste extract from tea factory as corrosion inhibitor in acidic media on boiler quality steel." *Industrial Crops and Products* 151 (2020): 112468. doi:10.1016/j.indcrop.2020.112468

Péerez, Teresa ZAYAS, Gunther Geissler, and Fernando Hernandez. "Chemical oxygen demand reduction in coffee wastewater through chemical flocculation and advanced oxidation processes." *Journal of Environmental Sciences* 19, no. 3 (2007): 300–305.

Pruvot-Woehl, Solene, Sarada Krishnan, William Solano, Tim Schilling, Lucile Toniutti, Benoit Bertrand, and Christophe Montagnon. "Authentication of Coffea arabica Varieties through DNA Fingerprinting and its Significance for the Coffee Sector." *Journal of AOAC International* 103, no. 2 (2020): 325–334. doi:10.1093/jaocint/qsz003.

Pua, Fei-Ling, Mohamad SyahmiSubari, Lee-WoenEan, and Shamala Gowri Krishnan. "Characterization of biomass fuel pellets made from Malaysia tea waste and oil palm empty fruit bunch." *Materials Today: Proceedings* 31 (2020): 187–190. doi:10.1016/j.matpr.2020.02.218

Rajput, Darshana, Samrat Paul, and Annika Gupta. "Green Synthesis of Silver Nanoparticles Using Waste Tea Leaves." *Advanced Nano Research* 3, no. 1 (2020): 1–14. doi:10.21467/anr.3.1.1-14.

Rijo, Bruna, Ana Paula Soares Dias, Marta Ramos, Nicole de Jesus, and Jaime Puna. "Catalyzed pyrolysis of coffee and tea wastes." *Energy* 235 (2021): 121252. doi:10.1016/j.energy.2021.121252

Rodríguez, N., J. R. Sanz, C. E. Oliveros, and C. A. Ramírez. "Beneficio del café en Colombia: Prácticas y estrategias para elahorrousoeficiente del agua y el control de la contaminaciónhídricaenelproceso de beneficiohúmedo del café." (2015): 1–35.

Sampaio, Armando, Giuliano Dragone, Mar Vilanova, José M. Oliveira, José A. Teixeira, and Solange I. Mussatto. "Production, chemical characterization, and sensory profile of a novel spirit elaborated from spent coffee ground." *LWT-Food Science and Technology* 54, no. 2 (2013): 557–563. doi:10.1016/j.lwt.2013.05.042

Sayar, Nihat Alpagu, SelçenDurmazŞam, Orkun Pinar, DamlaSerper, Berna SarıyarAkbulut, Dilek Kazan, and Ahmet Alp Sayar. "Techno-economic analysis of caffeine and catechins production from black tea waste." *Food and Bioproducts Processing* 118 (2019): 1–12. doi:10.1016/j.fbp.2019.08.014

Sealy, J. Robert. "A revision of the genus Camellia."_A Revision of the Genus Camellia (1958). Royal Horticultural Society, London.

Sharma, Kanishka, Singh Swati and T. Kumud. "Recognition and evaluation of authenticity of tea and coffee." International Journal of Advanced Research in Science, Communication and Technology 11 (2020): 1–8.

Silva, M. A., S. A. Nebra, MJ Machado Silva, and C. G. Sanchez. "The use of biomass residues in the Brazilian soluble coffee industry." Biomass and Bioenergy 14, no. 5–6 (1998): 457–467.

Suarez AgudeloJM. Aprovechamiento de los residuossólidosprovenientes del beneficio de café, enelmunicipio de Betania Antioquia: Usosyaplicaciones. Caldas: CorporaciónUniversitariaLasallista; 2012

Sui, Wenjie, Ying Xiao, Rui Liu, Tao Wu, and Min Zhang. "Steam explosion modification on tea waste to enhance bioactive compounds' extractability and antioxidant capacity of extracts." Journal of Food Engineering 261 (2019): 51–59. doi:10.1016/j.jfoodeng.2019.03.015.

Tang, Jun, Ju-Sheng Zheng, Ling Fang, YongxinJin, Wenwen Cai, and Duo Li. "Tea consumption and mortality of all cancers, CVD and all causes: a meta-analysis of eighteen prospective cohort studies." British Journal of Nutrition 114, no. 5 (2015): 673–683. doi:10.1017/ S0007114515002329.

Thanarasu, Amudha, Karthik Periyasamy, Kubendran Devaraj, PremkumarPeriyaraman, Shanmugam Palaniyandi, and Sivanesan Subramanian. "Tea powder waste as a potential co-substrate for enhancing the methane production in anaerobic digestion of carbon-rich organic waste." Journal of Cleaner Production 199 (2018): 651–658. doi:10.1016/j.jclepro.2018.07.225

Torres-Valenzuela, Laura Sofía, Johanna Andrea Serna-Jiménez, and Katherine Martínez. "Coffee by-products: Nowadays and perspectives." Coffee-Prod. Res (2020): 1–18. doi:10.5772/intechopen. 89508 2020

Torres-Valenzuela, Laura Sofía, Ana Ballesteros-Gomez, Alejandra Sanin, and Soledad Rubio. "Valorization of spent coffee grounds by supramolecular solvent extraction." Separation and Purification Technology 228 (2019): 115759.

Tripathi, Manoj, Jaya Narayan Sahu, and P. Ganesan. "Effect of process parameters on production of biochar from biomass waste through pyrolysis: A review." Renewable and Sustainable Energy Reviews 55 (2016): 467–481.

Uddin, M. N., K. Techato, M. G. Rasul, N. M. S. Hassan, and M. Mofijur. "Waste coffee oil: A promising source for biodiesel production." Energy Procedia 160 (2019): 677–682. doi:10.1016/j.egypro.2019.02.221

van den Brandt, Piet A. "Coffee or Tea? A prospective cohort study on the associations of coffee and tea intake with overall and cause-specific mortality in men versus women." European Journal of Epidemiology 33, no. 2 (2018): 183–200. doi:10.1007/s10654-018-0359

Vaverková, Magdalena Daria. "Landfill impacts on the environment." Geosciences 9, no. 10 (2019): 431. doi:10.3390/geosciences9100431.

Vázquez-Sánchez, Kenia, Nuria Martinez-Saez, Miguel Rebollo-Hernanz, Maria Dolores Del Castillo, Marcela Gaytán-Martínez, and Rocio Campos-Vega. "In vitro health promoting properties of antioxidant dietary fiber extracted from spent coffee (Coffee arabica L.) grounds." Food Chemistry 261 (2018): 253–259. doi:10.1016/j.foodchem.2018.04.064

Velissariou, Maria, Raymond Jay Laudano, Paul Martin Edwards, Stephen Michael Stimpson, and Rachel Lorna Jeffries. "Beverage derived from the extract of coffee cherry husks and coffee cherry pulp." U.S. Patent 7,833,560, issued 16 November 2010.

Verster, Joris C., and Juergen Koenig. "Caffeine intake and its sources: A review of national representative studies." Critical Reviews in Food Science and Nutrition 58, no. 8 (2018): 1250–1259.

Wachira, Francis N., S. Kamunya, S. Karori, R. Chalo, and T. Maritim. "The tea plants: botanical aspects." Tea in health and disease prevention. Elsevier (2013): 3–17. doi:10.1016/B978-0-12-384937-3.00001-X

Willson, Cyril. "The clinical toxicology of caffeine: A review and case study." Toxicology reports 5 (2018): 1140–1152.

Wu, Fangfang, Long Chen, Peng Hu, Yunxiao Wang, Jie Deng, and Baobin Mi. "Industrial alkali lignin-derived biochar as highly efficient and low-cost adsorption material for Pb (II) from aquatic environment." Bioresource Technology 322 (2021): 124539.

Xia, En-Hua, Hai-Bin Zhang, Jun Sheng, Kui Li, Qun-Jie Zhang, Changhoon Kim, Yun Zhang et al. "The tea tree genome provides insights into tea flavor and independent evolution of caffeine biosynthesis." Molecular Plant 10, no. 6 (2017): 866–877. doi:10.1016/j.molp.2017.04.002. PMID 28473262.

Xue, Zihan, Jingya Wang, Zhongqin Chen, Qiqi Ma, Qingwen Guo, Xudong Gao, and Haixia Chen. "Antioxidant, antihypertensive, and anticancer activities of the flavonoid fractions from green, oolong, and black tea infusion waste." *Journal of Food Biochemistry* 42, no. 6 (2018): e12690. doi:10.1111/jfbc.12690.

Yang, C. J., I. Y. Yang, D. H. Oh, I. H. Bae, S. Goo Cho, I. G. Kong, D. Uuganbayar, I. S. Nou, and K. S. Choi. "Effect of green tea by-product on performance and body composition in broiler chicks." *Asian-Australasian Journal of Animal Sciences* 16, no. 6 (2003): 867–872.

Yang, Doudou, Jin Liang, Yunsheng Wang, Feng Sun, Hong Tao, Qiang Xu, Liang Zhang, Zhengzhu Zhang, Chi-Tang Ho, and Xiaochun Wan. "Tea waste: an effective and economic substrate for oyster mushroom cultivation." *Journal of the Science of Food and Agriculture* 96, no. 2 (2016): 680–684. doi:10.1002/jsfa.7140

Yun, BeomYeol, Hyun Mi Cho, Young Uk Kim, Sung Chan Lee, Umberto Berardi, and Sumin Kim. "Circular reutilization of coffee waste for sound absorbing panels: A perspective on material recycling." *Environmental Research* 184 (2020): 109281. doi:10.1016/j.envres.2020.109281

Yuwono, SudarmintoSetyo, Kiki Fibrianto, Laila Yum Wahibah, and Aswin Rizky Wardhana. "Sensory attributes profiling of dampitrobusta coffee leaf tea (Coffea canephora)." *Carpathian Journal of Food Science & Technology* 11, no. 2 (2019): 165–176.

Zhang, Liang, Chi-Tang Ho, Jie Zhou, Jânio Sousa Santos, Lorene Armstrong, and Daniel Granato. "Chemistry and biological activities of processed Camellia sinensis teas: A comprehensive review." *Comprehensive Reviews in Food Science and Food Safety* 18, no. 5 (2019): 1474–1495. doi: 10.1111/1541-4337.12479

9 By-products from Malting, Brewing, and Distilling

Priyanka,[a] Ajay Singh,[a] Kritika Rawat,[a,*] and
Abhimanyu Kalne[b]

aDepartment of Food Technology, MataGujri College Fatehgarh Sahib,
Punjab, India
bDepartment of Agricultural Processing and Food Engineering,
Indira Gandhi Krishi Vishvavidyalya, Raipur, Chattisgarh, India
*kritikarawat8@gmail.com

CONTENTS

DOI: 10.1201/9781003207689-9

9.1   INTRODUCTION TO MALTING

Malting is an ancient method in biotechnology used since time immemorial for germination of grains in a controlled manner. The malting process ensures a desirable change in biochemical and physical properties inside grain, which is furthermore stabilized by drying of grains (Gupta et al. 2010). The foremost purpose behind malting is to develop hydrolytic enzymes namely α-amylase, β-amylase, dextrinase, and α-glucosidase, which does not exist in ungerminated cereal grain (Manna, 1995). Malt is prepared from a varied range of cereal grains as well as beans, peas, vetches, and pulse sprouts, but the uses of malt obtained from pulse sprout in food quality is rare, and is therefore not considered. Barley is the significant primarily cereal grain for preparation of malt for brewing, and is therefore, well-thought-out in details (Daniels, 1998). The malting process enhances the nutritional quality of cereal grains by decreasing the antinutritional factors (phytic acid, oxalic acid) present in whole grain.

The malting process involves three steps (Figure 9.1) i.e.

(1)   **Steeping:** steeping process involves immersing the whole grains in water. Dipping in water removes all the dirt, dust, and floating grains and ultimately helps in the germination process.

(2)   **Germination:** the second phase of the malting process starts when the grains soaks adequate quantity of water. The germination process is dependent upon temperature and time of germination, moisture content of soaked grains and availability of optimum oxygen, which facilitates starch hydrolysis and enzyme production (hydrolytic enzymes, amylases, proteases). The process can last upto 2–6 days and occurs speedily between 20°C and 30°C.

(3)   **Drying:** drying is the last step in malting. This process is done to decrease the moisture content, water activity, and chiefly prevent further growth occurring in whole grains. Grains are oven dried at 50°C for 24 hours, and further milled to form activated flour.

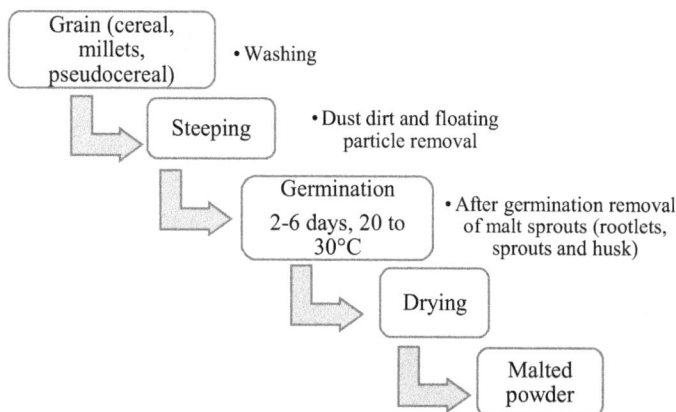

FIGURE 9.1   Flow chart depicting the malting process with separation of by-product.

9.2 BY-PRODUCTS OF MALTING

Floating grains removed during the steeping process is one of the by-products of malting, it is further sold at low cost as animal food (Papageorgiou and Skendi, 2018). Another by-product is malted sprouts, which includes rootlets, sprouts, and malt husk. It is separated during the drying process and characterized as a protein source.

9.2.1 MALTED SPROUTS

Malt sprouts are the by-products formed during germination of grains and separated during the drying process. It is the mixture of sprouts, rootlets, and husk (Samala et al., 1997). In the case of sorghum, an incomplete germination process leads to the formation of malted sorghum sprouts (Oke et al., 2017). Malt sprout is nutritious, cheap, and therefore an easily affordable ingredient for incorporation in different food product formations. Crude protein in barley sprout and sorghum ranged between 20 to 21.8 g/110g and 16.37 g/100g (Table 9.1). On performing the deep study on amino acid profile, the highest amino acid found in barley malt sprout is glutamic acid followed by aspartic acid and lysine (Salama et al., 1997). Malt sprouts are higher in macro- and micro-minerals than any grain (Zbstov et al., 2017). When combining wheat and barley malt sprouts, the mineral composition of barley malt sprouts is comparatively higher than wheat malt sprout. Maximum calcium content is observed in malted barley sprouts (Table 9.1). By including 6 g of malted wheat sprout in the human diet, fulfils 15% of the daily norm as per RDA (Eremina et al., 2021). Currently, due to its high nutrients, malt sprouts are largely used in feeding animals or utilization as a base for

TABLE 9.1
Composition of Malt Sprouts Obtained from Barley, Wheat, and Sorghum

	Malt sprouts' composition				
	Barley			Wheat	Sorghum
Parameters	Salama et al., 1997	Zubtsov et al., 2017	Creasy et al. 2001	Eremina et al., 2021	Oke et al. 2017
Moisture (%)	7.20	Ns	ns	Ns	Ns
Crude protein	21.8	Ns	20	Ns	16.37
Crude fat	25.5	Ns	ns	Ns	Ns
Crude fiber	16.33	Ns	ns	Ns	10.75
Ash	7.27	Ns	ns	Ns	Ns
Reducing sugar	2.9	Ns	ns	Ns	Ns
Non-reducing sugar	0.95	Ns	ns	Ns	Ns
Starch	26.7	Ns	ns	Ns	Ns
Calcium	1.19	0.39	0.19	0.08	0.9
Phosphorus	0.75	0.60	1.17	0.41	1.1
Potassium	0.60	1.364	0.69	0.81	Ns
Sodium	0.11	Ns	ns	Ns	Ns
Copper	0.0005	0.0001	0.0009	Ns	Ns
Iron	0.025	0.01	Ns	0.007	Ns
Magnesium	0.0025	0.19	0.17	0.15	Ns
Zinc	0.0060	0.0004	0.0006	0.005	Ns
Polyphenols	0.387	Ns	ns	0.25	Ns

Notes: The data is given in g/100g.

ns: not studied.

microorganism growth (Cejas et al., 2017). The food application of malted spouts is limited, but the nutritive properties present in them will definitely attract researchers for using malt sprouts as an ingredient in the formulation of food.

9.2.2 ROOTLETS

Rootlets contain a high proportion of nutrients; therefore, nowadays they have obtained an integral importance in research as well as their nutritional study in humans. During the germination stage of malting, rootlets are produced and must be separated from the malted barley after the kilning stage of malting. Germ/rootlets form during the malting process (germination stage). Of the total amount of barley, 3–5% of germ/rootlets can be obtained (Spiller and Amen, 1975; Werpy and Petersen, 2004). The removal of rootlets from the kilned malt is often referred to as the deculming process (Boulton, 2013). Anciently, this process was performed by crushing of kilned malt and passing it through a sieve, which allows broken rootles to pass and be further collected in piles (Kunze, 2004). At present, two machines are used for this process, namely, malt deculming and deculming screw machine (Daniels, 1998; Boulton, 2013). In a malt deculming machine, kilned malt moves into an air stream, where it is forced into an upright cylinder-shaped vessel, which leads to the breaking of rootlets from kilned malt. The malt Is heavier, and therefore passes through the air steam and is further collected, on the other hand, rootlets pass through a cyclone separator and get collected (Kunze, 2004). Some rootlets may pass through the heating chamber and can be collected from the bottom of the kiln, but the nutritive property of these collected rootlets will be less than those collected from cyclone due to the exposure to high temperature in kiln (Daniels, 1998). Therefore, due to their darker color they are separated from others. In a deculming screw machine, kilned malt passes through a perforated wall with angle beaters, which result in breakage of rootlets from the kilned malt and collected. Rootlets mixed with floating grains from steeping and malt dust to form blend and pelletized, which is further forced to pass through a die and cut into pellets (Neylon et al., 2020). These pellets are used for animal feed (Karlovic et al., 2020) because of its inexpensiveness and extremely nutritiousness, furthermore comprise crude protein (24.5%), oil (2.9%), neutral detergent fiber (40%), starch (5.5%), and sugar (13%). If pellets contain only rootlets it can be used as an important food supplement for humans.

9.2.2.1 Composition of Rootlets

The most common rootlets considered are barley rootlets. On considering the nutritional property and application of barley rootlets, its constituents and quality is most important. The barley variety chosen for malting, the germination time, and drying temperature influence the composition of barley rootlets (Kunze, 2004; Thiel and Toit, 1965).

Protein percentage in barley rootlet ranges between 23.9–38.9% (Table 9.2), which is higher than barley grain, malted barley, and wheat (Waters et al., 2013), therefore it provides an evidence that barley rootlets act as an excellent protein source. On the other hand, the highest protein is observed in pearl millet (38.4) followed by sorghum and finger millet (Table 9.3). After barley rootlets' proteome analysis, due to anatomical complexity and numerous processes occurring during growth of roots, the protein composition of barley rootlets' content is more varied (Mahalingam, 2020). The highest amount of protein found in rootlets is glutelins, followed by globulins, albumins, and prolamins (Neylon et al., 2020). Protein is formed by bonding/combination of different amino acids, therefore it becomes important to study the amino acid profile in rootlets. Table 9.4 depicts the essential and non-essential amino acid profile in different malted rootlets. Barley, sorghum, pearl millet, and finger millet rootlets contain a maximum amount of glutamic acid and aspartic acid. Among different rootlets, pearl rootlets have the highest aspartic acid content (Table 9.4). Barley rootlets contain 5.29 g/16 g N of lysine (Salama et al., 1997), which is a limiting amino acid in all cereals and its products. Due to its in-vitro protein digestibility (81–83.29%), rootles can play a significant role in human nutrition.

TABLE 9.2
The Range of Composition of Barley Rootlets

Parameters	Value (per 100 gm)	References
Moisture content	8.5–12.6%	Hegazi et al., 1975; Salama et al., 1997; Aggelopoulos et al., 2013; Begea et al., 2017; Chiş et al., 2020
Protein	23.9–38.9%	Aborus et al., 2017; Hegazi et al., 1975; Salama et al., 1997; Aggelopoulos et al., 2013; Waters et al.,2013; Begea et al., 2017; Chiş et al., 2020
Fat	1.7–4.4%	Hegazi et al., 1975; Aggelopoulos et al., 2013; Waters et al., 2013; Begea et al., 2017; Chiş et al., 2020
Ash	2.9–8.7%	Aborus et al., 2017; Hegazi et al., 1975; Salama et al., 1997; Aggelopoulos et al., 2013; Waters et al., 2013;Begea et al., 2017; Chiş et al., 2020
Crude fiber	9.7–20.5%	Hegazi et al., 1975; Salama et al., 1997
Carbohydrate	46–60%	Aborus et al., 2017; Waters et al., 2013; Chiş et al., 2020
Starch	2.6–26.5%	Salama et al., 1997; Waters et al., 2013

TABLE 9.3
Composition of Sorghum and Millets' Rootlets after 72 Hours' Germination Time

Parameters	Sorghum	Pearl millet	Finger millet	Reference
Protein	28.7	38.4	18.6	Malleshi and Klopfenstein,
Fat	–	1.9	1.5	1998
Ash	8.0	5.6	5.3	
Dietary fiber	–	8.4	9.4	
Crude fiber	–	–	–	

Note: The results are in grams per 100 grams.

The range of fat % in barley rootlets varies from 1.7–4.4% (Table 9.2). Sorghum rootlets do not contain any fat, which gives us an idea that sorghum rootlets' incorporation can reduce the fat content in any food (Table 9.3). Furthermore, products rich in other nutrients and low in fat can serve as nutritious fat free food to humans. Rootlets contain linoleic and linolenic acid as pre-dominant fatty acids, followed by palmitic acid (Neylon et al., 2020). The ash content in barley rootlets ranges from 2.8–8.6% (Table 9.2), which is higher than barley grain, malted barley, acrospire, and husk (Waters et al., 2013; Neylon et al., 2020).

The crude fiber ranges between 9.7 to 20.5% in barley rootlets (Table 9.2), which is substantially higher from barley grain and malted barley (Waters et al., 2013). Due to its significant fiber count largely rootlets are still used as cattle feed. On the other side, sorghum, pearl millet, and finger millet contain no crude fiber and dietary fiber (Table 9.3). After a detailed study on the total fiber (insoluble and soluble) content in barley malt and rootlets, Waters et al., 2013, depicted that a large portion of fiber was composed of insoluble fiber than soluble fiber. It gives a prediction that because of the bulky effect in fiber it can be used in the human diet. One of the important fibers present in rootlets is Arabinoxylan, which constitutes around one third of barley rootlet fiber (Waters et al., 2013). Furthermore, cellulose and lignin are also present in rootlets, which make rootlets a high source of fiber, which can further be used in the fortification of food, prebiotic and supplementation in the coming future. A Japanese invention involves the manufacturing of food products and beverages

TABLE 9.4
Amino Acid Range of Rootlets from Barley, Sorghum, Pearl Millet, and Finger Millet (g/100 g Protein)

	Waters et al., 2013	Malleshi and Klopfenstein, 1998		
Amino acid	Barley rootlets	Sorghum rootlets	Pear millet rootlets	Finger millet rootlets
Essential amino acid				
Threonine	0.05	4.13–4.40	4.75–4.89	4.76–5.10
Methionine	0.10	1.76–1.83	1.34–1.59	2.22–2.46
Tryptophan	0.02	Ns	Ns	Ns
Phenylalanine	0.87	3.32–4.51	3.54–3.77	3.90–4.36
Isoleucine	1.05	2.93–3.08	2.93–3.11	3.39–3.65
Leucine	1.45	6.92–7.14	6.22–6.82	7.99–8.14
Lysine	Ns	5.24–6.29	5.28–5.92	5.75–6.01
Non-essential amino acid				
Aspartic acid	2.61	12.05–20.59	17.52–21.22	11.81–12.14
Glutamic acid	3.02	12.73–16.41	12.83–13.62	12.06–13.44
Asparagine	0.43	Ns	Ns	Ns
Serine	0.88	5.27–5.91	5.30–5.51	5.34–5.70
Histidine	7.58	3.52–3.76	3.36–3.56	3.53–3.66
Glycine	0.47	4.79–5.35	4.36–4.87	5.35–5.57
Arginine	1.11	4.94–6.02	5.81–6.68	4.63–5.43
Alanine	1.19	7.02–7.71	6.95–9.18	7.64–8.33
Tyrosine	0.61	2.86–3.64	3.00–3.13	3.44–3.77
Valine	1.33	4.54–5.00	4.62–4.83	5.40–5.66
Proline	Ns	3.96–5.26	4.26–4.74	5.60–7.06
Cystine	Ns	0.65–0.71	1.03–1.29	0.81–1.11

Note: Ns=not studied.

comprising soluble dietary fiber by use of partially degraded hemicellulose extracted from plant germ (Karlovic et al., 2020).

Starch contents have been reported in the range of 2.6–26.5% (Table 9.1), which is much lower than barley grain. Monosaccharides, disaccharides, reducing, and non-reducing sugar have been reported in the range of 3.4–13.6% (Waters et al., 2013; Neylon et al, 2020; Hegazi et al., 1975). A maximum amount of monosaccharide i.e. glucose and fructose are found in barley rootlets, with a minimal amount of oligosaccharides i.e. sucrose, maltose and maltotriose (Waters et al., 2013).

However, apart from the nutritional properties of germ, an unwanted taste persists, which can affect the usage of rootlets in food products. Consequently, in future many researches will be conducted to find technologies to incorporate rootlets in manufacturing of different food products without effecting the taste of that product (Karlovic et al., 2020). Kondo et al. (2016) performed a study on bad taste of germ and concluded that the rootlets' particle size is directly proportional to the bad taste, as the rootlets' particle size increases, a decrease in bad taste is observed.

9.2.2.2 Application of Rootlets in Food

Rootlets can be considered as an important part of the human diet as they are high in protein, fiber, and low in fat, which make it a perfect ingredient for incorporation in preparation of different food products chiefly "low-fat high fiber diet".

The fiber present in barley rootlets is predominantly insoluble, and therefore, can act as roughage in the colon and provide health benefits to those who are suffering from diseases like piles, diabetes etc. (Kendall et al., 2010). Salama et al. (1997) observed that incorporation of barley rootlets boosted the nutritional quality of breads (flattened breads and pan breads). Barley rootlets are cheap and are mostly used as animal feed, therefore, utilization of rootlets as a food ingredient in the food industry can be a cost-effective way to increase nutrition in the human diet. After studying the functional properties, it was observed that barley rootlets possessed absorption and emulsification capability, which can play a specific role while using barley rootlets in food (Neylon et al., 2020). Bread is a staple food, therefore, enhancing bread with nutritional properties would be the best way to provide nutrition to consumers. Waters et al. (2013) prepared bread by substituting wheat with barley rootlets and fermented barley rootlets in 5,10,15 and 20% level, and estimated that incorporation of barley rootlets enhanced the nutritional properties, specifically amino acid and fiber. Due to substitution a decline in fat percentage (saturated) was observed, which made the bread low fat, high fiber. However, the increasing percentage level of both barley rootlets (fermented and non-fermented) reduced the volume of bread, produced dark color with more hardness, therefore, the best formulated bread was at the 5% level for both fermented and non-fermented barley rootlets.

Rootlets can be the best substitution from plant-based food with increased protein (amino acid) and fiber (Hoehnel et al., 2020), however to confirm this several investigations may be needed. To study the effect of barley rootlets' volatiles on biscuits Chiş et al. (2020) prepared biscuits by addition of barley rootlets and estimated that as the incorporation level over 15 % increased the aroma, an aldehydic unpleasant taste and flavor intensified. Moreover, barley rootlets at different percentage levels were utilized as a binder in sausage preparation and concluded that the 10% incorporation level was sensorily accepted and also concluded that barley rootlet usage reduced cooking losses and is a cost-effective way to increase nutrition (Salama et al., 1997). Studies on the food application of barley rootlet or any other rootlets are limited, but on the whole, the usage of barley rootlets in the preparation of food products can have a promising benefit to the food industry. However, with proper incorporation their inclusion in food may have a maximum point due to higher levels of insertion limit, because an increase in limit can directly affect taste of food. Due to its nutritional profile, in the coming future it can be used as a high fiber food or food fortifier in food industries.

9.2.3 MALT HUSK AND ACROSPIRES

Malt husk and acrospires are removed with malt sprouts. Table 9.5 given below shows the chemical constituents of husk and acrospire. Protein content in acrospires was highest in comparison to husk. However, husk possessed a maximum amount of lysine (7.5 g/16 gN), which limits amino acid in cereal products (Salama et al., 1997). As per Salama et al.'s (1997) investigation on malt sprouts

TABLE 9.5
Composition of Husk and Acrospires

Parameters	Acrospires	Husk
Moisture content (%)	10.30	7.60
Protein (%)	30.2	12.4
Crude fiber (%)	4.6	24.9
Ash (%)	5.8	7.9
Reducing sugar (%)	3.4	2.8
Non-reducing sugar (%)	1.1	0.9
Starch (%)	27.6	1.7
Polyphenol	0.3	0.4

and its constituents, he espied that maximum recovery of crude fiber is 60% from husk followed by acrospires and his studies also concluded that highest crude was depicted in malted husk. He also evaluated the functional properties of husk as functional properties are the key criteria that determine the quality as well as quantity of protein and crude fiber present in it. As per the study, the husk possessed good water and oil absorption capacity and the lowest foam and emulsion capacity. The lowest foam capacity and foam stability can be because of high fiber and low protein. There is very limited research on husk and acrospires in preparation of different food. They are cheap and presently used as animal feed. In the future, lysine and other fibers can be extracted from husk and further used in the food industry.

9.3   INTRODUCTION TO BREWING PROCESS

The process of brewing requires a large amount of water and rigorous energy. Beer manufacturing comprises blending of malt extract, hops, and sugar with water, followed by yeast (*Saccromycesserivecie*) fermentation. The brewing process is accomplished by certain steps namely malting, wort production, yeast fermentation [primary and secondary (maturation of fresh beer)], finishing (filtering), and packaging.

By-products are formed at different stages of the brewing process, to understand the formation of by-products, a brief brewing process is explained (Figure 9.2). Initially, barley malt is mixed with water and gradually the temperature is increased from 37 to 78°C. The upsurge in temperature will stimulate enzymatic hydrolysis and lead to formation of sugars, namely maltose and matotriose (fermentable) and dextrin (non-fermentable). This enzymatic conversion stage (mashing) produces wort (sweet liquid), which is further filtered. At this stage the residual solid mass (by-product) left after removal of wort is called brewer spent grains (Mussatto et al., 2006). The filtered wort is boiled for an hour with hops' addition in cooper (brewing kettle), further cooled and the extract obtained is separated. Hops' addition enhances the aroma, taste, flavor, and bitterness in beer. The hop residues

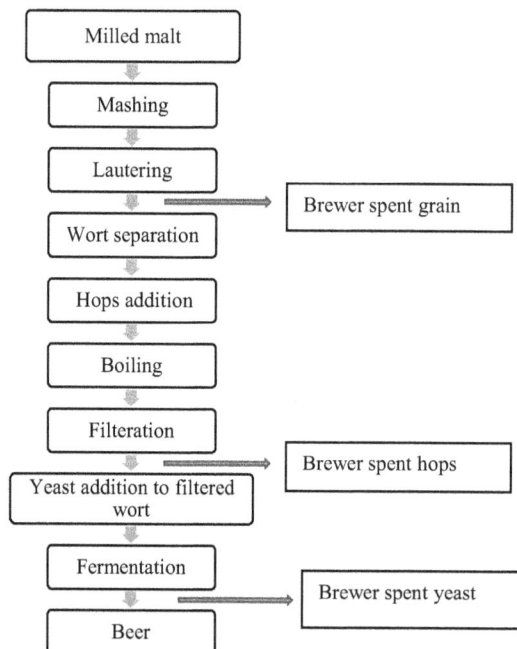

FIGURE 9.2   Flow chart of beer production with formation of different by-product.

left after extract separation is a by-product called spent hops (Keukeleire. 2000). The extracted hop wort is further fermented by addition of yeast in a fermented vessel. The fermentation process leads to conversion of sugars into ethanol and carbon dioxide. After the end of fermentation maximum cells of yeast are collected, which is again a by-product of beer called spent yeast or surplus yeast (Keukeleire, 2000). Finally, the brewing industry requires lots of water for the production of beer, therefore, extra water collected can also be called a by-product of the brewing industry.

9.4 BY-PRODUCTS OF BREWING

As discussed above the by-products of brewing are brewer spent grain, spent hops, spent yeast, and water. A detailed discussion is given below.

9.4.1 BREWER SPENT GRAIN

Out of the entire by-products produced in the brewing industry, brewer spent grain acquires the topmost position as it is collected in a surplus amount, approximately 85%. As discussed above, it is produced during the mashing and wort separation process. According to Kunze (2004), brewer spent grain comprises materials, which are not soluble namely cellulose, starch (non-digestible by enzymes), protein (high molecular weight), and additional composites, which are left over during the lautering process. Spent grain solid mass is left after wort separation forms three layers at the first top layer, which is fine and composed of protein and husk, the second intermediate layer is thick and comprises a maximum amount of spent grain mass, lastly the bottom layer, which consists of particles that are dense and huge (Kunze, 2004). Spent grain solid mass when removed is washed numerous times in hot water to recuperate the extract, which is entrapped in spent grain. The wet spent grain obtain after washing persist high moisture percentage, which fluctuates between 70–80% (Robertson et al., 2010). The spent grain removed can be used itself in wet form, which can be further sold to farmers at a low price as it has a high chance of getting spoiled. Spent grain can also be dried and its moisture content ranges between 10 to 12%. Drying helps in the reduction of the size of the mass and increases the shelf life upto months (Stroem et al., 2009; Tang et al., 2004). The foremost efficient drying technique used to dry spent grain is pressurized-steam fluid-bed drying, which is faster than other drying methods and helps in increasing productivity, sterilization, and reducing the probability of fire and explosion (Stroem et al., 2009).

9.4.1.1 Composition of Brewer Spent Grain

The chemical composition of spent grain can vary, depending on the quality of barley or other cereals used in beer production, but it can also depend on other factors such as harvesting time, malting germination conditions, and the quality of unsweetened raw materials.

Brewer spent grain contains a high amount of protein, which depends on species or type of beer produced, but in general it ranges from 12 to 24%. Protein chiefly present in barley is hordein, which is commonly called prolamin, which constitutes 50% of whole protein in brewer spent grain (Connolly et al., 2019; Tapia-Hernández et al., 2019). Glutelin is also present in spent grain. The presence of different protein during the extraction process differs and is extremely vital. Brewer spent grain when alkali treated 21.4% of yield was extracted with 60% purity of protein furthermore, combining alkali pre-treatment with diluted acid, 95% of extracted was estimated Du et al. (2020). A better result of protein extraction was reported by using carboxylate salt-urea DES (Wahlstrom et al., 2017), but with a disadvantage, on another hand, that urea is present in the extracted protein, therefore it cannot be used in food formulation for humans. Using ultrasound for protein extraction can be a promising technology and by this method the efficacy of separation of protein is improved by 68%, with a time reduction by 56% (Yu et al., 2020). Arabinoxylan is a polysaccharide consisting of two pentose sugars: xylose and arabinose (Izydorczyk and Biliaderis, 1995). Among other hemicelluloses, cellulose, and lignin, it is part of the dietary fiber found in BSG. It can bind to polyphenols such as ferulic acid and p-cumaic acid. Arabinoxylans can be recovered by ultrasound-assisted extraction (Reis et al., 2015), microwave-assisted extraction (Coelho et al., 2014) or HCl and ethanol extraction (after previous protein extraction) (Vieira et al., 2014). Studies show that supercritical extraction of CO_2 with ethanol 60% v/v at 35 MPa, 40 °C at an extraction time of 240 min allows a good recovery of phenolic or flavonoid fractions (Spinelli et al., 2016). Ultrasound-assisted extraction achieved similar productivity, after 30 min of treatment, in comparison to enzyme hydrolysis (Alonso-Riano et al., 2020). Birsan et al. compared conventional maceration, microwave-, and ultrasound-assisted extraction, using BSG from light and dark beer as well as their mixtures. Microwave and ultrasound extraction did not improve the total polyphenol yield when compared to the conventional maceration method (Birsan et al., 2019).

9.4.1.2 Application of Brewer Spent Grain in the Food Industry

Brewer spent grain can itself be again used for production of beer. As per Muller (1991) the spent grain mass extracted from the top layer after wort filtration contain 5% of undigested starch, therefore it can be further used for preparation of beer. Brewer spent grain contains a high amount of fiber and protein, therefore, it can be combined with different flour to increase the nutritional property of conventional or ready-to-eat food product. On combining durum wheat with brewer spent grain the fiber content amplifies to 135% with an increase in antioxidant value up to 19% (Nocente). The best optimum organoleptic properties were observed when 10% of brewer spent grain was added to prepare pasta (Nocente et al., 2019). In bread preparation, water absorption capacity is influenced by addition of brewer spent grain flour, which affects the texture and volume of the end product (Steinmacher et al., 2012). Brewer spent grain also influences the quality of cookies when incorporated with wheat flour. Kirjoranta et al. (2016) reported that incorporating brewer spent grain flour in wheat flour in a 1:4 ratio reduces the glycemic index of cookies. Apart from baked products brewer spent grain can also be used in meat production as study reveals that the addition of brewer spent grain in meat preparation reduced the fat content and increases the fiber content in meat (Ozvural et al., 2009). Brewer spent grain can further be used in preparation of high protein and high fiber protein food like bread, biscuits, snacks.

9.4.2 Spent Hops

Hops are dried flowers added in small quantities to wort while cooking as it enhances the taste, develops characteristic flavor, imparts bitterness, and possesses antimicrobial properties (Lewis et al., 2005). Spent hops are removed after completion of cooking of wort. It is demonstrated that only 15% of the components present in hops is absorbed in beer and the remaining 85% become a by-product that is spent hop (Huige, 2006). The chemical composition of spent hop is given in the Table 9.7.

Spent hop is a material with high amounts of nitrogen free extract, fibers, and proteins. Spent hop's protein can be isolated by supercritical acid carbon dioxide extraction (Daenicke et al., 1991). Spent hops are used as a biological pesticide as they contain a high amount and a different variety of essential oils (Bedini et al., 2015). These essential oils can be extracted by using steam distillation. Spent hop extract shows a promising effect on reducing platelet reactivity and also persist anticoagulating properties (Luzak, 2016). 85% of the polyphenolic compounds are present in spent hops therefore, spent hops are further utilized for extraction of polyphenolic compounds. Spent hops are chiefly used as a fertilizer. Apart from other by-products of beer preparation spent hops cannot be used as animal feed because of the bitter taste (O'Rourke, 1999). Spent hop possesses a higher fiber content than spent grain, but the energy value of spent hop is 59% less than spent grain (Huige, 2006), therefore to discard spent hop, animal feed manufacturers add 5% of spent hop in spent grain.

TABLE 9.6
Composition of Brewer Spent Grain in % db

Parameters								References								
	1	2	3	4	5	6	7	8	9	10	11	12	13	14	15	16
Moisture fresh	78–80	na	na	na	na	na	na	na	na	na	na	na	na	na	na	na
Moisture dried	7.07–8.06	na	na	na	na	na	na	na	na	na	na	na	na	na	na	na
Protein (%)	12.91–18.58	24	na	na	na	na	na	na	na	na	na	na	na	na	na	na
Ash (%)	2.63-2.96	2.4	1.2	4.6	4.6	3.3	3.3	na	na	2.27	0.76	1.1	4.8	2.48	4.1	4.1
Fat (%)	6.4-6.7	na	na	na	na	na	na	na	na	na	na	na	na	na	na	na
Total dietary fibre	33–38.52	na	na	na	na	na	na	na	na	na	na	na	na	na	na	na
Carbohydrates	24.58–34.57	na	na	na	na	na	na	na	na	na	na	na	na	na	na	na
Lignin	na	11.9	21.7	16.9	27.8	na	11.5	20–22	12.6	13–17	56.74	na	19.4	9.19	19.6	na
Cellulose	na	25.4	21.9	25.3	16.8	0.3	12	31–33	45.9	na	40.2	26	21.73	60.64	45	51
Hemicellulose	na	21.8	29.6	41.9	28.4	22.5	40	na	na	22–29	na	22.2	19.27	na	na	51
Lipid	na	10.06	na	na	na	na	13	6–8	na	na	2.50	na	na	6.18	na	9.4
Phenols	7.3–9.55	na	na	na	na	1	2	1–1.5	na	na	0.28	na	na	na	na	na
Starch	na	na	na	na	na	na	2.7	10–12	7.8	2–8	na	na	na	na	na	na

Sources: 1. Bravi et al. (2021), 2. Kanauchi et al. (2001), 3. Carvalheiro et al. (2004), 4.Silva et al. (2004), 5. Mussatto and Roberto (2006), 6. Celus et al. (2006), 7. Xiros et al. (2008), 8. Jay et al. (2008), 9. Treimo et al. (2009), 10. Robertson et al. (2010), 11. Khidzir et al. (2010), 12. Waters et al. (2012), 13. Meneses et al. (2013), 14. Sobukola et al. (2012), 15. Kemppainen et al. (2016), 16. Yu et al. (2020).

Note: na; not analyzed.

TABLE 9.7
Range Depicting Composition of Spent Hops

Components	References	
	Bravi et al., 2021	Karlovic et al., 2020
Moisture (%)	5.30–5.41	na
Total nitrogen (%)	6.35–8.33	40
Protein (%)	39.67–52.02	22–23
Ash (%)	2.11–2.33	6-6.5
Fat (%)	1.06–1.23	na
Total dietary fiber (%)	12.19–12.38	na
Crude fiber (%)	na	22–23
Carbohydrates (%)	26.78–39.94	na
Sesquiterpene hydrocarbons (%)	37	na
Monoterpene hydrocarbons (%)	27	na
Non-terpene derivatives (%)	18	na
Oxygenated sesquiterpene (%)	8	na
Oxygenated monoterpenes (%)	4	na
Lipids (%)	4.5	na

TABLE 9.8
Chemical Composition of Spent Yeast from Different Sources

References	Components							
	Moisture	Protein	Ash	Fat	Carbohydrates	Soluble fiber	Insoluble fiber	Lignin
Vieira et al., 2016 (g/100 g dw)	7.70	64.1	14	1.32	12.29	na	na	na
Cabellero-Cordoba and Sgarbieri, 2000 (%)	na	14.79	8.55	3.53	21.55	9.65	2.57	na
Saksinchai et al., 2001 (%w/w)	na	na	13.3	na	26.8	na	na	na
Marson et al., 2020 (% dw)	na	14.8	7	0.21	na	6.66	na	na
Mathias et al., 2015 (% dw)	na	14.6	5.9	na	na	na	na	na
Jacob et al.,2019 (g/100 g)	6.8	74.3	13.5	0.67	14.7	na	na	na
Karlovic et al., 2020 (%)	na	15-24	5	na	25–35	na	na	8-28

Note: na= not analyzed.

9.4.3 SPENT YEAST

Spent yeast (*Saccharomyces pastorianus* and *Saccharomyces cerevisiae*) is removed after completion of the fermentation of beer. A maximum amount of yeast is removed when the yeast is settled on the surface or bottom of the tank, furthermore the yeast that remains is separated through centrifugation and filtration. Spent yeast is available in inactive form in the market. It is rich in proteins, minerals, vitamins, nitrogen, and enzymes table. Spent yeast is chiefly used in the production of animal or fish food (Ferreira et al., 2010). It is a decent source of cysteine, glycine, glutamic acid, and nicotinic acid. In recent studies, spent brewer's yeast can also be used in human food formulation as it contains a high percent of protein, vitamins, and minerals. It comprises protein in a range of 45–60% and is generally regarded as safe (GRAS). 30% supplementation with spent yeast in meat did not alter the nutritional quality of meat (Ferreira et al., 2010). Spent yeast can also be used in the preparation of functional food as the yeast can provide a probiotic prebiotic combination in food (Rakin et al., 2004). Spent yeast contains polysaccharides, which are prepared from β-glucans, mannoproteins, and glycogen. Spent yeast can further be used for production of β-glucan, which is used in the food processing industry as a thickening, emulsifying, and stabilizing agent (Thammakiti et al., 2004). Worrasinchai et al. (2016) depicted that spent grain extracted β-glucan can be used as a fat replacer in mayonnaise and explored the possibility of replacing the fat in mayonnaise with β-glucan, derived from spent brewer's yeast. In a recent investigation, spent yeast was used to produce energy bars, which possessed a high amount of protein and phytic acid (Stojceska et al., 2008).

9.5 DISTILLATION AND ITS METHOD OF PRODUCTION

Distillation is a process whereby a liquid made of two or more parts is broken up into smaller parts of desired purity by the adding and elimination of heat from the mixture. The vapors/liquids distilled will be separated on the basis of boiling points. Distilled spirits are produced from agricultural raw materials such as grapes, other fruit, sugar-cane, molasses, potatoes, cereals, etc.

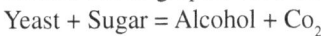

Yeast + Sugar = Alcohol + Co_2

9.5.1 METHODS

9.5.1.1 Alembic or Pot Distilling

Pot distilling is considered the most primitive form of distilling equipment. It is simple in design, consisting of a kettle to hold the mash, a steam coil and a condenser. This technique is used for single-stage batch distillation.

9.5.1.2 Column Distilling

Today, the majority of distillers employ continuous stills, in this design fermented liquor enters the upper end of the column and flows downward via an overflow fall and downpipe, criss-crossing a series of perforated trays. Steam is introduced from the bottom and rises with sufficient velocity to strip the volatile components. Vapors recondense and rise upward in the column depending upon their volatilities. Thus, liquids can be drawn off or recycled at the various plates. If higher efficiency is desired, rectifying stills can be added, which will concentrate volatile impurities within a narrow region of the still. Heat exchangers are usually coupled with continuous stills to save energy by preheating the incoming beer with outgoing stillage.

9.6 DISTILLED SPIRIT HEALTH BENEFITS AND THEIR TYPES

Distilled spirit products (Figure 9.3) result from distillation of plant or fruit juices (grapes, sugarcane, or agave), seeds (anise) or fruits and fruit marcs (grape or blackberry) after previous fermentation. These are popular strong alcoholic beverages served alone or as ingredients of various alcoholic drinks (Caldas et al., 2009; Barciela et al., 2008).

As can be seen, major, minor, and trace elements in these alcoholic beverages are of primary (natural) and secondary sources. The primary sources of elements are production region, environmental

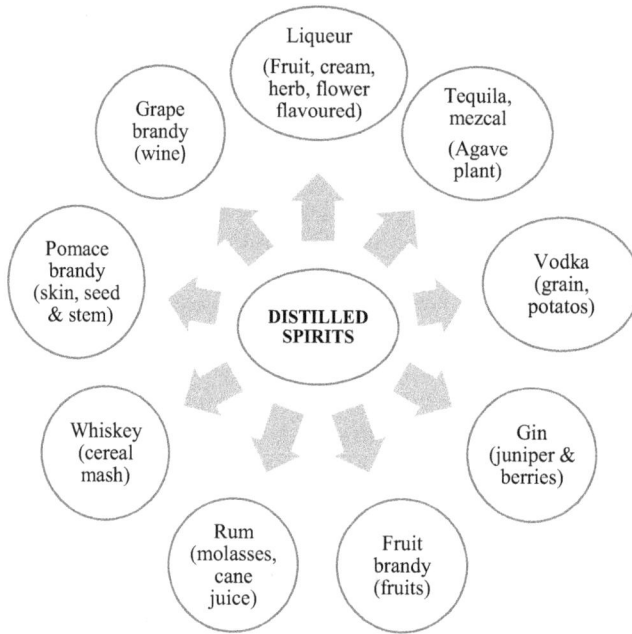

FIGURE 9.3 Different varieties of distilled spirits with their key fermentative ingredient.

conditions, and other natural activities affecting growth of plants and fruits and grains and the quality of raw materials (Rodriguez-solana et al., 2014; Barciela et al., 2008; Bonic et al., 2013). The secondary sources of elements in distilled spirit products are activities related to cultivation of plants, fruits, and grains (e.g., fertilization, agrochemical protection, and crop treatment) and production of spirits (e.g., materials and equipment used for maceration, fermentation, distillation and maturation, fabrication and storage, and, finally, technology and manufacturing processes, Figure 9.4).

Despite huge consumption of distilled liquid products, it seems that inorganic profiles of some of them remain inadequately distinct and need more extensive investigations (Nascimento et al., 1999; Navarro-Alarcon et al., 2007). This lack of clarity may hinder the establishment of appropriate standards of market quality and safety of products and badly affect their characteristics and marketing (Hernández-Caraballo et. Al., 2008). Greater attention to product authenticity and origin, quality control (QC), and suitable reference parameters of essential composition could certainly improve the safety of products and benefit the economy of regions dealing with production and trade of distilled spirits. The presence of selected trace elements (i.e., Al, As, Cd, Cu, Cr, Hg, Ni, Pb, Sb, and Zn) contributes to overall quality of alcoholic beverages and determines their suitability and safety for consumption in sight of the toxicity. Analysis of distilled spirits is required to assess potential health effects while the information about concentrations of certain elements is an important parameter of the QC of these products.

To avoid toxic effects on human health from these elements and to preserve the high quality and safety of final distillates, the composition of these products is controlled by national legislations and stringent regulations (Caraballo et al., 2003). Accordingly, permissible levels of selected trace elements in products are 10 mg L^{-1}, 10mg L^{-1}, 0.5 mg L^{-1}, and 2mg L^{-1}, respectively, for As, Cu, Pb, and Zn (Bonic et al., 2013) in Serbia, 0.1 mg L^{-1}, 5 mg L^{-1} and 0.2 mg L^{-1}, respectively, for As, Cu, and Pb (Penteado et al., 2009; Miranda et al., 2010) in Brazil, 10 mg L^{-1}, 1 mg L^{-1}, and 10 mg L^{-1}, respectively, for Cu, Pb, and Zn (Rodríguez et al., 2006) in Spain, and 4 mg L^{-1} and 2 mg L^{-1}, respectively, for Cu and Fe (Capote et al., 1999) in Venezuela. Apart from verification of quality and safety of distilled spirits, their reliable elemental analysis is significant from a marketing point

FIGURE 9.4 Flow chart representing production steps of distilled spirits.

TABLE 9.9
Composition of Distillers' Feed Dried grains with Solubles (Corn)

Composition	Percentage (%)	References
Moisture	9.0	Maisch et al., 2014; DISCUS Facts Book, 2012
Protein	27.0	Miller, 2013; Solomons, 2017
Fat	8.0	DISCUS Facts Book, 2012; Miller, 2013
Fibre	8.5	Maisch et al., 2014; DISCUS Facts Book, 2012
Ash	4.5	Maisch et al., 2014; DISCUS Facts Book, 2012

of view [i.e., to guarantee the products are genuine and of superior quality, particularly when these products are declared as possessing a certified brand of origin (CBO) label (Farinas et al., 2008)].

9.6.1 TYPES OF DISTILLED SPIRITS

9.6.1.1 Vodka

Vodka is a neutral, unaged distilled spirit. Vodka is prepared from fermentation of starchy substances, and the most commonly used is corn or milo. The yield of vodka is solely dependent on its inoculum used (% yeast) and time temperature combination of fermentation. Being a spirit, it is distilled at or above 190° proof and further distillate obtained can be treated with charcoal for its neutral taste and odor.

9.6.1.2 Whiskey

Whiskey is prepared from cereal mash and most commonly used is rye or corn malt. It is usually distilled at no more than 160° proof and distillates contain abundant flavor compounds. To prepare the rich flavor of whiskey it is kept for aging in charred white oak barrels, which specifically produce a desired mellowing effect. The changes that occur during aging can be monitored chemically. A significant loss of ethanol occurs during aging due to barrel soakage and some evaporation. Scotch,

TABLE 9.10
Quantitative Analysis of Bourbon Whiskey

Specific gravity 20°C/20°C	0.93
Proof	100.00
pH	4.1
Extract	136.0
Tannins	42.0
Furfural	1.3
Aldehydes	2
Esters	38.7
Others congeners	218.0
Total acids as acetic	54.0

Note: All values given are in grams per 100 liters.

Canadian and American light are a type of whiskey that may claim age for storage in reused cooperage while bourbon whiskey must by law be aged in new barrels. Table 9.10 presents quantitative data for the different classes of compounds found in bourbon whiskey (Maisch et al., 2014).

9.6.1.3 Tequila

Tequila is a distinctive product of the Jalisco area of Mexico and is produced from agava cactus.

9.6.1.4 Rum

Rum is distillate from molasses or fermented sugar cane juice. It is distilled between 80° and 190° proof.

9.6.1.5 Brandy

Brandy is also distilled at less than 190° proof from fermented fruit or fruit juices. Extensive chemical analyses have been made on aged as well as unaged distillates.

9.7 BY-PRODUCTS OF DISTILLED BEVERAGES

Fermentation and distillation of alcohol from grains has been recognized for at least 10,000 years (Brooke, 1992). The earliest distilling of whisky was reported in Scotland (Hertrampf and Piedad-Pascual, 2000). The recovery of materials from grains, which have undergone fermentation (distillers' feed), was developed by the beverage distilling industry.

Originally distillers' feeds were by-products from malted barley only. Nowadays all kind of grains are used for transforming starch into spirit. There are four different products of distillers' feed (Figure 9.5):

- Distillers' dried grains (DDG).
- Distillers' dried solubles (DDS).
- Distillers' dried grains with solubles (DDGS).
- Condensed distillers' solubles (CDS).

9.7.1 Manufacture and Processing

Distillers' grains are basically spoken "grains deficiency starch". They are obtained from de-alcoholized fermentation residues, which remain after grains have been fermented by the yeast

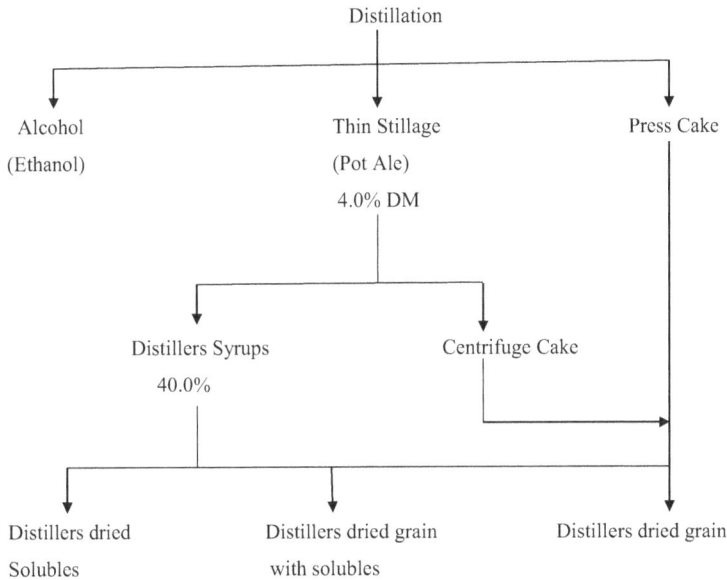

FIGURE 9.5   Diagram of the processing of distillers' dried feeds.

species *Saccharomyces cerevisiae*. All the essential starchless nutrients are recovered as distillers' grains by condensing remaining nutrients after starch removal. Because of the high moisture content fresh distillers' grains cannot be stored particularly under hot weather conditions. To make them an easy-to-handle feedstuff they have to be drum-dried and preferably pelletized.

"Light" **distillers' dried grains** (DDG) are obtained when whole stillage is screened for removal of coarser particles, pressing out excess water and drying. "Dark" distillers' dried grains (DDG) are derived from processing whole stillage.

The **"thin stillage"** is very watery but contains large quantities of water-soluble material (protein, vitamins). The dry matter content amounts to 4.0%. Thin stillage is also named "pot ale" if it comes from a malt distillery and "spent wash" if the source is from a grain distillery.

Condensed distillers' solubles (distillers' syrup) (CDS) are derived from condensing material to a dry matter content of 40 to 50%. Distillers' dried solubles (DDS) is then spray or drum dried "pot ale syrup" and "evaporated spent wash", respectively (Figure 9.4).

Distillers' dried solubles (DDS) is superior feedstuff. Since its processing is costly and the distilling industry adds back condensed distillers' soluble to the press cake prior to drying (Hughes, 1986). "Distillers' dried grains with soluble" (DDGS) is a now prevailing distillery by-product. It represents over 95% of the distillers' feeds used in animal feed such as in pig feed formulation.

9.7.2 WHISKY BY-PRODUCTS

A chief source of barley distillery by-products is production of malt whisky, which is generally based on malted barley (Russell et al., 2014). The production of grain whisky may use malted barley together with non-malted grains (maize, wheat, rye) yielding by-products containing up to 15% barley residues (Gizzi et al., 2001). The quality of barley distillery by-products depends on the malting and extraction process that malt undergoes in the distillery. The general process is as follows:

1. Germination: enzymes reduce starch into soluble and fermentable sugars.
2. Kilning: germinated grain is dried and converted into malt.

3. Mashing: ground malt is immersed in hot water and then filtered so that soluble sugars are extracted yielding a sugar-rich liquid (wort).
4. The wort is cooled and distilled, which yields alcohol. A liquid by-product the pot ale is evaporated to produce pot ale syrup.

Malt whisky production yields several by-products used in animal feeding.

9.7.2.1 Malt Distillers

They are also called draff, which are wet residues of malted barley obtained after mashing. Unlike maize or wheat distillers resulting from production of grain whisky, malt distillers are produced before distillation. Wet product is sold directly from the distilleries to farms or mixed with other by-products at a later stage. Excess draff that is not consumed by livestock is ensiled often with grass during summer for the following winter (Bell et al., 2012). ·

9.7.2.2 Pot Ale Syrup

Pot ale syrup is a concentrated liquid residue of distillation. It contains yeast, yeast residue, soluble proteins, and carbohydrates. It is medium to dark brown viscous syrup that resembles molasses. It is rich in protein and before 1960 it was used on its own, but now it is generally mixed with malt distillers (Crawshaw, 2004).

9.7.2.3 Malt Distillers' Dark Grains

It is also called barley dark grains, which are obtained by mixing draff and pot ale syrup followed by drying and pelleting the mix. They have a variable composition but characteristically richer in protein than draff due to the addition of the pot ale syrup. In the United Kingdom, drying plants mix materials from different distilleries (Crawshaw, 2004).

9.7.3 Ethanol as By-products

Barley distillers from fuel ethanol production follow a similar process as other grains. The ground grains mixed with water and enzymes (amylases) to produce a mash where starch hydrolysis occurs (liquefaction stage). Enzymes are added to the mash to transform starch into dextrose (saccharification stage). After saccharification, yeast is added to start the fermentation process, which produces a "beer" and CO_2. The beer passes through a continuous distillation column to yield alcohol at the top of the column. The product that remains at the bottom (whole stillage) is centrifuged and yields wet grains and thin stillage. Wet grains may be fed to livestock directly or can be dried to produce dried distillers' grain. Thin stillage can be sold as high-moisture feed or can be

TABLE 9.11
Distillery Feed By-products and Use

Spirit	Product	Type	Dry matter (%)	Livestock
Malt	Draff	Wet	23	Cattle, Sheep
Malt	Pot ale syrup	Liquid	45	Cattle, Pigs
Malt	Malt DDGS	Pellet	90	Cattle
Grain	Grain moist feeds	Wet	23	Cattle
Grain	Spent wash syrup	Liquid	45	Cattle, Sheep
Grain	Wheat/Maize DDGS	Pellet	90	Cattle, Sheep

Note: DDGS – distillers' dark grains with solubles.

dehydrated to produce condensed distillers' solubles. Condensed distillers' solubles and distillers' grain are often blended together to prepare wet or dried distillers' grain and solubles (WDGS or DDGS) (Mosier et al., 2006).

However, while maize and wheat grains are commonly used as ethanol feedstocks the use of barley in ethanol production has been limited due to its low starch and high fiber contents (which causes flow problems), the abrasive nature of its hull (high silica content) and presence of β-glucans (which increase viscosity of mash) (Hicks et al., 2005). Experiments with enzyme addition, hull and bran removal before fermentation and high-starch varieties have demonstrated that barley could be more widely used as a potential substrate for ethanol. Scientific literature on barley distillers often refer to blends or to experimental products (Kalscheur *et al.*, 2012).

9.7.4 BY-PRODUCTS OF TEQUILA

Agave bagasse is residual fiber remaining after cooked agave heads are shredded, milled, and sugars water-extracted. The bagasse is primarily rind and fibro vascular bundles dispersed throughout the interior of the agave head. It represents about 40% of total weight of milled agave on a wet weight basis. Bagasse is available all year in only two of the main regions of tequila producing areas in Mexico: Tequila region and Jalisco Highlands. Bagasse is composed of fiber and pith. The fiber is thick walled and long (10–12 cm). Some bagasse is mixed with clay and used to make bricks. other bagasse are separated from mechanical Pitt and sun-drying into mattresses, furniture, and packing materials but most of it is treated as waste and returned to the fields. In recent years, the tequila market has grown and gained international recognition. Market growth is expected to continue due to recently recognized "tequila origin denomination" by the European Union, which means more tequila will be produced with a substantial improvement of the process and tequila quality.

Research work on the use of agro-industrial waste is needed because of the serious economic and environmental problems caused by the disposal of these resources. The possibility of utilizing at least part of agave bagasse as an efficient ruminant feed could have a major impact on livestock production practices in areas where tequila is produced. The availability of energy to ruminant animal from this source as well from other lignocellulosic agricultural by-products is limited by close physical and chemical association between structural carbohydrates and lignin and crystalline arrangement of the cellulose polymer in plant cell walls (Scalbert et al., 1985). Lignin is the most important factor limiting degradation of cellulose by microorganisms (Gould, 1984). There is also an increasing trend towards use of lignocellulosic agricultural by-products in useful products such as sound-proofing, thermal insulation, and false ceilings (Sampathrajan et al., 1992; Raj and Kokta, 1991). The bagasse fiber could also be used to produce a wide variety of other products such as filters, sorbents, geotextiles, fiberboards, packaging, molded products, or in combination with other resources. The purpose of the present study was to investigate the use of agave bagasse pith as a possible animal feed and the fiber for fiberboard production.

9.7.5 BY-PRODUCTS OF WINE

The winemaking process is based on ancestral procedures being more an art than a science. As a general rule, many artisanal practices are strongly rooted in traditional vinification processes as well as limited human resources or physical infrastructures during the production operations limiting the update of the technological advances addressed to minimize waste production in several wine industries (Musee et al., 2007). Therefore, implementation of waste management in the wine industry is a challenging task making the development of innovative and effective valorization procedures necessary. In this sense, growing demand of final products and urgency of avoiding environmental footprints of this agro-industrial activity have encouraged a tough legal framework to ensure efficiency of the processes and to support the improvements of recovery and recycling procedures.

TABLE 9.12

Chemical Composition of Various Distilleries' By-products

Parameters (% db)	Various by-products									Reference
	1	2	3	4	5	6	7	8	9	
Dry matter	24.1	90.7	48.3	4.7	35.2	89.0	92.2	89.9	90.6	Crawshaw, 2004; Bonnet et al., 2010; Liu et al., 2011
Protein	20.3	27.8	37.4	17.9	31.8	29.5	31.3	33.5	37.3	Briggs, 2015; Crawshaw, 2004; Liu et al., 2011
Lipid	8.2	8.5	0.2	9.2	13.0	11.1	7.9	9.4	5.0	Bryce et al., 2008; Crawshaw, 2004; Liu et al., 2011
Fiber	17.6	11.6	0.2	-	8.2	7.9	8.2	8.1	7.7	Crawshaw, 2004; Liu et al., 2011; Linko, 1989
NDF	65.1	39.7	0.6	12.5	39.0	34.2	-	38.5	34.0	Crawshaw, 2004; Liu et al., 2011; Bryce et al., 2008; Gibson, 2011
Starch	1.5	3.2	1.8	25.1	4.9	9.3	-	-	4.2	Crawshaw, 2004; Gibson, 2011; Kanauchi, 2013; Liu et al., 2011
Ash	3.3	5.8	9.5	6.3	3.8	5.4	5.4	4.5	5.9	Crawshaw, 2004; Linko, 1989; Liu et al., 2011
Phosphorus (g/kg db)	3.3	9.7	19.0	-	8.2	7.9	8.0	7.4	9.1	Crawshaw, 2004; Liu et al., 2011; Bryce et al., 2008; Linko, 1989
Iron (mg/kg db)	-	231.0	-	-	116.0	123.0	154.0	-	-	Crawshaw, 2004; Briggs, 2015; Liu et al., 2011

Notes:

By product: 1. Draff (barley distillers' grain) Fresh; 2. Malt distillers' dark grain, dried; 3. Pot ale syrup, fresh; 4. Corn, thin stillage, fresh; 5. Corn distillers' wet grains with solubles (DWGS); 6. Corn distillers' dried grains with solubles (DDGS); 7. Rye distillers' dried grains with solubles (DDGS); 8. Sorghum distillers' dried grains with solubles (DDGS) and 9. Wheat distillers' dried grains with solubles (DDGS)

The type of residues produced is closely dependent on specific vinification procedures, which also affect physic-chemical properties of residual material. The major residues from winemaking activity are represented by organic wastes (grape pomace, containing seeds, pulp and skins, grape stems, and grape leaves), wastewater, emission of greenhouse gases (CO_2, volatile organic compounds, etc.) and inorganic wastes (diatomaceous earth, bentonite clay, and perlite) (Oliveira et al., 2013). In this regard, it is estimated that in Europe alone 14.5 million tons of grape by-products are produced annually (Chouchouli et al., 2013).

The valorization of winemaking by-products is mainly represented by elaboration of soil fertilizers as well as a fermentation substrate for biomass production and livestock feeds (Table 9.12) (Harsha et al., 2013; Yilmaz et al., 2004). However, there are several constraints for the current available options for reusing these unprofitable materials. For example, certain polyphenols present in winery by-products are known to be phytotoxic and display antimicrobial effects during composting. Regarding their use in livestock feed, some animals show intolerance to certain components such as condensed tannins, which negatively affect digestibility (González-Centeno et al., 2014). Hence, their valorization as a source of bioactive phytochemicals of application in pharmaceutical, cosmetic, and food industries might constitute an efficient, profitable, and environment-friendly alternative for residues (Makris et al., 2007).

9.8 CONCLUSION AND FUTURE PERSPECTIVES

The residues obtained after malting, brewing, and distillery by-products are commonly used as livestock feed. These by-products contain high levels of protein as well as fiber. There is worldwide need for protein as food in the human diet, which is fulfilled by fish and meat products, therefore the demand for meat and fish products is rising.

To reduce the demand of non-vegetarian food products and to develop a balance between environment and the human diet these by-products can be used as a protein alternative. These by-products are waste and can furthermore be incorporated into different food products to provide a vegan diet to consumers. Protein from distillery by-products could be used as a replacement for fish meal or soybean meal in feed formulations. Apart from protein brewing and malting by-products contain a high amount of fiber. The fiber present can be isolated or the residues can be dried to obtain flour, which ultimately helps in rising the fiber and protein content in food. New technologies should be developed to utilize these wastes and incorporate them into the human diet or use them as a part of the environment to benefit not only the food sector but every sector.

REFERENCES

Aborus, N. E., CanadanovićBrunet, J., Ćetković, G., Šaponjac, V. T., Vulić, J., & Ilić, N. (2017). Powdered barley sprouts: composition, functionality and polyphenol digestibility. *International Journal of Food Science & Technology*, *52*(1), 231–238.

Aggelopoulos, T., Bekatorou, A., Pandey, A., Kanellaki, M., & Koutinas, A. A. (2013). Discarded oranges and brewer's spent grains as promoting ingredients for microbial growth by submerged and solid-state fermentation of agro-industrial waste mixtures. *Applied Biochemistry and Biotechnology*, *170*(8), 1885–1895.

Alonso-Riaño, P., Sanz Diez, M. T., Blanco, B., Beltrán, S., Trigueros, E., & Benito-Román, O. (2020). Water ultrasound-assisted extraction of polyphenol compounds from brewer's spent grain: Kinetic study, extract characterization, and concentration. *Antioxidants*, *9*(3), 265.

Barciela, J., Vilar, M., García-Martín, S., Peña, R. M. & Herrero, C. (2008). Study on different pre-treatment procedures for metal determination in Orujo spirit samples by ICP-AES. *Analytica Chimica Acta*, *628*(1), 33–40.

Bedini, S., Flamini, G., Girardi, J., Cosci, F., & Conti, B. (2015). Not just for beer: Evaluation of spent hops (Humulus lupulus L.) as a source of eco-friendly repellents for insect pests of stored foods. *Journal of Pest Science*, *88*(3), 583–592.

Begea, M., Sirbu, A., Constantin, U., & Pitesti, B. (2017). Pilot technology to obtain a bio-based product from barley malt rootlets. *Int. J. Food Biosyst. Eng.*, *3*, 66–72.

Bell, J., Morgan, C., Dick, G. & Reid, G. (2012). Distillery feed by-products briefing. An AA211 Special Economic Study for the Scottish Government, SAC Consulting

Birsan, R. I., Wilde, P., Waldron, K. W., & Rai, D. K. (2019). Recovery of polyphenols from brewer's spent grains. *Antioxidants*, *8*(9), 380.

Bonaui, C., Dumoulin, E., Raoult-Wack, A. L., Berk, Z., Bimbenet, J. J., Courtois, F., & Vasseur, J. (1996). Food drying and dewatering. *Drying Technology*, *14*(9), 2135–2170.

Bonic, M., Te šević, V., Nikićević, N., Cvejic, J., Milosavljević, S. M., Vajs, V. & Jovanic, S. (2013). The contents of heavy metals in Serbian old plum brandies. *Journal of the Serbian Chemical Society*, *78*(7), 933–945.

Bonnet, A. & Willm, C. (2010). Wheat mill for getting proteins and ethanol [crusher Victory]. [French]. *Industries des Cereales*.

Boulton, C. (2013). *Encyclopaedia of brewing*. John Wiley & Sons.

Bravi, E., Francesco, G. D., Sileoni, V., Perretti, G., Galgano, F., & Marconi, O. (2021). Brewing By-Product Upcycling Potential: Nutritionally Valuable Compounds and Antioxidant Activity Evaluation. *Antioxidants*, *10*(2), 165.

Briggs, D. E. (1998). *Malts and malting*. Springer Science & Business Media.

Briggs, D. E. (2015). Barley for animal and human food. In *Barley* (pp. 492–525). Springer.

Brooke, M. (1992). A wee drop of Scotch. *Silver Kris*, *19*(**9**), 19–24.

Bryce, J.H., Piggott, J.R. & Stewart, G.G. (2008). *Distilled Spirits: Production, Technology and Innovation*. Nottingham University Press.

Caballero-Córdoba, G. M., & Sgarbieri, V. C. (2000). Nutritional and toxicological evaluation of yeast (Saccharomyces cerevisiae) biomass and a yeast protein concentrate. *Journal of the Science of Food and Agriculture*, *80*(3), 341–351.

Capote, T., Marcó, L. M., Alvarado, J., & Greaves, E. D. (1999). Determination of copper, iron and zinc in spirituous beverages by total reflection X-ray fluorescence spectrometry. *Spectrochimica Acta Part B: Atomic Spectroscopy*, *54*(10), 1463–1468.

Caldas, N. M., Raposo Jr, J. L., Neto, J. A. G. & Barbosa Jr, F. (2009). Effect of modifiers for As, Cu and Pb determinations in sugar-cane spirits by GF AAS. *Food Chemistry*, *113*(4), 1266–1271.

Carvalheiro, F., Esteves, M. P., Parajó, J. C., Pereira, H., & Gírio, F. M. (2004). Production of oligosaccharides by autohydrolysis of brewery's spent grain. *Bioresource Technology*, *91*(1), 93–100.

Cejas, L., Romano, N., Moretti, A., Mobili, P., Golowczyc, M., & Gómez-Zavaglia, A. (2017). Malt sprout, an underused beer by-product with promising potential for the growth and dehydration of lactobacilli strains. *Journal of Food Science and Technology*, *54*(13), 4464–4472.

Celus, I., Brijs, K., & Delcour, J. A. (2006). The effects of malting and mashing on barley protein extractability. *Journal of Cereal Science*, *44*(2), 203–211.

Chiş, M. S., Pop, A., Păucean, A., Socaci, S. A., Alexa, E., Man, S. M., & Muste, S. (2020). Fatty acids, volatile and sensory profile of multigrain biscuits enriched with spent malt rootles. *Molecules*, *25*(3), 442.

Chouchouli, V., Kalogeropoulos, N., Konteles, S. J., Karvela, E., Makris, D. P. & Karathanos, V. T. (2013). Fortification of yoghurts with grape (Vitis vinifera) seed extracts. *LWT-Food Science and Technology*, *53*(2), 522–529.

Coelho, E., Rocha, M. A. M., Saraiva, J. A., & Coimbra, M. A. (2014). Microwave superheated water and dilute alkali extraction of brewers' spent grain arabinoxylans and arabinoxylo-oligosaccharides. *Carbohydrate Polymers*, *99*, 415–422.

Connolly, A., Cermeño, M., Crowley, D., O'Callaghan, Y., O'Brien, N. M., & FitzGerald, R. J. (2019). Characterisation of the in vitro bioactive properties of alkaline and enzyme extracted brewers' spent grain protein hydrolysates. *Food Research International*, *121*, 524–532.

Crawshaw, R. (2004). Co-product feeds: animal feeds from the food and drinks industries. Nothingham University Press

Creasy, M. E., Gunter, S. A., Beck, P. A., & Weyers, J. S. (2001). Malt sprouts as a supplement for forage fed beef cattle. *Journal of Applied Animal Research*, *20*(2), 129–140.

Daenicke, R., Rohr, K., & Engling, F. P. (1991). Influence of brewers' spent hops silage in diets for dairy cows on digestion and performance variables. In *Proceedings of the, VDLUFA Congress: Umweltaspekte der Tierproduktion, Ulm*, *33*, 539–44.

De Boer, F. & Bickel, H. (1988). Livestock feed resources and feed evaluation in Europe: present situation and future prospects. EAAP Publication, Netherlands.

De Keukeleire, D. (2000). Fundamentals of beer and hop chemistry. *Quimica Nova*, *23*, 108–112.

Daniels, R. (1998). *Designing great beers: The ultimate guide to brewing classic beer styles*. Brewers Publications.

DISCUS Facts Book (2012). Distilled Spirits Council U.S., Washington, D.C.

Du, L., Arauzo, P. J., Meza Zavala, M. F., Cao, Z., Olszewski, M. P., & Kruse, A. (2020). Towards the properties of different biomass-derived proteins via various extraction methods. *Molecules*, *25*(3), 488.

Eremina, O. Y., Seregina, N. V., Ivanova, T. N., Shuldeshova, N. V., & Zaugolnikova, E. V. (2021). Micronutrient value and antioxidant activity of malt wheat sprouts. In *IOP Conference Series: Earth and Environmental Science*, *677*(2), 022107.

Farinas, M. V., García, J. B., Martín, S. G., Crecente, R. P. & Latorre, C. H. (2008). Determination of Cr and Ni in Orujo spirit samples by ETAAS using different chemical modifiers. *Food Chemistry*, *110*(1), 177–186.

Ferreira, I. M. P. L. V. O., Pinho, O., Vieira, E., & Tavarela, J. G. (2010). Brewer's Saccharomyces yeast biomass: characteristics and potential applications. *Trends in Food Science & Technology*, *21*(2), 77–84.

Gibson, G. (2011). Malting plant technology. In: Palmer, G.H. (Ed.), Cereal Science and Technology. *Aberdeen Univ. Press*, UK: 52.

Gizzi, G. & Givens, D. I. (2001). Distillers' dark grains in ruminant nutrition. Nut. Abstr. Rev. Series B, *71*(10), 1R–19R

González-Centeno, M. R., Knoerzer, K., Sabarez, H., Simal, S., Rosselló, C. & Femenia, A. (2014). Effect of acoustic frequency and power density on the aqueous ultrasonic-assisted extraction of grape pomace (Vitis vinifera L.)–a response surface approach. *Ultrasonics Sonochemistry*, *21*(6), 2176–2184.

Gould. J.M. (1984). Studies on the mechanism of alkaline peroxide delignification of agricultural residues. *Biotechnology and Bioengineering*, *27*, 225–231.

Gupta, M., Abu-Ghannam, N. and Gallaghar, E. (2010). Barley for brewing: Characteristic changes during malting, brewing and applications of its by-products. *Comprehensive Reviews in Food Science and Food Safety*, *9*, 318–328.

Harsha, P. S., Gardana, C., Simonetti, P., Spigno, G. & Lavelli, V. (2013). Characterization of phenolics, in vitro reducing capacity and anti-glycation activity of red grape skins recovered from winemaking by-products. *Bioresource Technology*, *140*, 263–268.

Hegazi, S. M., Ghali, Y., Foda, M. S., & Youssef, A. (1975). Nutritive value of barley rootlets, a by-product of malting. *Journal of the Science of Food and Agriculture*, *26*(8), 1077–1081.

Hernández-Caraballo, E. A., Avila-Gómez, R. M., Capote, T., Rivas, F. & Pérez, A. G. (2003). Classification of Venezuelan spirituous beverages by means of discriminant analysis and artificial neural networks based on their Zn, Cu and Fe concentrations. *Talanta*, *60*(6), 1259–1267.

Hernández-Caraballo, E. A., de Hernández, R. M. Á., Rivas-Echeverría, F. & Capote-Luna, T. (2008). Discrimination of Venezuelan spirituous beverages by a trace element-radial basis neural network approach. *Talanta*, *74*(4), 871–878.

Hertrampf, J. W. & Piedad-Pascual, F. (2000). Distillery by-products. In *Handbook on Ingredients for Aquaculture Feeds* (pp. 115–124). Springer.

Hicks, K. B., Flores, R. A., Taylor, F., McAloon, A. J., Moreau, R. A., Johnston, D. B., Senske, G. E., Brooks, W. S. & Griffey, C. A. (2005). Current and potential use of barley in fuel ethanol production. 2005 EWW/SSGW Conference, May 9–12, Bowling Green, KY

Hoehnel, A., Bez, J., Petersen, I. L., Amarowicz, R., Juśkiewicz, J., Arendt, E. K., & Zannini, E. (2020). Enhancing the nutritional profile of regular wheat bread while maintaining technological quality and adequate sensory attributes. *Food & Function*, *11*(5), 4732–4751.

Hughes, S.G. (1986): Replacement of distillers' dried grains with solubles compared for salmon diet formulation. *Feedstuffs*, *58*(53), 10.

Huige, N. J. (2006). Brewery by-products and effluents. In *Handbook of brewing* (pp. 670–729). CRC Press.

Izydorczyk, M. S., & Biliaderis, C. G. (1995). Cereal arabinoxylans: advances in structure and physicochemical properties. *Carbohydrate Polymers*, *28*(1), 33–48.

Jacob, F. F., Hutzler, M., & Methner, F. J. (2019). Comparison of various industrially applicable disruption methods to produce yeast extract using spent yeast from top-fermenting beer production: influence on amino acid and protein content. *European Food Research and Technology*, *245*(1), 95–109.

Jay, A. J., Parker, M. L., Faulks, R., Husband, F., Wilde, P., Smith, A. C., ... & Waldron, K. W. (2008). A systematic micro-dissection of brewers' spent grain. *Journal of Cereal Science*, *47*(2), 357–364.

Kalscheur, K. F., Garcia, A. D., Schingoethe, D. J., Diaz Royón, F. & Hippen, A. R. (2012). Feeding biofuel co-products to dairy cattle. In: Makkar, H. (Ed.), Biofuel co-products as livestock feed: Opportunities and challenges, Chapter 7: 115–154

Kanauchi, M. (2013). *SAKE* Alcoholic beverage production in Japanese food industry (Chapter 3). In: Muzzalupo, I. (Ed.), *Food Industry*. Intech.

Kanauchi, O., Mitsuyama, K., & Araki, Y. (2001). Development of a functional germinated barley foodstuff from brewer's spent grain for the treatment of ulcerative colitis. *Journal of the American Society of Brewing Chemists*, *59*(2), 59–62.

Karlovic, A., Jurić, A., Ćorić, N., Habschied, K., Krstanović, V., & Mastanjević, K. (2020). By-Products in the malting and brewing industries – Re-usage possibilities. *Fermentation*, *6*(3), 82.

Kemppainen, K., Rommi, K., Holopainen, U., & Kruus, K. (2016). Steam explosion of Brewer's spent grain improves enzymatic digestibility of carbohydrates and affects solubility and stability of proteins. *Applied Biochemistry and Biotechnology*, *180*(1), 94–108.

Kendall, C. W., Esfahani, A., & Jenkins, D. J. (2010). The link between dietary fibre and human health. *Food Hydrocolloids*, *24*(1), 42–48.

Khidzir, K. M., Abdullah, N., & Agamuthu, P. (2010). Brewery spent grain: Chemical characteristics and utilization as an enzyme substrate. *Malaysian Journal of Science, 29*(1), 41–51.

Kirjoranta, S., Tenkanen, M., & Jouppila, K. (2016). Effects of process parameters on the properties of barley containing snacks enriched with brewer's spent grain. *Journal of Food Science and Technology, 53*(1), 775–783.

Kondo, K., Nagao, K., & Yokoo, Y. (2016). *U.S. Patent No. 9,326,542.* Washington, DC: U.S. Patent and Trademark Office.

Kunze, W. (2004). Brewing Malting. *Berlin: Vlb*, 18–152.

Lewis, M. J. (2005). Celiac disease, beer, and brewing. *Tech. Q. Master Brew. Assoc. Am, 42*, 45.

Linko, P. (1989). The twin-screw extrusion cooker as a versatile tool for wheat processing (Chapter 22). In: Pomeranz, Y. (Ed.), Wheat Is Unique. *Amer. Assoc. of Cereal Chemists* Inc., St. Paul, MN, USA

Liu, K. & Rosentrater, K.A. (2011). Distillers Grains: Production, Processing, and Utilization. CRC Press, Boca Raton, FL, USA.

Luzak, B., Golanski, J., Przygodzki, T., Boncler, M., Sosnowska, D., Oszmianski, J. & Rozalski, M. (2016). Extract from spent hop (Humulus lupulus L.) reduces blood platelet aggregation and improves anticoagulant activity of human endothelial cells in vitro. *Journal of functional Foods, 22*, 257–269.

Mahalingam, R. (2020). Analysis of the barley malt rootlet proteome. *International Journal of Molecular Sciences, 21*(1), 179.

Maisch, W. F., Sobolov, M. & Petricola, A. J. (2014). Distilled beverages. In *Microbial technology* (pp. 79–94). Academic Press.

Makris, D. P., Boskou, G. & Andrikopoulos, N. K. (2007). Polyphenolic content and in vitro antioxidant characteristics of wine industry and other agri-food solid waste extracts. *Journal of Food Composition and Analysis, 20*(2), 125–132.

Malleshi, N. G., & Klopfenstein, C. F. (1998). Nutrient composition, amino acid and vitamin contents of malted sorghum, pearl millet, finger millet and their rootlets. *International Journal of Food Sciences and Nutrition, 49*(6), 415–422.

Manna, K. M., Naing, K. M. and Pe, H. (1995). Amylase activity of some roots and sprouted cereals and beans. *Food and Nutrition Bulletin* 16, 1–4.

Marson, G. V., Saturno, R. P., Comunian, T. A., Consoli, L., da Costa Machado, M. T., & Hubinger, M. D. (2020). Maillard conjugates from spent brewer's yeast by-product as an innovative encapsulating material. *Food Research International, 136*, 109365.

Mathias, T. R. D. S., Fernandes de Aguiar, P., de Almeida e Silva, J. B., Moretzsohn de Mello, P. P., & CamporeseS érvulo, E. F. (2017). Brewery waste reuse for protease production by lactic acid fermentation. *Food Technology and Biotechnology, 55*(2), 218–224.

Meneses, N. G., Martins, S., Teixeira, J. A., & Mussatto, S. I. (2013). Influence of extraction solvents on the recovery of antioxidant phenolic compounds from brewer's spent grains. *Separation and Purification Technology, 108*, 152–158.

Miller, D. L. (2013). *Biotechnol. Bioeng.* Symp. 5, 345–352.

Miranda, K., Dionísio, A. G. & Pereira-Filho, E. R. (2010). Copper determination in sugar cane spirits by fast sequential flame atomic absorption spectrometry using internal standardization. *Microchemical Journal, 96*(1), 99–101.

Mosier, N. & Ileleji, K. (2006). How fuel ethanol is made from corn. Purdue University, ID-328. Department of Agricultural and Biological Engineering

Muller, R. (1991). The effects of mashing temperature and mash thickness on wort carbohydrate composition. *Journal of the Institute of Brewing, 97*(2), 85–92.

Musee, N., Lorenzen, L. & Aldrich, C. (2007). Cellar waste minimization in the wine industry: a systems approach. *Journal of Cleaner Production, 15*(5), 417–431.

Mussatto, S. I., & Roberto, I. C. (2006). Chemical characterization and liberation of pentose sugars from brewer's spent grain. *Journal of Chemical Technology & Biotechnology: International Research in Process, Environmental & Clean Technology, 81*(3), 268–274.

Mussatto, S. I., Dragone, G., & Roberto, I. C. (2006). Brewers' spent grain: generation, characteristics and potential applications. *Journal of Cereal Science, 43*(1), 1–14.

Nascimento, R. F., Bezerra, C. W., Furuya, S. M., Schultz, M. S., Polastro, L. R., Neto, B. S. L. & Franco, D. W. (1999). Mineral profile of Brazilian cachaças and other international spirits. *Journal of Food Composition and Analysis, 12*(1), 17–25.

Navarro-Alarcón, M., Velasco, C., Jodral, A., Terrés, C., Olalla, M., Lopez, H. & Lopez, M. C. (2007). Copper, zinc, calcium and magnesium content of alcoholic beverages and by-products from Spain: Nutritional supply. *Food additives and contaminants*, *24*(7), 685–694.

Neylon, E., Arendt, E. K., Lynch, K. M., Zannini, E., Bazzoli, P., Monin, T., & Sahin, A. W. (2020). Rootlets, a malting by-product with great potential. *Fermentation*, *6*(4), 117.

Nocente, F., Taddei, F., Galassi, E., & Gazza, L. (2019). Upcycling of brewers' spent grain by production of dry pasta with higher nutritional potential. *LWT*, *114*, 108421.

O'Rourke, T. (1999) Mash separation. *Brewer's Guardian, 7*, 48–50.

Oke, F. O., Oso, A. O., Oduguwa, O. O., Jegede, A. V., Südekum, K. H., Fafiolu, A. O., & Pirgozliev, V. (2017). Growth, nutrient digestibility, ileal digesta viscosity, and energy metabolizability of growing turkeys fed diets containing malted sorghum sprouts supplemented with enzyme or yeast. *Journal of Animal Physiology and Animal Nutrition*, *101*(3), 449–456.

Oliveira, D. A., Salvador, A. A., Smânia Jr, A., Smânia, E. F., Maraschin, M., & Ferreira, S. R. (2013). Antimicrobial activity and composition profile of grape (Vitis vinifera) pomace extracts obtained by supercritical fluids. *Journal of Biotechnology*, *164*(3), 423–432.

Özvural, E. B., Vural, H., Gökbulut, İ., & Özboy-Özbaş, Ö. (2009). Utilization of brewer's spent grain in the production of Frankfurters. *International Journal of Food Science & Technology*, *44*(6), 1093–1099.

Özvural, E. B., Vural, H., Gökbulut, İ., & Özboy-Özbaş, Ö. (2009). Utilization of brewer's spent grain in the production of Frankfurters. *International Journal of Food Science & Technology*, *44*(6), 1093–1099.

Papageorgiou, M., & Skendi, A. (2018). Introduction to cereal processing and by-products. In *Sustainable Recovery and Reutilization of Cereal Processing By-Products* (pp. 1–25). Woodhead Publishing.

Penteado, J. C. P. & Masini, J. C. (2009). Multivariate analysis for the classification differentiation of Brazilian sugarcane spirits by analysis of organic and inorganic compounds. *Analytical letters*, *42*(17), 2747–2757.

Raj, R.G. & Kokta, B.V. (1991). The effect of processing conditions and binding material on the mechanical properties of bagasse fiber composites. *Eur. Polym. J. 27*(10), 1121–1123.

Rakin, M., Baras, J., & Vukasinovic, M. (2004). The influence of brewer's yeast autolysate and lactic acid bacteria on the production of a functional food additive based on beetroot juice fermentation. *Food Technology and Biotechnology*, *42*(2), 109–113.

Reis, S. F., Coelho, E., Coimbra, M. A., & Abu-Ghannam, N. (2015). Improved efficiency of brewer's spent grain arabinoxylans by ultrasound-assisted extraction. *Ultrasonics Sonochemistry*, *24*, 155–164.

Robertson, J. A., I'Anson, K. J., Treimo, J., Faulds, C. B., Brocklehurst, T. F., Eijsink, V. G., & Waldron, K. W. (2010). Profiling brewers' spent grain for composition and microbial ecology at the site of production. *LWT-Food Science and Technology*, *43*(6), 890–896.

Rodríguez Madrera, R., Suárez Valles, B., García Hevia, A., García Fernández, O., Fernández Tascón, N.,& Mangas Alonso, J. J. (2006). Production and composition of cider spirits distilled in "Alquitara". *Journal of Agricultural and Food Chemistry*, *54*(26), 9992–9997.

Rodríguez-Solana, R., Salgado, J. M., Domínguez, J. M., & Cortés, S. (2014). Assessment of minerals in aged grape marc distillates by FAAS/FAES and ICP-MS. Characterization and safety evaluation. *Food Control*, *35*(1), 49–55.

Russell, I. & Steward, G. (2014). Whisky: Technology, production and marketing. Academic Press, 2nd Edition, 414 p

Saksinchai, S., Suphantharika, M., & Verduyn, C. (2001). Application of a simple yeast extract from spent brewer's yeast for growth and sporulation of Bacillus thuringiensis subsp. kurstaki: a physiological study. *World Journal of Microbiology and Biotechnology*, *17*(3), 307–316.

Salama, A. R. A., ElSahn, M. A., Mesallam, A. S., & Shehata, A. M. E. T. (1997). The chemical composition, the nutritive value and the functional properties of malt sprout and its components (acrospires, rootlets and husks). *Journal of the Science of Food and Agriculture*, *75*(1), 50–56.

Sampathrajan, A., Vijayaraghavan, N.C., & Swaminathan, K.R. (1992). Mechanical and thermal properties of particle boards made from farm residues. *Biores. Tech.*, *40*, 249–251.

Scalbert, A., Monties, B., Lallemand, J.Y., Guittet, E., & Rolando, C. (1985). Ether linkages between phenolic acids and lignin fractions from wheat straw. *Phytochemistry*, *24*, 1359–1362.

Silva, J. P., Sousa, S., Rodrigues, J., Antunes, H., Porter, J. J., Gonçalves, I., & Ferreira-Dias, S. (2004). Adsorption of acid orange 7 dye in aqueous solutions by spent brewery grains. *Separation and Purification Technology*, *40*(3), 309–315.

Sobukola, O. P., Babajide, J. M., & Ogunsade, O. (2013). Effect of brewers spent grain addition and extrusion parameters on some properties of extruded yam starch-based pasta. *Journal of Food Processing and Preservation, 37*(5), 734–743.

Solomons, G. L. (2017). *Materials and Methods in Fermentation.* Academic Press.

Spiller, G. A., & Amen, R. J. 1975 Dietary fiber in human nutrition. CRC *Critical Reviews in Food Science and Nutrition, 7,* 39–70.

Spinelli, S., Conte, A., Lecce, L., Padalino, L., & Del Nobile, M. A. (2016). Supercritical carbon dioxide extraction of brewer's spent grain. *The Journal of Supercritical Fluids, 107,* 69–74.

Steinmacher, N. C., Honna, F. A., Gasparetto, A. V., Anibal, D., & Grossmann, M. V. (2012). Bioconversion of brewer's spent grains by reactive extrusion and their application in bread-making. *LWT-Food Science and Technology, 46*(2), 542–547.

Stojceska, V., Ainsworth, P., Plunkett, A., & İbanoğlu, S. (2008). The recycling of brewer's processing by-product into ready-to-eat snacks using extrusion technology. *Journal of Cereal Science, 47*(3), 469–479.

Stroem, L. K., Desai, D. K., & Hoadley, A. F. A. (2009). Superheated steam drying of Brewer's spent grain in a rotary drum. *Advanced Powder Technology, 20*(3), 240–244.

Tang, Z., Cenkowski, S., & Muir, W. E. (2004). Modelling the superheated-steam drying of a fixed bed of brewers' spent grain. *Biosystems Engineering, 87*(1), 67–77.

Tapia-Hernández, J. A., Del-Toro-Sánchez, C. L., Cinco-Moroyoqui, F. J., Juárez-Onofre, J. E., Ruiz-Cruz, S., Carvajal-Millan, E., & Rodríguez-Felix, F. (2019). Prolamins from cereal by-products: Classification, extraction, characterization and its applications in micro-and nanofabrication. *Trends in Food Science & Technology, 90,* 111–132.

Thammakiti, S., Suphantharika, M., Phaesuwan, T., & Verduyn, C. (2004). Preparation of spent brewer's yeast β-glucans for potential applications in the food industry. *International Journal of Food Science & Technology, 39*(1), 21–29.

Thiel, P. G., & Toit, P. D. (1965). The chemical composition of a brewery waste. *Journal of the Institute of Brewing, 71*(6), 509–514.

Treimo, J., Westereng, B., Horn, S. J., Forssell, P., Robertson, J. A., Faulds, C. B., & Eijsink, V. G. (2009). Enzymatic solubilization of brewers' spent grain by combined action of carbohydrases and peptidases. *Journal of Agricultural and Food Chemistry, 57*(8), 3316–3324.

Vieira, E., Rocha, M. A. M., Coelho, E., Pinho, O., Saraiva, J. A., Ferreira, I. M., & Coimbra, M. A. (2014). Valuation of brewer's spent grain using a fully recyclable integrated process for extraction of proteins and arabinoxylans. *Industrial Crops and Products, 52,* 136–143.

Wahlström, R., Rommi, K., Willberg-Keyriläinen, P., Ercili-Cura, D., Holopainen-Mantila, U., Hiltunen, J., & Kuutti, L. (2017). High yield protein extraction from brewer's spent grain with novel carboxylate salt-urea aqueous deep eutectic solvents. *ChemistrySelect, 2*(29), 9355–9363.

Waters, D. M., Kingston, W., Jacob, F., Titze, J., Arendt, E. K., & Zannini, E. (2013). Wheat bread biofortification with rootlets, a malting by-product. *Journal of the Science of Food and Agriculture, 93*(10), 2372–2383.

Werpy, T., & Petersen, G. (2004). Top value-added chemicals from biomass: volume I--results of screening for potential candidates from sugars and synthesis gas (No. DOE/GO-102004-1992). National Renewable Energy Lab., Golden, CO (US).

Worrasinchai, S., Suphantharika, M., Pinjai, S., & Jamnong, P. (2006). β-Glucan prepared from spent brewer's yeast as a fat replacer in mayonnaise. *Food Hydrocolloids, 20*(1), 68–78.

Xiros, C., Topakas, E., Katapodis, P., & Christakopoulos, P. (2008). Hydrolysis and fermentation of brewer's spent grain by Neurospora crassa. *Bioresource Technology, 99*(13), 5427–5435.

Yilmaz, Y., & Toledo, R. T. (2004). Major flavonoids in grape seeds and skins: antioxidant capacity of catechin, epicatechin, and gallic acid. *Journal of Agricultural and Food Chemistry, 52*(2), 255–260.

Yu, D., Sun, Y., Wang, W., O'Keefe, S. F., Neilson, A. P., Feng, H., & Huang, H. (2020). Recovery of protein hydrolysates from brewer's spent grain using enzyme and ultrasonication. *International Journal of Food Science & Technology, 55*(1), 357–368.

Yu, D., Sun, Y., Wang, W., O'Keefe, S. F., Neilson, A. P., Feng, H., & Huang, H. (2020). Recovery of protein hydrolysates from brewer's spent grain using enzyme and ultrasonication. *International Journal of Food Science & Technology, 55*(1), 357–368.

Zubtsov, Y. N., Eremina, O. Y., & Seregina, N. V. (2017). The micronutrient value of by-products of malting barley. *Voprosypitaniia, 86*(3), 115–120.

10 Pectate as a Gelling Agent in Foods

Gauri Harish Athawale,[1,] Shweta Jeevan Raichurkar,[1]
and Rajkumar Arjun Dagadkhair[2]*

[1]MIT school of Food Tech, MITADT University, Pune, India
[2]Scientist, Directorate of cashew research (ICAR), Puttur, Karnatka, India
*Corresponding author: athawalegauri@gmail.com

CONTENTS

10.1 INTRODUCTION

Pectin belongs to the family of polysaccharides, which is generally found in the primary cell wall and the middle layer of higher plants. The highest pectin content is found in dicotyledonous and monocotyledonous of plants (Mohnen, 2008). Pectin biosynthesis happens in the Golgi vesicles (Xiao, 2013). The released pectin is highly methylated and can be modified by the activity of pectin methylesterase. Certain areas of the pectin are involved in the formation of calcium ions to observe protein cross-links. Pectins are very important for the growth and development of plants, due to their impact on the rigidity and integrity of plant tissues (Sila et al. 2009; Voragen et al. 2009). They play a role in ion transport, determine the porosity of the cell wall, and affect the activation of the plant's immune system (Baldwin et al. 2014). Many functions of different pectins depend on their structure and concentration in the cell wall, and varies greatly owing to the enzymatic and chemical modifications, which occur during the growth and ripening, storage, and processing of the fruit (Schols et al. 2002; Voragen et al. 2009). Pectins are used in food design, due to its versatile gelling property. Common applications are found in the processing of jelly, jam, fruit juice, confectionery products, etc. (Voragen et al. 1995; Mierczyńska et al. 2015). Pectins act as a gelling

DOI: 10.1201/9781003207689-10

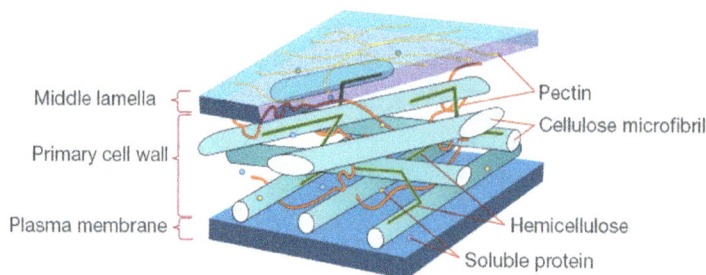

FIGURE 10.1   Schematic demonstration of the structural design of the cell wall.

Source: adapted from Antony et al. 2017.

agent, thickening agent, stabilizer, and emulsifier (Rascón, 2016). Pectins are pro-health due to the presence of essential constituent i.e. soluble dietary fiber (Kosmala et al. 2013). It has been reported that pectins can lower down the cholesterol level in blood (Behall, 1986), bind lead and mercury in the gastrointestinal tract, and can play an anti-cancer role (Glinsky, 2009).

Pectin polysaccharides are diverse and variable, it is generally believed that the function of pectin in food design depends on its structure and characteristics. Due to the variability of the source of plant pectin, it is extremely difficult to control the structure of pectin extracted from plants for specific purposes in various experiments. Comparison of different experimental results of different plants and different stages of maturity is unclear. Association of gelling properties of pectin towards its stability and its reaction with other natural ingredients is still a challenge to a researcher.

The term pectin in this chapter will be used to refer to water-soluble galacturonic acid polysaccharides with variable methyl ester content and degree of neutralization, which can form gels under suitable conditions. Pectin can be classified according to its degree of esterification (DE) and the percent of carboxyl groups esterified with methanol (Figure 10.1). Pectins with DE > 50% are high-methoxyl pectins (HM-pectins); those with DE < 50% are low-methoxyl pectins (LM-pectins).The degree of amidation (DA) indicates the percent of carboxyl groups in the amide form.

In the 1930s, the Californian fruit growers' exchange described a process for treating solid plant material under mild alkaline conditions to de-esterify the pectin present in it to produce a crude insoluble material. Insoluble material was later called a pectate pulp. Sodium pectate can be extracted from this pulp by boiling it with alkaline phosphate sequestrant and further can be purified by alcohol precipitation. Pectate pulp can also be used directly in any application. Crude pulp obtained from citrus waste contains 20–30% of pectin and on heating in the presence of phosphate sequestrants, pectate can be extracted and a dispersion of predominantly cellulose material can be obtained in a pectate solution. If calcium ions are introduced into the dispersion then calcium pectate gel can be formed. Pectate pulps also have gained interest in the market as they are a very cheap gelling agent. Pectate pulps can be made into gels in neutral pH canned foods specifically meat products. Its application in the pet industry is growing as it could be used as an alternative source to carrageenan for developing canned products consisting of a meat in a polysaccharide gel. It finds application as a stabilizer in frozen products, in reformed fruit and meat chunks, yoghurts, dietetic jam, and deserts.

10.2   STRUCTURE

Pectin is a structural polysaccharide present in the cell wall of higher plants. It is both multimolecular and polydisperse, that is, it is heterogeneous in terms of chemical structure and molecular weight. In any pectin sample, the number and percentage of individual monomeric unit types will vary, the average composition and distribution of molecular weights will also vary with the source, the conditions used for isolation, and any subsequent treatments. Since all of these parameters define physical properties, various products can be produced by controlling the source, isolation procedure, and subsequent processing to increase its functionality.

Pectin is composed of α-1,4-D-galacturonic acid residues. Different domains of pectin can be illustrated as follows: homogalacturonan (HG), rhamnogalacturonan I (RGI), rhamnogalacturonan II (RGII), xylogalacturonan (XGA), apiogalacturonan (AGA), arabinan, galactan, arabinogalactan I (AGI) and arabinogalactan II (AGII). The most abundant domain of the pectin macromolecule is homogalacturonan (Figure 10.2), which represents approximately 60% of all pectins in cell walls (Voragen, 2009). The HG molecule consists of a linear chain of α-1,4-D-galacturonic acid units, some of which are esterified with methanol or/and acetyl groups are present at O-2 or/and O-3. For example, apple, sugar beet, and citrus have 72–100 D-galacturonic acid residues in the HG backbone. The degree of methylation (DM) is expressed as the ratio of methyl-esterified carboxyl groups to the total number of galacturonic acid units. DM is an important parameter connected to gelling capability and which can be measured by infrared spectroscopy. It is calculated based on the areas under the peaks at $1600–1630\ cm^{-1}$ and $1740\ cm^{-1}$ related with the antisymmetric stretching vibrations of COO^- and the C=O stretching vibration of the ester. The technical classification of pectin is based on the degree of methylation. More than 50% of the carboxyl groups of high methoxy pectin (HM) are esterified with methanol, and low methoxy pectin (LM) is between 5% and 50% and for pecticacid it is less than 5% (Baron, 2016).

Rhamnogalacturonan I have repeating α-1,4-D-galacturonic acid and α-1,2-L-rhamnose residues (Figure 10.2). The length of the RGI backbone is approximately 100–300 repeating units after isolation from the suspension-cultured cells. RGI can be shorter and may consist of approximately 20 residues of this disaccharide, which has been reported in sugar beet. Galactose and/or arabinose residues are attached in a side chain at C-4 of the rhamnose residues. The side chain is made up of a single sugar residue or combined chains of arabinans, galactans, or arabinogalactans. The proportion of branched rhamnose units is 20–80% depending on the source of the polysaccharide. It is

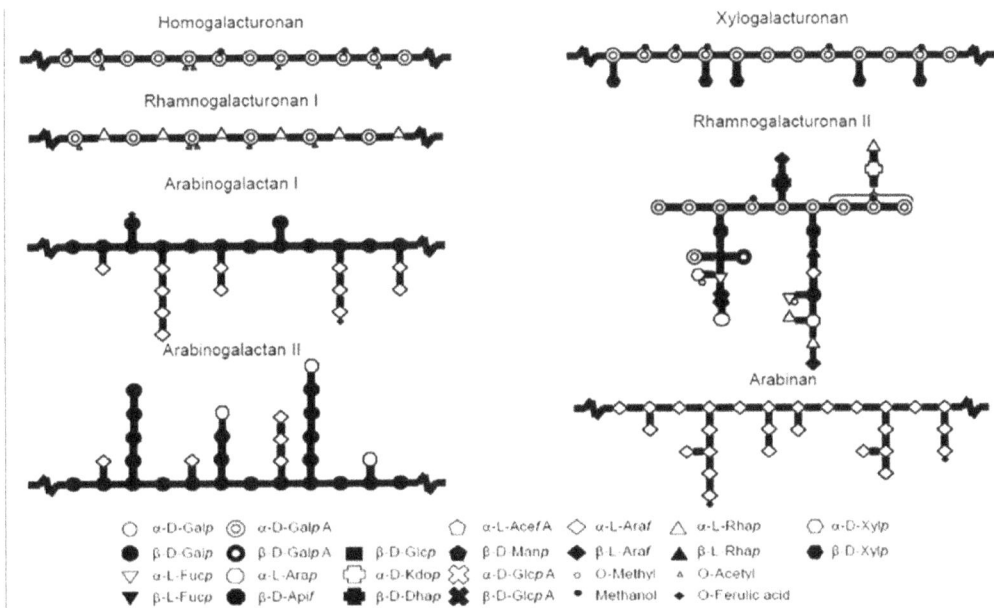

FIGURE 10.2 Schematic demonstration of pectin structural elements.

Source: adapted from Voragen et al., 2009.

presumed that RGI and HG are covalently linked (Coenen et al. 2007). These linkage can be broken by endopolygalacturonase enzyme or chromatographic fractionation method by chain cleavage.

The rhamnogalacturonan II backbone comprises 7–9 galacturonic acid residues branched with four side chains (at C-2 and C-3) having fucose, arabinose, apiose, rhamnose, aceric acid, galactose, galacturonic acid, glucuronic acid, 3-deoxy-lyxo-2-heptulosaric acid (DHA), 2-O-methyl-xylose, 3-deoxy-manno-2-octulosonic acid (KDO), or 2-O-methyl-fucose (Figure 10.2). RGII isolated from suspension-cultured pea stems and sycamore cells are found to be covalently bound to boron (O'Neill et al. 1996). It can affect the formation of the complex structures in the cell wall, which contains pectin polysaccharides such as HG, RGI and RG II. Some vitro studies have shown that RGII dimers are formed through apiosyl residues.

Xylogalacturonanhave homogalacturonan backbone with xylose residues attached to O-3 of some galacturonic acid units. Presence of the number of branches and the degree of methylation of HG may fluctuate noticeably subject to the source of polysaccharides. In the apiogalacturonan macromolecule, one or two apiosyl residues connected with each other are linked to galacturonic acid units. Arabinans contain a chain of α-1,5-L-arabinofuranosyl units with attached α-arabinofuranosyl residues at C-2 or/and C-3 in the arabinose molecule. Mostly arabinans are side chains of rhamnogalacturonan I found in primary cell walls (Albersheim et al., 1996).

Galactans posses β-D-galactose units linked by β-1,4-glycosidic bonds. They are the most common type of side chain of rhamnogalacturonan I in the case of potato (Jarvis et al. 1981).

Arabinogalactan I have a backbone containing β-1,4-D-galactopyranosyl residues with attached short chains of α-1,5-arabinose residues at C-3 in the galactose molecule. Nevertheless β-galactos emerges to the galactan chain at O-6 position (Schols, 2002). Arabinogalactan II have chains of β-D-galactopyranosyl units, where glycosidic linkages are found to be at C-1, C-3, and C-6 in galactose molecules (Leivas et al. 2016). This form of arabinogalactan is extremely branched. It may have L-arabinopyranosyl residue on the end of the chain of β-1,6-D-galactopyranosyl units. A longer chain of α-1,3-arabinose residues can be attached to the backbone of the β-1,3-D-galactopyranosyl unit. In addition, arabinofuranosyl residues or arabinopyranosyl residues possibly will be attached at C-6 in the galactose molecule of the backbone. Arabinogalactans I and II may be rhamnogalacturonan I side chains.

The key feature of pectin, being a D-galacturonic acid polymer, is the linear chain of (1›4)-linked α-D-galactopyranosyluronic acid units, which makes it an α-D-galacturonan [a poly (α-D-galactopyranosyluronic acid) or an α-D-galacturonoglycan]. In all natural pectins, some carboxyl groups are in the form of methyl ester. Depending on the isolation conditions, the remaining free carboxylic acid groups can be partially or completely neutralized, i.e. partially or completely present as sodium, potassium, or ammonium carboxylate. Ratio of esterified D-galacturonic acid units to total D-galacturonic acid determines the degree of esterification (DE). DE strongly affects the solubility, gelling temperature, gel forming ability, gel properties, and gelation conditions.

In some sources of pectin, some units appear as 0-2 or 0-3 acetates. This esterification makes gelling difficult, so that when one of the eight galactopyranoic acid units is monosterified with acetic acid at 0-2 or 0-3, gelation is completely inhibited. Therefore the presence of acetyl groups, as some potential sources of commercial pectin, e.g. sugar beet, sunflower, and potato, is less desirable. There are also neutral sugars mainly L-rhamnose. However, there are differences in their distribution in the linear chain. A group of workers (Rees, 1971) reported that, in citrus, apple, and sunflower pectins, L-rhamnopyranosyl units are more or less uniformly inserted into the galacturonan chain in the following manner: 0-α-D-GalpA-(1›2)-0-L-Rhap-(1›4)-0-α-D-GalpA. The configuration "of the L-rhamnopyranosyl linkage" is not known, but calculations show that it should be beta to provide the necessary degree of kinking in the structure. It was also reported that the L-rhamnopyranosyl units were unevenly distributed with in the galacturonan backbone. According to the researcher, apple pectin consisted of "smooth" areas [poly (α-D-galactopyranosyluronic acid) regions] and "hairy" regions. Where the latter regions consisted of rhamnogalacturonan sequences, which contained highly branched arabinogalactan side chains and galacturonan sequences with short side chains composed of D-xylose. Data from the former group (Powell, 1982) showed that L-rhamnopyranosyl units appeared in the chain. The length of the poly (α-D-galactopyranosyluronic acid) sequences between L-rhamriopyranose interruptions (whether single units or blocks of units) was found to be constant and the length of the sequence was approximately 25 units in each pectin studied (citrus, apple, sunflower). The total content of neutral sugars varies depending on the source, the extraction conditions performed, and the subsequent processing, e.g. ferulic acid was esterified to the neutral side chains of spinach sugar and sugar beet.

10.3 CLASSIFICATION OF PECTIN

One of the main applications of pectin to form gel is very attractive. The gelation mechanism of pectins can be controlled by their degree of esterification. Thus the gelling mechanism of HM (high methoxy) and LM (low methoxy) pectins are different (Sundar, 2012). The gelling process of pectins is directly affected by external and internal factors. These parameters include the average molecular weight, degree of methylation, temperature, pH, charge distribution along the backbone, ionic strength, and the presence of co-solutes.

The degree of esterification of HM pectin falls in the range of 50–80%. It requires precise conditions to form gel, such as low pH (2.5–3.5) and the presence of soluble solids; mainly sucrose (55–75%) or other similar co-solutes (e.g. sorbitol, ethylene glycol) (Sundar, 2012). The role of sugar in gel development is to reduce water activity by promoting hydrophobic interactions to stabilize the binding area (junction zone). The role of sugar also depends on the molecular geometry of the sugar and its interaction with adjacent water molecules. The formed gel has a two-dimensional network of pectin molecules in which a solvent (water) containing the co-solutes sugar and acid are immobilized. The buildup of the 3D network depends on the formation of junction zones stabilized by hydrogen bonding between carboxyl and secondary alcohol groups and by hydrophobic interactions between methyl esters (Sharma, 2006). HM-pectin gels are thermally reversible. HM-pectins are soluble in hot water and often contain a dispersing agent such as dextrose to prevent lumping (Sundar, 2012). HMP (high ester/high methoxy pectin) are extracted at acidic solutions and high temperature. HMP having soluble solids' content higher than 60% and a pH between 2.8 and 3.6 will produce hydrogen bonds and hydrophobic interactions that connect the individual pectin chains together. Such bonds are formed when water will combine with sugar and then ultimately pectin strands will be forced to stick together to form a 3D network. This gelling mechanism can be called as a low-water activity gel or sugar-acid-pectin gel. Most commercial pectins are HMP.

Less than 50% of the total carboxyl groups are esterified in LM pectins. Regardless of the sugar content, they can form gels and are more chemically stable to moisture and heat compared to HM pectin (Sharma, 2006). LM pectins can be obtained in a wider pH range as they are more resistant to pH compared to HM pectin (Sundar, 2012). LM pectins can form gel in the existence of divalent cations (usually calcium (Ca^{2+})), which can be easily reversed by adding monovalent ions like sodium (Na^+) and potassium (K^+) (Munarin, 2013). Here gelation is due to the formation of intermolecular bonding regions between pairs of carboxyl groups in the homogalacturonic acid and smoothing regions of different chains in close contact. The structure of such a connection zone is generally attributed to the binding process called an "egg box" (Figure 10.3) (Morris, 1982). Initial strong association of two polymers into a dimer is followed by the formation of weak interdimer aggregation, mainly governed by electrostatic interactions and ionic bonding of carboxyl groups (Sundar, 2012). Gelling capacity of LM pectins is inversely proportional to the degree of methylation. LM pectins have more distributed free carboxyl groups very sensitive to low calcium levels. Plants such as beets and potatoes, have acetyl groups in their structure, which can prevent calcium ions from forming gels, but give pectin emulsion stabilizing properties (Sharma, 2006). The texture of the low methoxy pectin gel can be controlled by adjusting the ratio of calcium to pectin. Gels produced with high levels of pectin and relatively little calcium will be elastic in nature whereas gels produced with more calcium and minimal pectin will be brittle in nature and in some cases it may lead to syneresis. Though sugar is not compulsory for gel formation in low methoxyl pectins, still small amounts of sugar (10–20%) if added may tend to decrease syneresis and can give desirable firmness to the gel, also the presence of sugar will reduce the amount of calcium required to form gel. However, high concentrations of sugar interfere with gel formation because the dehydration of the sugar favors hydrogen bonding and decreases cross-linking by divalent ion forces. During jam manufacturing, where the situation of lower soluble solids and higher pH-values than high-ester pectins comprises it is mandatory to add calcium salt (dicalciumphosphate) along with LM to form a gel. LMP (low ester/low methoxyl pectin) containing pectic acids are extracted at lower temperatures with less acidic solutions, but in the presence of other chemical compounds. Generally, LM pectins form gels with a range of pH from 2.6 to 7.0 and with 10–70% of soluble solids content.

Non-esterified galacturonic acid units can be free acids (carboxyl groups) or a salt formed with sodium, potassium, or calcium. The salts of partially esterified pectins are called pectinates. If the degree of esterification is below 5% the salts are called pectates, the insoluble acid form, pectic acid. Plants of pear, potato, and beet contain pectin with acetylated galacturonic acid in addition to methyl esters. Acetylation prevents gel formation, but increases stabilization and emulsification of pectin.

Amidated pectin is an altered and improved form of pectin. Here conversion of galacturonic acid takes place into carboxylic acid amide with the help of ammonia. They are more tolerant to varying calcium concentrations that occurs when in use (Belit, 2004). https://en.wikipedia.org/wiki/Pectin - cite_note-22To make a pectin-gel, the ingredients are heated to dissolve the pectin. When cooled underneath gelling temperature, gel begins to form. A gel that is too strong will result in syneresis or a granular texture, while a weak gel will result in a gel that is too soft. Amidated pectin gels are thermally reversible (they can be heated and then solidified again after cooling), while traditional pectins will remain liquid.

High-ester pectins solidify at higher temperatures compared to low-ester pectins. The relationship between the degree of esterification and gelation reaction with calcium is inversely proportional to each other. Whereas lower pH or higher soluble solids (usually sugars) will lead to escalation of gelation rate. Therefore, jam and jelly, or higher-sugar confectionery jellies with higher sugar content, can choose suitable pectin.

10.4 PECTIN SOURCES

Plant sources used to extract pectin, galacturonic acid content, and reported yield are summarized in Table 10.1. At commercial scale, pectin is generally prepared from apple pomace and citrus peel through acid extraction, filtration, and alcohol precipitation. Pectin content of apple pomace, sugar beet, and citrus peel is about 10–15%, 10–20%, and 20–30% of the dry matter, respectively.

10.5 PROPERTIES

10.5.1 Physical Properties

Pectins are soluble in pure water, but they are insoluble in aqueous solutions and thus will form gel when dissolved at a higher temperature. Monovalent cation (alkali metal) salts of pectinic and pectic acids are typically soluble in water whereas di and trivalent cation salts are weakly soluble or insoluble. Though pectins are not used as a thickening agent still its solutions exhibit a non-Newtonian pseudoplastic behavior characteristic. Similar to solubility, viscosity of a pectin solution is related to the DE (degree of esterification), molecular weight, and concentration of the formulation, as well as the pH value of the solution and the existence of counter ions. For example, the addition of monovalent cations will cause the viscosity to decrease. As the DE (degree of esterification) decreases more the decrease in viscosity is found. Whereas an opposite effect is seen when salts of divalent

TABLE 10.1
Galacturonic Acid Content and Yield of Pectin Depending on the Source and Method of Extraction

Pectin source	Galacturonic acid content (%)	Yield (%)	Extraction method	References
Apple pomace	~21–44	~10–17	Subcritical water (t = 5 min, T = 130–170 °C, solid/liquid ratio 1:30)	Wang et al. 2014
Black currant	37.1	-	Citric acid (t = 30 min, T = 90 °C, pH = 2.5, solid/liquid ratio 1:50)	Mierczyńska et al. 2017
Black mulberry pomace	~29–43	~9–14	Hydrochloric acid and microwave-assisted extraction (Irradiation time 10–30 min, pH = 2, Power 300–900 W, solid/liquid ratio between 1:15 and 1:30)	Mosayebi 2015
Cacao pod husks	~60 (of total sugar content)	~11	Nitric acid (t = 30 min, T = 100 °C, pH = 3.5)	Vriesmann, 2017
Carrot	16.5	-	Citric acid (t = 30 min, T = 90 °C, pH = 2.5, solid/liquid ratio 1:50)	Mierczyńska et al. 2017
Gold kiwifruit pomace	~82–85 (of total non-starch polysaccharides)	~4	Citric acid (t = 1 h, T = 50 °C, pH = 2.2, pomace/acid solution ratio 1:3 w/v), water (t = 30 min, T = 5 °C, pomace/water ratio 1:3 w/v) and enzymatic extraction (t = 30 min, T = 25 °C, enzyme: Celluclast 1.5 L)	Yuliarti et al. 2015
Mango peel	~29–35 (T = 20 °C) ~52–53 (T = 80 °C)	2 17	Citric acid, conventional extraction (t = 2 h, pH = 2.5, solid/liquid ratio 1:40) and ultrasound-assisted extraction (t = 15 min, pH = 2.5, solid/liquid ratio 1:40)	Wang et al. 2016
Okra (Abelmoschuses culentus L.)	~43–63	11–15	Aqueous extraction, phosphate buffer (t = 1 h, T = 80 °C, pH = 6.0, solid/liquid ratio 1:15)	Kpodo et al. 2017
Opuntiaficusindica cladodes	~69	~19	Acidified water, ultrasound-assisted extraction (t = 70 min, T = 70 °C, pH = 1.5, solid/liquid ratio 1:30)	Bayar et al. 2017
Orange juice wastes	~46–74	~1–11	Hydrochloric acid, ohmic extraction (T = up to 90 °C, pH = 1.5–4, voltage gradient 5–30 V/cm, solid/liquid ratio between 1:10 and 1:40)	Saberian et al. 2018
Peach	26	-	Citric acid (t = 30 min, T = 90 °C, pH = 2.5, solid/liquid ratio 1:50)	Mierczyńska et al. 2017
Plum	23.8	-	Citric acid (t = 30 min, T = 90 °C, pH = 2.5, solid/liquid ratio 1:50)	Mierczyńska et al. 2017
Pomegranate peel	~70–82	~3–9	Citric acid (t = 40–150 min, T = 70–90 °C, pH = 2–4)	Pereira et al. 2016
Raspberry	23.1	-	Citric acid (t = 30 min, T = 90 °C, pH = 2.5, solid/liquid ratio 1:50)	Mierczyńska et al. 2017
Strawberry	33.9	-	Citric acid (t = 30 min, T = 90 °C, pH = 2.5, solid/liquid ratio 1:50)	Mierczyńska et al. 2017
Sugar beet pulp	~54–64	~1–9	Hydrochloric acid (t = 1 h, T = 80 °C, pH = 1.5, solid/liquid ratio 1:20)	Guo et al. 2016

and trivalent cations are added. Solubility, viscosity, and gelation are inter-related e.g. factors that increase gel strength will increase gelling tendency, decrease solubility, and increase viscosity, and vice versa.

Physical properties of pectin are the function of its structure, that is, the structure of linear polyanions (polycarboxylates). Therefore, the monovalent cation salt of pectin is highly ionized in the solution, and due to Coulombic repulsion, the distribution of ionic charges in the entire molecule tends to extend it. Coulombic repulsion between carboxylate anions prevents aggregation of polymer chains. Every polysaccharide chain, especially every carboxylate group, will be highly hydrated. The monovalent pectin salt solution shows a stable viscosity because each polymer chain is hydrated, extended, and independent.

Because the commercial importance of pectin is mainly due to its unique ability to form a spreadable gel in the presence of a dehydrating agent (sugar) at a pH 3 or close to 3 in the presence of calcium ion (jams, jelly, and marmalades made from fruit juices or whole fruit), this is the most researched and focused attribute. The factors that decide whether gelation occurs and affect the characteristics of the gel are the concentration of co-solute (sugar), the pH value, the concentration and type of cation, temperature and the concentration of pectin. Factors that affect gelation depends on the following molecular properties of a particular pectin: molecular weight, degree of esterification (DE), degree of amidation (DA), presence of acetate esters, and heterogeneity. All these parameters are interdependent. Generally speaking, under similar conditions, the degree of gelation, gelation temperature, and gel strength are generally proportional, and each property is generally proportional to molecular weight and inversely proportional to DE. As the pH value decreases (the concentration of hydrogen ions in the solution increases) the ionization of carboxylate groups is inhibited. Highly hydrated carboxylate groups will only convert to carboxylic acid groups when slightly hydrated. Due to the loss of some charge, the polysaccharide molecules no longer repel each other over their entire length; due to the loss of some water of hydration, they can combine to form a gel for part of their length. Apparent pK (a measure of the strength of an acid on a logarithmic scale) values (the pH value at 50 % dissociation) vary with the DE of the pectin (Plaschina et al., 1978); the apparent pK value of 65% DE is 3.55 and the apparent pK value for 0% DE is 4.10. At slightly higher pH values pectins will gel with increasing degrees of methylation due to fewer carboxylate anions at any given pH value. The influence of pH on the gel texture is greater than the influence on the strength of the gel.

It is believed that gels are formed when pectin molecules (usually highly hydrated by water molecules) loses some of the water of hydration due to the competitive hydration of co-solute molecules. Contact between pectin chains increases due to reduction in hydration leading to formation of binding zones (mainly through hydrogen bonds) and resulting in a network of polymer chains, which entraps water and solute molecules. Sugar acts as co-solute in jams, jellies, and marmalades. High methoxyl pectins will gel only in the presence of large concentrations (at least 55% w/w) of sugar whereas low methoxypectin will gel in the absence of sugar (if divalent cations are present), but increasing the soluble solids will increase the gel temperature and gel strength. Increase in the concentration of divalent cations (calcium ions) leads to an increase in gel temperature and gel strength. Divalent cations are not required for gelation of high methoxypectin due to the low number of carboxylate groups to be bridged and due to the formation of hydrophobic regions parallel to the helix axes by columnar stacking of methyl ester groups (Walkinshaw et al. 1981). Any process containing pectin at potential gelling circumstances (i.e. pH, necessary concentration of an appropriate pectin, concentration of divalent cations and concentration of co-solutes) should be prepared at a temperature above the gelling temperature (temperature at which gelation occurs). When the hot pectin solution is cooled, the thermal energy of the molecules decreases and their tendency to form junction zones upon collision increases. LM pectin forms gel rapidly whereas HM pectin forms gel slowly. Concentration of pectin required for gel formation is inversely related to the concentration of soluble solids, for in general, increasing the concentration of co-solutes, i.e.

decreasing the water activity, increases the size and number of junction zones. Increase in the concentration of pectin increases the gel strength due to increase in the number of junction zones, when other factors are held constant.

Gel strength of HM pectin increases with increasing DE, as does the rate of gelation but at constant pH. As the DE of a HM pectin is lowered, a lower pH is required for gelation e.g. pH required to produce a given percentage of carboxyl groups in the carboxylic acid form increases owing to an increase in apparent pK, when DE is lowered (Plaschina et al., 1978). Even the absolute number of adjoining carboxylate groups, which must be converted into unionized carboxylic acid groups increases before the chain association. As the DE of HM-pectin decreases a greater percentage of carboxylate groups must be converted into free carboxylic acid groups to effect gelation. Commercial LM-pectins who have the lowest DE values have the highest gelling temperatures and the greatest requirement for divalent cations (for cross bridging). Amidation results in a higher gelling temperature and a decreased need for a divalent cation. The distribution of carboxyl/carboxylate can affect gelation. Pectins with blocks of methyl ester and carboxyl groups (as opposed to a random distribution) usually produces weaker gels and requires a greater content of divalent cations.

10.5.2 CHEMICAL PROPERTIES

Soluble pectins can undergo deesterification and depolymerization in aqueous systems. The maximum pH stability for pectin is around 4. For pH values both above and below 4, deesterification and depolymerization occur simultaneously, where the rate of deesterification will be greater than the rate of depolymerization. Presence of solutes can lower the activity of water and the rate of both reactions. Deesterification can be caused by normal acid and basic catalysis of ester hydrolysis. Depolymerization occurs by means of acid-catalyzed hydrolysis of glycosidic bonds at low pH values (BeMiller, 1967). Acid-catalyzed hydrolysis happens especially at the L-rhamnopyranosyl glycosidic bonds. Galacturonoglycan chains with a degree of polymerization of about 25 is produced after hydrolysis of these linkages (Powell et al. 1982). Side chains, containing L-arabinofuranosyl units, should also be favorably removed by hydrolysis because of the inherent stability to acid-catalyzed hydrolysis of glycuronosyl glycosidic bonds and the inherentnon stability of furanosyl glycosidic bonds (BeMiller, 1967). If these side chains are attached to rhamnogalacturonan sequences, it would not be possible to convert "hairy" regions to "smooth" regions by treatment with acid because the lability of the L-rhamnopyranosyl bonds would result in concurrent depolymerization of the main chain.

Pectin solutions are stable at room temperature at pH values around 5–6. Rise in temperature, leads to cleavage of pectin chains by beta elimination reaction, and a reaction, which is generally stimulated by organic anions. Deesterification of pectin proceeds concurrently with the beta-elimination depolymerization reaction, which happens only at monosaccharide units, which are esterified. Deesterification and depolymerization reactions occur rapidly at pH above 6 even at room temperature, these rates of each reaction increase with an increase in pH.

Hydroxyl-group reactions, such as esterification, etherification, oxidation, and acetylation can be done in the same way as they are on other polysaccharides. Esterification of carboxyl groups and interactions with cations, including polycations such as proteins below their isoelectric pH, occur as they do with other glycuronoglycans. Reduction of carboxyl groups to hydroxyraethyl groups was done with diborane (Smith et al., 1960) and by borohydride treatment of methyl and hydroxyethyl esters.

Some of the methyl carboxylate groups are converted into carboxamide groups when ammonia is used to prepare low methoxypectin from high methoxypectin thus producing "amidated pectin". Molecule becomes less hydrophilic in the presence of amide groups in LM pectin, which increases their tendency to form gels, and they will be less sensitive to calcium ions. Overall gelling temperature can increase with an increase in DA. Reaction of pectins with primary and secondary alkylamines can also make pectin amides.

Several types of enzymes can act on the pectin molecule. Enzymes produced by plants plays a significant role in the processes (like ripening, storage, and processing) leading to structural changes in fruits and vegetables. Controlling their action is important for pectin production. Many industries like juice industry uses fungal enzyme to improve juice clarity and juice yield.

Pectinesterases causes and speeds up the hydrolysis process of the methyl ester group. Action of pectinesterases on methyl α-D-galactopyranosyluronate unit adjacent to a non-esterified α-D-galactopyranosyluronic acid unit produces pectins, which contain blocks, rather than a random distribution, of carboxyl groups. Such reactions are undesirable for commercial use. Fungal pectinesterases produced LM pectin, showing similar results like acid or base deesterified pectin in terms of gelling capacity and sensitivity to calcium ions. Many other enzymes are also known like lyases or transeliminases, which perform similar work. All catalyze depolymerization by a beta-elimination reaction in the same way, which occurs during base-catalyzed depolymerization. Pectinlyases (endo-enzymes), catalyzes beta-eliminations at esterified D-galacturonic acid units. Pectate lyases (exo-enzymes) catalyzes beta-eliminations at non-esterified D-galacturonic acid units. Polygalacturonases can depolymerize pectin by catalyzing hydrolysis of glycosidic bonds. As DE is inversely proportional to the rate of depolymerization it suggests a requirement for a non-esterified D-galacturonic acid unit. Endo-polygalacturonases can attack randomly whereas exo-polygalacturonases releases mono or disaccharides fromnon-reducing termini.

10.6 PECTIN-METAL ION INTERACTION AND THEIR EFFECT ON RHEOLOGY OF FOOD

Many divalent and trivalent cations can also bind to pectin chains like calcium.Mierczyńska et al. (2015) studied the influence of the addition of Ca^{2+}, Mg^{2+}, and Fe^{2+} on the rheological properties of the modified cell wall polysaccharide matrix (MPS) obtained from apple (having a good amount of pectin). The result showed that the presence of ions like Ca^{2+} and Fe^{2+} led to an increase in the viscosity of MPS solutions. It was also observed that the addition of Mg^{2+} to MPS solutions led to a decline in viscosity. The cause of the increase in viscosity of MPS solutions in the presence of calcium ions can be connected with the cross-linking of low-methoxy pectins with calcium ions agreeing with the "egg-box" model (Cybulska et al. 2012). Addition of iron ion also led to the increase in viscosity due to interaction of pectins with Fe^{2} the same as that like the interaction between pectin and Ca^{2+}. Binding of Fe^{2+} to citrus pectins was recently investigated by Celus et al. (2017). On the basis of adsorption isotherms, the authors noted that the structural properties of pectins such as DM (degree of methylesterification) and DB_{abs} (degree of blockiness) plays an influential role on iron ions' binding capacity. DM mainly determines the maximum adsorption capacity while DB_{abs} mostly influences the pectin-iron ions' interaction energy. They suggested that chemically de-methylesterified pectins with DM in the range of 20–67% and DB_{abs} in the range of 5–44%, can interact with iron ions according to the "egg-box" model. The degree of methylesterification (DM) of pectin, which is the percentage of GalA (galacturonic acid) units of HG that are methylesterified at C6, is of great importance in determining its polyanionic nature and associated functional properties (Voragen et al., 2009). Assifaoui et al. (2015) stated that Zn^{2+} can interact with both the charged carboxylic group as well as with hydroxyl groups of GalA while Ca^{2+} binding can occur only via carboxylate groups. Thus, interactions between divalent cations and pectin, either with high DM, either at low pH, might occur as well via cross-linking mechanisms. Debon and Tester anticipated that Fe^{3+} can interact with the carboxyl groups of the galacturonic acid residue in acidic conditions (Debon et al. 2001).

The interactions between Ba^{2+}, Ca^{2+}, $Mg^{2+}$and Zn^{2+} and polygalacturonate can be studied using measurements of isothermal titration calorimetry, viscosity, turbidity, and molecular dynamics

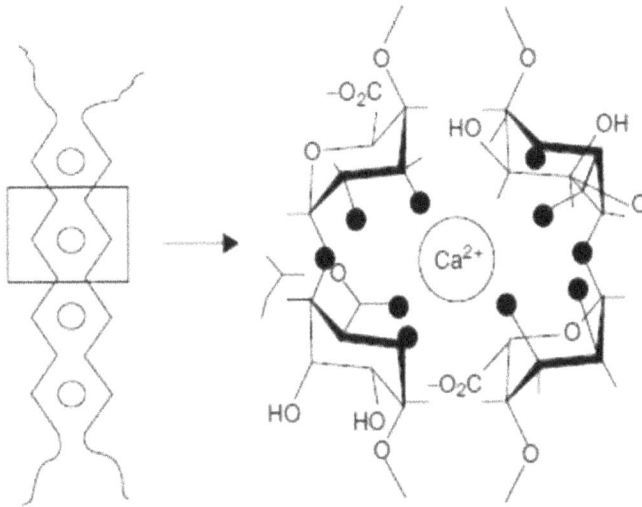

FIGURE 10.3   Structure of pectin of low methoxyl as call "the egg box" model.

Source: adapted from Axelos et al. 1991.

simulations (Huynh et al. 2016). Binding of Mg^{2+} to polygalacturonate is found to be weak due to the high affinity of Mg^{2+} with water whereas Ca^{2+}, Ba^{2+}, and Zn^{2+} interacts very well with polygalacturonate to form a network. Binding mode of Ba^{2+} and Ca^{2+} to carboxylate ions is found to be bidentate (two oxygen atoms of the carboxylate ion are involved in this process), while Zn^{2+} was bound in monodentate coordination (by one oxygen atom of the carboxylate ion). Huynh et al.(2016) suggested a two-step binding mechanism of Ba^{2+}, Ca^{2+}, and Zn^{2+} with polygalacturonate, where in the first step monocomplexes and point-like cross-links are to be formed while the second step is connected with the formation of dimers. Fang et al. (2008) proposed a similar binding mechanism of Ca^{2+} to low methoxy pectins. Huynh et al. (2016) suggested interaction of Mg^{2+} polygalacturonatevia polycondensation when specific conditions are provided.

Wellner et al. (1998) investigated the interactions between potassium pectate and different divalent cations. They observed similarity in band shifts in the FT-IR spectra for ions such as Cu^{2+}, Sr^{2+}, Ni^{2+}, Ca^{2+}, Cd^{2+}, Pb^{2+}, and Zn^{2+}. The band shifts were found in the range of 1200–950 cm^{-1}, especially the shift of the 1018 cm^{-1} band (the main maximum was connected with C–C and C–O vibrations of the pyranoid ring). Their results suggested that these cations are linked with the pectate chain in a similar way. The Mg^{2+} band at 1004 cm^{-1} showed a lower intensity than the bands of other cations in pectate solution.

Debon and Tester (2001) scrutinized the binding ability of Zn^{2+} and Ca^{2+} with low methoxy by equilibrium dialysis in acidic and neutral media. At pH 1 (strong acidic solutions) these ions not bound by pectins may be due to the presence of protonated carboxyl groups, which prevented electrostatic interactions. In neutral solutions (like water) the binding of Zn and Ca ions with pectin was evident due to electrostatic interactions between charged carboxyl groups of pectins and these ions. The binding process of Zn^{2+} and Ca^{2+} with pectins was also studied by Schlemmer (1989). Degree of binding was found to be the highest at low ionic strength and pH from 5.0–7.5 in diluted solutions of sodium hydroxide/hydrochloric acid. For the same pH range but at higher ionic strength and in the bicarbonate and carbon dioxide buffering system, pectin did not bind or only bound to a limited extent with Zn^{2+} and Ca^{2+}. Furthermore studies are required to determine the interaction mechanism of pectin systems and metal ions. It is probably that the binding of divalent and trivalent metal ions by pectin is similar to the above-described "egg-box" model. Nevertheless, the differences in the atomic properties of elements can have a big impact on their binding mode to pectin. Application of various metal ions in the gelation process of low methoxy pectins can modify the structure and properties of gels and therefore food products.

10.6.1 CONFORMATION

Gels are formed when polymeric molecules interact over part of their length to form a network of solute molecules and solvent. The connection regions resulted from these chain interactions is of limited size, if they are too large, a precipitate will form rather than a gel. The inserted L-rhamnopyranosyl units can provide the required irregularities' fold lines in the structure and limit the size of the junction regions (Figure 10.4). The presence of "hairy" regions can also be a limiting factor in the degree of chain association. Junction regions are formed between regular, unbranched pectin chains when hydration of the molecules is reduced (addition of a co-solute), the negative charges on the carboxylate groups are removed (addition of acid) and/or polymer chains are bridged by divalent cations (calcium ions).

Analysis of proton NMR (nuclear magnetic resonance) spectra and mathematical model building suggests that individual α-D-galactopyranosyluronic acid units have the $^4C_1$ conformation (Rees, 1971), Sodium and calcium pectates, pectic acid, and pectinic acid all occur in the solid state (fibers) as right-handed (3_1) helices with a three-fold screw axis (trisaccharide repeat) (Palmer, 1945; Sathyanarayana 1973; Walkinshaw, 1981). In solid pectinic acid, the polymer molecules pack so that the chains remain parallel to each other whereas in pectates they pack as corrugated sheets of anti-parallel chains (Walkinshaw, 1981).

It is further suggested that junction zones in pectinic acid (HM-pectin plus sucrose) gels are formed by a columnar stacking of methyl ester groups to form cylindrical hydrophobic areas parallel to the helix axes. Two models in calcium pectate (LM-pectin) gels have been proposed for

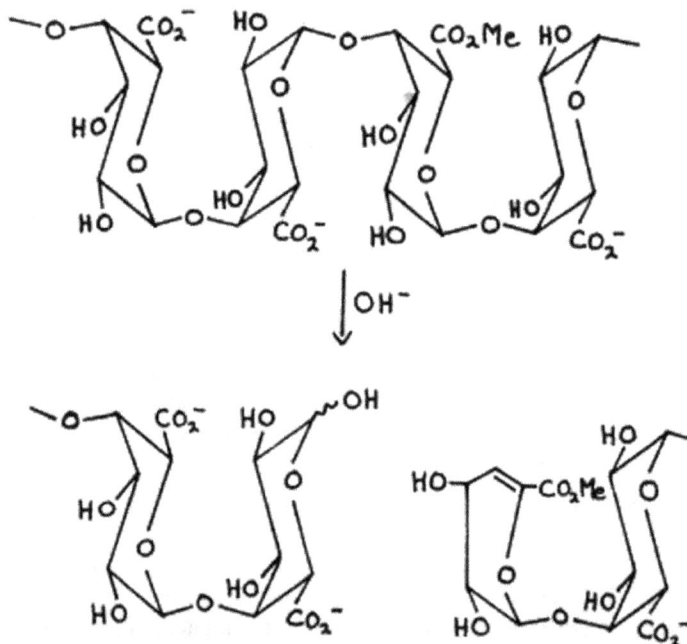

FIGURE 10.4   Alkaline depolymerization of a sequence of a pectinic acid of DE 25.

Source: adapted from James 1986.

the formation of junction zones. One suggests an aggregation of chains by a cross-linking of carboxylate anions with calcium ions to form a structure similar to that of the corrugated sheets of anti-parallel helices (three–six chains in an average junction zone) found in solid calcium pectate (Walkinshaw, 1981). The other is the "egg box" model used to describe the formation of calcium alginate gels (Morris, 1978). This model is proposed because of the close similarity between (1›4)-linked poly (α-D-galactopyranosyluronic acid) segments of pecticacid and (1›4)-linked poly (α-L-gulopyranosyluronic acid) segments of alginic acids, segments that are mirror images except for the configuration at C-3. From circular dichroism and equilibrium dialysis studies, it has been concluded that interchain association of hydrated pectinic acid molecules, in the presence of swamping levels of monovalent counterions, is limited to the formation of dimers of chains of 2_1 helical symmetry with specific site-binding of calcium ions along one face of each participating chain (Gidley, 1979; Morris, 1982). When Ca^{2+} is the sole counterion, these dimers further aggregate without rearrangement, leading to an approximate doubling of the amount of Ca^{2+} bound cooperatively (Gidley, 1979; Morris, 1982). Morris (1982) concluded that drying of a calcium pectinate gel effects a polymorphic phase transition in which associated, regular, buckled chains with two-fold symmetry ("egg box") as found in L-guluronoglycan chain segments are converted into associated chains with three-fold symmetry as found in solid state calcium pectinate (Palmer, 1945; Sathyanarayana, 1973; Walkinshaw, 1981). It should be noted that the axial-axial linkages in a chain of aldohexopyranosyl units linked $1\rightarrow4$ gives a buckled conformation naturally.

10.7 HEALTH BENEFITS OF PECTIN

Pectins perform various functions in the human body. They are considered some of the safest food additives for which no acceptable daily intake (ADI) has been stipulated. Pectins have also been included into a group of completely safe substances and are referred to as GRAS – generally regarded as safe (Gasik, 2013). Pectins bind and remove excess fat supplied with food, thereby reducing the amount of cholesterol, reducing the incidence of atherosclerosis, and preventing liver disorders. Indirectly, they also protect against heart attack (Brouns, 2012). By binding cholesterol and bile salts from food, apple pectins enforce the body to use fat tissue reserves in metabolic processes, thereby reducing the amount of total cholesterol that circulates in the blood and improving the ratio between concentrations of HDL and LDL cholesterol fractions. Pectins reduce the levels of phospholipids, triglycerides, and free fatty acids in plasma and tissues to modify the distribution of lipoproteins, wherein their force is dependent on the degree of their methylation and amidation. The most positive effects have been attributed to high-methoxylated pectins, including their contribution to weight loss, while low-methoxyl pectins have no such clear effect on cholesterol and lipid metabolism. Thus, the high-methoxylated pectins are functional in the treatment of diseases (such as obesity, diabetes, or hypercholesterolemia) and prevention, whereas the low-methoxyl pectins can be used in diets to prevent these types of disorders (Wikiera, 2014). Pectin lowers blood glucose and insulin levels after consumption of foods with carbohydrates (Gasik, 2013), by coating the mucosa they hamper the contact of glucose with the mucosa, which slows down the absorption of sugar in the stomach. Due to the slower postprandial absorption of sugar, there is no large demand for insulin, which is compulsory for patients with high diabetes. Kim (2005) also reported similar results, where the addition of pectin to a glucose solution increased its viscosity and reduced glucose absorption in the small intestine. Pectins have also anti-carcinogenic properties. Modified citrus pectins (MCP) have anti-metastatic effects (McCarty, 2006), these are derivatives of pectins treated with a base and then with an acid, which leads to their transformation into lower molecular weight unbranched, rich in galactose polysaccharides with improved absorption. Studies on the effect of the modified citrus pectin on the development of breast and prostate cancer have demonstrated that low concentrations of modified citrus pectin 0.25–1.0 (mg/ml) did not stop the growth of cancer cells, however such an effect was confirmed after combining MCP with two newly tested phytocompounds responsible for the health of breast and prostate (Jiahua, 2013). In clinical trials, the MCPs were investigated for their effect on the time of doubling the level of prostate-specific antigen (PSA) in men, in whom the PSA was still detectable in the blood and increased after local therapy for prostate cancer. The study did not allow to clarify the extent to which the therapeutic measures affected the suppression of metastasis, cancer cell growth or directly affected the cancer cells (McCarty, 2006).

Human digestive enzymes do not digest pectin and the colonic microflora decomposes them into short chain fatty acids. When they are absorbed into the bloodstream, they are metabolized in the liver, which is why they are used in dietetic formulas. The capability of pectin to bind food components in the digestive tract by pectins results in a slower process of digestion. Owing to the binding capacity of significant amounts of water and the ability to form coatings, which cover food in the stomach, the contact with digestive enzymes is impaired and digestion is slowed down. The swelling of pectins in the digestive tract after ingestion increases the feeling of satiety, which is exploited in slimming diets used in the treatment of obesity and overweight (Gasik, 2013).

Pectins are used in a therapeutic pectin diet for infants during diarrhea and constipation. In the case of diarrhea, the action of lactic fermentation bacteria is inhibited, they cannot grow and are removed with feces. During the treatment of diarrhea, pectins occurring in herbs (e.g. berberis fruit, leaves of wild strawberries, or blackberries) protect gastrointestinal mucosa. In addition, pectins have a positive effect on peristalsis, thereby regulating constipation and helping excrete undigested food residues (Nowak and Mitka, 2004).

10.8 INDUSTRIAL APPLICATIONS OF PECTIN

Pectin is extensively used in the food industry as a supplement with the properties of thickener, gelling agent, stabilizer, and as a carrier of other substances. They are coated on the surface of food products and used as bioactive ingredients (the soluble fiber fraction). The selection of a suitable pectin depends on the type of finished product.

10.8.1 APPLICATION OF PECTIN IN FRUIT PRODUCTS AND DAIRY

Use of pectin in jams is one of the best-known uses and remains one of the most important markets for pectin. Pectin is present in fruit by default and thus the added pectin complements it. Pectin holds a unique position under permitted gelling agent. Normally 0.2–0.4% of pectin needs to be added, depending on the type and amount of fruit under processing (jam, jelly, marmalade, etc.). Lemon or grapefruit marmalade required crucial attention if the fruit content needs to be increased in it. Too much pectin from the fruit can lead to strong gel formation or even to pre-gelation and syneresis effect. Adjustment of texture and rate of setting can be done by increasing the pH leading to a controlled condition. If fruit pectin does not form gel then gelation can be accomplished by adding an amidated low-methoxyl pectin in a mixture capable of gelling at the high pH. HM pectins are useful in production of jams where soluble solids are 60 %. Many countries now permit jams with less sugar content perhaps having 30–55% soluble solids, or even lower. In such cases LM pectins are used at 0.75–1.0% to provide excellent gelling results. For the production of jam, jelly, or marmalade selection of the correct pectin and fruit concentration is required, if possible more calcium-sensitive pectin and lower soluble solids should be used. At very low solids, it may be necessary to add a calcium salt to get the best gelling result. Jam makers also produce a wide variety of jam and other fillings for the bakery and allied industries using pectin as a thickening or gelling agent. It can be made with HM pectin in liquid or powder form, or special textures can be made using LM as they are heat-reversible and thus will melt and gel on cooling. Cookie jams are made to have a low water activity so they do not transfer water to the cookies.

Careful formulation of LM pectin and calcium sequestrants such as diphosphate should be done to develop pasty but stable products having 65% sugar solids, which can be diluted and melted later to restore a layer of clear icing and shine for cakes. Many pectin manufacturers offer a mixture of pectin and pectate for yogurt and similar products. Many times the modified starch is used as a thickener to ensure even distribution with a pumpable texture. Although starches being inexpensive, a drawback of their usage is that they can mask the subtle flavors of the fruit and impart a mushy texture (Gasik, 2013). HM pectins can be used at the sugar content 30–60 % whereas LM pectins are selected according to the solid level to use effectively. Amidated pectins are less critical of exact conditions of use, and give a greater degree of thixotropy, but conventional LM pectins give a more acceptable texture.

Gelatin has been used in dessert jellies, including snack foods, fruit jellies, and many more such products. Its melt in the mouth property provides good flavor release compared to some gelling alternatives like carrageenan. Here the low melting point causes retail products to melt during transportation under hot climatic condition. Slow build-up of texture is a drawback on production lines where one wants to put a layer of cream, custard, or other fillings on top of the jelly. LM pectin gels made from amidated pectin at precisely controlled melting points can be successfully deposited at set temperatures on jellies. The manufacturing process of amidated pectin can be held in a stirred tank slightly below their setting point, later the set point will be suppressed by gentle shearing and then finally it will set immediately after depositing.

Agglomeration of casein when heated to pH below 4.3 can be prevented by addition of HM pectin. It is used as a stabilizer for blends of milk and fruit juices and UHT-treated drinkable yogurts. It can also stabilize whey acidified dairy products and soy drinks where whey protein precipitation is prevented. Spoonable yogurts are thickened by adding very low concentrations of frozen methoxyl pectin before cultivation.

Addition of pectin into a fruit base, with careful formulation, can have the thickened fruit yogurt. Pectin jellies can be manufactured to desired final solids' content and requires no heating to remove excess moisture. At sugh high sugar levels, it is essential to use slow-setting pectin with a mixture of sugars, including glucose syrup to prevent crystallization of the sugars. Proper sourness as well as the necessary tight pH control are attained by addition of a fruit acid buffer system. Unless the product is continuously calcined and deposited, it is necessary to store an additional portion of the acid for up to a short time before placing it in a suitable mold. Cooking is done at a pH 4.5, at which pectin is stable but not gelling, and the pH drops to 3.6 only when encapsulation is required. Once the acid has been added, the time available to settle the product in the mold is limited, as setting depends on both time and temperature. If a non-acidic product is required, such as Turkish Delight (confectionary item based on a gel of starch and sugar), LM pectin can be used at a pH around 4.5. This would result in a high sugar content in confectionery, but the addition of the diphosphate mixture can control the pH and isolate a controlled amount of added calcium so that the setting rate could be precisely controlled. Such control is essential for modern continuous production processes.

10.8.2 Application of Pectin in Beverages, Sauces, and Ice Pops

Low level of pectins (usually of controlled viscosity) are used in low calorie soft drinks as they are thin and lack the characteristic mouth feel provided by sugar in conventional soft drinks to improve the texture. Pectins are used to control ice crystal size in both sorbets and ice pops as well as to reduce the tendency for flavor and color to be sucked out of the ice structure. Chutneys and sauces depend on the natural pectins from vegetables and fruits for their texture. It is possible that sufficient development of gel does not take place due to the presence of a lesser amount of pectin in natural fruits and vegetables, thus it may be required to add additional pectin either LM or HM depending on the solids' level and other factors to improve the texture and also the uniformity of gel.

10.8.3 Application of Pectin in Pharmaceuticals

In addition to its growing use in food, pectin has found numerous applications in the pharmaceutical field as well. A traditional use of pectin is found in anti-diarrheal mixtures in combination with kaolin and sometimes bismuth compounds. Pectin is also used to preserve the viscosity of some syrups. The pharmaceutical industry uses low-methoxy pectin as a carrier of drugs to the gastrointestinal tract, e.g. matrix tablets, gel granules, and film dosing of drugs. Pectin hydrogel is also used in tablet formulation, as a binding agent and used in matrix tablet formulations of controlled-release drugs. By varying the degree of esterification of the less oxidized pectin, Sriamornsak and Nunthanid (1998) altered the form of drug release from the calcium pectinate gel granules. Calcium pectinate is considered to be an insoluble hydrophilic coating applied to form a sustained release during surface complexation (the process of making an atom or compound form a complex with another). Spherical granules containing calcium acetate, obtained by extrusion, were then coated with a pectin solution. The seeds are surrounded by insoluble and homogeneous coatings of calcium pectinate gels (Sriamornsak, 2003).

10.8.4 Use of Pectin in Edible Coating and Packaging

Pectin ability to form gels and films clarifies its use in the development of biodegradable food packaging or edible coatings for food preservation (Vayghan et al. 2020). The use of food packaging and edible coatings based on materials derived from petroleum sources are related to the depletion of natural resources (Gaona et al. 2021). In contrast, pectin coatings have been studied to extend the shelf life of food products in the last years, principally due to the fact they are renewable, biodegradable, and biocompatible (Table 10.2). In the food products, the coating has a significant effect to control the water loss and reduce decay in fruits, maintaining their firmness and extending their shelf life. Food packaging is responsible for millions of tons of waste in landfills every year, causing a severe impact on the environment (Souza et al. 2020). In this sense, the growing concerns about safety and the demand for natural and ecological technologies have been motivating research into sustainable materials for use in food packaging. Edible pectin coatings change the atmosphere around the fruits, modifying the oxygen levels inside the fruits and, therefore, limiting physiological degradation. In addition, it also decreases the quality degradation induced by fruit ripening in terms of texture or loss of bioactive compounds during storage.

10.8.5 Use of Pectin as an Emulsifying Agent

Use of pectin as an emulsifying agent is favored by its molecular characteristics (protein portion, acetyl group, acetylation position, ferulic acid content, degree of esterification, neutral sugar side chain, and average molecular weight) and environmental conditions (pectin concentration and pH of the solution) (Yang et al., 2018; Chen et al., 2016). Proteins play the main role to confer emulsifying capacity: the droplet size of pectin-stabilized emulsions decreased (4.12 to 1.5 mm) as the protein content of pectin from beetroot pulp increased (0.5 wt% to 3 wt%) (Freitas et al., 2020). Pectin with the highest protein content exhibited good emulsifying capacity. Anchor formation by the protein portion allows the adsorption of pectin on the oil drop surface, decreasing the interfacial tension at the oil/water interface, enhancement and improvements on the emulsion stability due to the protein–oil/water interface binds, allowing the pectin to form a thicker stabilizing layer protecting droplets against aggregation. Use of pectin increases the viscosity of the continuous aqueous phase of emulsions, stimulating the restriction of the mobility of the droplets dispersed in the oil, inhibiting or minimizing its tendency to migrate and coalesce. Presence of a low amount of pectin (about 1.5%) can act as an emulsifier compared to other polysaccharides (4% of soy-soluble polysaccharides and up to 10% of gum Arabic). Pectins can be used in emulsified oils and emulsion based foods like low-fat mayonnaise, fatty dairy products, ice cream, and emulsified meat products.

TABLE 10.2
Pectin Applications

Use of pectin	Goal	Results	Reference
Films	Films of apple pectin (2%), low-acyl gellan (0.5%) and glycerol (2.2%) incorporated with nisin were produced and the kinetics release was studied (at 5 and 30 °C) from brain heart infusion (BHI) broth and nutrient agar (food simulant) used for cultivation of L. monocytogenes food model to determine the period of film bioactivity and its potential use as an antimicrobial film	At equilibrium time (72 h), 83% of initial nisin was released at 30 °C while it was 81% at 5 °C; Release patterns of nisin from the pectin-gellan films indicate that these materials can be used as anti-listerial films for prolonging shelf life of packed food systems; The incorporation of nisin formulation into the films led to more plasticized films	Rivera et al. 2020
Films	Citrus high-methoxyl pectin (PEC) films activated by nanoemulsions (NE) of copaiba oil (CP) at 1%, 3%, and 6% (wt. of water) and Tween 80 (1 wt.% of CP) to ultrapure water were evaluated of chemical, morphological, thermal, mechanical and microbial properties	The addition of CP-NE into the films improved their physico-chemical and antimicrobial properties, and increased the film's biodegradation profile; The active pectin film showed great potential to be used as food packaging	Norcino et al. 2020
Coating	Edible coating of HM sunflower pectin (1% w/v), with sweeteners: sucrose (10%), stevia or saccharin (concentration 400× lower than sucrose) and their possible combination with two modified atmosphere packs (MAP) to extend the shelf life of strawberries	The coatings, formed only with calcium and stevia or saccharin, prolonged the shelf life of strawberries up to 12 days compared to uncoated fruits by reducing microbial growth, maintaining the fruit firmness, and constant mass loss; When combined with MAP (10% CO_2, 85% N_2, and 5% O_2), the same edible coatings have extended the shelf life of strawberries up to 23 days	Muñoz et al. 2021
Coating	Edible coatings of fish gelatin (3% w/v), orange peel HM pectin (3% w/v) and glycerol (15%) applied in packaged ricotta cheese stored under refrigeration for 7 days	The coating provided an improvement in the physical-chemical and textural properties of the cheese during storage; The coating increases the microbial stability of the cheese during refrigerated storage and offers health-promoting benefits to consumers	Jridi et al. 2020

Emulsified pectin oils as a fat substitute can be used to develop products with low-fat and/or low-salt content, improving the nutritional quality of food (Naqash, 2017; Kim, 2016).

REFERENCES

Albersheim P., Darvill A.G., O'Neill M.A., Schols H.A., & Voragen A.G.J.1996. A hypothesis: The same six polysaccharides are components of the primary cell walls of all higher plants. In: Visser J., Voragen A.G.J., editors. Pectins and Pectinases. Elsevier Science, B.V.; Amsterdam, the Netherlands: 47–55.

AllwynSundarraj A., & Ranganathan, T.V. 2017. A review - pectin from agro and industrial waste. *International Journal of Applied Environmental Sciences*. 12(10): 1777–1801.

Assifaoui, A., Lerbret, A., Uyen, H. T., Neiers, F., Chambin, O., Loupiac, C., & Cousin, F. 2015. Structural behaviour differences in low methoxy pectin solutions in the presence of divalent cations (Ca 2+ and Zn 2+): a process driven by the binding mechanism of the cation with the galacturonate unit. *Soft Matter*, 11(3): 551–560.

Axelos, M.A.V., & Thibault, J.F.1991. The chemistry of low-methoxyl pectin gelation. In The Chemistry and Technology of Pectin; Walter, R.H., Ed.; Academic Press: San Diego, CA, USA. 109–118.

Baldwin L., Domon J.M., Klimek J.F., Fournet F., Sellier H., Gillet F., Pelloux J., Lejeune-Hénaut I., Carpita N.C., & Rayon C. 2014. Structural alteration of cell wall pectins accompanies pea development in response to cold. *Phytochemistry*. 104: 37–47.

Baron A., Turk M., & Quéré J.M.L. 2016. From fruit to fruit juice and fermented products.Handbook of Food Science and Technology, Food Biochemistry and Technology. ISTE Ltd.; London, UK.231–273

Bayar N., Bouallegue T., Achour M., Kriaa M., Bougatef A., & Kammoun R. 2017. Ultrasonic extraction of pectin from Opuntiaficusindica cladodes after mucilage removal: Optimization of experimental conditions and evaluation of chemical and functional properties. *Food Chemistry*. 235: 275–282.

Behall K., & Reiser S. 1986. Effects of pectin on human metabolism. In: Fishman M.L., Jen J.J., editors. Chemistry and Function of Pectins. American Chemical Society; Washington, DC, USA. 48–265.

Belit H.-D., Grosch W., & Schieberle P. 2004. Food Chemistry. Springer, Berlin.

BeMiller J. N. 1967. Advance Carbohydrate Chemistry. 22: 25.

BeMiller. J. N. (1986). An Introduction to Pectins: Structure and Properties.

Brouns F., Theuwissen E., Adam A., Bell M., Berger A., & Mensink R.P. 2012. Cholesterol-lowering properties of different pectin types in mildly hyper-cholesterolemic men and women. *European Journal of Clinical Nutrition*. 66: 591–599.

Celus M., Kyomugasho C., Kermani Z.J., Roggen K., van Loey A.M., Grauwet T., & Hendrickx M.E. 2017. Fe^{2+} adsorption on citrus pectin is influenced by the degree and pattern of methylesterification. *Food Hydrocolloids*. 73:101–109.

Chen, H., Qiu, S., Gan, J., Liu, Y., Zhu, Q., & Yin, L. 2016. New insights into the functionality of protein to the emulsifying properties of sugar beet pectin. *Food Hydrocolloids*. 57: 262–270.

Coenen G.J., Bakx E.J., Verhoef R.P., Schols H.A., & Voragen A. 2007. Identification of the connecting linkage between homo- or xylogalacturonan and rhamnogalacturonan type I. *Carbohydrate Polymer*. 70: 224–235.

Cybulska J., Pieczywek P.M., & Zdunek A. 2012. The effect of Ca^{2+} and cellular structure on apple firmness and acoustic emission. *European Food Research and Technology*. 235: 119–128

De Vries, J. A., Rombouts, F. M., Voragen, A. G. J., & Pilnik, W. 1982. *Carbohydrate Polymer*. 2: 25

Debon S.J.J., & Tester R.F. 2001. In vitro binding of calcium, iron and zinc by non-starch polysaccharides. *Food Chemistry*. 73: 401–410.

Fang Y., Al-Assaf S., Phillips G.O., Nishinari K., Funami T., & Williams P.A. 2008. Binding behavior of calcium to polyuronates: Comparison of pectin with alginate. *Carbohydrate Polymer*. 72: 334–341.

Freitas, C.M.P., Sousa, R.C.S., Dias, M.M., & Coimbra, J.S. 2020. Extraction of pectin from passion fruit peel. *Food Engineering Reviews*. 12: 460–472

Gaona-Sánchez, V.A., Calderón-Domínguez, G., Morales-Sánchez, E., Moreno-Ruiz, L.A., Terrés-Rojas, E., de la Salgado-Cruz, M.P., Escamilla-García, M., & Barrios-Francisco, R. 2021. Physicochemical and superficial characterization of a bilayer film of zein and pectin obtained by electrospraying. *Journal of Applied Polymer Science*. 138: 1–15

Gasik A, & Mitek M. 2013. The use of pectin preparation. *Overview Bakery and Pastry*. 61(3):18–21.

Gidley, M. J., Morris, E. R., Murray, E. J., Powell, D. A., & Rees, D. A. 1979. Spectroscopic and stoicheiometric characterisation of the calcium-mediated association of pectate chains in gels and in the solid state. *Journal of Chemical Society and Chemical Communications*. (22): 990–992.

Glinsky V.V., & Raz A. 2009. Modified citrus pectin anti-metastasic properties: One bullet, multiple targets. *Carbohydrate Research*. 344: 1788–1791.

Guo X., Meng H., Tang Q., Pan R., Zhu S., & Yu S. 2016. Effects of the precipitation pH on the ethanolic precipitation of sugar beet pectins. *Food Hydrocolloid*. 52: 431–437.

Hilz H. 2007. Characterization of cell wall polysaccharides in billberries and black currants. Doctoral Thesis, Wagen in genUniversity

Huynh U.T.D., Lerbret A., Neiers F., Chambin O., & Assifaoui A. 2016. Binding of divalent cations to polygalacturonate: A mechanism driven by the hydration water. *The Journal of Physical Chemistry B.* 120: 1021–1032.

Jarvis M.C., Hall M.A., Threlfall D.R., & Friend J. 1981. The polysaccharide structure of potato cell walls: Chemical fractionation. *Planta.* 152: 93–100.

Jiahua J, Eliaz I, & Sliva D. 2013. Synergistic and additive effects of modified citrus pectin with two polybotanical compounds, in the suppression of invasive behavior of human breast and prostate cancer cells. *Integrative Cancer Therapies.* 12(2): 145–152.

Jridi M., Abdelhedi, O., Salem, A., Kechaou, H., Nasri, M., & Menchari, Y. 2020. Physicochemical, antioxidant and antibacterial properties of fish gelatin-based edible films enriched with orange peel pectin: Wrapping application. *Food Hydrocolloids.* 103: 105688.

Kim M. 2005. High-methoxyl pectin has greater enhancing effect on glucose uptake in intestinal perfused rats. *Nutrition Research.* 21: 372–377

Kim, H.W., Lee, Y.J., & Kim, Y.H.B. 2016. Effects of membrane-filtered soy hull pectin and pre-emulsified fiber/oil on chemical and technological properties of low fat and low salt meat emulsions. *Journal of Food Science and Technology.* 53: 2580–2588.

Kosmala M., Milala J., Kołodziejczyk K., Markowski J., Zbrzeźniak M., & Renard C.M.G.C. 2013. Dietary fiber and cell wall polysaccharides from plum (Prunusdomestica L.) fruit, juice and pomace: Comparison of composition and functional properties for three plum varieties. *Food Research International.* 54: 1787–1794.

Kpodo F.M., Agbenorhevi J.K., Alba K., Bingham R.J., Oduro I.N., Morris G.A., & Kontogiorgos V. 2017. Pectin isolation and characterization from six okra genotypes. *Food Hydrocolloids.* 72: 323–330.

Leivas C.L., Iacomini M., & Cordeiro L.M.C. 2016. Pectic type II arabinogalactans from starfruit (Averrhoacarambola L.) *Food Chemistry.* 199: 252–257.

McCarty M., & Block K. 2006. Toward a core nutraceutocal program for cancer management. *Integrative Cancer Therapies.* 5(2): 150–171.

Mierczyńska J., Cybulska J., Sołowiej B., & Zdunek A. 2015. Effect of Ca^{2+}, Fe^{2+} and Mg^{2+} on rheological properties of new food matrix made of modified cell wall polysaccharides from apple. *Carbohydrate Polymer.* 133: 547–555.

Mierczyńska J., Cybulska J., & Zdunek A. 2017. Rheological and chemical properties of pectin enriched fractions from different sources extracted with citric acid. *Carbohydrate Polymer.* 156: 443–451.

Mohnen D. 2008. Pectin structure and biosynthesis. *Current Opinion in Plant Biology.* 11: 266–277.

Morris, E. R., Powell, D. A., Gidley, M. J., & Rees, D. A. 1982. Conformations and interactions of pectins. *Journal of Molecular Biology.* 155(4): 507–516.

Morris, E. R., Rees, D. A., Thorn, D., & Boyd J. 1978. Chiroptical and stoichiometric evidence of a specific, primary dimerisation process in alginate gelation. *Carbohydrate Research.* 66(1): 145–154.

Mosayebi V., & Yazdi F.T. 2015. Optimization of microwave assisted extraction (MAE) of pectin from black mulberry (Morusnigra L.) pomace. *Journal of Food and Bio Process Engineering.* 1:34–48

Munarin, F., Tanzi, M.C., & Petrini, P. 2013. Corrigendum to "Advances in biomedical applications of pectin gels". *International Journal of Biological Macromolecules.* 55: 681–689.

Muñoz-Almagro, N., Herrero-Herranz, M., Guri, S., Nieves, C., Antonia, M., & Mar, V. 2021. Application of sunflower pectin gels with low glycemic index in the coating of fresh strawberries stored in modified atmospheres running title: Strawberries coated with sunflower pectin and stored in modified atmospheres. *Journal of the Science of Food and Agriculture.* 101(14), 5775–5783.

Naqash, F., Masoodi, F.A., Rather, S.A., Wani, S.M., & Gani, A. 2017. Emerging concepts in the nutraceutical and functional properties of pectin – A Review. *Carbohydrate Polymers.* 168: 227–239

Norcino, L.B., Mendes, J.F., Natarelli, C.V.L., Manrich, A., Oliveira, J.E., & Mattoso, L.H.C. 2020. Pectin films loaded with copaiba oil nanoemulsions for potential use as bio-based active packaging. *Food Hydrocolloids.* 106: 105862

Nowak K., & Mitka K. 2004. Pectins - polysaccharides of natural origin (2). *Fermentation Industry and Fruit-Vegetable.* 48(09): 24–25.

O'Neill M.A., Warrenfeltz D., Kates K., Pellerin P., Doco T., Darvill A.G., Albersheim P. 1996. Rhamnogalacturonan-II, a pectic polysaccharide in the walls of growing plant cell, forms a dimer that is covalently cross-linked by a borate ester. In vitro conditions for the formation and hydrolysis of the dimer. *Journal of Biological Chemistry.* 271: 22923–22930.

Palmer, K. J., & Hartzog, M. B. 1945. An X-Ray diffraction investigation of sodium pectate. *Journal of the American Chemical Society*. 67: 2122–2127.

Pereira P.H.F., Oliveira T.Í.S., Rosa M.F., Cavalcante F.L., Moates G.K., Wellner N., Waldron K.W., & Azeredo H.M.C. 2016. Pectin extraction from pomegranate peels with citric acid. *International Journal of Biological Macromolecules*. 88: 373–379.

Plaschina, I. G., Braudo, E. E., & Tolstoguzov, V. B.1978. Circular dichlorism studies of pectic solutions. *Carbohydrate Research*. 60(1): 1–8.

Powell, D. A., Morris, E. R., Gidley, M. J., & Rees, D. A.1982. Conformations and interactions of pectins: II. 1982. Influence of residue sequence on chain association in calcium pectate gels. *Journal of Molecular Biology*. 155(4): 517–531

Rascón-Chu A., Díaz-Baca J.A., Carvajal-Millán E., López-Franco Y., Lizardi-Mendoza J. 2016. New use for an "old" polysaccharide: Pectin-based composite materials. In: Handbook of Sustainable Polymers: Structure and Chemistry. Pan Stanford Publishing Pte. Ltd.; Singapore: 72–107

Rees, D. A., & Wight, A. W. *Journal of the Chemical Society*. B 1971, 1366.

Rivera-Hernández, L., Chavarría-Hernández, N., delLópez Cuellar, M.R., Martínez-Juárez, V.M., & Rodríguez-Hernández, A.I. 2020. Pectin-gellan films intended for active food packaging: Release kinetics of nisin and physico-mechanical characterization. *Journal of Food Science and Technology*. 58: 2973–2981

Saberian H., Hamidi-Esfahani Z., AhmadiGavlighi H., Banakar A., & Barzegar M. 2018. The potential of ohmic heating for pectin extraction from orange waste. *Journal of Food Processing and Preservation*. 42: e13458.

Sakooei-Vayghan, R., Peighambardoust, S.H., Hesari, J., & Peressini, D. 2020. Effects of osmotic dehydration (with and without sonication) and pectin-based coating pretreatments on functional properties and color of hot-air dried apricot cubes. *Food Chemistry*. 311: 125978.

Sathyanarayana, B. K., & Rao, V. S. R. 1973. *Current Sciences*. 42: 773.

Schlemmer U. 1989. Studies of the binding of copper, zinc and calcium to pectin, alginate, carrageenan and gum guar in HCO_3–CO_2 buffer. *Food Chemistry*. 32: 223–234.

Schols H.A., & Voragen A.G.J. 2002. The chemical structure of pectins. In: Seymour G.B., Knox J.P., editors. Pectins and Their Manipulation. Blackwell Publishing; Oxford, UK. 1–29.

Sharma, R., Naresh, L., Dhuldhoya, N., Merchant, S., & Merchant, U. 2006. An overview on pectins. *Times Food Process Journal*. 23(2), 44–51.

Sila D.N., Van Buggenhout S., Duvetter T., Fraeye I., de Roeck A., van Loey A., Hendrickx M. 2009. Pectins in processed fruits and vegetables: Part II – Structure–function relationships. *Comprehensive Reviews in Food Science and Food Safety*. 8: 86–104.

Smith, F., & Stephen, A.M. 1960. *Tetrahedron Lett*. 7: 17.

Souza, V.G.L., Pires, J.R.A., Rodrigues, C., Coelhoso, I.M., & Fernando, A.L. 2020. Chitosan composites in packaging industry-current trends and future challenges. *Polymers*. 12: 417.

Sriamornsak P, & Nunthanid J. 1998. Calcium pectinate gel beads for controlled release drug delivery: I. Preparation and in vitro release studies. *International Journal of Pharmacology*. 160: 207–212.

Sriamornsak P. 2003. Chemistry of pectin and its pharmaceutical uses: A review. Silpakorn*University International Journal*. 3(1–2): 206–228.

Sundar, A., Rubila, S., Jayabalan, R., & Ranganathan, T.V. 2012. A review on pectin: Chemistry due to general properties of pectin and its pharmaceutical uses. *Science Reports*.1: 1–4.

Voragen A.G.J., Coenen G.-J., Verhoef R.P., & Schols H.A. 2009. Pectin, a versatile polysaccharide present in plant cell walls. *Structural Chemistry*. 20: 263–275.

Voragen A.G.J., Pilnik W., Thibault J.-F., Axelos M.A.V., & Renard C.M.G.C. 1995. Chapter 10 Pectins. In: Stephen A.M., editor. Food Polysaccharides and Their Applications. Marcel Dekker, Inc.; New York, NY, USA. 287–339.

Vriesmann L.C., & de Oliveira Petkowicz C.L. 2017. Cacao pod husks as a source of low-methoxyl, highly acetylated pectins able to gel in acidic media. *International Journal of Biological Macromolecules*. 101: 146–152.

Walkinshaw, M. D., & Arnott, S. 1981. Conformations and interactions of pectins: I. X-ray diffraction analyses of sodium pectate in neutral and acidified forms. *Journal of Molecular Biology*. 153(4): 1055–1073.

Wang M., Huang B., Fan C., Zhao K., Hu H., Xu X., Pan S., & Liu F. 2016. Characterization and functional properties of mango peel pectin extracted by ultrasound assisted citric acid. *International Journal of Biological Macromolecules*. 91: 794–803.

Wang X., Chen Q., & Lü X. 2014. Pectin extracted from apple pomace and citrus peel by subcritical water. *Food Hydrocolloids*. 38: 129–137.

Wellner N., Kačuráková M., Malovíková A., Wilson R.H., & Belton P.S. 1998. FT-IR study of pectate and pectinate gels formed by divalent cations. *Carbohydrate Research*. 308: 123–131.

Wikiera A, Irla M, & Mika M. 2014. Health promoting properties of pectin. *Advances in Hygiene and Experimental Medicine*. 68: 590–596 (in Polish).

Xiao C., & Anderson C.T. 2013. Roles of pectin in biomass yield and processing for biofuels. *Front. Plant Science*. 4: 67.

Yang, J.S., Mu, T.H., & Ma, M.M. 2018. Extraction, structure, and emulsifying properties of pectin from potato pulp. *Food Chemistry*. 244: 197–205.

Yuliarti O., Goh K.K.T., Matia-Merino L., Mawson J., & Brennan C. 2015. Extraction and characterization of pomace pectin from gold kiwifruit (Actinidiachinensis). *Food Chemistry*. 187: 290–296.

11 Dairy Waste and Its Valorization

Richa Singh, Mayur Suresh Dhale, and Anusha Kishore
ICAR-National Dairy Research Institute, Karnal, Haryana, India
Corresponding author: richasingh.ndri@gmail.com

CONTENTS

11.1 INTRODUCTION

Waste produced from food processing industries causes several environmental problems due to the generation of wastewater with high biological oxygen demand (BOD) and chemical oxygen demand (COD). The volume of effluent generated in the dairy industry is very high when compared to any other food processing industries. The water required for washing and cleaning operations of a dairy plant is about to 2–5 liters of water per liter of processed milk and 0.2–10L of effluent per liter of processed milk. The dairy industry generates a huge amount of wastewater, as the water is used throughout different processes such as cleaning and sanitization of equipments, cleaning of floor, heating and cooling processes. Proper disposal of dairy wastewater is challenging as it contains a high concentration of organic materials (lactose, proteins, lipids, sulphates, and chlorides), and a high concentration of suspended and dissolved solids.

Different physical, chemical, and biological methods are used for the treatment of dairy wastes to decrease organic waste and thereby minimize environmental issues. Physical and chemical treatments remove the organic load of dairy wastewater to a certain extent. Chemical compounds such as aluminium sulphate, ferric chloride, and ferrous sulphate form precipitants with protein and fat present in whey, the efficiency of the precipitant formed is the success of this method. These reagents are costly and removal of soluble COD is less effective, due to these reasons biological methods are more preferred. The waste obtained from food industries can be classified into four as follows

11.2   SOLID WASTES

These refer to the waste that is obtained in solid form as a result of processing or are generated during maintenance. Examples are as follows: minute particles in the exhaust air coming out of a drier, solid ghee residue in ghee, ash from a boiler, (if solid fuel is used), packaging section, where packaging materials may be disposing in the form of low density polyethylene films, cartoons, bottles etc, Wastes generated by adverse events (explosions, fire etc.), Damaged equipment, contaminated equipment, and contaminated soil.

11.3   LIQUID WASTES

It includes wastes, which are obtained in liquid form as a result of processing, cleaning, flushing etc. The wastes derived from processing may include unreacted raw materials, impurities, or by-products generated due to operational deterioration, e.g., suppose the milk is not pasteurized at the right time and if it becomes sour then it will be dumped in the drain. Acid and lye used in clean in place (CIP), and water used in cleaning purposes come under these wastes.

11.4   OILY WASTES

It includes the waste that is the result of leakage in compressors, hydraulic machines, crankcase, and coolant leakage and motors where oil is needed for lubrication. These wastes result from the leakage in compressors, hydraulic machines, crankcase, and coolant leakage and motors where oil is required for lubrication. These wastes need a different disposal method, therefore, they are differentiated from liquid wastes.

11.5   GASEOUS WASTES/WATER VAPORS

It includes the wastes that are released in the air in the form of gases or volatile vapors. The emissions from chimneys are gaseous waste and it pollutes the environment to a great extent. The fumes of chimneys consist of many gases like CO_2 and CO. It also includes the refrigerant leakage from the compressor's pipelines.

11.6   DIFFERENT TYPES OF WASTE GENERATED FROM DAIRY INDUSTRIES

Dairy industries are growing owing to the rising demand of milk and milk products. Dairy products' processing result in a huge amount of effluents. These effluents cause a major environmental impact. Table 11.1 represents the sources of waste in different dairy processes.

TABLE 11.1
Source of Dairy Industry Waste

Dairy process	Source of waste
Preparation stages	
Milk receiving – storage	Poor drainage of tankers. Spills and leaks from hoses and pipes. Spills from storage and silos and tanks. Foaming and cleaning operation
Pasteurization-ultraheat treatment	Liquid losses/leaks. Recovery of downgraded products, cleaning operations. Foaming and deposits on surface of pasteurization and heating equipment
Homogenization	Liquid losses/leaks and cleaning operations
Separation-clarification (centrifuge, RO)	Pipe leaks. Foaming and cleaning operations
Product processing stages	
Market milk	Foaming. Product washing, cleaning operations, over filling. Poor drainage, Sludge removal from clarifiers and separators. Leaks damaged milk packages. Cleaning of filling machinery
Cheese making	Overfilling vats. Incomplete separation of whey from curds, using salt in cheese making. Spills and leaks and cleaning operations
Butter making	Cleaning operations and vacreation and salt use
Powder manufacture	Spills of powder handling, start up and shut down losses. Stack losses. Plant malfunction. Cleaning of evaporators and driers. Bagging losses

Source: Yonar et al. 2018.

Solid waste and liquid waste generate two categories of waste in the dairy industry. Recyclable materials, such as out of date products included in solid waste and whey, wastewater, and sludge are included in liquid waste. The most common types of effluents produced by the dairy processing industry are suspended solids and organic matter, oil and greases, residue of cleaning products, by-products such as whey concentrates etc. Dairy wastewater is mostly white in color (whey has yellowish-green color) with unpleasant odor and turbid character and mostly contains a high concentration of dissolved organic components, like lactose, minerals, fat, and whey protein. The composition of raw dairy wastewater differs mainly according to the type of products and operations. In dairy sludge, casein, fat, lactose, valuable nitrogen, phosphorus, potassium, and organic matter are present.

11.7 WHEY

Whey is considered as a main pollutant in milk processing wastewater due to its high volumetric load and organic content. It accounts for approximately 85–95% of the milk volume and 55% of the milk components. Whey contains carbohydrates (4–5%), mostly lactose. Proteins and lactic acid account for less than 1% of the total, lipids for 0.4–0.5%, and salts range from 1% to 3%. After removal of curd the yellow-greenish translucent liquid fraction of milk that remains is whey. Sweet whey and acidic whey are generated based on the type of cheese production. Globally 94% of the total production is sweet whey. Protein content of sweet whey is higher than acidic whey. BOD and COD are strongly connected to lactose content; for example, sweet whey has about 7 g/L more lactose than acidic whey, making it more polluting. Whey has a BOD of 30 to 50 g/L and a COD of 60 to 80 g/L as a result of its large production and high organic content, the liquid whey is creating great difficulty for the dairy industry in terms of their disposal and is commonly regarded as the "environmental problem". Table 11.2 shows the compounds of whey in a different milk origin.

TABLE 11.2

Compounds of Whey in a Different Milk Origin (Species)

Compound	Cow		Goat		Sheep	
	Sweet	Acidic	Sweet	Acidic	Sweet	Acidic
Ash (%w/w)	0.54±0.09	0.62±0.03	0.51±0.04	0.70±0.03	0.55±0.01	0.82±0.02
BOD_5 (g/L)	41.35±3.60	33±5.80	n.a.	n.a.	n.a.	n.a.
COD (g/L)	67.50±6.20	61.20±3.80	n.a.	68±2	n.a.	76.30±3.80
Lactose (% w/w)	0.48±0.00	0.45±0.28	0.48±0.03	0.35±0.25	0.44±0.06	0.42±0.02
Oils & fats (% w/w)	0.56±0.08	0.44±0.10	0.60±0.23	0.25±0.19	1.45±0.41	0.99±0.01
pH	5.91±0.08	4.73±0.04	6.06±0.12	4.10±0.48	5.87±0.30	4.72±0.20
Protein (% w/w)	0.95±0.05	0.81±0.20	0.88±0.20	0.85±0.06	2.52±0.12	1.64±0.02
Total solids (% w/w)	6.94±0.07	6.68±1.30	6.93±0.19	6.30±0.20	9.39±0.47	6.69±0.01

Source: Kaminarides et al. (2018). Secchi et al. (2018). Chwialkowska et al. (2019).

11.8   WASTE WATER

Dairy requires a huge quantity of water in order to run, resulting in a large amount of effluent. Dairy wastewater is white in color, has an unpleasant odor, and is turbid. Approximately 2.5 L of wastewater is generated in the processing of 1 L of milk. Contaminated water, including sanitary operations, accounts for 50–80% of total water utilized in the dairy plant. The volume and qualities of the wastewater are heavily influenced by the factory's size, technology used, the efficacy and complexity of CIP techniques, and good manufacturing practices (GMP) etc. The implementation of GMP, on the other hand, has the potential to lower the world's wastewater mean volume from 0.5–37 to 0.5–2 m³ of effluent per m³ of processed milk. The intended volumetric load is currently 1 m³ of effluent per tonne of produced milk.

11.8.1   Processing Water

Processing water is formed in the cooling of milk in special coolers and condensers, as well as condensates from the evaporation of milk or whey. Milk and whey drying produces vapors, which form the cleanest effluent after condensation although it may contain volatile substances as well as milk or whey droplets from evaporators. In general, processing waters lack pollutants and, after minimal pretreatment, they can be reused or discharged. Water reusage is possible for installations that are not in direct contact with derived products. For example, the water from the cooling of products during pasteurization after the last rinse of bottles and condensates generated in vacuum installations from secondary vapors can be utilized for room cleaning, lawn irrigation, etc.

11.8.2   Cleaning Wastewater

Cleaning wastewater usually comes from washing equipment, which is in direct contact with milk or dairy products. It also includes milk and product spillage, whey, pressing and brine, CIP effluents or equipment malfunction, and even operational errors. These effluents are in large quantities and are highly polluted, thus requiring further treatment.

11.9   SLUDGE

Dairy sludge contains large quantities of casein, lactose, fat and inorganic salts, besides detergents, sanitizers etc., and is valuable source of nitrogen and phosphorus. Dairy sludge has considerably

TABLE 11.3
Physicochemical Properties of Dairy Sludge

Compound	Dairy sludge levels
Ash (%w/w)	22.50±3.35
Cadmium (mg/kg)	0.07±0.01
Calcium (g/kg)	48.52±9.40
Carbon (% w/w)	38.40±2.51
Chromium (mg/kg)	10.79±0.44
Copper (mg/kg)	10.04±0.46
Iron (g/kg)	27.70±1.50
Magnesium (g/kg)	2.90±0.43
Mercury (mg/kg)	0.03±0.00
Nickel (mg/kg)	6.80±5.10
Nitrogen (%w/w)	5.30±0.12
Phosphorus (g/kg)	33.04±0.25
Potassium (g/kg)	7.10±0.40
Sulphur (g/kg)	6.10±3.20
Zinc (mg/kg)	82.41±2.99

higher fertilizer value than municipal sludge. Dairy industries emit 5 to 25 % of the total volume of treated wastewater as dairy sludge. Its composition varies depending on the environmental conditions and wastewater treatment process (Ganju and Gogate, 2017; Sharmila et al., 2020).In dairy sludge, the BOD and COD amount to 1.1–1.14 g/L, and 3.7–4.02 g/L, respectively (Bhattacharya et al., 2019) and its pH is approximately 7–8. Dairy sludge has a higher concentration of heavy metals like copper and zinc. Table 11.3 shows physicochemical properties of dairy sludge.

11.10 MANAGEMENT OF DAIRY WASTE

Different physical, chemical, and biological methods are there for treating dairy waste. Primarily biological methods such as trickling filters, activated sludge process, sequencing batch reactor (SBR), aerated lagoons, aerobic bioreactor, up-flow anaerobic filters, up-flow anaerobic sludge blanket (UASB) reactor and bio-coagulation are used for dairy waste treatment. Aerobic treatment of dairy wastewater is high energy intensive. Anaerobic treatment reflects very poor nutrient removal and this effluent needs additional treatment. Physical and chemical methods such as coagulation and flocculation can also be used for treatment of dairy waste.

Overall, dairy wastewater treatment is divided into three processes: primary, secondary, and tertiary processes. Primary treatment consists of physical screening and sedimentation of dairy waste to remove large particles or debris if any, flow and composition balancing to stabilize effluent, addition of chemicals to control pH, and dissolved air floatation (DAF) to remove fat, oil, and grease (FOG). Secondary treatment removes organic materials and it is of two types: biological and physico-chemical treatments. Biological treatment can be aerobic and anaerobic, which removes the organic materials present. Large quantities of dairy processing sludge (DPS) are produced during secondary treatment and pollutants can be absorbed into it. Aerobic biological treatments include activated sludge process, bio-towers, sequencing batch reactors, and membrane bioreactors and are carried out using dissolved oxygen. Aerobic treatment method is reliable and cost-effective in producing a high-quality effluent, but it results in high DPS generation and costly disposal problems. DPS is relatively a new waste type and it is much cleaner and a valuable fertilizing product than biosolids, which are derived from sewage sludge. Anaerobic biological treatments involve anaerobic lagoons, membrane anaerobic reactor systems, up-flow anaerobic sludge blankets, and completely

TABLE 11.4
Composition of Milk Processing Effluents

Milk processing effluent	Active reaction (pH)	Concentration/(g/L)							
		BOD$_5$	COD	FOG	TS	TSS	TN	TP	Alkalinity as CaCO$_3$
Mixed dairy	4–11	0.24–5.9	0.5–10.4	0.02–1.92	0.71–7	0.06–5.80	0.01–0.66	0–0.6	0.32–1.2
Milk reception	7.18	0.8	2.54	1.06	-	0.65	-	-	-
Dairy/sewage = 7:3	9.1±6.7	1.08–2.81	2.04–4.73	0.24–0.29	-	0.53–1.13	-	0.02–0.03	-
Fluid milk	5–9.5	0.5–1.3	0.95–2.4	-	-	0.09–0.45	-	-	-
Yoghurt	4.53	-	6.5	-	-	-	-	-	-
Butter	12.08	0.22–2.65	8.93	2.88	-	0.7–5.07	-	-	-
Ice cream	5.1–6.96	2.45	5.2	-	3.9	3.1	-	0.014	0.22
Cheese	3.38–9.5	0.59–5	1–63.3	0.33–2.6	1.92–53.2	0.19–2.5	0.018–0.83	0.005–0.28	-
Cottage cheese	7.83	2.6	17.65	0.95	-	3.38	-	-	-
Cheese whey	3.92–6.5	27–60	50–102.1	0.9–14	55–70.9	1.27–22.15	0.2–1.76	0.12–0.53	-
Hard cheese whey	5.8	9.48	73.45	0.99	-	7.15	-	-	-
Soft cheese whey	5.35	26.77	58.55	0.49	-	8.31	-	-	-
Cottage Cheese whey	4.5	-	79	-	68	-	2	-	-
Cheese whey wastewater	4.6	35	-	0.8	-	-	-	0.64	-
Whey processing effluent	5–9	0.59–1.21	1.07–2.18	-	-	0.08–0.44	-	-	-
Milk permeate	5.55–6.52	-	52.94–57.46	-	11.61–15.39	1.94–3.4	0.3–0.4	0.35–0.45	2.5
Condensate	8.3	-	-	-	-	-	0.0006	0.0001	-
Washing wastewater	10.37	3.47	14.64	3.11	-	3.82	-	-	-

Notes: BOD$_5$=biological oxygen demand for 5 days, COD=chemical oxygen demand, FOG=fat, oil, and grease, TS=total solids, TSS=total suspended solids, TN=total nitrogen, TP=total phosphorus stirred tank reactors and less DPS is generated. In tertiary treatment, phosphorus is removed by using chemicals like aluminium (Al) and/or iron (Fe) salts, before final discharge. In an enhanced biological phosphorus removal (EBPR) process, there is no need for precipitants. The EBPR process can be done through an activated sludge process by recirculating sludge through anaerobic and aerobic conditions. Table 11.4 gives the composition of milk processing effluents.

11.11 TREATMENT OF DAIRY WASTE

Wastewater characterization plays a major role while designing the wastewater treatment system. The concentration of COD for dairy wastewater varies considerably. Pollution loads of wastewater producing dairy industries are very different. Cheese producing plants and yoghurt producing plants have different COD concentration. The following are different treatments used in dairy industries as per their requirements:

11.12 WETLAND/NATURAL PROCESS

Wetlands are a type of sustainable effective wastewater treatment, having the similar function as that of conventional treatment and are more economical, environment friendly, and energy-efficient forms. A wetland uses microbial communities for the wastewater improvement, therefore, this is considered as a natural process. This treatment requires simple construction and lacks sludge recycling and uses wetland plants (hydophytes), aggregate materials, and microorganisms for the construction. Aerobic condition is used to treat dairy wastewater. Heavy loaded dairy wastewater treatment in facultative wetland results in 85% reduction in BOD in five days at 20°C. Wetlands were successfully used for the treatment of dairy wastewater in many countries.

11.12.1 CHEMICAL PRECIPITATION

Chemical precipitation involves the addition of chemicals to separate the suspended and dissolved solids by sedimentation process. Phosphorus and heavy metal can be removed by this primary settling process. Precipitants such as alum, ferric sulphate, ferrous sulphate are used. These are used primarily for treatment of metallic cations, anions, detergents, organic molecules, and oily emulsions. Coagulation/flocculation processes are directly applied to raw wastewater to separate suspended, dissolved, and colloidal contents. Coagulation involves the addition of coagulants such as iron or aluminium to overcome the factors that promote stability of the system. Flocculation allows the destabilized particles to come together and separate through gravity settling.

11.12.2 COAGULATION

Coagulation or flocculation is one among the most important physico-chemical treatment steps, which helps in the treatment of industrial wastewater. This step reduces the suspended and colloidal particles that are responsible for turbidity of water and helps in the reduction of organic substances responsible for the increase in COD and BOD contents. Addition of coagulant results in destabilization of particulate matter, followed by collision of particle and floc formation this finally results in flotation or sedimentation. Natural coagulation in dairy wastewater can be achieved by using certain lactic acid bacteria. These bacteria ferment lactose and convert it into lactic acid, which denatures the milk protein in waste water. The lactic acid bacteria in combination with chitosan results in COD reduction of 49–82%, while when carboxymethylcellulose (CMC) was used, COD reduction of 65–78% was achieved. Treatment with powdered activated charcoal (PAC) after chitosan treatment helps in removing complete color and odor from the wastewater. An oxidation pretreatment with $FeSO_4$

and H_2O_2 resulted in up to 80% removal of the fat contained in the cheese wastewater. Plant-based coagulant, tannin can be directly used and have better performance than inorganic coagulants, the generated waste is biodegradable, and it can be applicable over a wide of pH range. The combined action of tannin and polyaluminium chloride can be used in the coagulation of dairy wastewater. The use of tannin as a coagulant is most useful for treating wastewater from the dairy industry.

11.12.3 ADSORPTION PROCESS

Adsorption involves the removal of organic compounds from waste water. Adsorbents such as activated carbon, synthetic polymeric, and silica-based adsorbents are commonly used. Activated carbon is the most useful among this due to its ability to adsorb a wide range of organic compounds and is cost effective. Adsorption can be classified into two: physical and chemical adsorption. In physical adsorption, van der Waals forces are used and activated carbon is the best example. A chemical reaction takes place between adsorbate and adsorbent and this reaction does not have a wide application in wastewater treatment. Adsorption onto solid surfaces removes organic compounds, chemicals, and heavy metals. Fly ash, bagasse fly ash, rice husk ash, and activated carbon are low-cost adsorbents.

11.12.4 MEMBRANE PROCESSES

Membrane processes such as microfiltration, ultrafiltration, nanofiltration, dialysis, electrodialysis, and reverse osmosis are very promising methods. Membrane filtration is the separation or removal of particulate and colloidal substances from a liquid, which acts as a selective barrier, having typically 0.0001–1.0 μm pore size. The membrane technology is a nonthermal environment friendly technology, which minimizes the adverse effects caused by the temperature rise such as change in phase, denaturation of proteins, and change in sensory attributes of the food product.

- Micro-filtration: micro-filtration (MF) is a low pressure driven membrane filtration process and is used for reduction of bacteria and fat removal in milk as well as whey.
- Ultra-filtration: ultra-filtration (UF) is a medium pressure driven process and is used in decalcification of permeates from the dairy industry wastewater and in reduction of lactose concentration. In UF process, the solutes and suspended effluents vary in their size, which triggers movement due to hydrostatic forces. In the dairy industry,UF process is commonly used for the separation of whey from dairy waste.
- Reverse osmosis: reverse osmosis (RO) is a high pressure driven membrane treatment process. Treatment of dairy wastewater by RO is used for generating water, which can be reused and also to reduce the effluent volume. Large volume of effluents produced during the starting, equilibrating, rinsing, and stopping of the dairy processing units. RO of wastewater produces purified water, which can be re-used as boiler make-up water and for cooling purposes in the dairy industries. Treatment of 100 m³/day of wastewater with 540 m² RO units produced 95% of water recovery.
- Nano-filtration: nano-filtration (NF) membrane separation process is medium to high pressure driven process. NF membrane treatment process operates at lower pressure, lower energy consumption, and shows higher permeate recoveries than RO. NF is becoming a viable alternative to the conventional treatment over RO.Tertiary treatment of wastewater effluents by NF showed an efficient COD removal of 98% while the total nitrogen and phosphorus removal of 86% and 89%, respectively. The NF permeates showed a COD of 87 mg/L and calcium concentration of 3.2 mg/L. This permeate can be used for washing floors, trucks and all external areas that require low quality water.

11.12.5 ELECTROCHEMICAL PROCESS

Electrolysis is the process of degrading organic or inorganic substances by applying electrical charge. Oxidation and reduction reactions occur in the electrolytic cell, which contains both the anode and cathode. When electricity is applied to a cell, negative ions will migrate towards the anode and positive ions will migrate towards the cathode and cations get reduced and anions get oxidized at both electrodes. Electrochemical processes such as electrocoagulation (EC), electroflotation, and anodic oxidation are some examples used for dairy waste treatment. EC is mostly effective for a wide range of pollutants such as heavy metals, organic compounds, microorganisms etc., so EC is considered as one of the most promising water remediation techniques. EC is a primary wastewater treatment for inducing the controlled electrogeneration of flocculants/coagulants on site, usually under the application of a constant current. EC is a complex process involving several physical and chemical phenomena with the formation of aluminium or iron cations from the dissolution of the corresponding sacrificial anode(s) and the simultaneous production of OH− anions by cathodic reduction of water. The polymeric metal hydroxides thus formed act as excellent coagulating agents, which favor the removal of suspended, dissolved, or colloidal matter, and ultimately yield great percentages of removal of turbidity and color. Coagulation mainly occurs by the process destabilization, once the metal cation combines with the negatively charged particles moving to the anode by electrophoretic motion.

11.13 BIOLOGICAL TREATMENT

Biological treatment is the most preferred method for dairy wastewater, and includes processes such as trickling filters, aerated lagoons, activated sludge, up-flow anaerobic sludge blanket (UASB), anaerobic filters, sequential batch reactor (SBR), etc. It is the most promising method for removing organic material from dairy waste. However, the formation of sludge during aerobic biodegradation may cause serious disposal problems and is costly, including the sewage sludge treatment. Ability of sludge to adsorb organic material and even toxic heavy metals may be a serious problem. However, the biological system has the ability to microbiologically transform complex organics and even to adsorb heavy metals if suitable microorganisms are present. The biological treatment for purification of effluents assimilates all the dairy wastewater components and is one of the most reliable methods for dairy effluent treatment. The biological treatment of dairy wastewater includes both aerobic as well as anaerobic treatment processes. Aerobic and anaerobic treatment of the organic effluents are effective due to its performance for COD and BOD removal, even though there are a few drawbacks like aerobic processes consuming high energy whereas anaerobic treatments degrade nutrients partly. All conventional biological processes available for treatment of dairy industry wastewater may not be very feasible due to large land requirements and high operational cost. Based on oxygen requirement, biological treatment can be divided into two branches: aerobic and anaerobic methods.

11.13.1 AEROBIC TREATMENT

The aerobic method is used in most of the dairy wastewater treatment plants, it has reduced efficiency and is due to the rapid acidification and filamentous growth, resulted from the low water buffer capacity and high lactose level, respectively. Aerobic biological treatment depends on the microorganisms, which are grown in the presence of oxygen-rich environment and oxidize organic compounds into carbon dioxide, water, and cellular material. Nitrogen from ammonia (NH_3) is easily degraded in aerobic method, while effectiveness is less in the case of phosphorus removal. Aerobic bacteria are less effective than anaerobic bacteria. Complete-mix processes are more sensitive to high organic load problem i.e. bulking sludge, thus, they are less effective than plug flow systems. In continuous mode, synthetic dairy wastewater treatment gives good results under aerobic biological system with over 96% of degradation achieved and with COD, total Kjeldahl nitrogen (TKN) and pH of 4 g/L,1g/L, and 11.5g/L respectively. Percolating filters or conventional trickling filters are the oldest biological methods, which result in high quality final effluents. In high strength dairy effluents treatment (rich in fat, oil, and grease), aerobic filter usage is limited. Problems like biomass loss and filter fouling arise with high fat and heavy biofilm blockage, resulting in reduction in the efficiency. Organic loading for wastewater should not be more than 0.28–0.30 kg BOD m−3 and recirculation should be performed. Sequencing batch reactor (SBR) can be used for dairy wastewater treatment, its different loading capabilities and effluent flexibility. SBR is a single tank fills and draw unit, which is used for aeration, settling, withdrawing effluent, and recycling of solids. Purification of milk effluents with SBR reduces the COD (91–97%), TS (63%), volatile solids (VS) (66%), TKN (75%), and total nitrogen (TN) (38%). Intermittently, aerated sequencing batch reactors (IASBR) have improved the remediation limitation of nutrients (mainly nitrogen and phosphorous) by aerobic biotechnology when high organic load is present in the wastewater, as well as in the dairy effluents. The IASBR technology was applied with three different aeration rates (0.4–0.8 L/min). High efficiency of nutrient removal was reached (above 90% of orthophosphates and ammonium nitrogen removal in synthetic dairy wastewater) when the reactor operated with 0.6L/min of aeration rate.

11.13.2 Anaerobic Treatment

Anaerobic treatments are more suitable than an aerobic treatment system, mainly due to the cost effectiveness. Anaerobic treatment is for treating wastewater with high organic content. The dairy wastewater treatment can be performed at low cost scale by using anaerobic and facultative systems but is less effective. For the treatment of dairy wastewater, up-flow anaerobic sludge blanket (UASB) reactors, hybrid digester, and anaerobic sequencing batch reactors (ASBR) are also used. Up-flow anaerobic sludge blanket (UASB) has been used for treating dairy effluents. UASB reactors in dairy wastewater treatment have the organic loading rate of up to 6.2 g COD per day and could be increased up to 7.5 g COD per day. The HRT and loading range of UASB reactors are lies in range of 2.4–13.5 kg COD at HRT of 3–12 h and COD reduction ranged from 95.6% to 96.3% in 3 h. A UASB reactor shows limitation in wastewater treatment due to accumulation of FOG, and subsequently increases the time of hydrolysis. To overcome these problems, a modified UASB reactor with a scum extraction device and lamella settlers are available. The modified UASB reactor is efficient in operational unit and is cost effective.

Enterobacter aerogens and methanogenic bacteria produce methane and hydrogen gas from dairy wastewater. The different concentration of dairy wastewater showed maximum biomass growth rate (0.21 per hour) at 75% concentration. The production of methane (190 CH4 ml/gCOD and 0.59 LCH4 per liter) and hydrogen up to 105 ml H2/g-COD and 0.562 L-H2 per liter was reported from dairy wastewater. A hybrid anaerobic digester used in combination with up-flow sludge blankets and fixed bed designs at an influent substrate concentration of COD in dairy effluent, the COD removal rate obtained up to 90–97% at an OLR between 0.82 and 6.11 kg COD/(m3 day) at HRT range of 4.1–1.7. When anaerobic digester used for the treatment of dairy effluent, the methane removed up to 0.354 m₃ CH4/kg at HRT of 1.7 days.

ASBR enhance the efficiency of dairy effluent treatment in nonfat dry milk (NFDM) processing and removes COD and BOD up to 62% and 75%, respectively at HRT of 6 h at low temperature. The change in temperature from 5 to 20 °C reduces 62–90% of COD and 75–90% of BOD at HRT between 6 and 24 h for soluble organic loads. ASBR also provides 26–44% volatile solid removal in two stage thermophilic ASBR while removal of volatile solids in mesophilic ASBR systems ranges from 26% to 50% for dairy effluents.

The advantage of anaerobic treatment over aerobic treatment is that there is no need of aeration and a relatively low area is needed to carry out this process. Anaerobic treatment is of two types: single-phase anaerobic treatment and two-phase anaerobic treatment.

- Single-phase anaerobic treatment: the single-phase anaerobic treatment employs use of filter reactors for low concentration of suspended solids. Removal of COD ranges from 78% to 92% by the use of a laboratory-scale plastic medium anaerobic filter reactor. In order to treat very dilute dairy wastewater, an up-flow anaerobic filter reactor (UAF) is used. A pilot-scale UAF provides more than 85% COD and 90% BOD removal. The conventional single-phase anaerobic treatment is now being replaced by two-phase anaerobic treatment of waste-water in which the performance of an acid phase (acidogenic) reactor is of paramount importance.
- Two-phase anaerobic treatment: the two-phase anaerobic treatment system is especially used for the removal of high concentrations of suspended organic solids from wastewaters generated from food and agricultural industries. In a two-phase anaerobic treatment system, the first phase is an acidogenic reactor and the second phase is a methanogenic reactor. Anaerobic fermentation of wastewaters generated from the cheese-making process showed up to 19% of initial sugar, which is converted to volatile fatty acids. Biodegradation of whey generates n-butyric acid, which can used further. About 95% of carbohydrates, 82% of proteins, and 41% of lipids can be degraded by acidogenesis of dairy wastewaters.

To obtain the effluent discharge limits of dairy industry wastewaters, anaerobic treatment process is used in combination with aerobic treatment process. This process aims at reducing BOD by more than 90% and COD removal by 85%. The industrial scale treatment facility of certain factories consists of an anaerobic equalization tank, followed by an UASB and aerated lagoons.

11.13.3 Ultrasonication

Ultrasound technology utilizes mechanical waves at a frequency greater than 16 kHz (above human hearing threshold). These waves can be classified into two based on the frequency ranges: (1) low-energy (high frequency, low power) utilizing frequencies higher than 100 kHz at intensities below 1 W cm^{-2} and (2) high-energy (low frequency, high power) utilizing frequencies between 20 and 500 kHz and intensities higher than 1 W cm^{-2}. Low energy ultrasound is applicable in nondestructive analytical technique, which measures the physicochemical properties of food such as acidity, firmness, protein interactions, ripeness, sugar content, etc., whereas high energy ultrasound is applicable in changing physical and chemical properties of food and inactivation of microorganisms in foods. Ultrasonication treatments can be done directly and indirectly. Direct application involves immersion of the ultrasonic probe directly into liquid medium. The major drawback of this method is the possibility of metal contamination from the detached metal pieces from the probe. The indirect application involves ultrasonication bath. The ultrasonic baths are not powerful devices. In ultrasound application, cavitation occurs near the tip of the ultrasonication probe and at the transducer surface in the bath. In both applications, cavitation intensity decreases exponentially as the distance increases from the probe tip and transducer surface. Extensive research done on the application of ultrasound in the field of dairy; however, due to its limitations such as metal contamination, poor spatiotemporal distribution, low energy efficiencies, poor transmission of ultrasound energy in large volume sample, and higher operational cost, this is not widely used.

Application of ultrasound in a dairy waste water treatment is an emerging trend in wastewater management. The large amount of FOG content is repressive for anaerobic treatment due to flotation of sludge deposition of fat scum layer on the reactor surface. Solidification of fats at low temperature leads to clogging and even develops an obnoxious odor. This is due to the low hydrolysis in the anaerobic reactor. So, pretreatment is required to remove oils and grease from wastewater prior to the biological treatment process. Hydrolysis and dissolution of fats in pretreatment processes, boosts biological degradation of fat present in dairy wastewater, thus accelerating the process and reducing the time consumed. The physico-chemical pretreatment techniques did not render a good COD removal and the reagents used are costly and this method required harsh reaction conditions with a subsequent generation of problematic sludge. For this reason, biochemical techniques like using specific enzymeslipases acquired consideration owing to the stringent environmental regulations, clean and friendly approach of enzymes, mild reaction conditions, high specificity, and easy availability. Lipases break down the triglycerides to free fatty acids (FFA) and glycerol in the presence of the oil/water interface. This enzyme helps in degrading FOGs in dairy wastewater from dairies, slaughterhouses, and the pet food industry. The enzymatic prehydrolysis of fat under the influence of ultrasound drastically reduces the reaction time from 24 h to 40 min as compared to conventional stirring with improved yield.

11.14 GREEN WASTE MANAGEMENT: *RECOVERY OF VALUABLE INGREDIENTS FROM DAIRY WASTE*

Whey is the principal by-product of the dairy industry, which is formed during the manufacture of casein and cheese during milk coagulation. Due to the huge production of whey and its high organic content, exhibiting a BOD ranging from 30 to 50 g/L and a COD ranging from 60 to 80 g/L, and is regarded as the "environmental problem" and is creating a great difficulty for their proper disposal. Whey contains a number of nutrients such as lipids (0.4–0.5% w/v), soluble proteins (0.6–0.8% w/v), lactose (4.5–5% w/v), and mineral salts (8–10% of dried extract) thus whey can be exploited as a resource of a number of valuable end products rather than as a waste stream. Whey can be utilized as a cheap source of lactose and protein and it can be further used in food, dairy, and pharmaceutical industries. Lactose can be either directly fermented or it can be hydrolyzed to produce glucose and galactose. Proteins possess high nutritional value and versatile functional properties. Hence, the recovery of lactose and protein helps to reduce the BOD and COD loading of whey and also helps in solving the problem of environmental pollution caused by the disposal of whey. The proteins can be recovered efficiently by ultrafiltration with high yield and purity. Nanofiltration is used to remove ions from the feed solution and can be used to concentrate valuable components of whey. Thus, various membrane processes are used for the efficient recovery of proteins. Ultrafiltration using hollow fiber module separates lactose and protein with high yield and purity from whey. Ultrafiltration in a diafiltration mode improves the yield of protein in the retentate, and this retentate can be freeze-dried. Nanofiltration of the permeate stream from ultrafiltration to concentrate the lactose part and this can also be freeze-dried. Approximately 90% of lactose and 80% of protein can be recovered in the form of freeze-dried products. The recovered whey proteins and lactose can be converted to high value end products such as whey protein concentrate (WPC) and bioethanol.

11.15 CONCLUSION

The dairy business is a large food processing sector that consumes a lot of water and produces a lot of dairy effluent. The content of trash varies by industry, and information on the composition of wastewater streams is sparse. It has a high organic content, BOD and COD, and temperature in general. If not adequately handled and put directly onto the soil, it can cause major environmental concerns and even damage humans, aquatic species such as fish, and crops. For dairy waste treatment, physicochemical, and biological treatment methods are used.

REFERENCES

Adulkar T V and Rathod V K. "Ultrasound assisted enzymatic pre-treatment of high fat content dairy wastewater". *Ultrasonics Sonochemistry* 21. 2014. pp.1083–1089.

Ahmad T, Aadila R M, Ahmeda H, Rahmana U, Soaresb B C.V, Souzab S L Q. Pimentelc T C, Scudinod H, Guimarãesd J T, Esmerinod E A, Freitasd M Q, Almadab R B, Vendramelb S M R, Silvab M C and Cruzb A G. "Treatment and utilization of dairy industrial waste: A review". *Trends in Food Science & Technology*. 2019. 88.

Ashekuzzaman S M, Forrestal P, Richards K and Fenton O. "Dairy industry derived wastewater treatment sludge: Generation, type and characterization of nutrients and metals for agricultural reuse". *Journal of Cleaner Production* 230. 2019. Pp.1266–1275.

Bhattacharya R, Kundu P, Mukherjee J and Mukherjee S. "Kinetics study of a suspended growth system for sustainable biological treatment of dairy wastewater". *Environmental biotechnology for soil and wastewater implications on ecosystems,* pp. 15–20.

Chwialkowska J, Duber A, Zagrodnik R, Walkiewicz F, Łężyk M and Oleskowicz-Popiel P. "Caproic acid production from acid whey via open culture fermentation – Evaluation of the role of electron donors and downstream processing". *Bioresource Technology,* 2019, pp. 74–83,

Das B, Sarkar S, Sarkar A, Bhattacharjee S and Bhattacharjee C. "Recovery of whey proteins and lactose from dairy waste: A step towards green waste management". *Process Safety and Environmental Protection* 10(1). 2016. pp.27–33

Environment Protection Act (1997) A1997-92 Republication No 59 Effective: 31 August 2017

Feng G L, Letey J, Chang A C and Mathews M C. "Simulating dairy liquid waste management options as a nitrogen source for crops". *Agriculture, Ecosystems and Environment* 110. 2005. pp. 219–229.

Ganju S and Gogate P R. "A review on approaches for efficient recovery of whey proteins from dairy industry effluents". *Journal of food engineering.* 2017, pp. 84–96.

International Dairy Federation, 1993. Environmental influence of chemicals used in the dairy industry which can enter dairy wastewater, Bulletin of the IDF, No. 288.

Kaminarides S, Aktypis A, Koronios G, Massouras T and Papanikolaou S. "Effect of *in situ* produced bacteriocin thermophilin T on the microbiological and physicochemical characteristics of Myzithra whey cheese, 2017, pp. 213–222.

Karpati I, Bencze L and Borszerki J. "New process for physicochemical pretreatment of dairy effluents with agricultural use of sludge produced". *Water Science and Technology* 22(9). 1995. pp.93–100.

Orhon D, Gorgun E, Germirli F and Artan N. "Biological treatability of dairy wastewaters", *Water Research* 27(4). 1993. pp.625–633.

Roufou S, Griffin S, Katsini L, Polanska M, VanImpe J F M and Valdramidis V P. "The (potential) impact of seasonality and climate change on the physicochemical and microbial properties of dairy waste and its management". *Trends in Food Science & Technology* 116. 2021. pp.1–10.

Sarkar B, Chakrabarti P P, Vijaykumar A and Kale V. "Wastewater treatment in dairy industries – possibility of reuse". *Desalination.* 195(1–3). 2006. pp. 141–152.

Secchi N, Fadda C, Piccinini M, Pinna I, Piga A, Catzeddu P and Fois S. "The effects of ovine whey powders on durum wheat-based doughs". *Journal of food quality.* 2018.

Sharmila V G, Angappane S, Gunasekaran M, Kumar G and Bhanu R. "Immobilized ZnO nano film impelled bacterial disintegration of dairy sludge to enrich anaerobic digestion for profitable bioenergy production: Energetic and economic analysis". *Bioresource Technology.* I. 2020. p. 123276.

Shi W, Healy M G, Ashekuzzaman S M, Daly K, Leahy J J and Fenton O. "Dairy processing sludge and co-products: A review of present and future re-use pathways in agriculture". *Journal of Cleaner Production.* 314. 2021. p. 128035.

Sutariya S, Sunkesula V, Kumar R and Shah K. "Emerging applications of ultrasonication and cavitation in dairy industry: a review". *Cogent Food & Agriculture.* 4. 2018. p. 1549187.

Vourch M, Balannec B, Chaufer B and Dorange G. "Treatment of dairy industry wastewater by reverse osmosis for water reuse", *Desalination.* 219(1–3). 2008. pp. 190–202.

Wildbrett G. "Bewertung von reinigungs- und desinfektionsmittelnimabwasser". *Dtsch. Milchwirtschaft.* 39. 1988. pp. 616–620.

Yonar T, Sivrioğlu Ö, Özengin N. "Physico-chemical treatment of dairy industry wastewaters: A review". *Technological approaches for novel applications in dairy processing.* 2018. pp. 179.

12 Use of Microbiological Agents in Upgrading Waste for Feed and Food

Neha Rani Bhagat[1] and Arup Giri[2,*]
[1]DRDO-Defence Institute of High Altitude Research (DIHAR),
Ladakh UT, India
[2]Baba Mastnath University, Asthal Bohar, Rohtak, Haryana, India
*Corresponding author: arupsatadal@gmail.com

CONTENTS

12.1   INTRODUCTION

Every year, the world produces 2.01 billion tonnes of municipal solid garbage, which is dumped without any treatment. The average amount of garbage generated per person each day is 0.74 kg, which ranges from 0.11 to 4.54 kg. All these wastes may reduce the environmental quality, and this may be enhanced by the global population substantially increasing.

DOI: 10.1201/9781003207689-12

Many studies have discovered that waste material from various sectors can be converted to beneficial stuff using microbial treatment. Some research found that food waste has the potency to be converted into animal feedstuff like fish feed or broiler feed after microbial treatment. Microbial biotransformation has proved to be an imperative tool for centuries in boosting the production of various chemicals used in food, pharmaceutical, agrochemical, and industrial products. This is being used in the transformation of various pollutants or compounds, including hydrocarbons, pharmaceutical substances, and metals.

Animal feed, single cell protein, amylase, bioethanol, biobutanol, indole-3-acetic acid, protease production, astaxanthin (pigment), bioherbicide, biosorbents, bleomycin, ergosterol, laccase, lactic acid, meroparamycin, neomycin, oxytetracycline, poly (3-hyrdroxybutyric acid), protein, rifamycin, xanthan, bioactive phenolic compounds, fibrinolytic enzyme, pectin lyase, citric acid, acetic acid, fumaric acid, biosurfactant, wine (anti-oxidant rich), cellulose, lycopene, polygalactouronase, vanillic acid and vanillin, proanthocyanidins, anthocynidins, phenolic acids, vitamin E and oryzanol, ferulic, p-coumaric, sinapic and syringic, lipase, nisin, beta-galactosidase, pullulanase, xylanase, oligogalacturonides, sophorolipids, bacteriocins, ellagic acid, anti-oxidant peptides (preservatives), caratenoids, biopolymers (pha) are all food and food gradients being developed from agriculture residue – wheat bran, sea bass (*Dicentrarchus labrax*) processing wastes, wine industry residue-gape marc, whey, food waste mixture (cooked rice, meat, vegetables, bakery by-products, barley bran, wheat bran, and broiler poultry litter), dairy waste and kitchen based wastages, pineapple waste, potato/mash waste, banana waste, starchy food waste, cassava fibrous residue, wheat bran, rice bran, soybean meal, brewery waste, dairy waste (whey), peels of citrus fruits, soybean meal, mixed restaurant food waste and bakery waste, orange peel, rice husk, sugarcane baggase, rice bran, citrus peel waste, fish skin, viscera, muscle, orange peel waste, turbot skin waste, wheat bread waste, pressed juice from oil palm, wheat bran and rape seed meal, whey soy bean and rapeseed oil, corn oil, spent coffee, grounds oil, rice straw, molasses, sugarcane bagasse, cheese whey, etc with proper microbial treatment (Okorie and Asagbra, 2005; Okorie and Asagbra, 2008; Yang and Yuan, 1990; Ezejiofor *et al.*, 2012).

This book chapter reveals the microbial treatment of different wastes for animal feed development.

12.2 VARIOUS MICROBIOLOGICAL AGENTS FOR BIOTRANSFORMATION OF WASTES INTO FOOD OR FEED

The processing of waste using living organisms is a simple, cost-effective, and environmentally safe alternative to physico-chemical methods. In this context, microbial-based bio-treatment is emerging as a reasonable approach for waste conversion into valuable end-products as microbes have the unique ability to metabolize organic matter present in wastes for their growth and development, which further helps in waste biotransformation (Lee *et al.*, 2019). In addition, microbial biotransformation has proved to be an imperative tool for centuries in boosting the production of various chemicals used in food, pharmaceutical, agrochemical, and industrial products.

Already, microbial biotransformation is widely being utilized in the transformation of various pollutants or compounds, including hydrocarbons, pharmaceutical substances, and metals (Karigar *et al.*, 2011; Smitha *et al.*, 2017). The microbial agents involved in such a process include bacteria, filamentous fungi, algae, yeast, and actinomycetes, and it leads to biotransformation through fermentation. This fermentation process is regarded as an emerging approach for waste conversion to valuable end-products. In addition, the type of fermentation for the bioconversion vastly depends on the type of waste used as the crude feedstock. Based on the crude feedstock, i.e., substrate, there are two types of fermentation: solid state fermentation (SSF) involving the solid substrate, and submerged fermentation (SmF) involving the liquid substrate (Sadh *et al.*, 2018; Ng *et al.*, 2020).

Further, the common microbes involved in the biotransformation of waste into value-added products are discussed below.

12.2.1 BACTERIA

Bacteria, the basic biological fundamentals in waste treatment, have diverse biochemical activity, making it possible for them to metabolize most of the organic compounds present in bio-wastes (Adebayo and Obiekezie, 2018). On the basis of metabolizing activity, bacteria include two major groups: first, the bacteria utilizing the organic compounds, and second, the bacteria utilizing the lysed products released by the organic compound utilizing bacteria. The bacteria utilizing the organic compounds in the waste are the most important group and will determine the biotransformation of the organic substrate. Moreover, the various bacterial agents involved in the biotransformation of waste into feed or bioactive compounds, such as lactic acid bacteria (*Lactobacillus sp.*, *Pediococcus sp.*, *Enterococcus sp.*, *Streptococcus sp.*, *Bacillus sp.*, *Xanthomonas sp.*, *Streptomyces sp.*, *Penicillium sp.*, *Pseudomonas sp.*, *Amycolatopsis Mediterranean*, *Rheinheimera sp.*, *Lysinibacillus sp.*, *Burk holderiacenocepacia etc.* are discussed in Table 12.1.

12.2.2 FUNGI AND YEASTS

Further, fungi and yeast also play an important role in the stabilization of organic wastes. Like bacteria, fungi can metabolize almost every type of organic compound found in waste. The fungi have the potential ability to predominate over the bacteria, but they do not except under unusual environmental conditions. Under normal environmental conditions, fungi will be present and will aid in the stabilization of organic matter. *Aspergillus sp.*, *Trichoderma sp.*, *Cryptococcus albidus sp.*, *Rhizopus sp.*, *Candida sp.*, *Saccharomyces sp.*, *Chaetomium sp.*, *Phanerocheate chrysosporium*, *Alternaria alternata*, *Rhodotorula mucilaginosa*, and *Neurospora intermedia* are the common examples utilized in biotransformation of wastes and are discussed in Table 12.1.

12.3 MICROBIOLOGICAL WASTE MANAGEMENT FOR FISH FEED DEVELOPMENT

A lot of study has been carried out for the production of fish feed from the food waste and wastages of the fish industry (Table 12.2). Meantime, using microbes like single cell protein, microscopic zooplankton, yeast, bacteria, algae, etc. are being tested for their efficiency to produce fish feed from waste material (Spalvins and Blumberga, 2018).

12.4 MICROBIOLOGICAL TREATMENT OF AGRO-INDUSTRIAL WASTE FOR FEED/FOOD

For the most part, industries from two major sectors, agriculture and food, generate enormous amounts of residues every year, either as solid waste or effluents (Sadh *et al.*, 2018). Agro-industrial wastes are generally known as the wastes generated from the agricultural field and industrial residues such as apple pomace, coconut husk, lime peel, molasses, pistachio shell, potato waste, orange peel, rice bran, wheat bran, etc. (Sadh *et al.*, 2018; Kee *et al.*, 2020). If these residues are directly released into the environment, they can cause a variety of health and environmental issues, making them a serious global concern (RodríguezCouto 2008; Okonko *et al.*, 2009; Sadh *et al.*, 2018). For instance, around 147.2 million metric tonnes of fiber sources are found around the world, while wheat straw residues and rice straws were estimated to be 709.2 and 673.3 million metric tonnes in the 1990s, respectively (Belewu and Babalola 2009; Sadh *et al.*, 2018). These agro-industrial residues, on the

TABLE 12.1
Various Valuable Products Produced through Microbiological Biotransformation of Waste

Value added-products	Microorganism	Substrate	References
Animal feed	*Candida utilis*	Agriculture residue- wheat bran	Yunus *et al.*, 2015
	Rhizopus oligosporus		
	Enterococcus gallinarum, Streptococcus sp.,	Sea bass (*Dicentrarchus labrax*) processing	Ozyurt *et al.*, 2017
	Pediococcus acidilactici Lactobacillus plantarum	wastes	
	Lactobacillus brevis		
	Aspergillus oryzaeTrichoderma reesei	Wine industry residue-gape marc	Zepf and Jin, 2013
	Lactobacillus bulgaricus	Whey	Reddy *et al.*, 1976
	Lactic acid bacteria	Food waste mixture (cooked rice, meat,	Yang *et al.*, 2006
	Lactobacillus salivarius	vegetables, bakery by-product, barley bran,	
		wheat bran and broiler poultry litter)	
Single cell protein	*Lactobacillus acidophilus*	Dairy waste and kitchen based wastages	Khan *et al.*, 2010; Gaur *et al.*, 2017; Mensahand Twumasi, 2017
	Saccharomyces cerevisiae Candida tropicalis		
	Saccharomyces sp.		
Bioethanol	*Saccharomyces cerevisiae Aspergillus niger*	Pineapple waste, potato/ mash waste, banana waste	Hossain and Fazliny, 2010; Dhabekar and Chandak, 2010
Indole-3-acetic acid	*Bacillus subtilis*	Cassava fibrous residue	Swain and Ray, 2008
Protease production	*Lactobacillus delbrueckii sp.*	Wheat bran, rice bran, soybean meal, brewery	Jarun *et al.*, 2008; Maghsoodi *et al.*, 2013; Mathias *et al.*, 2017; de Castro *et al.*, 2015
	Bacillus licheniformis	waste	
	Aspergillus niger		
Astaxanthin (pigment)	*Yamadazyma guilliermondii Yarrowia lipolytica*	Wheat waste, olive pomace, bakery waste	Dursun and Dalgic, 2016; Haque *et al.*, 2016
	Xantophylomyces dendrorhous Sporidiobolus salmonicolor		
	Monascus purpureus		
Bioherbicide	*Phoma sp.*	Soybean bran, bagasse and corn steep liquor	Klaic *et al.*, 2017
Biosorbents	*Aspergillus niger*	Apple pomace	Dhillon *et al.*, 2017
Bleomycin	*Streptomyces mobaraensis*	Date syrup	Radwan *et al.*, 2010
Ergosterol	*Cryptococcus albidus sp.*	Dairy waste (whey)	Németh and Kaleta, 2015

Product	Microorganism	Waste substrate	Reference
Laccase	*Rheinheimera sp.* *Lysinibacillus sp.* *Trametes versicolor*	Peels of citrus fruits, soybean meal, tofu dreg, Brewer's spent grain	Sharma *et al.*, 2017 Dhillon *et al.*, 2012
Lactic acid	*Lactobacillus sp.* *R. oryzae* *Aspergillus awamori* *Aspergillus oryzae* *Lactobacillus rhamnosus*	Mixed restaurant food waste and bakery waste brewery spent grain and brewery yeast	Hitha *et al.*, 2014; Ranjit and Srividya, 2016; Radosavljević *et al.*, 2019
Meroparamycin	*Streptomyces sp. strain MAR01*	Rice, wheat bran, quaker, bread, and ground corn	El-Naggar *et al.*, 2009
Neomycin	*Streptomyces fradiae NCIM 2418*	Apple pomace, cotton seed meal, soy bean powder and wheat bran	Vastrad and Neelagund, 2011
Oxytetracycline	*Streptomyces Rimosus* *S. vendagensi* *S. speibonae*	Groundnut shell, sweet potato residues, cassava peels, cocoyam peels	Okorie and Asagbra, 2005 Okorie and Asagbra, 2008; Yang and Yuan, 1990; Ezejiofor *et al.*, 2012
Poly (3-Hyrdroxybutyric Acid)			
Protein	*Bacillus subtilis*	Orange peel	Sukan *et al.*, 2014
	Chaetomium sp. *Aspergillus niger*	Orange peel	Yalemtesfa *et al.*, 2010
Rifamycin	*Amycolatopsis Mediterranean*	Coconut oil cake, groundnut oil cake, ground nut shell and rice husk	Vastrad and Neelagund, 2012
Xanthan	*Xanthomonas citri*	Potato peel	Vidhyalakshmi *et al.*, 2012
Bioactive phenolic compounds	*Aspergillus fumigates* *A. terreus* *A. wentii* *Penicillium citrinum* *P. granulatum* *P. expansum*	Wheat straw, rice straw, corn cob, pea pod, sugarcane baggase	Chandra and Arora, 2016
Fibrinolytic enzyme	*Bacillus halodurans*	Banana peel, black gram husk, paddy straw, rice bran, and wheat bran	Vijayaraghvan *et al.*, 2016
Pectin lyase	*Aspergillus brasiliensis*	corn steep liquor and orange peel	Pili *et al.*, 2017

(continued)

TABLE 12.1 (Continued)
Various Valuable Products Produced through Microbiological Biotransformation of Waste

Value added-products	Microorganism	Substrate	References
Citric acid	*Aspergillus niger NRRL. 2001 Aspergillus ornatus Alternaria alternata Yarrowia lipolytica*	Apple pomace, brewer's spent grain, citrus waste, sphagnum peat moss; peanut shell,cassava bagasse, coffee husk, moasmi peel, pineapple peel	Ali and Vidhale, 2013; Ali *et al.*, 2016; Vandenberghe *et al.*, 2000; Kumar *et al.*, 2003; Prabha and Rangaiah, 2014 Shojaosadati and Babaeipour, 2002; Dhillon *et al.*, 2011; Sagar *et al.*, 2018
Acetic acid	*Acetobacter aceti Saccharomyces cerevisiae*	Papaya peel, pineapple peel	Raji *et al.*, 2012; Vikas and Mridul, 2014; Sagar *et al.*, 2018
Fumaric acid	*Rhizopus oryzae 1526*	Apple pomace; pulp and paper solid waste	Das *et al.*, 2015 Das *et al.*, 2016
Biosurfactant	*Bacillus subtilis ANR 88*	Potato peels, orange peels, banana peels, and bagasse	Rane *et al.*, 2017
Wine (antioxidant rich)	*Saccharomyces cerevisiae (NCIM 3206)*	Potato, pumpkin and carrot peels	Chakraborty *et al.*, 2017
Cellulase	*Trichoderma viride Bacillus cereus*	Wheat bran, rice bran, corn husks	Vintila *et al.*, 2009; Grover *et al.*, 2012
Lycopene	*Aspergillus niger*	Tomato waste	Parveen *et al.*, 2016
Polygalactouronase	*Aspergillus niger*	Wheat bran, coffee pulp	Maldonado and Strasser de Saad, 1998
Vanillic acid and vanillin	*A. niger Pycnoporus cinnabarinus*	Pineapple canary waste	Lun *et al.*, 2014
Proanthocyanidins, anthocynidins, phenolic acids, vitamin E and oryzanol	-	Rice bran	Huang and Lai, 2016
Ferulic, p-coumaric, sinapic and syringic	*Aspergillus oryzae Rhizopus oryzae*	Rice bran	Razak *et al.*, 2017
Lipase	*Penicillium simplicissimum Pseudomonas aeruginosa Burk holderiacenocepacia*	Castor bean waste; Jatropha curcas seed cake; Sugarcane bagasse, sunflower seed and olive oil	Godoy *et al.*, 2011 Joshi *et al.*, 2011 Liu *et al.* 2016 Ferrarezi *et al.*, 2014

Nisin	*Lactococcus lactis*	Date by product	Khiyami *et al.*, 2008
Biopolymer (PHA)	*Cupriavidus necator* *Thermus thermophiles HB8* *Cupriavidus necator H16* *Bacillus firmus* *Burkholderia sp.* *B. megaterium* *Methylobacterium sp.* *H. pseudoflava*	Pressed juice from oil palm, wheat bran and rape seed meal, whey soy bean and rapeseed oil, corn oil, spent coffee, grounds oil, rice straw, molasses sugarcane bagasse, cheese whey	Zahari *et al.*, 2012; Kachrimanidou *et al.*, 2016 Pantazaki *et al.*, 2009; Taniguchi *et al.*, 2003; Chaudhry *et al.*, 2011; Obruca *et al.*, 2014 Sindhu *et al.*, 2013; Chaudhry *et al.*, 2011; Lopes *et al.*, 2014 Obruca *et al.*, 2011 Nath *et al.*, 2008 Koller *et al.*, 2007

Source: adapted from Sadh *et al.*, 2018; Sindhu *et al.*, 2019; Ng *et al.*, 2020; Arun *et al.*, 2020.

TABLE 12.2
Production of Fish Feed from Different Waste Precursors

Sl. No.	Waste description	Developed fish feed	References
1.	Fishery waste material like fish processing waste, trash fish waste	Silver pompano fish feed	Tugiyono *et al.*, 2020
2.	Food waste	Feed pellets	Mo *et al.*, 2019a
3.	Food waste	Feed pellets	Wong *et al.*, 2016
4.	Food waste from local hotels and restaurants	Feed pellets	Cheng *et al.*, 2014
5.	Plant extract with food waste	Feed pellets	Mo *et al.*, 2016
6.	Food waste with trash fish	Feed pellets	Man *et al.*, 2020
7.	Fermented food waste	Feed mixture	Mo *et al.*, 2019b

other hand, contain a wide variety of carbohydrates, minerals, and proteins, making them extremely high in nutrients and bioactive compounds (Graminha *et al.*, 2008; Sadh *et al.*, 2018). Thus, they can be utilized as "raw material" rather than "wastes" for various biotechnological applications (Sadh *et al.*, 2018; Kee *et al.*, 2020). Indeed, such nutrient availability in these residues could provide appropriate environments for the microorganism's growth and will be a promising carbon source for the microorganism to produce valuable by-products through fermentation (Sadh *et al.*, 2018; Kee *et al.*, 2020). Furthermore, reusing such waste will establish a cost-effective waste management system while also ensuring its long-term profitability (Mihai and Ingrao, 2016; Sadh *et al.*, 2018). In fact, numerous reports have also emphasized the nutritional and nutraceutical properties found in agro-industrial biowastes, which can be fortified by fermentation to develop efficient animal-feed supplements and value-added compounds (Ajila *et al.*, 2012; Kee *et al.*, 2020). Remarkably, animal feed production from agro-industrial residues is one of the most cost-effective and long-term solutions for meeting the ongoing and massive demand for animal feed (Ajila *et al.*, 2012). In this context, this section reveals common agro-industrial bio-wastes that are being re-utilized as substrates for various microbiological agents to produce valuable products.

12.4.1 POTATO WASTE

The potato industry generates a large amount of waste of about 12 million tonnes annually in the form of potato pulp, peel, waste, and distillery waste (Boushy *et al.*, 1994; Kot *et al.*, 2020). These generated wastes are detrimental to the environment and, as a result, their disposal has become an issue of concern in the potato-starch industry. Currently, for managing these wastes, microbial agents are becoming an alternative solution for re-utilizing these wastes for the production of various value-added products. As a result, after microbial fermentation, potato wastes are repurposed as animal and poultry feed, reducing waste discharge to the environment (Wang *et al.*, 2010; Ajila *et al.*, 2012). For instance, Wang and his colleagues evaluated the nutritional quality of feed developed by fermentation of potato wastes and studied its influence on poultry farming, consequently confirming its safety as feed for poultry (Wang *et al.*, 2010). Furthermore, recently, Patelski and his colleagues (2020) also evaluated potato pulp waste for cultivating yeast, *Pichia stipites*, and *Candida guilliermondii*, and consequently suggested it as a promising raw material for obtaining yeast single cell protein (Patelski *et al.*, 2020). They also stated that the potato pulp's local availability and low prices as well as its easy conversion process into single cell protein could be a promising alternative for meeting the ever-rising demand for animal feed. Furthermore, various reports have highlighted the use of potato wastes in the production of organic acids, enzymes, and bioethanol by the action of microbes such as *Rhizopus oryzae*, *Pleurotus ostreatus*, and *Sacchromyces cerevisiae* (Oda *et al.*, 2002; Mabrouk and El Ahwany, 2008; Izmirlioglu and Demirci, 2012; Mladenovi *et al.*, 2016).

12.4.2 Sugarcane Bagasse

Globally, about 510 million tonnes of sugarcane bagasse are produced, while around 94 and 93 million tonnes are produced from two countries, Brazil and India, respectively (Birru, 2016; Millati *et al.*, 2019). Sugarcane bagasse has been used as a fuel in the paper and pulp industries, in the manufacturing of structural materials, and in agriculture. This sugarcane bagasse comprises 46.42% of cellulose, 23.97%of hemicellulose, and 28.09% of lignin and is also rich in carbon, which makes it an ideal substrate for microbes to produce various value-added products and feeds (Loh *et al.*, 2013; Candido *et al.*, 2017; Millati *et al.*, 2019). On the basis of such a composition, bagasse can be further exploited for animal feed production by employing certain microbes (Millati *et al.*, 2019). For instance, various reports have determined the animal-feed production from bagasse by fermentation using basidiomycetes, *P. sajorcaju* and*Pleurotussp.* (Nigam *et al.*, 1987; Nigam, 1990; Puniya *et al.*, 1996; Ajila *et al.*, 2012).

12.4.3 Citrus Waste

Citrus fruits, such as grapefruits, lemons, limes, oranges, mandarins, and tangerines, are among the world's most commonly grown fruits (Sharma *et al.*, 2017). One-third of these citrus fruits are processed, resulting in 50 to 60% of organic waste (Satari and Karimi, 2018). Every year, the citrus processing industry generates massive quantities of waste, with citrus peel waste accounting for nearly half of the wet fruit mass (Sharma *et al.*, 2017). These wastes are of immense economic worth because they contain various bioactive compounds, viz., ascorbic acid, carotenoids, dietary fiber, essential oils, flavonoids, sugars, organic acids, and polyphenols etc. having nutraceutical and therapeutic properties such as anti-oxidative, anti-cancerous, and anti-inflammatory properties (Sharma *et al.*, 2017; Satari and Karimi, 2018). These wastes often contain high levels of sugars that can be bio-transformed into bioethanol by fermentation.

For instance, citrus wastes such as orange peel can also be fermented to produce singlecell protein, which can be used as a component in ruminant animal feed (Ajila *et al.*, 2012). Recently, orange peel is also being utilized for enzymes (pectinases, xylanases etc.) production through solid state fermentation by fungisuch as *Aspergillus*, *Rhizopus*, *Fusarium oxysporum*, *Neurospora crassa*, and *Penicillium decumbens* (El-Bakry *et al.*, 2015; Mamma *et al.*, 2008; Satari and Karimi, 2018).

12.4.4 Pomace of Grapes

Grape pomace is a residue from wineries, and most wineries in developing countries dispose of this waste in landfills, resulting in environmental issues (Kumanda *et al.*, 2019). However, these wastes are found to be rich in bioactive compounds, having potential applications in food, pharmaceutical, and cosmetic factories (Pintać *et al.*, 2019). Therefore, reutilization of grape pomace as a feed would provide an environmentally friendly, lower-cost alternative that could help to improve food and nutrition security (Kumanda *et al.*, 2019). These wastes also contain phenolic compounds, which have anti-microbial and anti-oxidant properties (Kumanda *et al.*, 2019). For instance, Kumanda and colleagues (2019) utilized this grape pomace waste as a feed ingredient in broiler chickens and found that grape pomace has the potential as feed and will boost feed utilization efficiency while providing health benefits to consumers of poultry (Kumanda *et al.*, 2019).

12.4.5 Rice Industry Waste

Rice provides about 21% of global human food energy (per capita) and 15 per cent of protein (per capita) (Kee *et al.*, 2020). According to statistics, China produced 148.5 million metric tonnes (MMT) in 2018/2019, followed by India and Indonesia (116.42 MMT and 36.7 MMT, respectively) (Kee *et al.*, 2020). Along with 660 million tonnes of rice production, 650 to 975 million dry tonnes of straw, 800 million tonnes of husk, and some quantity of bran (10% of rice) are produced (Domnguez-Escribá and Porcar, 2010; Santos *et al.*, 2017; Peanparkdee and Iwamoto, 2019). These residues are rich sources of cellulose, hemicelluloses, lignin, and silica, which can be exploited as raw material for the production of high value-added products such as organic acids, biofuels, bioenergy, and biomaterials, etc. (Santos *et al.*, 2017). These residues are also considered high in bio-active compounds, such as essential amino acids, flavonoids, phenolics, γ-oryzanol, and vitamin E (Peanparkdee and Iwamoto, 2019). Therefore, researchers are currently focusing on the develop-ment of value-added compounds from these wastes as well as their application as feed or food (Peanparkdee and Iwamoto, 2019).

For instance, Bisaria and labmates used the fungus *Pleurotus sajor-caju* for the bioconversion of rice straw and wheat straw and consequently found an increase in the protein level of treated waste (Bisaria *et al.*, 1997; Ajila *et al.*, 2012). In addition, various reports have also highlighted the use of some fungi for bioethanol production from these residues, such as *Aspergillus*, *Rhizopus*, *Monilia*, *Neurospora*, *Fusarium*, *Trichoderma*, and *Mucor* (Santos *et al.*, 2017). Also, Karimi and his labmates (2014) identified certain bioactive compounds in rice straw varieties and found that rice straw contained phenolic compounds and flavonoids (Karimi *et al.*, 2014; Peanparkdee and Iwamoto, 2019). Furthermore, rice husk was also evaluated for its nutritive value and found to be composed of various bioactive compounds (Peanparkdee and Iwamoto, 2019). Furthermore, rice commonly utilized as animal feed is also found to be a rich source of phenolic acids, flavonoids, vitamin E, and γ-oryzanol (Peanparkdee and Iwamoto, 2019). Other than nutritive value, these residues are also known for their anti-inflammatory and anti-cancerous properties (Peanparkdee and Iwamoto, 2019). Recently, Matrawy and his labmates utilized a rice husk-based medium for the pro-duction of xylanase enzymes through perfentation by an isolate of *Thermomyces lanuginosus* strain A3-1 DSM 105773 (Matrawy *et al.*, 2021). They also stated that inoculation of fungi, Thermomyces lanuginosus, helped in recycling of the accumulated rice husk waste and also caused xylanase pro-duction through a cost-effective process (Matrawy *et al.*, 2021).

12.4.6 OLIVE OIL WASTE

Olive oil production is expected to be 2.9 million tonnes per year, with 15 million tonnes of olive mill waste produced per year (Nasopoulou and Zabetakis, 2013). Agro-industrial waste from olive production involves olive pomace, olive skin, pulp, and olive mill waste water, which possess det-rimental environmental issues, and therefore their bio-management is necessary (Nasopoulou and Zabetakis, 2013; Berbel and Posadillo, 2018). Olive pomace is the most frequently produced residue from the olive industry, where for every 100 kg of olives, 40–80 kg of olive pomace is produced. As these olive waste residues are rich sources of nutrients, bioactive and phenolic compounds, they can be exploited further for the production of value-added compounds or can also be re-utilized as feed and food (Nasopoulou and Zabetakis, 2013; Berbel and Posadillo, 2018). Therefore, a number of bacteria, fungi, yeasts, and mushrooms are being utilized for the processing of these wastes into enzymes, organic acids, biopolymers, biosurfactants, biofuels, nutraceuticals, and pharmaceuticals (Darvishi, 2012).

For instance, recently, Medouni-Haroune and his colleagues (2018) utilized olive pomace as a substrate for *Streptomyces sp.* S1M3I for submerged fermentation to upgrade the nutritional value for its application as livestock feed (Medouni-Haroune *et al.*, 2018). In addition, another olive industry waste, olive oil mill wastewater was also used as a substrate for a fungal consortium (i.e., *Pleurotus sp.* namely *P. floridae*, *P. eryngii*, *P. ostreatus*, *P. sajor-caju* (basidiomycete fungi); *Saccharomyces cerevisiae* (yeast), Kluyveromyces lactis (yeast); *Oidodendronsp.* (filamentous fungi), and *Penicillumsp.* (filamentous fungi)) to obtain a potentially valuable microbial biomass by the waste bioremediation (Laconi *et al.*, 2006). Laconi and his labmates (2006) also suggested the possible utilization of this microbiologically treated waste as an additive to animal feed (Laconi *et al.*, 2006).

12.4.7 WHEY

Whey is an important by-product of the food and dairy industry (Obruca *et al.*, 2014). Cheese whey constitutes around 80 to 90% of the total milk used. Whey, in fact, retains 55% of total milk nutrients as well as large amounts of liquid waste with a high organic content (Obruca *et al.*, 2014). About 50% of total world-generated cheesewhey is treated and converted into numerous food products, while the remaining 50% is discarded as a waste, which can pose environmental problems (Marwaha and Kennedy, 1988; Obruca *et al.*, 2014). Therefore, to overcome such disposal problems, it is more reasonable to utilize whey as a substrate for the production of high value-added products rather than as a waste (Obruca *et al.*, 2014). Indeed, whey is considered to be rich in fermentable nutrients and can be exploited as a substrate for the production of various valuable products (Obruca *et al.*, 2014). Moreover, many compounds viz. antibiotics, biopolymers, carotenoids, vitamins, organic acids, polysaccharides, etc. are also found to be produced from this biowaste by employing various microbiological agents as mentioned in Table 12.1.

12.4.8 MOLASSES

Molasses is a by-product of sugar-rich crops and grain industries, such as soya bean molasses, sugarcane molasses, beet molasses, and so on (Preston *et al.*, 1986). Molasses is a rich carbon-source for the production of various value-added products as it mainly comprises carbohydrates, proteins, lipids, minerals, and ash (de Oliveira *et al.*, 2020). These wastes are already being exposed for the production of malic acid, lipases, butanol, and bioethanol (de Oliveira *et al.*, 2020) and can be further exploited for supplementation of animal feed due to their nutritive value. Recently, Abdel-Rahman and his colleagues (2021) also utilized beet molasses for lactic acid production through fermentation with *Enterococcus hirae* and consequently found improved productivity as well as yield of lactic acid (Abdel-Rahman *et al.*, 2021).

12.5 FOOD PROCESSING WASTE TO EXTRACT FOOD-GRADE BIOACTIVE COMPOUNDS

Food waste is characterized as both precooked agricultural waste and kitchen leftovers that are discarded by a variety of industries, including food processing, viz., fruit and vegetable processing industries, meat and poultry industries, dairy industries, marine industries, grain processing industries, and households as waste. Such industries involve the use of huge quantities of fruits, vegetables, dairy, meat, poultry, marine, brewing, and grains, subsequently producing a huge amount of waste after processing. In fact, this waste generation varies with the source and each source constitutes 21.3% of the waste fraction from the ice cream and dairy industry, 26% from the drink industry, 14.8% from the vegetable and fruit industry, 12.9% from grains and starchy products, 8% from meat processing, 3.9% from vegetable and animal oils and fats industries, and 0.4% from the fish and fish products industries (Lee *et al.*, 2019). According to the food and agriculture organization, food processing industries are known to produce about 1.3 billion tonnes of waste (FAO, 2012; Lee *et al.*, 2019). Moreover, globally, food waste has become a major issue of concern as the wastage is estimated to be around 1.6 billion tons, constituting a loss of around 1 trillion US dollars (Dahiya *et al.*, 2018; Lee *et al.*, 2019). This loss is estimated to be worth approximately 900 billion USD in social and environmental terms. It has been further estimated that urban food waste will increase from 278 million tonnes in 2005 to 416 million tonnes in 2025 (Melikoglu *et al.*, 2013; Lee *et al.*, 2019). Therefore, management of such hugely generated waste is definitely required to avoid environmental issues.

Food bio-waste is often disposed of as municipal waste or transferred to landfills, incineration plants, or composting facilities, where only a portion of the garbage is reprocessed (Lee *et al.*, 2019; Ng *et al.*, 2020). However, due to a lack of disposal sites, high transportation costs, significant economic losses, environmental concerns, and burgeoning food shortage conditions, the focus has shifted to a more efficient use of these bio-wastes for human or animal consumption rather than dumping them in an environmentally friendly and sustainable manner (Lee *et al.,* 2019; Ng *et al.,* 2020). The bulk of food waste is still disposed of in landfills, resulting in severe environmental contamination as well as direct or indirect greenhouse gas emissions, all of which contribute to significant air pollution (Karthikeyan *et al.,* 2017). The re-use of bio-wastes is recognized as the most cost-effective strategy to combating waste-related environmental degradation and losses in the long run, in order to address such environmental difficulties caused by the accumulation of food waste. However, due to the diverse and unknown compounds included in bio-waste, this application is limited, complicating the overall waste management system. As a result, various studies have been conducted in recent years to find innovative and cost-effective bioprocesses for converting food waste into valuable bio-products. In this context, microorganisms are being used to manage food waste by converting it into a valuable bi-product that can be consumed as food or used as a bioactive molecule (Lee *et al.,* 2019). Food wastes, in fact, are frequently regarded as a superior crude feedstock for microbial fermentation to produce a variety of bioactive compounds (Ng *et al.,* 2020). By reducing the requirement for food waste treatment, biotransformation of food waste into value-added bio-products with higher functionality can reduce production costs and pollution (Dursun and Dalgic, 2016; Lee *et al.,* 2019, Ng *et al.,* 2020). Alternatively, utilizing food waste as a crude feedstock, microbial fermentation can produce a variety of bio-products such as proteins, enzymes, anti-oxidants, and pigments (Sadh *et al.,* 2018; Ng *et al.,* 2020). Fermentations used in the biotransformation process not only create useful products at a low cost, but they also have significant nutritional value. Sindhu *et al.* (2019), for example, unquestionably specified food waste as a source of numerous value-added goods by focusing on bioactive compound production via chemical, enzymatic, and microbiological processes using food waste as a raw material (Sindhu *et al.,* 2019).

Although organic and inorganic components, such as carbohydrates, proteins, fats, lipids, minerals, phenolic compounds, and others, make up the majority of food and kitchen waste, their composition might vary depending on the source of trash (Sindhu *et al.,* 2019; Lee *et al.,* 2019). Furthermore, food waste from various sources has varying chemical compositions, but it is a great sustainable raw material for the development of bio-products of interest. Furthermore, big food processing enterprises produce trash, which leads to significant losses and requires management. The following subheadings go through each of these.

12.6   FRUIT AND VEGETABLE PROCESSING INDUSTRIES

Food industries process the substrate for juices, jams, and jellies, etc., and usually produce effluents as well as solid waste in huge amounts constituting discarded fruits, vegetables, peels, seeds, etc., whereas the effluents contain liquid waste from juice and wash water (Sadh *et al.,* 2018). Globally, the fruit and vegetable processing industries' productivity varies in different regions of the world, i.e., 85 million tonnes in North America, Oceania, and Sub-Saharan Africa; 145 million tonnes in North Africa, West and Central Asia; 175 million tonnes in Latin America; 200 million tonnes in Europe; 310 million tonnes in South and Southeast Asia, and 640 million tonnes in industrialized Asia (Lee *et al.,* 2019). Out of all the countries, China, the USA, the Philippines, and India are considered the major contributors to fruit and vegetable processing industries' generated waste, with about 32, 15, 6.5, and 1.8 million tonnes of waste, respectively (Wadhwa and Bakshi, 2013; Lee *et al.,* 2019). Remarkably, food wastage per capita in Sub-Saharan Africa and South and South East Asia varies between 120-170 kg/year, where the losses are almost two times those in Europe and North America (Gustavsson *et al.,* 2011; Lee *et al.,* 2019). In addition, the Koyambedu market in Chennai, India, Asia's largest vegetable, fruit, and flower market, produces approximately 80 tonnes of solid waste per day (Kameswari *et al.,* 2007). Moreover, as the focus is getting towards re-utilizing the wastes for consumption or bioactive compounds' production, various countries that facea shortage of feed for livestock, for example, a deficiency of 25 million tonnes of concentrates, 159 million tonnes of green forages, and 117 million tonnes of crop residues in India, a shortage of 10 million tonnes of protein feed, 30 million tonnes of energy feed, and 20 million tonnes of aquatic feed in China, respectively, are intending on utilizing wastes as feed (Chen, 2012; Gorti *et al.*, 2012; Sadh *et al.*, 2018).

For instance, potato peel, fruit and vegetable wastes can be utilized as substrates for organic acid production through microbial fermentation (Jawad *et al.*, 2013; Ng *et al.*, 2020). Another example is the carbohydrate-rich waste generated after food processing that could prove to be a good feedstock for bioethanol production through fermentation by the yeast, *Saccharomyces cerevisae* (Parmar and Rupasinghe, 2013; Ng *et al.*, 2020). In addition, these wastes are also being re-utilized to produce various bioactive compounds such as bioethanol, singlecell protein, biopolymers, organic acids, enzymes, nutraceuticals, pharmaceuticals, and pigments, such as mentioned in Table 12.1, where microbes are facilitating these bio-waste transformations.

12.6.1 MEAT AND POULTRY INDUSTRIES

Meat and meat products are essential components of the human diet because they include bioactive molecules and nutrients (Pogorzelska-Nowicka *et al.*, 2018; Bycrs *et al.*, 2002; Alao *et al.*, 2017; Sadh *et al.*, 2018; Ng *et al.*, 2020). As a result, there has been a huge growth in the demand for animal-origin foods. The impact of diverse livestock businesses on a country's GDP is growing significantly, accounting for more than 40% of the overall agricultural sector and more than 12% of GDP (Jayathilakan *et al.*, 2012). Because by-products are underused, their output has expanded in lockstep with the growth of the meat and poultry sectors. However, the processing of animals for food results in the development of a variety of wastes, including carcasses, viscera, fat, meat trimmings, hoofs, hides, feathers, heads, bones, blood, urine, and meat, skins, viscera, blood, oil, by-catch, off-spec, cuts, bones, and debris (Lee *et al.*, 2019).

Every year, 1.8 million tonnes of feathers are produced in the chicken industry, which are high in structural proteins, particularly keratin. Furthermore, the chicken sector creates a considerable volume of shells, resulting in 1.5 million tonnes of trash every year (Jayathilakan *et al.*, 2012; Dong *et al.*, 2014; Lee *et al.*, 2019). These egg shells contain eggshell membrane (up to 11% of the egg weight) and are a valuable source of proteins, polypeptides, and polysaccharides that can be used in food and nutraceuticals (Jain and Anal, 2016; Lee *et al.*, 2019). Furthermore, the meat and poultry processing industries have 16 to 45 million tonnes of bones and vast quantities of skin, which contain vital minerals like calcium, lipids, and proteins and can be used as protein sources. Furthermore, meat industry waste contains high levels of nitrogen, phosphate, and grease. Because it includes a high quantity of protein and iron, animal blood, a type of meat by-product, is a significant culinary by-product (Wan *et al.*, 2002). Bah and his colleagues explain how animal blood gathered from various slaughterhouses is becoming a new source of bioactive chemicals (Bah *et al.*, 2013). As a result, increased or improved utilization of meat and poultry sector by-products could create capital with minimal expenditure while also addressing environmental challenges caused by waste decomposition in the meat business.

12.6.2 DAIRY INDUSTRIES

The dairy industries constitute raw milk processing into various products like butter, cheese, milk, condensed milk, ice cream, yoghurt and milk powder through chilling, pasteurization, and homogenization processes. Milk and milk-related products are one of the basic necessities, and therefore, dairy industries are growing continually (Raghunath *et al.*, 2016). Likewise, other bio-wastes, dairy processing related wastes such as water effluents, milk sludge, whey and curd, also cause serious environmental problems (Brião and Tavares, 2007; Sadh *et al.*, 2018). These dairy-related wastes, on the other hand, are high in nitrogen, minerals, dissolved sugars, proteins, and fats and may be repurposed as a substrate for the production of various bioactive compounds and biopolymers (Watkins and Nash, 2010; Sadh *et al.*, 2018).

12.6.3 MARINE INDUSTRIES

Seafood such as fish and shellfish related industries constitute the marine industries, which produce processed products for consumption by humans as food (Qin *et al.*, 2005). India is considered the second biggest producer of fish globally. These marine industries produce an enormous amount of bio-waste, where only India generates more than 2 million metric tonnes of waste residue (Nurdiyana and Mazlina, 2009). These wastes include solid wastes like fish carcasses, viscera, skin, heads and liquid effluents generated from water discharges, blood water from drained fish tanks, brine etc. (Michail *et al.*, 2006; Sadh *et al.*, 2018). These bio-wastes are rich in nutraceuticals and bioactive compounds (Helkar *et al.*, 2016).

12.6.4 GRAIN PROCESSING INDUSTRIES

Cleaning, grading, drying, seed processing, conveying and out loading, storage, vegetable oil processing, aspiration, and filtering are all processes used in these sectors to create all sorts of grain, including seed, granules, vegetable oil, and other bulk commodities. India's grain processing industry is said to handle more than 200 million tonnes of diverse grains each year. However, during the oil extraction process, these companies produce a large amount of oil cakes as waste, as well as some wastewater and air outflows after each phase of grain processing (Sadh *et al.*, 2018). These waste leftovers from cereal, pulse, and other grain processing are known to be a promising food source, containing bioactive substances such as phytochemicals, and could be used for their positive nutraceutical qualities.

12.6.5 BREWERY INDUSTRIES

Brewery industries are one of the leading water consumers, where large amounts of water are utilized for the production of the brew, and to wash, clean, and sanitize different units after each finished cluster (Olajire, 2012). This brew processing leads to the generation of organic wastes such as brewer's spent grains, residual brewing yeast, and trub called wet brewery wastes (Mathias *et al.*, 2014). And the disposal and management of such large volumes of waste pose significant challenges. Brewer's spent grain and brewer's yeast constitute approximately 20% and 1.5 to 3% of the total volume of brewery produced (Sosa-Hernández *et al.* 2016). These solid wastes are elucidated as a rich source of fiber, protein, vitamins B, lipids, fatty acids, carbohydrates, minerals (K, Mg, Ca, P, Fe), and enzymes, which can be utilized as feed for the animal feed industry (Cooray and Chen 2018; Dos Santos Mathias *et al.*, 2014; Merino *et al.* 2016; Radosavljević *et al.*, 2019). Thus, these organic wastes can be directly utilized as animal feed for all livestock, such as sheep, horses, rabbits, and carps, and as a feed supplement for poultry, and can also be used for the improvement of soil (Thomas and Rahman, 2006). Moreover, malt can also be utilized for the production of bio-ethanol and various bioactive compounds (Gencheva et al., 2012). For instance, Radosavljević and his colleagues elucidated that brewer's spent grain, malt rootlets, brewer's yeast, and soy lecithin can be utilized as raw materials and bio-transformed into lactic acid by *Lactobacillus rhamnosus* stimulated fermentation (Radosavljević *et al.*, 2019).

12.7 MUNICIPAL SOLID WASTE TO ANIMAL FEEDSTUFFS

The world's population is increasing more rapidly in urban areas than in rural areas. Relatively, a population burst is occurring, and in the meantime, the municipal waste (MW) load in the environment is also increasing. It is estimated that by 2050, MW will be 3.40 BT (9.32 MT/day) (Kaza *et al.*, 2018). Therefore, there is an urgent need to convert MW with microbial treatment into a beneficiary platform. Some of the research articles found that MW has the capability for the development of animal feedstuff, but some research found that MW should be converted to animal feed-stuff after proper purification. A lot of study should have to be carried out in this aspect so that our environment may remain in a less polluted condition.

12.8 CONCLUSION

All the findings in this book chapter lead us to conclude that fruit and vegetable processing industries, meat and poultry industries, dairy industries, and marine industries, grain processing industries, brewery industries, and municipal solid waste refineries have the great potency to develop animal feedstuffs with microbial treatment. However, most of these technologies belong at the lab level. There is an urgent need to excel this technology at the social level so that food waste-related pollution may remain under control.

REFERENCES

Abd Razak, D. L., Abd Rashid, N. Y., Jamaluddin, A., Sharifudin, S. A., Abd Kahar, A., & Long, K. (2017). Cosmeceutical potentials and bioactive compounds of rice bran fermented with single and mix culture of *Aspergillus oryzae* and *Rhizopus oryzae*. *Journal of the Saudi Society of Agricultural Sciences, 16*(2), 127–134. https://doi.org/10.1016/j.jssas.2015.04.001

Abdel-Rahman, M. A., Hassan, S. E. D., Alrefaey, H. M., El-Belely, E. F., Elsakhawy, T., Fouda, A.,& Khattab, S. M. (2021). Subsequent improvement of lactic acid production from beet molasses by *Enterococcus hirae* ds10 using different fermentation strategies. *Bioresource Technology Reports, 13*, 100617.https://doi.org/10.1016/j.biteb.2020.100617

Adebayo, F. O., & Obiekezie, S. O. (2018). Microorganisms in waste management. *Research Journal of Science and Technology, 10*(1), 28–39. https://doi.org/10.5958/2349-2988.2018.00005.0

Ajila, C. M., Brar, S. K., Verma, M., Tyagi, R. D., Godbout, S., & Valéro, J. R. (2012). Bio-processing of agro-by-products to animal feed. *Critical reviews in biotechnology, 32*(4), 382–400. https://doi.org/10.3109/07388551.2012.659172

Alao, B. O., Falowo, A. B., Chulayo, A., & Muchenje, V. (2017). The potential of animal by-products in food systems: Production, prospects and challenges. *Sustainability, 9*(7), 1089. https://doi.org/10.3390/su9071089

Ali, S. R., Anwar, Z., Irshad, M., Mukhtar, S., & Warraich, N. T. (2016). Bio-synthesis of citric acid from single and co-culture-based fermentation technology using agro-wastes. *Journal of Radiation Research and Applied Sciences, 9*(1), 57–62. https://doi.org/10.1016/j.jrras.2015.09.003

Ali, S. S., & Vidhale, N. N. (2013). Protease production by Fusarium oxysporum in solid-state fermentation using rice bran. *American Journal of Microbiological Research, 1*(3), 45–47. Retrieved from https://citeseerx.ist.psu.edu/viewdoc/download?doi=10.1.1.968.1940&rep=rep1&type=pdf

Arun, K. B., Madhavan, A., Sindhu, R., Binod, P., Pandey, A., Reshmy, R., & Sirohi, R. (2020). Remodeling agro-industrial and food wastes into value-added bioactives and biopolymers. *Industrial Crops and Products, 154*, 112621.https://doi.org/10.1016/j.indcrop.2020.112621

Bah, C. S., Bekhit, A. E. D. A., Carne, A., & McConnell, M. A. (2013). Slaughterhouse blood: an emerging source of bioactive compounds. *Comprehensive Reviews in Food Science and Food Safety, 12*(3), 314–331. https://doi.org/10.1111/1541-4337.12013

Belewu, M. A., & Babalola, F. T. (2009). Nutrient enrichment of waste agricultural residues after solid state fermentation using Rhizopus oligosporus. *J Appl Biosci, 13*, 695–699. Retrieved from https://www.cabdirect.org/cabdirect/abstract/20093214461

Berbel, J., & Posadillo, A. (2018). Review and analysis of alternatives for the valorisation of agro-industrial olive oil by-products. *Sustainability, 10*(1), 237. https://doi.org/10.3390/su10010237

Birru, E. (2016). Sugar cane industry overview and energy efficiency considerations. KTH School of Industrial Engineering and Management, 1,1e6. Retrieved from https://www.diva-portal.org/smash/get/diva2:905929/FULLTEXT02.pdf

Bisaria, R., Madan, M., & Vasudevan, P. (1997). Utilisation of agro-residues as animal feed through bioconversion. *Bioresource Technology*, *59*(1), 5–8. https://doi.org/10.1016/S0960-8524(96)00140-X

Boushy El-, A.R.Y. and A.F.B. Van der Poel. (1994). Poultry Feed from Waste Processing and Use. Chapman andhall (Ed).Charmley, E., Nelson, D. and Zvomuya, F. 2006. Nutrientcycling in vegetable processing industry: Utilization of potato peel by-products. *Canadian Journal of Soil Science, 86*, 612–629.

Brião, V. B., & Tavares, C. R. (2007). Effluent generation by the dairy industry: preventive attitudes and opportunities. *Brazilian Journal of Chemical Engineering*, *24*(4), 487–497. https://doi.org/10.1590/S0104-66322007000400003

Byers, T., Nestle, M., McTiernan, A., Doyle, C., Currie-Williams, A., Gansler, T., Thun, M., & American Cancer Society 2001 Nutrition and Physical Activity Guidelines Advisory Committee (2002). American Cancer Society guidelines on nutrition and physical activity for cancer prevention: Reducing the risk of cancer with healthy food choices and physical activity. *CA: ACancer Journal for Clinicians*, *52*(2), 92–119. https://doi.org/10.3322/canjclin.52.2.92

Candido, R. G., Godoy, G. G., & Goncalves, A. R. (2017). Characterization and application of cellulose acetate synthesized from sugarcane bagasse. *Carbohydrate Polymers*, *167*, 280–289. https://doi.org/10.1016/j.carbpol.2017.03.057

Chakraborty, K., Raychaudhuri, U., & Chakraborty, R. (2017). Optimization of bioprocess parameters for wine from household vegetable waste production by employing response surface methodology. *International Food Research Journal*, *24*(1). Retrieved from http://ifrj.upm.edu.my/24%20(01)%202017/(4).pdf

Chandra, P., & Arora, D. S. (2016). Production of antioxidant bioactive phenolic compounds by solid-state fermentation on agro-residues using various fungi isolated from soil. 20,1–8. https://doi.org/10.3923/ajbkr.2016.8.15

Chaudhry, W. N., Jamil, N., Ali, I., Ayaz, M. H., & Hasnain, S. (2011). Screening for polyhydroxyalkanoate (PHA)-producing bacterial strains and comparison of PHA production from various inexpensive carbon sources. *Annals of Microbiology*, *61*(3), 623–629. https://doi.org/10.1007/s13213-010-0181-6

Chen, J. (2012). Aquatic feed industry under tension in world and China's grain supply and demand. *Chin. Fish*, *6*, 32–34.

Cheng, Z., Mo, W. Y., Man, Y. B., Nie, X. P., Li, K. B., & Wong, M. H. (2014). Replacing fish meal by food waste in feed pellets to culture lower trophic level fish containing acceptable levels of organochlorine pesticides: health risk assessments. *Environment International*, *73*, 22–27. https://doi.org/10.1016/j.envint.2014.07.001

Cooray, S. T., & Chen, W. N. (2018). Valorization of brewer's spent grain using fungi solid-state fermentation to enhance nutritional value. *Journal of Functional Foods*, *42*, 85–94. https://doi.org/10.1016/j.jff.2017.12.027

Dahiya, S., Kumar, A. N., Sravan, J. S., Chatterjee, S., Sarkar, O., & Mohan, S. V. (2018). Food waste biorefinery: Sustainable strategy for circular bioeconomy. *Bioresource technology*, *248*, 2–12. https://doi.org/10.1016/j.biortech.2017.07.176

Darvishi, F. (2012). Microbial biotechnology in olive oil industry. *Olive Oil–Constituents, Quality, Health Properties and Bioconversions*, 309–330. https://www.intechopen.com/chapters/27040

Das, R. K., Brar, S. K., & Verma, M. (2015). A fermentative approach towards optimizing directed biosynthesis of fumaric acid by Rhizopus oryzae 1526 utilizing apple industry waste biomass. *Fungal Biology*, *119*(12), 1279–1290. https://doi.org/10.1016/j.funbio.2015.10.001

Das, R. K., Brar, S. K., & Verma, M. (2016). Potential use of pulp and paper solid waste for the bio-production of fumaric acid through submerged and solid state fermentation. *Journal of Cleaner Production*, *112*, 4435–4444. https://doi.org/10.1016/j.jclepro.2015.08.108

de Castro, R. J. S., Ohara, A., Nishide, T. G., Bagagli, M. P., Dias, F. F. G., & Sato, H. H. (2015). A versatile system based on substrate formulation using agroindustrial wastes for protease production by Aspergillus niger under solid state fermentation. *Biocatalysis and Agricultural Biotechnology*, *4*(4), 678–684. https://doi.org/10.1016/j.bcab.2015.08.010

de Oliveira, J. M., Michelon, M., & Burkert, C. A. V. (2020). Biotechnological potential of soybean molasses for the production of extracellular polymers by diazotrophic bacteria. *Biocatalysis and Agricultural Biotechnology*, *25*, 101609.https://doi.org/10.1016/j.bcab.2020.101609

Dhabekar, A., & Chandak, A. (2010). Utilization of banana peels and beet waste for alcohol production. *Asiatic J. Biotech. Res*, *1*, 8–13. https://doi.org/10.3923/ajbkr.2012.1.14

Dhillon, G. S., Brar, S. K., Verma, M., & Tyagi, R. D. (2011). Enhanced solid-state citric acid bio-production using apple pomace waste through surface response methodology. *Journal of Applied Microbiology, 110*(4), 1045–1055. https://doi.org/10.1111/j.1365-2672.2011.04962.x

Dhillon, G. S., Kaur, S., & Brar, S. K. (2012). In-vitro decolorization of recalcitrant dyes through an eco-friendly approach using laccase from Trametes versicolor grown on brewer's spent grain. *International Biodeterioration & Biodegradation, 72*, 67–75. https://doi.org/10.1016/J.IBIOD.2012.05.012

Dhillon, G. S., Rosine, G. M. L., Kaur, S., Hegde, K., Brar, S. K., Drogui, P., & Verma, M. (2017). Novel biomaterials from citric acid fermentation as biosorbents for removal of metals from waste chromated copper arsenate wood leachates. *International Biodeterioration & Biodegradation, 119*, 147–154.

Domínguez-Escribá, L., & Porcar, M. (2010). Rice straw management: the big waste. *Biofuels, Bioproducts and Biorefining, 4*(2), 154–159. https://doi.org/10.1002/bbb.196

Dong, X. B., Li, X., Zhang, C. H., Wang, J. Z., Tang, C. H., Sun, H. M., ... & Chen, L. L. (2014). Development of a novel method for hot-pressure extraction of protein from chicken bone and the effect of enzymatic hydrolysis on the extracts. *Food Chemistry, 157*, 339–346. https://doi.org/10.1016/j.foodchem.2014.02.043

dos Santos Mathias, T. R., de Mello, P. P. M., & Servulo, E. F. C. (2014). Solid wastes in brewing process: A review. *Journal of Brewing and Distilling, 5*(1), 1–9. https://doi.org/10.5897/JBD2014.0043

Dursun, D., & Dalgıç, A. C. (2016). Optimization of astaxanthin pigment bioprocessing by four different yeast species using wheat wastes. *Biocatalysis and Agricultural Biotechnology, 7*, 1–6. https://doi.org/10.1016/j.bcab.2016.04.006

El-Bakry, M., Abraham, J., Cerda, A., Barrena, R., Ponsá, S., Gea, T., & Sánchez, A. (2015). From wastes to high value added products: novel aspects of SSF in the production of enzymes. *Critical Reviews in Environmental Science and Technology, 45*(18), 1999–2042.

El-Naggar, M. Y., El-Assar, S. A., & Abdul-Gawad, S. M. (2009). Solid-state fermentation for the production of meroparamycin by Streptomyces sp. strain MAR01. *Journal of Microbiology and Biotechnology, 19*(5), 468–473. https://doi.org/10.1080/10643389.2015.1010423

Ezejiofor, T. I. N., Duru, C. I., Asagbra, A. E., Ezejiofor, A. N., Orisakwe, O. E., Afonne, J. O., & Obi, E. (2012). Waste to wealth: Production of oxytetracycline using streptomyces species from household kitchen wastes of agricultural produce. *African Journal of Biotechnology, 11*(43), 10115–10124. https://doi.org/10.4014/jmb.0807.457

FAO. (2012). Towards the Future We Want: End hunger and make the transition to sustainable agricultural and food systems.

Ferrarezi, A. L., Ohe, T. H. K., Borges, J. P., Brito, R. R., Siqueira, M. R., Vendramini, P. H., ...& Gomes, E. (2014). Production and characterization of lipases and immobilization of whole cell of the thermophilic Thermomucor indicae seudaticae N31 for transesterification reaction. *Journal of Molecular Catalysis B: Enzymatic, 107*, 106–113. https://doi.org/10.1016/j.molcatb.2014.05.012

Gaur, S., Mathur, N., Singh, A., & Bhatnagar, P. (2017). Characterization of dairy waste and its utilization as substrate for the production of single cell protein. *J. Biotechnol. Biochem, 3*, 73–78. Retrieved from www.iosrjournals.org/iosr-jbb/papers/Volume%203,%20Issue%204/Version-2/N0304027378.pdf

Gencheva, P., Dimitrov, D., Dobrev, G., & Ivanova, V. (2012). Hydrolysates from malt spent grain with potential application in the bioethanol production. *Journal of BioScience and Biotechnology*, 135-141. https://bit.ly/3xFzbXO

Godoy, M. G., Gutarra, M. L., Castro, A. M., Machado, O. L., & Freire, D. M. (2011). Adding value to a toxic residue from the biodiesel industry: production of two distinct pool of lipases from Penicillium simplicissimum in castor bean waste. *Journal of Industrial Microbiology and Biotechnology, 38*(8), 945–953. https://doi.org/10.1007/s10295-010-0865-8

Gorti, R. K., Suresh, K. P., Sampath, K. T., Giridhar, K., & Anandan, S. (2012). *Modeling and forecasting livestock and fish feed resources: Requirement and availability in India* (Doctoral dissertation, National Institute of Animal Nutrition & Physiology).

Graminha, E. B. N., Gonçalves, A. Z. L., Pirota, R. D. P. B., Balsalobre, M. A. A., Da Silva, R., & Gomes, E. (2008). Enzyme production by solid-state fermentation: Application to animal nutrition. *Animal Feed Science and Technology, 144*(1–2), 1–22. https://doi.org/10.1016/j.anifeedsci.2007.09.029

Grover, S., Kathuria, R. S., & Kaur, M. (2012). Energy values and technologies for non woody biomass: as a clean source of energy. *IOSR Journal of Electrical and Electronics Engineering, 1*(2), 10–14. www.iosrjournals.org/iosr-jeee/Papers/vol1-issue2/I01201014.pdf?id=2686

Gustavsson, J., Cederberg, C., Sonesson, U., Van Otterdijk, R., & Meybeck, A. (2011). Global food losses and food waste. *Food and Agriculture Organization of the United Nations, Rome*. 1-38. Retrieved from https://www.fao.org/fileadmin/user_upload/suistainability/pdf/Global_Food_Losses_and_Food_Waste.pdf

Haque, M. A., Kachrimanidou, V., Koutinas, A., & Lin, C. S. K. (2016). Valorization of bakery waste for biocolorant and enzyme production by Monascus purpureus. *Journal of Biotechnology*, *231*, 55–64. https://doi.org/10.1016/j.jbiotec.2016.05.003

Helkar, P. B., Sahoo, A. K., & Patil, N. J. (2016). Review: Food industry by-products used as a functional food ingredients. *International Journal of Waste Resources*, *6*(3), 1–6. https://doi.org/10.4172/2252-5211.1000248

Hitha, C. S., Hima, C. S., Yogesh, B. J., Bharathi, S., & Sekar, K. V. (2014). Microbial utilization of dairy waste for lactic acid production by immobilized bacterial isolates on sodium alginate beads. *Int. J. Pure App. Biosci*, *2*(4), 55–60. Retrived from www.ijpab.com/form/2014%20Volume%202,%20issue%204/ IJPAB-2014-2-4-55-60.pdf

Hossain, A. B. M. S., & Fazliny, A. R. (2010). Creation of alternative energy by bio-ethanol production from pineapple waste and the usage of its properties for engine. *African Journal of Microbiology Research*, *4*(9), 813–819. https://doi.org/10.5897/AJMR.9000206

Huang, Y. P., & Lai, H. M. (2016). Bioactive compounds and antioxidative activity of colored rice bran. *Journal of Food and Drug Analysis*, *24*(3), 564–574. https://doi.org/10.1016/j.jfda.2016.01.00

Izmirlioglu, G., & Demirci, A. (2012). Ethanol production from waste potato mash by using Saccharomyces cerevisiae. *Applied Sciences*, *2*(4), 738–753. https://doi.org/10.3390/app2040738

Jain, S., & Anal, A. K. (2016). Optimization of extraction of functional protein hydrolysates from chicken egg shell membrane (ESM) by ultrasonic assisted extraction (UAE) and enzymatic hydrolysis. *LWT-Food Science and Technology*, *69*, 295–302. https://doi.org/10.1016/j.lwt.2016.01.057

Jarun, C., Sinsupha, C., Yusuf, C., & Penjit, S. (2008). Protease Production by Aspergillus oryzae in solid-state fermentation using agroindustrial substrate. *Journal of Chemical Technology and Biotechnology*, *83*(7), 1012–1018. https://doi.org/10.1002/jctb.1907

Jawad, A. H., Alkarkhi, A. F., Jason, O. C., Easa, A. M., & Norulaini, N. N. (2013). Production of the lactic acid from mango peel waste–Factorial experiment. *Journal of King Saud University-Science*, *25*(1), 39–45. https://doi.org/10.1016/j.jksus.2012.04.001

Jayathilakan, K., Sultana, K., Radhakrishna, K., & Bawa, A. S. (2012). Utilization of by-products and waste materials from meat, poultry and fish processing industries: a review. *Journal of Food Science and Technology*, *49*(3), 278–293. https://doi.org/10.1007/s13197-011-0290-7

Joshi, C., Mathur, P., & Khare, S. K. (2011). Degradation of phorbol esters by Pseudomonas aeruginosa PseA during solid-state fermentation of deoiled Jatropha curcas seed cake. *Bioresource Technology*, *102*(7), 4815–4819. https://doi.org/10.1016/j.biortech.2011.01.039

Kachrimanidou, V., Kopsahelis, N., Vlysidis, A., Papanikolaou, S., Kookos, I. K., Martínez, B. M.,& Koutinas, A. A. (2016). Downstream separation of poly (hydroxyalkanoates) using crude enzyme consortia produced via solid state fermentation integrated in a biorefinery concept. *Food and Bioproducts Processing*, *100*, 323–334.

Kameswari, K. S. B., Velmurugan, B., Thirumaran, K., & Ramanujam, R. A. (2007,September). Biomethanation of vegetable market waste untapped carbon trading opportunities. In *Proceedings of the International Conference on Sustainable Solid Waste Management* (pp. 415–420).

Karigar, C. S., & Rao, S. S. (2011). Role of microbial enzymes in the bioremediation of pollutants: a review. *Enzyme Research*, *2011*. https://doi.org/10.4061/2011/805187

Karimi, E., Mehrabanjoubani, P., Keshavarzian, M., Oskoueian, E., Jaafar, H. Z., & Abdolzadeh, A. (2014). Identification and quantification of phenolic and flavonoid components in straw and seed husk of some rice varieties (Oryza sativaL.) and their antioxidant properties. *Journal of the Science of Food and Agriculture*, *94*(11), 2324–2330. https://doi.org/10.1002/jsfa.6567

Karthikeyan, O. P., Mehariya, S., & Wong, J. W. C. (2017). Bio-refining of food waste for fuel and value products. *Energy Procedia*, *136*, 14–21. https://doi.org/10.1016/j.egypro.2017.10.253

Kaza, S., Yao, L., Bhada-Tata, P., & Van Woerden, F. (2018). *What a waste 2.0: A global snapshot of solid waste management to 2050*. World Bank Publications.

Kee, S. H., Chiongson, J. B. V., Saludes, J. P., Vigneswari, S., Ramakrishna, S., & Bhubalan, K. (2020). Bioconversion of agro-industry sourced biowaste into biomaterials via microbial factories–a viable domain of circular economy. *Environmental Pollution*, 116311. https://doi.org/10.1016/j.envpol.2020.116311

Khan, M., Khan, S. S., Ahmed, Z., & Tanveer, A. (2010). Production of single cell protein from Saccharomyces cerevisiae by utilizing fruit wastes. *Nanobiotechnica Universale*, *1*(2), 127–132.

Khiyami, M., Aboseide, B., & Pometto, A. (2008). Influence of complex nutrient sources: dates syrup and dates Pits on Lactococcus lactis growth and nisin production. *Journal of Biotechnology*, *136*, S736.

Klaic, R., Sallet, D., Foletto, E. L., Jacques, R. J., Guedes, J. V., Kuhn, R. C., & Mazutti, M. A. (2017). Optimization of solid-state fermentation for bioherbicide production by Phoma sp. *Brazilian Journal of Chemical Engineering*, *34*, 377–384. https://doi.org/10.1590/0104-6632.20170342s20150613

Koller, M., Hesse, P., Bona, R., Kutschera, C., Atlić, A., & Braunegg, G. (2007). Potential of various archae- and eubacterial strains as industrial polyhydroxyalkanoate producers from whey. *Macromolecular Bioscience*, *7*(2), 218–226.

Kot, A. M., Pobiega, K., Piwowarek, K., Kieliszek, M., Błażejak, S., Gniewosz, M., & Lipińska, E. (2020). Biotechnological methods of management and utilization of potato industry waste—a review. *Potato Research*, 1–17.

Kumanda, C., Mlambo, V., & Mnisi, C. M. (2019). From landfills to the dinner table: Red grape pomace waste as a nutraceutical for broiler chickens. *Sustainability*, *11*(7), 1931. https://doi.org/10.3390/su11071931

Kumar, D., Jain, V. K., Shanker, G., & Srivastava, A. (2003). Utilisation of fruits waste for citric acid pro- duction by solid state fermentation. *Process Biochemistry*, *38*(12), 1725–1729. https://doi.org/10.1016/s0032-9592(02)00253-4

Laconi, S., Molle, G., Cabiddu, A., & Pompei, R. (2006). Bioremediation of olive oil mill wastewater and production of microbial biomass. *Biodegradation*, *18*(5), 559–566. https://doi.org/10.1007/s10532-006-9087-1

Lee, J. K., Patel, S. K. S., Sung, B. H., & Kalia, V. C. (2019). Biomolecules from municipal and food industry wastes: An overview. *Bioresource Technology*, *298*, 122346.https://doi.org/10.1016/j.biortech.2019.122346

Liu, Y., Li, C., Meng, X., & Yan, Y. (2013). Biodiesel synthesis directly catalyzed by the fermented solid of Burkholderia cenocepacia via solid state fermentation. *Fuel Processing Technology*, *106*, 303–309.

Loh, Y. R., Sujan, D., Rahman, M. E., & Das, C. A. (2013). Sugarcane bagasse—The future composite material: A literature review. *Resources, Conservation and Recycling*, *75*, 14–22. https://doi.org/10.1016/j.resconrec.2013.03.002

Lopes, M. S. G., Gomez, J. G. C., Taciro, M. K., Mendonça, T. T., & Silva, L. F. (2014). Polyhydroxyalkanoate biosynthesis and simultaneous remotion of organic inhibitors from sugarcane bagasse hydrolysate by Burkholderia sp. *Journal of Industrial Microbiology and Biotechnology*, *41*(9), 1353–1363. https://doi.org/10.1007/s10295-014-1485-5

Lun, O. K., Wai, T. B., & Ling, L. S. (2014). Pineapple cannery waste as a potential substrate for micro- bial biotranformation to produce vanillic acid and vanillin. *International Food Research Journal*, *21*(3), 953.

Mabrouk, M. E., & El Ahwany, A. M. (2008). Production of 946-mannanase by Bacillus amylolequifaciens 10A1 cultured on potato peels. *African journal of Biotechnology*, *7*(8).

Maghsoodi, V., Kazemi, A., Nahid, P., Yaghmaei, S., & Sabzevari, M. A. (2013). Alkaline protease production by immobilized cells using B. licheniformis. *Scientia Iranica*, *20*(3), 607–610.

Maldonado, M. C., & De Saad, A. S. (1998). Production of pectinesterase and polygalacturonase by Aspergillus niger in submerged and solid state systems. *Journal of Industrial Microbiology and Biotechnology*, *20*(1), 34–38. https://doi.org/10.1038/sj.jim.2900470

Mamma, D., Kourtoglou, E., & Christakopoulos, P. (2008). Fungal multienzyme production on industrial by- products of the citrus-processing industry. *Bioresource Technology*, *99*(7), 2373–2383. https://doi.org/10.1016/j.biortech.2007.05.018

Man, Y. B., Mo, W. Y., Zhang, F., & Wong, M. H. (2020). Health risk assessments based on polycyclic aromatic hydrocarbons in freshwater fish cultured using food waste-based diets. *Environmental Pollution*, *256*, 113380.https://doi.org/10.1016/j.envpol.2019.113380

Marwaha, S. S., & Kennedy, J. F. (1988). Whey—pollution problem and potential utilization. *International Journal of Food Science & Technology*, *23*(4), 323–336. https://doi.org/10.1111/j.1365-2621.1988.tb00586.x

Mathias, T. R. D. S., Fernandes de Aguiar, P., de Almeida e Silva, J. B., Moretzsohn de Mello, P. P., Camporese Sérvulo, E. F. (2017). Brewery Waste Reuse for Protease Production by Lactic Acid Fermentation. *Food Technology and Biotechnology*, 55, 218–224. https://doi.org/10.17113/ftb.55.02.17.4378.

Matrawy, A. A., Khalil, A. I., Marey, H. S., & Embaby, A. M. (2021). Biovalorization of the raw agro-industrial waste rice husk through directed production of xylanase by Thermomyces lanuginosus strain A3-1 DSM 105773: A statistical sequential model. *Biomass Conversion and Biorefinery*, *11*(5), 2177–2189.

Medouni-Haroune, L., Zaidi, F., Medouni-Adrar, S., Kernou, O. N., Azzouz, S., & Kecha, M. (2018). Bioconversion of olive pomace by submerged cultivation of Streptomyces sp. S1M3I. *Proceedings of the National Academy of Sciences, India Section B: Biological Sciences*, *88*(4), 1425–1433.

Melikoglu, M., Lin, C. S. K., & Webb, C. (2013). Analysing global food waste problem: pinpointing the facts and estimating the energy content. *Central European Journal of Engineering*, *3*(2), 157–164. https://doi.org/10.2478/s13531-012-0058-5

Mensah, J. K., & Twumasi, P. (2017). Use of pineapple waste for single cell protein (SCP) production and the effect of substrate concentration on the yield. *Journal of Food Process Engineering*, *40*(3), e12478. https://doi.org/10.1111/jfpe.12478

Merino, D., Ollier, R., Lanfranconi, M., & Alvarez, V. (2016). Preparation and characterization of soy lecithin-modified bentonites. *Applied Clay Science*, *127*, 17–22. https://doi.org/10.1016/j.clay.2016.04.006

Michail, M., Vasiliadou, M., & Zotos, A. (2006). Partial purification and comparison of precipitation techniques of proteolytic enzymes from trout (Salmo gairdnerii) heads. *Food Chemistry*, *97*(1), 50–55. http://dx.doi.org/10.1016/j.foodchem.2005.03.022

Mihai, F. C., & Ingrao, C. (2018). Assessment of biowaste losses through unsound waste management practices in rural areas and the role of home composting. *Journal of Cleaner Production*, *172*, 1631–1638. http://dx.doi.org/10.1016/j.jclepro.2016.10.163

Millati, R., Cahyono, R. B., Ariyanto, T., Azzahrani, I. N., Putri, R. U., & Taherzadeh, M. J. (2019). Agricultural, industrial, municipal, and forest wastes: an overview. *Sustainable resource recovery and zero waste approaches*. https://doi.org/10.1016/B978-0-444-64200-4.00001-3

Mladenović, D., Pejin, J., Kocić-Tanackov, S., Stefanović, A., Đukić-Vuković, A., & Mojović, L. (2016). Potato stillage and sugar beet molasses as a substrate for production of lactic acid and probiotic biomass. *Journal on Processing and Energy in Agriculture*, *20*(1), 17–20. Retrieved from https://scindeks-clanci.ceon.rs/data/pdf/1821-4487/2016/1821-44871601017M.pdf

Mo, W. Y., Lun, C. H. I., Choi, W. M., Man, Y. B., & Wong, M. H. (2016). Enhancing growth and non-specific immunity of grass carp and Nile tilapia by incorporating Chinese herbs (Astragalus membranaceus and Lycium barbarum) into food waste based pellets. *Environmental Pollution*, *219*, 475–482. https://doi.org/10.1016/j.envpol.2016.05.055

Mo, W. Y., Man, Y. B., & Wong, M. H. (2019a). Evaluation of Nutritional Values of Food Waste-Based Feed Pellets and Common Feeding Materials for Culturing Freshwater Fish. In *Environmental Sustainability and Education for Waste Management* (pp. 305–321). Springer, Singapore.

Mo, W. Y., Man, Y. B., Zhang, F., & Wong, M. H. (2019). Fermented food waste for culturing jade perch and Nile tilapia: Growth performance and health risk assessment based on metal/loids. *Journal of Environmental Management*, *236*, 236–244. https://doi.org/10.1016/j.jenvman.2019.01.102

Nasopoulou, C., & Zabetakis, I. (2013). Agricultural and aquacultural potential of olive pomace a review. *Journal of Agricultural Science*, *5*(7), 116.

Nath, A., Dixit, M., Bandiya, A., Chavda, S., & Desai, A. J. (2008). Enhanced PHB production and scale up studies using cheese whey in fed batch culture of Methylobacterium sp. ZP24. *Bioresource Technology*, *99*(13), 5749–5755. https://doi.org/10.1016/j.biortech.2007.10.017

Németh, Á. R. O. N., & Kaleta, Z. O. L. T. Á. N. (2015). Complex utilization of dairy waste (whey) in Biorefinery. *WSEAS Trans. Environ. Dev*, *11*, 80–88. www.wseas.us/journal/pdf/environment/2015/a1872601-220.pdf

Ng, H. S., Kee, P. E., Yim, H. S., Chen, P. T., Wei, Y. H., & Chi-Wei Lan, J. (2020). Recent advances on the sustainable approaches for conversion and reutilization of food wastes to valuable bioproducts. *Bioresource Technology*, *302*, 122889.https://doi.org/10.1016/j.biortech.2020.122889

Nigam, P. (1990). Investigation of some factors important for solid-state fermentation of sugar cane bagasse for animal feed production. *Enzyme and Microbial Technology*, *12*(10), 808–811.

Nigam, P., & Singh, D. (1994). Solid-state (substrate) fermentation systems and their applications in biotechnology. *Journal of Basic Microbiology*, *34*(6), 405–423. https://doi.org/10.1002/jobm.3620340607

Nurdiyana, H., & Siti Mazlina, M. K. (2009). Optimization of protein extraction from fish waste using response surface methodology. *Journal of Applied Sciences*, *9*(17), 3121–3125. https://doi.org/10.3923/jas.2009.3121.3125

Obruca, S., Marova, I., & Certik, M. Biotechnological conversion of whey into high-value products. *Agricultural Research Updates*, 125.

Obruca, S., Marova, I., Melusova, S., & Mravcova, L. (2011). Production of polyhydroxyalkanoates from cheese whey employing Bacillus megaterium CCM 2037. *Annals of Microbiology*, *61*(4), 947–953. https://doi.org/10.1007/s13213-011-0218-5

Oda, Y., Saito, K., Yamauchi, H., & Mori, M. (2002). Lactic acid fermentation of potato pulp by the fungus Rhizopus oryzae. *Current Microbiology*, *45*(1), 1–4. https://doi.org/10.1007/s00284-001-0048-y

Okonko, I. O., Adeola, O. T., Aloysius, F. E., Damilola, A. O., & Adewale, O. A. (2009). Utilization of food wastes for sustainable development. *EJEAFChe*, *8*(4), 120–44.

Okorie, P. C., & Asagbra, A. E. (2008). Oxytetracycline production by mix culture of Streptomyces rimosus and S. vendagensis in solid-state fermentation of cassava peels. *J. Ind. Res. Technol*, *2*, 43–47.

Okorie, P.C., & Asagbra, A. E. (2005). Production of oxytetracycline by Streptomyces rimosus in solid –state fermentation of groundnut shells. *World Journal of Biotechnology*, *6*, 903–908.

Olajire, A. A. (2012) The brewing industry and environmental challenges. *Journal of Cleaner Production*. https://doi.org/10.1016/j.jclepro.2012.03.003

Olajire, A. A. (2020). The brewing industry and environmental challenges. *Journal of Cleaner Production*, *256*, 102817. https://doi.org/10.1016/j.jclepro.2012.03.003

Ozyurt, G., Ozkutuk, A. S., Boga, M., Durmus, M., & Boga, E. K. (2017). Biotransformation of seafood processing wastes fermented with natural lactic acid bacteria; the quality of fermented products and their use in animal feeding. *Turkish Journal of Fisheries and Aquatic Sciences*. *17*, 543–555. https://doi.org/10.4194/1303-2712-v17_3_11

Pantazaki, A. A., Papaneophytou, C. P., Pritsa, A. G., Liakopoulou-Kyriakides, M., & Kyriakidis, D. A. (2009). Production of polyhydroxyalkanoates from whey by Thermus thermophilus HB8. *Process Biochemistry*, *44*(8), 847–853. https://doi.org/10.1016/j.procbio.2009.04.002

Parmar, I., & Rupasinghe, H. V. (2013). Bio-conversion of apple pomace into ethanol and acetic acid: enzymatic hydrolysis and fermentation. *Bioresource technology*, *130*, 613–620. https://doi.org/10.1016/j.biortech.2012.12.084

Parveen, J., Iqrah, A., Yumi, Z., & Irwandi, J. (2016). Process development for maximum lycopene production from selected fruit waste and its antioxidant and antiradical activity. *Journal of Food Processing and Technology*, *7*(4), 1000576.

Patelski, P., Berłowska, J., Balcerek, M., Dziekońska-Kubczak, U., Pielech-Przybylska, K., Dygas, D., & Jędrasik, J. (2020). Conversion of potato industry waste into fodder yeast biomass. *Processes*, *8*(4), 453. https://doi.org/10.3390/pr8040453

Peanparkdee, M., & Iwamoto, S. (2019). Bioactive compounds from by-products of rice cultivation and rice processing: Extraction and application in the food and pharmaceutical industries. *Trends in Food Science & Technology*, *86*, 109–117.

Pili, J., Danielli, A., Nyari, N. L., Zeni, J., Cansian, R. L., Backes, G. T., & Valduga, E. (2018). Biotechnological potential of agro-industrial waste in the synthesis of pectin lyase from Aspergillus brasiliensis. *Food Science and Technology International*, *24*(2), 97–109. https://doi.org/10.1177/1082013217733574

Pintać, D., Četojević-Simin, D., Berežni, S., Orčić, D., Mimica-Dukić, N., & Lesjak, M. (2019). Investigation of the chemical composition and biological activity of edible grapevine (Vitis vinifera L.) leaf varieties. *Food chemistry*, *286*, 686–695.

Pintać, D., Majkić, T., Torović, L., Orčić, D., Beara, I., Simin, N., ...& Lesjak, M. (2018). Solvent selection for efficient extraction of bioactive compounds from grape pomace. *Industrial Crops and Products*, *111*, 379–390.

Pogorzelska-Nowicka, E., Atanasov, A. G., Horbańczuk, J., & Wierzbicka, A. (2018). Bioactive compounds in functional meat products. *Molecules*, *23*(2), 307. https://doi.org/10.3390/molecules23020307

Prabha, M. S., & Rangaiah, G. S. (2014). Citric acid production using Ananas comosus and its waste with the effect of alcohols. *Int. J. Curr. Microbiol. Appl. Sci*, *3*(5), 747–754.

Preston, T. R., Sansoucy, R., & Aarts, G. (1986). Molasses as animal feed: An overview. *Sugarcane as Feed, FAO Animal Production and Health Papers 72, Proceeding of an FAO Expert Consultation, Santo Domingo, Dominic Republic*.

Puniya, A. K., Shah, K. G., Hire, S. A., Ahire, R. N., Rathod, M. P., & Mali, R. S. (1996). Bioreactor for solid-state fermentation of agro industrial wastes. *Indian Journal of Microbiology*, *36*(3), 177–178. https://doi.org/10.1016/j.biortech.2010.11.126

Qin, G., Liu, C. C., Richman, N. H., & Moncur, J. E. (2005). Aquaculture wastewater treatment and reuse by wind-driven reverse osmosis membrane technology: a pilot study on Coconut Island, Hawaii. *Aquacultural Engineering*, *32*(3–4), 365–378. https://doi.org/10.1016/j.aquaeng.2004.09.002.

Radosavljević, M., Pejin, J., Pribić, M., Kocić-Tanackov, S., Romanić, R., Mladenović, D.,& Mojović, L. (2019). Utilization of brewing and malting by-products as carrier and raw materials in l-(+)-lactic acid production and feed application. *Applied Microbiology and Biotechnology*, *103*(7), 3001–3013. https://doi.org/10.1007/s00253-019-09683-5

Radwan, H. H., Alanazi, F. K., Taha, E. I., Dardir, H. A., Moussa, I. M., & Alsarra, I. A. (2010). Development of a new medium containing date syrup for production of bleomycin by Streptomyces mobaraensis ATCC 15003 using response surface methodology. *African Journal of Biotechnology*, *9*(33), 5450–5459.

Raghunath, B. V., Punnagaiarasi, A., Rajarajan, G., Irshad, A., & Elango, A. (2016). Impact of dairy effluent on environment—a review. *Integrated Waste Management in India*, 239–249.

Raji, Y. O., Jibril, M., Misau, I. M., & Danjuma, B. Y. (2012). Production of vinegar from pineapple peel. *International Journal of Advanced Scientific Research and Technology*, *3*(2), 656–666.

Rane, A. N., Baikar, V. V., Ravi Kumar, V., & Deopurkar, R. L. (2017). Corrigendum: agro-industrial wastes for production of biosurfactant by Bacillus subtilis ANR 88 and its application in synthesis of silver and gold nanoparticles. *Frontiers in Microbiology*, *8*, 878.https://doi.org/10.3389/fmicb.2017.00492

Ranjit, C., & Srividya, S. (2016). Lactic acid production from free and polyurethane immobilized cells of Rhizopus oryzae MTCC 8784 by direct hydrolysis of starch and agro-industrial waste. *International Food Research Journal*, *23*(6). Retrieved from http://ifrj.upm.edu.my/23%20(06)%202016/(47).pdf

Reddy, C. A., Henderson, H. E., & Erdman, M. D. (1976). Bacterial fermentation of cheese whey for production of a ruminant feed supplement rich in curde protein. *Applied and Environmental Microbiology*, *32*(6), 769–776. https://doi.org/10.1128/aem.32.6.769-776.1976

Rodríguez Couto, S. (2008). Exploitation of biological wastes for the production of value-added products under solid-state fermentation conditions. *Biotechnology Journal: Healthcare Nutrition Technology*, *3*(7), 859–870. https://doi.org/10.1002/biot.200800031

Sadh, P. K., Chawla, P., & Duhan, J. S. (2018). Fermentation approach on phenolic, antioxidants and functional properties of peanut press cake. *Food Bioscience*, *22*, 113–120.

Sadh, P. K., Kumar, S., Chawla, P., & Duhan, J. S. (2018). Fermentation: a boon for production of bioactive compounds by processing of food industries wastes (by-products). *Molecules*, *23*(10), 2560. https://doi.org/10.1016/j.fbio.2018.01.011

Sagar, N. A., Pareek, S., Sharma, S., Yahia, E. M., & Lobo, M. G. (2018). Fruit and vegetable waste: Bioactive compounds, their extraction, and possible utilization. *Comprehensive Reviews in Food Science and Food Safety*, *17*(3), 512–531. https://doi.org/10.1111/1541-4337.12330

Santos, F., Machado, G., Faria, D., Lima, J., Marçal, N., Dutra, E., & Souza, G. (2017). Productive potential and quality of rice husk and straw for biorefineries. *Biomass Conversion and Biorefinery*, *7*(1), 117–126.

Satari, B., & Karimi, K. (2018). Resources, Conservation & Recycling Citrus processing wastes: Environmental impacts, recent advances, and future perspectives in total valorization. *Resour. Conserv. Recycl.*, *129*, 153–167.

Sharma, A., Gupta, V., Khan, M., Balda, S., Gupta, N., Capalash, N., & Sharma, P. (2017). Flavonoid-rich agro-industrial residues for enhanced bacterial laccase production by submerged and solid-state fermentation. *3 Biotech*, *7*(3), 1–8. https://doi.org/10.1007/s13205-017-0836-0

Shojaosadati, S. A., & Babaeipour, V. (2002). Citric acid production from apple pomace in multi-layer packed bed solid-state bioreactor. *Process Biochemistry*, *37*(8), 909–914. https://doi.org/10.1016/S0032-9592(01)00294-1

Sindhu, R., Gnansounou, E., Rebello, S., Binod, P., Varjani, S., Thakur, I. S., & Pandey, A. (2019). Conversion of food and kitchen waste to value-added products. *Journal of Environmental Management*, *241*, 619–630. https://doi.org/10.1016/j.jenvman.2019.02.053

Sindhu, R., Silviya, N., Binod, P., & Pandey, A. (2013). Pentose-rich hydrolysate from acid pretreated rice straw as a carbon source for the production of poly-3-hydroxybutyrate. *Biochemical Engineering Journal*, *78*, 67–72. http://ir.niist.res.in:8080/jspui/handle/123456789/1160

Smitha, M. S., Singh, S., & Singh, R. (2017). Microbial biotransformation: a process for chemical alterations. *J Bacteriol Mycol Open Access*, *4*(2), 85. https://doi.org/10.15406/jbmoa.2017.04.00085

Sosa-Hernández, O., Parameswaran, P., Alemán-Nava, G. S., Torres, C. I., & Parra-Saldívar, R. (2016). Evaluating biochemical methane production from brewer's spent yeast. *Journal of Industrial Microbiology and Biotechnology*, *43*(9), 1195–1204.

Sousa, B. A., & Correia, R. T. P. (2012). Phenolic content, antioxidant activity and antiamylolytic activity of extracts obtained from bioprocessed pineapple and guava wastes. *Brazilian Journal of Chemical Engineering*, *29*(1), 25–30. https://doi.org/10.1590/S0104-66322012000100003

Spalvins, K., Zihare, L., & Blumberga, D. (2018). Single cell protein production from waste biomass: comparison of various industrial by-products. *Energy procedia*, *147*, 409–418. https://doi.org/10.1016/j.egypro.2018.07.111

Sukan, A., Roy, I., & Keshavarz, T. (2014). Agro-industrial waste materials as substrates for the production of poly (3-hydroxybutyric acid). *Journal of Biomaterials and Nanobiotechnology*, *5*(4), 229–240. https://doi.org/10.4236/jbnb.2014.54027

Swain, M. R., & Ray, R. C. (2008). Optimization of cultural conditions and their statistical interpretation for production of indole-3-acetic acid by Bacillus subtilis CM5 using cassava fibrous residue. Retrieved from http://nopr.niscair.res.in/bitstream/123456789/1809/1/JSIR%2067%288%29%20622-628.pdf

Taherzadeh, M., Bolton, K., Wong, J., & Pandey, A. (Eds.). (2019). *Sustainable resource recovery and zero waste approaches*. Elsevier.

Taniguchi, I., Kagotani, K., & Kimura, Y. (2003). Microbial production of poly (hydroxyalkanoate) s from waste edible oils. *Green Chemistry*, *5*(5), 545–548. https://doi.org/10.1039/B304800B

Thomas, K. R., & Rahman, P. (2006). Brewery wastes. Strategies for sustainability. A review. *Aspects of Applied Biology*, *80*, 1–11.

Tugiyono, T., Febryano, I. G., Yuwana, P., & Suharso, S. (2020). Utilization of Fish Waste as Fish Feed Material as an Alternative Effort to Reduce and Use Waste. *Pakistan Journal of Biological Sciences*, *23*(5), 701–707.

Vandenberghe, L. P., Soccol, C. R., Pandey, A., & Lebeault, J. M. (2000). Solid-state fermentation for the synthesis of citric acid by Aspergillus niger. *Bioresource Technology*, *74*(2), 175–178.

Vastrad, B. M., & Neelagund, S. E. (2012). Optimization of process parameters for rifamycin b production under solid state fermentation from Amycolatopsis mediterranean MTCC14. *Int J Curr Pharm Res*, *4*(2), 101–108.

Vastrad, B. M., & Neelagund, S. E. (2011). Optimization and production of neomycin from different agro industrial wastes in solid state fermentation. *International Journal of Pharmaceutical Sciences and Drug Research*, *3*(2), 104–111.

Vidhyalakshmi, R., Vallinachiyar, C., & Radhika, R. (2012). Production of xanthan from agro-industrial waste. *Journal of Advanced Scientific Research*, *3*(2), 56–59.

Vijayaraghavan, P., Vincent, S. G. P., Arasu, M. V., & Al-Dhabi, N. A. (2016). Bioconversion of agro-industrial wastes for the production of fibrinolytic enzyme from Bacillus halodurans IND18: Purification and biochemical characterization. *Electronic Journal of Biotechnology*, *20*, 1–8. http://dx.doi.org/10.1016/j.ejbt.2016.01.002

Vikas, O. V., & Mridul, U. (2014). Bioconversion of papaya peel waste in to vinegar using Acetobacter aceti. *Int. J. Sci. Res*, *3*(11), 409–411.

Vintila, T., Dragomirescu, M., Jurcoane, S., Vintila, D., Caprita, R., & Maniu, M. (2009). Production of cellulase by submerged and solid-state cultures and yeasts selection for conversion of lignocellulose to ethanol. *Romanian Biotechnological Letters*, *14*(2), 4275–4281.

Wadhwa, M., & Bakshi, M. P. S. (2013). Utilization of fruit and vegetable wastes as livestock feed and as substrates for generation of other value-added products. *Rap Publication*, *4*, 1–67.

Wan, Y., Ghosh, R., & Cui, Z. (2002). High-resolution plasma protein fractionation using ultrafiltration. *Desalination*, *144*(1–3), 301–306.

Wang, T. Y., Wu, Y. H., Jiang, C. Y., & Liu, Y. (2010). Solid state fermented potato pulp can be used as poultry feed. *British Poultry Science*, *51*(2), 229–234. https://doi.org/10.1080/00071661003781864

Watkins, M., & Nash, D. (2010). Dairy factory wastewaters, their use on land and possible environmental impacts-a mini review. *The Open Agriculture Journal*, *4*(1). https://doi.org/10.2174/1874331501004010001

Wong, M. H., Mo, W. Y., Choi, W. M., Cheng, Z., & Man, Y. B. (2016). Recycle food wastes into high quality fish feeds for safe and quality fish production. *Environmental Pollution*, *219*, 631–638. https://doi.org/10.1016/j.envpol.2016.06.035

Yalemtesfa, B., Alemu, T., & Santhanam, A. (2010). Solid substrate fermentation and conversion of orange waste in to fungal biomass using Aspergillus niger KA-06 and Chaetomium Spp KC-06. *African Journal of Microbiology Research*, *4*(12), 1275–1281.

Yang, S. S., & Yuan, S. S. (1990). Oxytetracycline production by Streptomyces rimosus in solid state fermentation of sweet potato residue. *World journal of Microbiology and Biotechnology*, *6*(3), 236–244. https://doi.org/10.1007/BF00327798

Yang, S. Y., Ji, K. S., Baik, Y. H., Kwak, W. S., & McCaskey, T. A. (2006). Lactic acid fermentation of food waste for swine feed. *Bioresource Technology*, *97*(15), 1858–1864. https://doi.org/10.1016/j.biortech.2005.08.020

Yunus, F. U. N., Nadeem, M., & Rashid, F. (2015). Single-cell protein production through microbial conversion of lignocellulosic residue (wheat bran) for animal feed. *Journal of the Institute of Brewing*, *121*(4), 553–557. https://doi.org/10.1002/jib.251

Zahari, M. A. K. M., Zakaria, M. R., Ariffin, H., Mokhtar, M. N., Salihon, J., Shirai, Y., & Hassan, M. A. (2012). Renewable sugars from oil palm frond juice as an alternative novel fermentation feedstock for value-added products. *Bioresource Technology*, *110*, 566–571. https://doi.org/10.1016/j.biortech.2012.01.119

Zepf, F., & Jin, B. (2013). Bioconversion of grape marc into protein rich animal feed by microbial fungi. *Chemical Engineering & Process Techniques*, *1*(2), 1011.

13 Uses of Enzymes in Food Industry Waste Utilization

Janifer Raj Xavier, Om Prakash Chauhan,* and
Anil Dutt Semwal

Defence Food Research Laboratory, Defence Research and
Development Organization, Siddharthanagar, Mysore, India
*Corresponding author: opchauhan@gmail.com

CONTENTS

13.1 INTRODUCTION

The food processing sector is growing rapidly in the modern world due to increasing demand for food and energy due to exponential population growth. Practices of intensification in agricultural and food production resulted in increased accumulation of solid and liquid wastes. The industrial food processing sector is a significant segment of food and produces a large volume of waste, which accounts up to 39.0 percent of total waste generated. General household waste, food retail units and food distribution losses accounts up to 42.0, 14.0, and 5.0 percent, respectively. Wastes are generated during different food processing operations, shipping, and storage. Deleterious environmental effects of organic wastes include microbial contamination by pathogens and putrefied odor owing to the presence of moisture and nutrient richness. Therefore, it is important to prevent, reduce, reuse, and recycle food processing waste throughout the processing line and consumption scenario (Papargyropoulou et al 2016). High value biomolecules derived from fruit and vegetable waste obtained by food processing industries include pigments, phenolic compounds, dietary fibers, sugar derivatives, organic acids, and minerals. Globally, 1.3 billion tonnes of food is wasted annually and the Food and Agriculture Organization (FAO) estimates on food waste and losses account for more than 10 percent of worldwide energy consumption (FAO 2019). Being in the tropical belt India, Brazil and China accounts for the major share in accumulation of food waste. Value added products using food industry wastes are a valuable source of additives and nutraceutical supplements, based on their nutritional value and health benefits. The concept of biorefinery to derive value added by-products from food processing industry generated wastes has emerged recently. Various eco-friendly, efficient, and sustainable methods and technologies has been reported in recent years for providing solutions to problems associated with waste accumulation. Processes such as fermentation, anaerobic, and aerobic composting, pyrolysis involving chemical, thermo chemical, and enzymatic transformations are being reported based on the biorefinery concept similar to refinery for crude fossil fuels. Paiho et al (2020) insisted the importance of considering social aspects related to the management of wastes for a successful implementation of the same. Enzymatic transformation of food processing derived biowaste and valorization into useful products has gained importance as a non-conventional environmentally friendly approach and are preferred over land filling and incineration. The highly specific nature of enzymes towards their substrate makes them efficient tools for hydrolysis of various types of plant substrates such as fiber, protein, lipids, and so on. Amylases, cellulases, xylanases, and pectinases act upon starch, cellulose, hemicelluloses, and pectin rich material present in fruit and vegetable processing waste. Enzymes are natural substances capable of increasing the rate of specific biochemical reactions in living systems without being used up during the process. Though enzymes catalyze various reactions in living organisms, they also act *in vitro* and find applications in various industrial applications. Their specificity, efficacy, safety, suitability, and ease of usage are the major reasons, which are attributed to the usage of enzymes in various food processing applications. The food enzymes' market size was estimated at US $1,944.8 million in 2018 and is expected to reach US $3,056.9 million by 2026, registering a CAGR of 5.6 percent from 2019 to 2026 (www.freedoniagroup.com/2018). Enzymes catalyze reactions by reducing the energy barrier for transformation of reactants into desirable products. Specific catalytic mechanisms of enzymes include approximation, covalent catalysis, general acid base, and molecular strain. Enzymes belong to broad categories such as oxido reductases, transferases, hydrolases, lyases, isomerases, and ligases find importance in volarization of food applications. Detailed understanding on the mechanism of action of various enzymes on a wide variety of substrates as well as their protein structure and specific functions may be useful for enzymatic hydrolysis of food processing industry waste. This chapter discusses the recent trends in the importance of managing food processing industry derived waste using enzymatic methods and their transformation into valuable by-products. We also discuss the types of food industry wastes generated and processes and principles related to sustainable approaches to repurpose those using enzymes/biocatalysts in terms of food waste biorefinery. Use of green chemistry for sustainable utilization of wastes generated from cereals, pulses, oilseed, and fruit and vegetable processing are highlighted in this chapter. Recent advances on enzyme catalysis for reutilization of plant based residues generated after industrial food processing operations into pigments, aromatic oils, dietary fiber, bioactive peptides, and antioxidant molecules are also discussed with various examples.

13.2   GLOBAL FOOD WASTE AND LOSSES AND THE IMPORTANCE OF FOOD WASTE MANAGEMENT

Recent years have seen an increased awareness in sustainable development, and waste management in the food industry is being given a major emphasis due to huge volumes of generated waste and problems encountered in their disposal. Post harvest food production losses alone account up to 14 percent of global food loss, which does not include losses at retail level. South and Central Asia contribute more than 20 percent while losses in Europe and North America are estimated up to 15 percent (FAO 2020). According to the food waste index report (2021) of the United Nations environment program, approximately 931000000 metric tons of food was wasted globally in 2019 (UNEP 2021). Food waste generated in developed nations were attributed to behavior of user or consumer of food, which accounts up to 30 percent of total food wasted amounting to annual loss of 161 billion USD. However, availability of adequate infrastructure facilities for food processing and preservation is related to food wastage in developing nations (USEPA 2018). Beverage and drink industries contribute 26 percent of total food waste generated while dairy waste accounts up to 21 percent; fruits and vegetable processing and cereal processing into finished products generate 14.8 and 12.9 percent of waste, respectively (Tsegaye et al 2021). In Indian context, fruit and vegetable losses alone account for 12 to 21 MT, respectively among the generated agricultural waste and are worth around 4.4 USD, out of the total waste worth 10.6 billion USD (NAAS 2019). Fruit and vegetable processing involves generation of effluents and food waste such as peels, seeds, fruit and vegetable pomace, residual starches, fibers, and are rich in soluble solids, starches, fibers, organic acids, etc. Plant based commodity viz. cereals, pulses, oilseeds, fruits and vegetables are related to generation of myriad wastes and are categorized into five major segments as follows:

1. Agricultural production losses pertaining to physical damage during harvest operations as well as during grading and spillage during loading and transportation.
2. Post harvest losses such as decay and rotting during storage of the fresh commodity as well as transportation losses due to long distances between farm and storage godowns or wholesale markets.
3. Cereals, pulses, oilseeds, fruit and vegetable processing losses at manufacturing and domestic operations, which includes rejection of commodity due to non-suitability of maturity index, spillage, and microbial degradation. Various food processing operations such as washing, peeling, slicing, dicing, grating, pulping, and spillage during such operations.
4. Distribution losses at wholesale and retail markets such as *mandis*, super markets, wet markets etc., during unloading, improper storage, and microbial spoilage of fresh commodity.
5. Consumption losses attributed to food waste during consumption at consumer level at commercial outlets and household.

Department of Economic and Social Affairs, United Nations adopted an agenda for sustainable development in 2015 and formulated Sustainable Development Goals (SDGs) with 17 major targets aimed to reduce the globally generated global food waste to half, at the producer and consumer end by the year 2030 under the goal titled responsible consumption and production. It also emphasizes reducing food losses during food processing operations as well as in supply chains to ensure sustainable production and consumption patterns. Furthermore, to attain the SDGs in food waste management, it is important to understand the kind of waste generated at various stages of food production, processing and consumption in order to formulate sustainable and practical strategies for waste management. Food loss index (FLI) and food waste index (FWI) are the two clearly defined indices given by FAO, to measure the progress in attaining the sustainable methods of food waste management. FLI indicates the food losses till post harvest production stage while FWI measures the food waste generated at customer level at retail outlets (UN 2015).

Pfaltzgraff et al (2013) described food waste into two major categories: (i) pre-consumer generated wastes (agricultural production activities, industrial food processing operations and distribution) and (ii) post-consumer generated wastes (food preparation activities and food consumption). Food loss denotes the food disposed along the food chain from production location such as farms, agricultural production units, and protected cultivation units till post harvest processing. It denotes the food wasted and not available for productive usage in the form of seed or fodder. Loss in quantity of food produced due to choices of raw material producers, suppliers, and processors are responsible for food loss. Wastage of food due to various reasons such as sorting and grading in supply chain, expiry of processed foods and leftovers at household and eateries, which are discarded without consumption. Thus, the food waste is the food unavailable for consumption and is affected by decisions of food service providers, household food usage and consumers. Food loss and food waste accounts for about 12 and 22 percent, respectively of 1/3 of total food produced but it remains unavailable for human consumption.

Globally, around 95 percent of food wastes generated at various stages of food processing from production to consumer are dumped in landfills and account for a major share of anthropogenic activity derived total methane emission i.e. an annual equivalent of 113 million tonnes of carbon dioxide. Considerable losses to environment in terms of pollution could be prevented by avoiding such huge losses of food waste, in the form of food in raw as well as processed form as well as losses during various food related operations. Promising changes are being observed recently on the pre-treatment of harmful industrial effluents before being let out into the environment due to stringent regulations by Government agencies. Furthermore, awareness on reusing and recycling wastes generated from food processing industries and changes in consumer perception on avoiding food waste are encouraging. Arizona State could reduce about 27.85 percent of household generated food waste through awareness campaigns on food waste prevention. Modern food waste reduction policies implemented by the French government aimed at the production of bio-plastic and bio-gas through volarization of food wastes. However, the sustainable approach to avoid/reduce food wastes and food losses points to circular bioeconomy, which would pave the right direction to achieve the SDGs related to food industry waste generation and utilization (Jain et al 2022).

13.3   FOOD WASTE BIOREFINERY: A SUSTAINABLE APPROACH FOR WASTE MANAGEMENT

Public institutions and private bodies suggest the consideration of the principles of the waste management hierarchy while taking actions on food waste management. Key terms in the hierarchy include prevent, reuse, recycle, recover, and dispose and indicates the importance of various ways and means in the management of generated food wastes. Food wastes are of different kinds and a multipronged approach involving various stakeholders is essential for their effective management in a dynamic and context based methods (Mattila et al 2018). Utilization of food and food processing waste for the generation of energy in terms of biofuels, has been projected as a venture with huge potential to attain energy security. Waste volarization of nutrient rich food industry generated wastes into useful value added by-products are replacing practices of traditional waste management such as land filling and incineration. Ubando et al (2021) reviewed various models of establishing a biorefinery using various ligno cellulosic substrates by technologies available recently for bioconversion of wastes into wealth. The importance of biorefinery strategies has been emphasized as a tool to realize a circular bioeconomy for sustainable development. Economic, social, and environmental aspects of the industrial sector in adopting methods of biorefinery have also been reviewed along with present challenges and future scope.

Biorefining i.e. conversion of waste into wealth using suitable bioprocesses represent an important waste management strategy, which leads to sustainable development by reduction on the dependence of fossil fuels as well as reduction of the emission of greenhouse gases. Processes of food waste biorefinery are categorized into three major classes based on substances that act up on the food and food industrial waste to be valorized. (1) Conversion of food wastes into useful products by use of enzymes or microbes termed as biological pathway. (2) Treating food wastes at high temperatures along with chemical based solvents termed the thermo chemical process. (3) Treatment of food wastes using solvents termed the chemical process, which catalyze conversion of waste into useful by-products. Greater efficiencies are being obtained on the use of the above methods in combination rather than singly. Efficient ecological and resource management, low carbon and methane foot prints, least dependance on conventional fossil fuels and integrated volarization of wastes from agriculture, fisheries and allied fields are the exclusive benefits obtained by practicing the concept of circular bioeconomy. Moreover, circular bioeconomy involving ideas of biorefinery could guarantee energy as well as environmental security. Food security is attained by environmental security thus eliminating hunger issues globally. Quality of life, health and wellness of mankind could be attained and maintained by food and environmental security. Therefore, lifelong wellness and prosperity would be assured without any adverse economic impacts on food, energy, and our ecosystem (Leong et al 2021).

Definitions of bioeconomy and circular economy have been clearly stated by the European bioeconomy council "Bioeconomy is the knowledge based production and use of biological resources to provide products, processes and services in all economic sectors within the frame of a sustainable economic system." "Circular economy is defined as the elimination/minimization of waste generations during the processing and production of products, materials and resources by maintaining the value of the product as long as possible." These concepts emphasize production of bio based materials and bio based compounds in a sustainable manner with zero waste (EU 2018). Kover et al (2022) described use of integrated biorefinery, which uses a combination of various bioprocesses to derive value added by-products from biowaste. A clear understanding of biochemical parameters of the type of waste to be used as raw material is considered as the most important step for successful biorefining process.

Galanakis (2016) elaborated the concept of reviving valuable and useful compounds by utilizing food wastes and termed it "universal recovery strategy." Goals of circular bioeconomy would be achieved by bioconversion of wastes into useful products such as industrial biocatalysts (enzymes), bioactive molecules, and biofuel followed by increased market demand for bio-derived eco-friendly by-products. Addressing the issues of environmental pollution by waste utilization and opening new avenues for sustainable economical growth, by the concept of circular bioeconomy requires the latest technological interventions. Advanced technologies viz. enzyme assisted extraction, immobilized enzymes in suitable substrates and bioreactors ensure conversion of food industry derived waste to zero-level decomposed state. Furthermore, the advanced techniques use green eco-friendly approaches for conversion of wastes into useful bioplastics, bioactives, and biofuels with the least carbon foot print (Sharma et al 2021).

A closed loop system of circular bioeconomy wherein utilization of waste to produce useful substances; diversified applications of derived biomolecules in various industrial and household processes is the need of the hour to achieve SDGs (Mak et al 2019). Ng et al (2020) emphasized the importance of sustainable approaches for food waste volarization into useful biomaterials. Accumulation of food waste is considered as a major segment for environmental pollution and natural product scarcity. Agro processing wastes rich in phytochemicals with desirable properties are considered as valuable sources of by-products viz. enzymes, nutraceutical molecules, cosmetics, biofuel, and animal feed. The increasing population leads to the increased accumulation of waste generated at global level. Wastes generated from crop based sources are suitable candidates for volarization due to their richness in content of bioactive compounds and extraction of enzymes and exopolysaccharides, for synthesis of novel molecules and biodegradable polymers. Integration of conventional and novel technologies for food waste volarization is proposed to be an efficient tool for sustainable economy. Sustainable methods for conversion of agriculture and food derived wastes into useful by-products and their reuse to meet the goal of circular bioeconomy were elaborated and various examples have been depicted (Table 13.1).

Software-based process as a decision-support tool and their use in case studies pertaining to two industrial wastes and their conversion in a sustainable manner was described (Martin-Rios et al 2018). A quality evaluation tool with nine different stages along with parameters to evaluate the nature of various types of industry generated wastes, which followed the evaluation of their economic, social, and environmental feasibility in comparison to traditional methods was used. Understanding the nature of generated waste and classifying them, is the foremost important step to take strategic decisions to alleviate their negative impact caused on environment and society. Ioannidou et al (2020) elaborated the importance of the availability of renewable crude resources to derive value added products using the concept of biorefinery. Fermentable sugars and fractionated molecules available in the European Union (EU) for production of useful bio based polymers and chemicals were emphasized. Sustainability indicator and end of life scenario of biowaste with an example of succinic acid production using solid municipal wastes as well as their techno-commercial feasibility covering raw material requirement and profitability of the venture has been studied in detail.

TABLE 13.1

List of Possible Substrates of Food Industry Waste and Volarized By-products

S. No.	Food industry substrate	By-products	Reference
1.	*Cucurbita pepo* and *Linum usitatissimum* seedcakes	Polyphenols	Ratz-Łyko and Arct (2015)
2.	Tomato waste	Carotenoids	Strati et al (2015)
3.	Waste cooking oil	Free fatty acids	Kochepka et al (2015)
4.	Sugarcane molasses	Levan and levansucrase	Xavier et al (2017)
5.	Wheat bran	Xylooligosaccharides	Xavier et al (2018)
6.	Onion skin	Pectic oligosaccharides	Baldassarre et al (2018)
7.	Food waste	Biofuel	Patel et al (2019)
8.	Fruit and vegetable by-products	Biologically active compounds	Lu et al (2019)
9.	Sunflower seed hulls	Levoglucosan and/or furfural from bio-oils obtained from pyrolyzed	Casoni *et al* (2019)
10.	Fruit, nut, cereal, and vegetable waste	Biofuel	Shehu et al (2019)
11.	Pomegranate peel	Pectin and punicalagin rich phenolics	Talekar et al (2019)
12.	Rice, corn and bakery	Biohydrogen, bioethanol, butanol, biogas and biocoal as biofuels, industrial valued enzymes, biomass, biofertilizer, proteins, organic acids and polysaccharides	Hassan et al (2021)
13.	Raspberry pomace	Antioxidant and potentially anti-inflammatory anthocyanin fractions	Szymanowska et al (2019)
14.	Rice bran	Poly glutamic acid	Xavier et al (2020)
15.	Tomato peels	Lycopene	Gu et al (2020)
16.	Fruits and vegetable waste	Enzymes, exopolysaccharides, bioplastics and biofuels	Esparza et al (2020)
17.	Vegetable food waste	Enzymes	Sabater et al (2020)
18.	Fruit and vegetable stems, leaves, peels, pulps, seeds, and roots	Antioxidants, oils, fiber, fatty acids, isoprenoids, lipids, proteins, saponins, and phytoestrogens	Jiménez-Moreno et al (2020)
19.	Whole safflower plant	Bioethanol	Hashemia et al (2020)
20.	Biomass waste	Monosaccharides, oligosaccharides,biofuels, bioactive molecules, nanocellulose, and lignin	Mou et al (2020)
21.	Fruit and vegetable by-products	Bioactive compounds	Trigo et al (2020)
22.	Pulse processing by-products	Primary and secondary plant metabolites	Rudraraju et al (2021)
23.	Fruits and vegetable peels, seeds or pomace	Natural pigments such as anthocyanins, betalains, carotenoids, and chlorophylls	1. Sharma et al (2021)
24.	Vanilla waste compounds	Wall material for bioactive compounds and biofuel	Peña-Barrientos et al (2021)
25.	Apple industrial waste residue	Poly-hydroxyalkanoates	Liu et al (2021)

Teigiserova et al (2020) reviewed factors responsible for the valorization of food from categories such as surplus, waste, and loss in the context of circular bioeconomy. Definitions related to the future of food waste management has been explained through a simple matrix by categorizing food wastes into six major categories viz. safe to eat, unfit for consumption naturally (pits), industry generated waste, unfit for consumption due to pest attack, inedible to wrong processing, and the rest of waste not categorized in the former. Six broad categories such as surplus only (I category), food waste (II, III, IV, and V category) and food loss (VI category) and a food waste pyramid explaining the food waste management hierarchy has also been elaborated to emphasize the importance of biorefineries in the future of waste management. Policies and initiatives of the EU to reach sustainable goals have also been briefed.

13.4 ENZYME ASSISTED FOOD INDUSTRY WASTE VALORIZATION: PROCESS AND PRINCIPLES

Non-conventional methods involving enzymes for biotransformation and extraction of bioactives by volarization of fruit and vegetable processing waste are gaining importance in the present scenario due to their robustness, specificity, and eco-friendly nature. The highly specific nature of enzymes towards their substrate makes them efficient tools for hydrolysis of various types of plant substrates such as fiber, protein, lipids, and so on. Recent reports on the volarization of food waste generated by processing industries indicate that the cost of production and price of derived product are important in deciding the feasibility of waste biomass conversion into useful by-products. Utilization of enzymes and microorganisms as important partners for eco-friendly bioconversion of wastes from source to end product. Process optimization of enzyme based methods during waste valorization of different types of food industry generated biowaste is important to make the methods economically viable and thereby profitable and easy adoption by industries (Usmani et al 2021).

Long term food waste processing technologies with two stage processes, for direct processing of food wastes, to derive value added products and second stage to process residues by use of non-conventional strategies are being recommended for successful volarization (Esparza et al 2020). Non-conventional methods involving enzymes for biotransformation and extraction of bioactives by the volarization of fruit and vegatble processing waste are gaining importance due to their eco-friendly nature. The highly specific nature of enzymes towards their substrate makes them efficient tools for hydrolysis of various types of plant substrates such as fiber, protein, lipids, and so on. Higher recovery rates, cost effectiveness, and their suitability to extract thermo labile phenolics compounds are the major advantages of newer processes. Greener and newer extraction strategies to improve efficiency, yield, and quality of bioactives extracted to replace older conventional methods using toxic solvents are looked upon in modern times. Furthermore, naturally obtained bioactives have a greater reach in the field of food and nutraceuticals as they are considered safer than their solvent extracted counterparts. Optimization of extraction time and moderate use of solvents during enzymatic extraction of biomolecules would improve performance ability of such processes at an industrially viable scale (Nadar et al 2018).

Extraction strategies play a major role on the quality of the extracted bioactive. Enzymes act upon a variety of substrates such as plant fiber, proteins, and oils making it an excellent ingredient for pre-treatment of plant biomass to enhance extraction of phytochemicals. Pham et al (2015) described general management methods for conversion of food waste into energy. Development of viable methods for food waste management is hindered by heterogenisity, higher water content, and lower calorific value of obtained food waste. Sowbaghya and Chitra (2010) extensively reviewed enzymatic extraction of flavors and colorants from phyto sources. Enzyme pre-treatment for extraction of phytochemicals from avocado, corn olives, and sunflower have been reported in the recent past. Proteases, lipases, and esterases are capable of acting up on plant components and aid in extraction of valuable bioactives. Water as well as thio-esters, amines, oximes, and alchocols are known substrates for these enzyme actions therefore extraction of flavors and colors from plant sources could result in obtaining natural compounds for food flavoring and coloring.

The concept of biorefinery, which describes conversion of renewable biomass to valuable products using advanced techniques of biotechnology, process chemistry, and engineering was described (Ragauskas et al 2000). Presently, the concept of biorefinery is linked with bioeconomy and circular bioeconomy, which strongly support restoration and regeneration by intentional design to support sustainable development. Separation of useful products from wastes using green methods has attracted attention based on a holistic biorefinery approach (Zuin and Ramin 2018). Greener methods for separation of bioactives such as solvent extraction, ultrasonication, supercritical fluid, ionic liquid, enzyme based, accelerated solvent, alkaline processing, and microemulsions are available to extract wealth from waste. These methods are applied singly or in combination based on the nature of raw material and output expected. Systematic reviews on biorefinery reported use of solvent extraction (25%), microwave assisted extraction (19%), supercritical fluid (13%) while usage of ionic liquids, enzymes and subcritical fluid accounts for up to (13%) for extraction of valuable bioactives. Three major biopolymers constitute a major proportion of agricultural and food processing waste viz. cellulose, hemicelluloses, and lignin, constituting 40–50, 20–30, and 20–35 percent, respectively. Jegannathan and Nielsen (2013) elaborated in detail the environmental impact of using enzymatic processes on products and systems. Lifecycle Impact Assessment (LCA) and Environment Impact Assessment (EIA) tools were used as indicators for the environmental impact to compare traditional methods with enzymatic methods. Generalized recommendations for execution of enzyme based methods in process systems were also provided based on extensive literature survey. Enzymatic processes and their benefits imparted on the environment over traditional processes has been agreed upon and reviewed in detail over the past decade by many researchers.

Requirement of pre-treatment before enzymatic hydrolysis of plant based wastes rich in lignocellulosic substances such as cellulose, hemicelluloses, pectin, and lignin found in primary and secondary cell wall of rigid plant cells has been emphasized for enhanced recovery. Norrrahim et al (2021) reviewed different types of pre-treatment such as mechanical, chemical, electrical, biobased, or combination technologies that are necessary for producing a porous material to release useful biomolecules from plant mass. Pre-treated biomass has the following major advantages over untreated ones during biorefining processes: formation of simple sugars on hydrolysis, reduced carbohydrate and by-product formation, as well as being profitable due to reduced processing duration.

Increasing permeability of plant cells and extraction of value added substances using enzymes is considered to be an eco-friendly approach. Feasibility of extraction at lower temperatures and replacement of strong extraction solvents, are considered to be the major advantages of involving enzymes, the biocatalysts for food waste valorization (Kover et al 2022). Among the various aspects of industrial biotechnology, enzyme technology and its application has been declared as the most sustainable and with the most potential in comparison to traditional methods of processing and volarization. Microbial production of enzymes using optimized liquid (submerged fermentation) and solid (solid state fermentation) media and their scale-up followed by downstream processing paved the way for large scale production of useful enzymes for industrial applications. Šelo et al (2021) reviewed the importance of solid state fermentation for valorization of agro-food industrial residues.

Natural products have been categorized into primary metabolites, structural components, and secondary metabolites. Bioactive molecules fall into types, such as flavanoids, alkaloids, terpenoids etc., and occur less than 0.01 percent of dry weight. They find a wide variety of applications as food additives, flavors, anti-microbials, nutraceuticals, biopreservatives, biopestiiceds, cosmetics, pharmaceuticals etc. Extraction procedures using enzymes were found to be robust from industrial food wastes resulting from food processing. Cipolatti et al (2019) reviewed in detail the state-of-the-art of enzymes involved chemical transformations and their implications in achieving sustainable processes. Agro-industrial residues waste such as fruits, cereals, and other lignocellulosic materials and their enzyme based volarization was reviewed in detail (Barcelos et al 2020).

Globally, management of food waste with the goal of converting the waste into various price worthy by-products for use in industry or food applications are the major goals of sustainable food waste volarization. Plazzotta and Manzocco (2019) elaborated in detail the principles and definitions of food waste volarization as well as the current limitations faced during commercialization of the same. Onu and Mbohwa (2021) elaborated sustainable industrial waste management strategies for developing countries and sub-Saharan African region with respect to agricultural waste utilization. Zhu et al (2020) reviewed the latest innovative methods and technologies such as high pressure, nanotechnology, ultrasound, electrical methods etc., for conversion of food wastes into useful products. Sodhi et al (2021) provided insights on recent trends and challenges involved in volarization of wastes derived from agricultural and food processing applications. Novel technologies such as protein engineering, immobilization of enzymes, editing of genomes, high pressure, sonication etc., and their application in conversion of biowaste to useful materials were elaborated.

13.5 VALORIZATION OF FOOD PROCESSING WASTE INVOLVING GREEN CHEMISTRY/ENZYME CATALYSIS

13.5.1 CEREAL PROCESSING

Development of novel products from cereal waste with useful properties were explored and described based on concepts of circular economy through a biorefinery approach in numerous literatures (Figure 13.1). Production and purification of enzymes used for volarization of cereal processing industry generated waste by solid state fermentation using residues are gaining popularity due to their eco-friendly nature. Hemicellulose xylan degrading xylanase production potential using wheat bran (*Triticum* sp.) and paddy straw (*Oryza sativa*) was promising and further the produced enzyme was used to derive useful molecules from crop residues of wheat and rice (Teigiserova et al 2021).

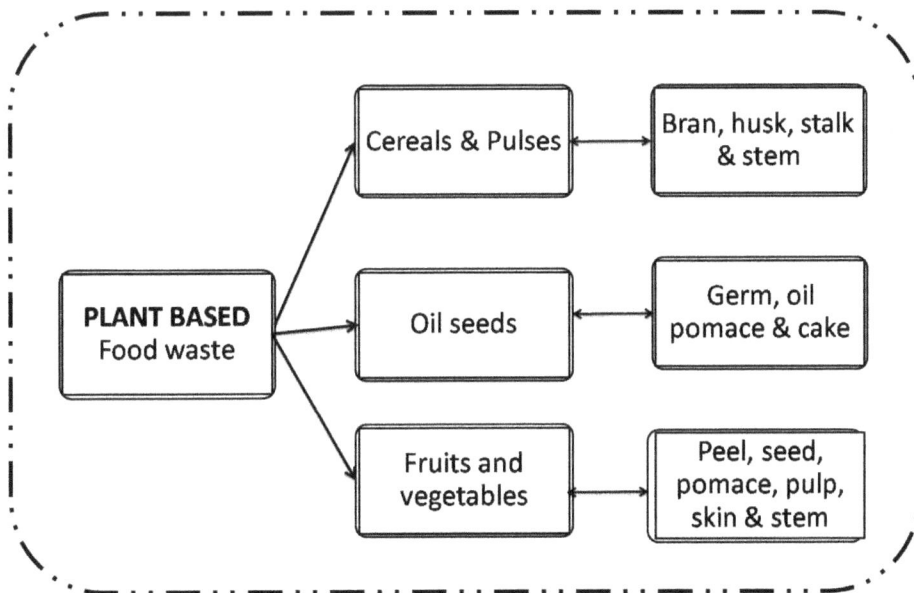

FIGURE 13.1 Types of plant-based food industry wastes/residues.

Skendi et al (2020) reviewed in detail cereal processing industry based waste obtained after wet and dry milling of grains. Bread making, brewing, and industrial starch production are the major players in generation of biowaste rich in sugars, dietary fibers, minerals, proteins, and antioxidant rich polyphenolic compounds. Melnichuk et al (2020) reported cost effective production and purification of alpha amylase through solid state fermentation using *Aspergillus oryzae* of soy husk (*Glycine max*) and wheat flour processing waste. Medium consisting of soybean husk (45.0%), flour mill waste (55.0%) at 30°C for 144 h yielded 47,000 U/g on dry weight basis of solid state substrate.

Wheat bran as the sole carbon substrate was found to be promising for production of ferulic acid esterase up to 4.18 ± 0.12 IU/ml using *Enterococcus lactis* strain SR1. Ferulic acid production from lignocellulosic waste was also reported. Sugarbeet pulp (*Beta vulgaris* subsp. *vulgaris*), maize bran (*Zea mays*), and rice bran were also found suitable for ferulic acid esterase production (Sharma et al 2020). Ferri et al (2020) reported ferulic acid production from wheat bran (Tritello) using a combination of enzymes for pre-treatment and final hydrolysis. Protein and sugars were removed using enzymes Alcalase® and Termamyl® followed by Pentopan® and feruloyl esterase, to hydrolyse phenols. Enzymatic process yielded higher levels of ferulic acid from wheat bran substrate in comparison to traditional acid or alkali extraction. Vangsøe et al (2019) reported the function of aleurone cells in wheat bran for production of arabinoxylan oligosaccharides. Enzymes, which act up on primary and secondary cell wall namely cellulose, β-glucanase, and xylanase, were used for treatment of wheat bran to hydrolyze wheat bran. Arabinoxylans for production of oligosaccharides were obtained from aleurone up to 164 g kg⁻¹ of wheat bran in comparison to fraction obtained from outer pericarp at 15 g kg⁻¹. Arabinoxylan: xylan ratio obtained after enzymatic hydrolysis has been found as a major factor to obtain the desired level of oligosaccahrides.

Tang et al (2021) reported pre-treatment of wheat straw using humic acid, a naturally derived surfactant and dilute sulphuric acid enhanced enzymatic hydrolysis of cellulase up to 92.9 percent. Natural surfactants play an important role in green processes of waste management. Hemicellulose degrading endo acting xylanases acted upon arabinoxylan derived from rice bran and yielded XOS and arabinoxylooligosaccharides (AXOS). Defatted rice bran was found to be the potential substrate and yielded up to 456.69 mg/g of rice bran arabinoxylan. Ultraflo Max® and Ultraflo L®, the commercially available xylanases were compared for their efficiency on enzymatic hydrolysis of rice processing wastes. Ultraflo L® converted 45.67 percent of rice arabinoxylan into short chain oligosaccharides while only 8.3 percent of waste could be converted into oligosaccharides using Ultraflo Max® (Truong and Rumpagaporn 2019).

Enzymatic hydrolysis and fermentative properties of barley bran (*Hordeum vulgare*) derived dietary fibers was reported (Karimi et al 2020). Enzymes such as α-amylase, glucoamylase, and protease were used to obtain dietary fibers and were analyzed for end products on fecal fermentation for the presence of glucans and other short chain fatty acids. α-amylase treatment produced dietary fibers containing the least pure glucan (approx. purity level 55.0 %) while combination of enzymes such as glucoamylase, α-amylase, and protease yielded glucans with moderate quality (approx. purity level 68.0 %). This study revealed the importance of enzyme combinations to be optimized to obtain a quality product.

13.5.2 Pulse and Oilseed Processing

Mirpoor et al (2021) reviewed in detail the suitability of seed oil cakes derived as a by-product of oil extraction industries as substrates for bioplastic production. Fifty percent of seed weight goes into seed oil cakes rich in bioactive secondary metabolites, fiber, and proteins. Seed cakes of four crops namely *Sesamum indicum* L. (Til), *Cannabis sativa* (Hemp), *Nigella sativa* (Cumin black), and *Cynara cardunculus* (Cardon) and their volarization to support circular economy has been elaborated. Traditional uses of seed oil cakes such as animal feed and recent developments as sustainable substrates for biodegradable plastics and biofuels and development of methods for modern biorefineries have been explained.

Achinivu et al (2021) reported the production of sinapic acid using a chemo-enzymatic method based volarization of mustard bran (*Brassica juncea*). Sinapine was successfully extracted from mustard bran followed by hydrolysis into sinapic acid up to 10.0 mg/g of bran. Chia (*Salvia hispanica* L.) expeller is obtained on extraction of chia seed oil and is widely non-utilized. Papain treatment of chia expeller produced protein hydolysates rich in antioxidants. Furthermore, antioxidant biopeptides are in demand in recent years due to their suitability for development of nutraceuticals as well as functional foods (Cotabarren et al 2019). Sesame or gingelly oil meal was hydrolyzed using enzymes such as alcalase, papain, and pepsin for production of peptides possessing aceytlcholinesterase (ACE) inhibitor properties. Protein hydrolysates obtained by enzymatic treatment were of superior quality and may find applications in development of different functional foods (Chatterjee et al 2015).

Cruz et al (2019) reported a direct recovery of lipids and soap from soap stock, a waste generated from vegetable oil refinering units. Traditionally concentrated acids were being used for the same purpose. Free fatty acids could be retracted from wastes by a combination of enzymatic treatment with hexane as a solvent. Casali et al (2021) reviewed volarization of soap stock obtained from oil processing industries. Acid oil is the by-product derived from soap stock derived out of vegetable oil processing. Industrial production of lipase as well as conversion of soap stock into biofuel, additives, surfactants, and plasticizers in continuous flow line was also discussed.

Enzymatic hydrolysis and fermentation employing two yeast species namely *Saccharomyces cerevisiae* and *Pichia stipitis* for bioethanol production from (*Elaeis guineensis*) empty palm fruit bunches after oil extraction. Alkaline pre-treatment enhanced the recovery of xylose and glucose, and a reduction of sugars up to 51.5 ± 4.1 g/L was obtained (Polprasert et al 2021). Nutongkaew et al (2020) reported the production of acetic acid and ethanol from oil palm trunk waste using fungi assisted submerged and solid state fermentation. *Trichoderma koningiopsis* TM3 obtained partially purified enzyme was used for maximum yield up to 0.48 reducing sugars per g of substrate, which were further converted into ethanol by *S. cerevisiae* TISTR5055 on addition of yeast extract to the medium. *Ceratocystis paradoxa* TT1 also produced carboxy methyl cellulase activity while *T. koningiopsis* TM3 showed xylanase activity. Acetic acid productivity increased 4.4-fold during single step fermentation while ethanol production increased by 1.55-fold during two step fermentation. Hence, the necessity of various kinds of enzymes, from different sources as well as optimal culture condition for production of valuable by-products need to be explored for efficient hydrolysis of complex biowastes. Bukhari et al (2020) reported use of enzymatic hydrolysis of effluent generated from palm oil mills for bioconversion into fermentable sugars and a biofloculant (BM 8).

Olive (*Olea europaea*) mill industrial wastes are rich in lignin content and contain polysaccharides in moderate quantities. Pomace and stones were obtained as waste after industrial production of olive pomace oil and virgin olive oil. Hot water treatment of olive pomace at 130 °C for 30 min yielded xylooligosaccharides (XOS). Commercially available enzymes such as Celluclast 1.5 L®, Saczyme Yield®, and Ultimase BWL 40® acted upon olive pomace and stones hydrolyzed it to glucan with conversion ratio of about 90 and 55 percent, respectively (Miranda et al 2019). Olive cake generated as waste by olive processing industries are bioconverted into protein rich substances with reduced anti-nutritional factors using fungi namely *Fusarium flocciferum* and *Rhizodiscina* cf. *lignyota* (Chebaibi et al 2019).

13.5.3 Fruit and Vegetable Processing

Fruit and vegetable processing generates 25–30 percent waste of total product (Kumar et al 2020). The major waste generated by fruit and vegetable processing units is peels accounting for up to 90–92 percent of the total waste generated. Volarization of useless peels into value added biomolecules with food and pharmaceutical applications using greener approaches would be essential for being cost effective, environmentally safe and help in attaining sustainable circular bioeconomy (Rifna et al 2021). Tropical fruit production is estimated as 84 million tonnes presently and around 50 percent of the production is being wasted throughout the supply chain. Eco-friendly green technologies for volarization of tropical fruit waste aiming at zero waste are being given importance to meet SDGs. Fruit processing waste is a rich source of metabolites with applications in the pharma, food, and energy sector. Impending purposes of compounds isolated from processing waste and biorefinery concepts pertaining to their generation and utilization were also elaborated (Villacís-Chiriboga et al 2020).

Dimou et al (2019) elaborated in detail the richness of fruit processing waste streams of bioactive compounds and their importance in extraction of various useful by-products such as oil, additives, and dietary fibers. Growing interest on non-thermal processing and its advantage in the recovery of bioactive molecules has also been described. Peels, seeds etc., are being used as fertilizers but fruit processing residues and their recyclability into highly value added compounds pave the way for a sustainable economy model. Sharma et al (2021) elaborated various methods for production of natural pigments from fruit and vegetable processing waste.

Ruviaro et al (2019) reported volarization of phenolic glycosides rich citrus (*Citrus* sp) juice processing wastes. Enzymes such as cellulose, β-glucosidase, pectinase, and tannase were studied individually and in combination for extraction of bioactives. Enzyme mediated biotransformation of citrus waste resulted in extraction of bioactives with higher antioxidant activity due to enhanced release of sugars from enzyme treated waste. β-glucosidase treatment at 20 U/g substrate for 24 h duration yielded maximum naringenin and hesperetin. Aglycon production was improved when a combination of β-glucosidase and tannase was used for hydrolysis of citrus peels. Citrus peel on enzymatic treatment for 1 h at pH 4.8 and temperature 60 °C using Viscozyme L® yielded 1590 of polyphenols in comparison to 1169.23 without enzyme (Nishad et al 2019).

Phenolic antioxidants were extracted from the peel of sweet lime, *Citrus limetta* Risso harvested and processed in Pakistan using enzyme assisted extraction for 30–120 minutes at a pH range of 5 to 8 at temperatures ranging from 30–75 °C. Enzyme assisted lycopene extraction from tomato peel was reported in Italy. Pectin and pectic oligosaccharide extraction from roots, florets, peels, pulps, and leaves obtained after processing of chicory, cauliflower, citrus, sugar beet, and endive in France and Finland using enzymes for 4 h at 50 °C. Statistical optimization based on response surface methodology (RSM) and artificial neural networking (ANN) of α-amylase production up to 12.19 U/ml by submerged fermentation using *Streptomyces* sp. KP314280 was reported. Orange waste (*Citrus sinensis*) derived from orange processing industry was used as a media component (Ousaadi et al 2021).

Citrus processing leads to residue generation of approximately 50 percent of the total weight of processed material. Phenolic glycosides including hesperetin and naringenin were extracted by enzymatic treatment of citrus fruit residues using 20 U g/l of β-glucosidase for 24 h incubation. Synergistic effect of β-glucosidase and tannase was also reported for aglycon production. Enhanced hesperetin content was reported on enzymatic intervention for releasing bioactive phenolics, which may be attributed to the hydrolysis of polysaccharides and increased release of polyphenols (Ruviaro et al 2020). Pomello (*Citrus maxima*) fruit waste was subjected to enzyme as well as ultrasound assisted extraction to extract phenolic molecules such as naringin and hesperidin with antimicrobial activity. High pressure extraction up to 600 MPa combined with enzymes were reported to extract bioactives from pomegranate (*Punica granatum*) peels (Alexandre et al 2018).

Wongkaew et al (2021) reported recovery of oligosaccharides from mango (*Mangifera indica*) peel of variety Thai Chok Anan using pectin transforming enzymes. Prebiotic potential of pectic oligosaccharides were tested on lactic acid bacterial namely *Bifidobacterium animalis* TISTR 2195 and *Lactobacillus reuteri* DSM 17938. Probiotic activity scores indicated that pectic oligosaccharides derived from mango peel waste comprised of glucose and fructose supported growth of tested bacteria. Supplementation of obtained prebiotics (4.0%) produced acetic and propionic acid on fermentation. Pectin derived from mango peel waste of variety chokanan was volarized to novel pectic oligosaccharides using pectinase enzyme for 2 days. The prebiotic nature of the derived oligosaccharides on *B. animalis* TISTR 2195 and *L. reuteri* DSM 17938 was also reported by their enhanced growth and survival on supplementation of prebiotic pectic oligosacchries.

Dranca and Oroian (2019) carried out research on apple (*Malus domestica*) (variety: Fălticeni) pomace waste utilization and succeeded in optimizing pectin extraction. Commercially available cellulose enzyme Celluclast 1.5L® based optimization was performed by statistical method. Enzyme hydrolysis using a concentration of 42.5 µL/g was performed for 18 h 14 min at temperature 48.3 °C and predicted 96.02 percent esterification of the side groups of the complex pectin molecule as well as a pectin yield of 6.76 percent. Moreover, characterization by FT-IR revealed similarity between pectin obtained from whole fruits of apple and citrus and apple pomace. Therefore, utilization of biomaterials considered as waste need to be explored to redirect fresh produce for mitigating hunger by increasing food availability in the supply chain.

Cellulase and tannase treated Syrah grape (*Vitis vinefera*) pomace yielded 0.74–0.76 mg gallic acid equivalent (GAE)/g of polyphenol on enzyme treatment at pH: 5 at 45 °C for 2 h (Meini et al 2019). Pomace derived from raspberry (*Rubus idaeus* L.) processing is rich in polyphenols, PUFA, and tocols. Alkaline protease (1.2U/100g substrate) acted upon the pomace for 2 h at pH 9.0 for conversion of waste into molecules rich in polyphenol with higher antioxidant activity has been reported (Saad et al 2019). Bilberry (*Vaccinium myrtillus* L.) processing waste in the form of pomace was subjected to Viscozyme L® enzyme assisted extraction of bioactive fractions. Viscozyme L® was used (2U/g) on pomace at pH of 4.5 at 46 °C for 1 h to obtain 56.15 g/100 g of extract on dry weight basis. Solid-liquid extraction using enzyme yielded fractions with higher antioxidant nature as indicated by oxygen radical absorbance capacity (ORAC), Cupric ion reducing capacity (CUPRAC) and 2-2'-Azinobis-(3-Ethyl benzothiazoline-6-sulfonic acid) assay (ABTS) in comparison to traditional method. Kitrytė et al (2017) reported enzymatic extraction of polyphenolic compounds from chokeberry (*Aronia* sp) pomace using Viscozyme L® and CeluStar XL® and extracted 15 mg GAE/g of polyphenols at 40 °C and pH of 3.5 for 7 h incubation.

Redondo-Gómez et al (2020) elaborated possible solutions for sustainable utilization of banana (*Musa* sp) processing waste with respect to Latin-American context. Valuable by-products were also highlighted, such as fibers (nano cellulose), biodegradable plastics, and biofuels, derived using a biorefinery approach. Furthermore, generation of bioenergy and treatment of waste water from banana waste was explained. Viayaraghavan et al (2019) reported production of pectinase and amylase using a mixture of peels of orange and banana. A three and two fold increase in production of pectinase and amylase was also reported with molecular weights of 37 kDa and 60 kDa, respectively. Carrot juice (*Dacus carota*) and orange juice were clarified and saccaharified using amylase and pectinase on the pulp of the fruits and vegetables during juice processing.

Pomegranate waste has been valorized for production of ellagic acid, a praised bioactive substance with promising nutraceutical potential. Solid state fermentation and resultant hydrolystic action by *Aspergillus niger* improved up to five fold in comparison to ultrasound, microwave, and solvent extraction of pure ellagic acid (Moccia et al 2019). El Barnossi et al (2021) reviewed in detail sustainable utilization of peels generated during processing of banana, pomegranate, and tangerine (*Citrus reticulate*) into useful by-products such as oil, food, cosmetics, and pharma grade substances using enzymatic treatments and other recent technologies.

Pineapple (*Ananas comosus*) waste includes crown, peels, trimmings, and stem, which account up to 60 percent of its weight. Enzymatic volarization yielded insoluble dietary fiber rich flour and chlorogenic, caffeic, and ferulic acid rich pineapple juice by-products (Campos et al 2020). Mora-Sandí et al (2021) reviewed in detail waste volarization of avocado (*Persea americana*) processing into valuable by-products such as pharmaceutical compounds and biodegradable plastics. Recent trends in nanotechnology for waste biomass volarization for scientific growth also have been detailed. Date seeds contain cellulose (20.63%), lignin (27.34%), and hemicelluloses (13.49%) and are promising substances to be hydrolyzed into useful fermentable sugars. Acid-enzyme based pre-treatment and hydrolysis obtained reducing sugars from date seed waste (Hamid et al 2020).

Cellulase treated seed waste for 6 h at temperature of 45 °C produced up to 31 g/l of pre-treated waste; combined treatments of acid and enzyme resulted in extraction of 81.0 percent of fermentable sugars.

Catalkaya and Kahveci (2019) reported an enzymatic pre-treatment prior to solvent extraction of lycopene pigment from tomato waste (*Solanum lycopersicum*). Highest lycopene revival, enhanced polyphenol recovery, and higher antioxidant nature as well as deep red color pigment was obtained by statistical optimization of the process to volarize tomato waste derived from food processing industries. Optimized conditions for hydrolysis of tomato waste using cellulase and pectinase at the rate of 0.2 ml/g for 5 h at 40 °C followed by enzyme pre-treated substrate to ethyl acetate at 5 ml/g resulted in 11.5 mg lycopene/g of oleoresin. Ethyl acetate solvent was used at 5 ml per gram of tomato waste, which aided in extraction of red colored oleoresin as a by-product with higher lycopene content and antioxidant activity. Lycopene extraction from tomato peel generated as a hybrid approach using an enzyme treatment consisting of 3.0% of pectinase and cellulase at 50ºC and pH 5.0 along with ultrasound at 20 kHz of 10 W (Ladole et al 2018).

Commercially available cellulase was used for hydrolysis of pre-treated Italian green pepper (*Capsicum annuum*) waste for production of fermentable sugars. Pre-treatment using hot water for 40 minutes at 180 °C enhanced yield of glucose up to 61.02 percent. This study conducted in Spain revealed the use of huge quantities of green pepper waste as a valuable feedstock for production of reducing sugars, which may find application in bioethanol production on further saccharification. Processing residue of red colored *Capsicum annuum* was liquefied using pectinase, Viscozyme L®, and cellulose at 60 °C for 1 h for extraction of carotenoids (Martín-Lara et al 2020). Natural dyes were extracted from cashew-nut (*Anacardium officinale*) husk using cellulase and pectinase in aqueous phase for 60–180 min at pH of 9.5 using Taghuchi statistical optimization. Highly bioactive phenolic compounds such as flavanones were extracted from residues of Brazilian citrus cultivar such as Hamlin, Valencia, Pera riu, and Pera Natal processing waste. Tannase, pectinase, and cellulase were used for 30 h at a temperature of 40 °C and proved good results (Madeira Jr et al 2015).

Endoxylanase obtained using *A. niger* MTCC 9687 was used to produce XOS from cauliflower stalk. Prebiotic efficiency of produced XOS was tested on lactic acid bacteria such as *B. bifidum*, *Lactobacillus delbrueckii* ssp. Helveticus, and *Lactiplantibacillus plantarum*. Antimicrobial and antioxidant nature of XOS was also reported as well as bone cancer cells (MG-63) lost their viability on treatment with XOS (Majumdar et al 2021). Jackfruit (*Artocarpus heterophyllus*) seed powder derived from jack fruit seeds after usage of fruit was hydrolysed using glucoamylase and amylase for production of L-lactic acid up to 109 g/L of medium (Nair et al 2016).

Microbial cultivation is essential for production of cellular and intercellular secreted microbial enzymes. Solid state and submerged cultivation requires cost effective substrates for sustainable production of food and industrial enzymes. Watermelon (*Citrullus lanatus*) peel waste has been substituted as universal fungal growth medium in lieu of potato dextrose agar, which is used for cultivation of yeast and molds. Taguchi optimized medium composition (g/L) comprises of homogenized fresh watermelon peel 500.0 and dextrose 10.0. *Fusarium oxysporum* and *A. niger* were cultivated successfully in the optimized media and an increase in dry biomass weight of up to 199 percent was also reported (Hasanin and Hashem 2020). Enzymatic hydrolysis of fruit and vegetable waste through solid state fermentation using *A. niger* and *Rhizopus oryzae* with addition of fructose and glucose yielded succinic acid up to 27 g/L or 1.18g/ g sugar (Dessie et al 2018). Redondo-Gómez et al (2020) reviewed the conversion of biowaste into lactic acid by the fermentation process and enzymes are found to enhance the rate of conversion of wastes into lactic acid. Carlos et al (2020) reviewed microbial fermentation based volarization methods used for conversion of fruit and vegetable processing waste into useful by-products. Lactic acid in its pure form could be obtained by fermentation using *Penicillium* species. Proteins were produced on waste fermentation by *Bacillus subtilis*, *Rhizopus oligosporus* etc. *Aspergillus*, *Trichoderma*, and *Yarrowia* based fermentation yielded bioactive carbohydrates in the form of prebiotic dietary fibers.

13.5.4 Miscellaneous

Tanruean et al (2021) reported enzymatic volarization of Thai based agricultural wastes such as corncobs, rice straw, wheat bran, peanut husks palm residue, red tea (*Camelia sinensis*) leaves etc. Solid state fermentation of *Thermoascus aurantiacus* SL16W, a thermophilic fungus was carried out using the agricultural residue as substrate for production of industrially important enzymes. Phytase (84.1 U/g), xylanase (162 U/g) and endoglucanase (19 U/g) production was maximum in substrates namely rice bran, red tea leaves and wheat bran, respectively. Glucans are biopolymers obtained from D-glucose molecules with medical applications. Hence, lignocellulosic agri-food biowaste are promising substrates for production of glucan (Abdeshahian et al 2021).

Cebin et al (2021) reported extraction of hemicellulose xylan from pea (*Pisum sativum*) pod and walnut (*Juglans regia*) shells by treatment with alkali and delignification. Endoxylanases belonging to glycosyl hydrolases 11 (GH 11) hydrolyzed hemicelluloses into prebiotic xylooligosaccharides (XOS) namely xylobiose and xylotriose. Pod husks of cocoa (*Theobroma cocoa*) were explored for pectin extraction using enzymatic methods. Statistical optimization using response surface methodology (RSM) produced pectin up to 10.20 g/100 g followed by 52.06 g and 5.31 galacturonic acid from pectin and pod husk substrate, respectively. Enzymatic volarization using cellulose was found to produce superior quality pectin in terms of physico-chemical and rheological parameters in comparison to sonication and chemical extraction methods (Hennessey-Ramos et al 2021). Hazel nut (*Corylus hazel*) shells were pre-treated using hot water, dilute acid, and alkali to recover glucose for further use. Studies indicated the necessity of partial removal of lignin and destruction of the crystalline structure of cellulose and hemicelluloses for effective enzymatic hydrolysis of lignocelluloses rich substrate. Therefore, pre-treatment of generated waste has been identified as an essential step in valorization of waste to meet targets of circular bioeconomy (Hoşgün et al 2021).

Brewer's spent grain, a solid waste generated from brewery are generally used as animal feed. Llimós et al (2020) reported production of enzymes by solid state fermentation using brewer's spent grain as substrate and also produced fermentable sugars using the produced enzymes. Furthermore, *Burkholderia cepacia* and *Cupriavidus necator* were cultivated for production of polyhydroxyalkanoates up to 9.0 ± 0.44 mg/g dry substrate thus ensuring two-step volarization using enzymatic hydrolysis. Kavalopoulos et al (2021) explored enzymatic saccharification of acid pre-treated brewer's spent grain using cellulase for an integrated biorefinery, which could produce three by-products namely biogas, bioethanol, and biodiesel.

Polyhydroxyalkanoates are gaining importance in the field of food packaging owing to their biodegradable nature. Food processing leftovers such as brewer's spent grain, olive and grape pomace were used as substrates for solid state fermentation and enzymatic action. Brewer's spent grain substrate could produce 0.33 g/kg of polyhydroxyalkanoates in one hour (Martínez-Avila et al 2021). Hence, sustainable substrates have to be explored for production of biodegradable microbial polymers, which would pave the way for reduction in the overall cost incurred in the production line.

Zaharudin et al (2018) reported volarization of used cooking oil using an immobilized enzyme system. Lipases act upon used cooking oils to derive useful by-products such as free fatty acids and glycerol. Okino-Delgado et al (2017) reported that lipase hydrolyzed waste soybean cooking oil was less toxic in comparison to untreated cooking oil. Orange waste was used to derive lipase enzyme capable of transforming lipids and used to treat cooking oil waste ensuring a circular bioeconomy process. Soybean cooking oil waste has been subjected to Lipozyme TL IM®, alipase enzyme assisted ultrasound hydrolysis for removal of formed free fatty acids due to oxidation reactions (Zenevicz et al 2016). Bakery waste was used as substrate for solid state fermentation of *Monascus purpureus* and bioconverted into biocolorant for use in textile and food applications. Initial substrate contained 5g/L of glucose, which was converted into color of concentration 24AU/g glucose. Glucoamylase (8U/g) and protease (117U/g) were also produced using the same substrate rich in sugars and amino acids (Haque et al 2016).

13.6 OPPORTUNITIES AND PROSPECTS

Food waste management practices require a holistic approach from understanding the kinds of wastes generated as well as adoption of mitigation strategies to achieve goals of sustainable development as per principles of circular bioeconomy. Scientific research needs to be taken up to provide substrate and region specific, economical and feasible methods for establishing biorefineries based on various food wastes derived along the food supply chain. In addition to valorization strategies, steps for reduction of food losses and food wastes occurring at various stages of production and consumption are the need of the hour. Sustainable goals to be achieved in food waste management entail a significant assessment and complete overhaul of food systems from the point of production until the consumer. Use of virgin feed stocks for production of eco-friendly alternatives for plastics and other fossil fuels invites lot of criticism as it contributes to loss of food and feed available for use. Hence, use of food processing derived waste and other lignocellulosic residues as substrates for production of value added by-products in the form of biofuels, bioactive molecules, bioactive peptides, biodegradable polymers etc., is a feasible and welcomed advancement. Furthermore, understanding the importance of prevention and reduction of food waste and conversion of generated wastes using green chemistry based enzymatic catalysis involving methods arekey strategies of the modern world. Environmentally benign technologies are gaining popularity due to the sustainable solutions they offer for problems related to environmental pollution. Behavioral changes through creation of awareness among producers and consumers, creation of social responsibility by corporate involved in agro-food industrial setup as well as bioconversion of waste into wealth, by incorporation of core principles of circular bioeconomy are the paramount measures to ease the burden of food waste on our planet based on win-win approaches.

REFERENCES

Abdeshahian, Peyman, Jesus Jimenez Ascencio, Rafael R. Philippini, **Felipe Antonio** Fernandes Antunes, Andre S. de Carvalho, Mojgan Abdeshahian, Julio Cesar dos Santos, and Silvio Silverio da Silva. "Valorization of lignocellulosic biomass and agri-food processing wastes for production of glucan polymer." *Waste and Biomass Valorization* 12, no. 6 (2021): 2915–2931.
Achinivu, Ezinne C., Amandine L. Flourat, Fanny Brunissen, and Florent Allais. "Valorization of waste biomass from oleaginous "oil-bearing" seeds through the biocatalytic production of sinapic acid from mustard bran." *Biomass and Bioenergy* 145 (2021): 105940.
Alexandre, Elisabete MC, Silvia A. Moreira, Luís MG Castro, Manuela Pintado, and Jorge A. Saraiva. "Emerging technologies to extract high added value compounds from fruit residues: Sub/supercritical, ultrasound-, and enzyme-assisted extractions." *Food Reviews International* 34, no. 6 (2018): 581–612.
Baldassarre, Stefania, Neha Babbar, Sandra Van Roy, Winnie Dejonghe, Miranda Maesen, Stefano Sforza, and Kathy Elst. "Continuous production of pectic oligosaccharides from onion skins with an enzyme membrane reactor." *Food Chemistry* 267 (2018): 101–110.
Barcelos, Mayara CS, Cintia L. Ramos, Mohammed Kuddus, Susana Rodriguez-Couto, Neha Srivastava, Pramod W. Ramteke, Pradeep K. Mishra, and Gustavo Molina. "Enzymatic potential for the valorization of agro-industrial by-products." *Biotechnology Letters* (2020): 1–29.
Bukhari, Nurul Adela, **Soh Kheang** Loh, **Abu Bakar** Nasrin, and **Jamaliah Md** Jahim. "Enzymatic hydrolysate of palm oil mill effluent as potential substrate for bioflocculant BM-8 production." *Waste and Biomass Valorization* 11, no. 1 (2020): 17–29.
Campos, Débora A., Tânia B. Ribeiro, José A. Teixeira, Lorenzo Pastrana, and Maria Manuela Pintado. "Integral valorization of pineapple (ananas comosus L.) by-products through a green chemistry approach towards added value ingredients." *Foods* 9, no. 1 (2020): 60.
Carlos, W. G., Dela Cruz, C. S., Cao, B., Pasnick, S., & Jamil, S. "Novel Wuhan (2019-nCoV) Coronavirus." *Am J Respir Crit Care Med*, (20220): 7–8.
Casali, Beatrice, Elisabetta Brenna, Fabio Parmeggiani, Davide Tessaro, and Francesca Tentori. "Enzymatic methods for the manipulation and valorization of soapstock from vegetable oil refining processes." *Sustainable Chemistry* 2, no. 1 (2021): 74–91.

Casoni, Andrés I., Patricia M. Hoch, María A. Volpe, and Victoria S. Gutierrez. "Catalytic conversion of furfural from pyrolysis of sunflower seed hulls for producing bio-based furfuryl alcohol." *Journal of Cleaner Production* 178 (2018): 237–246.

Catalkaya, Gizem, and Derya Kahveci. "Optimization of enzyme assisted extraction of lycopene from industrial tomato waste." *Separation and Purification Technology* 219 (2019): 55–63.

Cebin, Aleksandra Vojvodić, Marie-Christine Ralet, Jacqueline Vigouroux, Sara Karača, Arijana Martinić, Draženka Komes, and Estelle Bonnin. "Valorisation of walnut shell and pea pod as novel sources for the production of xylooligosaccharides." *Carbohydrate Polymers* 263 (2021): 117932.

Chatterjee, Roshni, Tanmoy Kumar Dey, Mahua Ghosh, and Pubali Dhar. "Enzymatic modification of sesame seed protein, sourced from waste resource for nutraceutical application." *Food and Bioproducts Processing* 94 (2015): 70–81.

Chebaibi, Salima, Mathilde Leriche Grandchamp, Grégoire Burgé, Tiphaine Clément, Florent Allais, and Fatiha Laziri. "Improvement of protein content and decrease of anti-nutritional factors in olive cake by solid-state fermentation: A way to valorize this industrial by-product in animal feed." *Journal of Bioscience and Bioengineering* 128, no. 3 (2019): 384–390.

Cho, Eun Jin, Ly Thi Phi Trinh, Younho Song, Yoon Gyo Lee, and Hyeun-Jong Bae. "Bioconversion of biomass waste into high value chemicals." *Bioresource Technology* 298 (2020): 122386.

Cipolatti, Eliane Pereira, Martina Costa Cerqueira Pinto, Rosana Oliveira Henriques, José Carlos Costa da Silva Pinto, Aline Machado de Castro, Denise Maria Guimarães Freire, and Evelin Andrade Manoel. "Enzymes in green chemistry: The state of the art in chemical transformations." *Advances in Enzyme Technology* (2019): 137–151.

Cotabarren, Juliana, Adriana Mabel Rosso, Mariana Tellechea, Javier García-Pardo, Julia Lorenzo Rivera, Walter David Obregón, and Mónica Graciela Parisi. "Adding value to the chia (Salvia hispanica L.) expeller: Production of bioactive peptides with antioxidant properties by enzymatic hydrolysis with Papain." *Food Chemistry* 274 (2019): 848–856.

Cruz, M., M. F. Almeida, J. M. Dias, and M. C. Alvim-Ferraz. "Adding value to soapstocks from the vegetable oil refining: Alternative processes." In *Wastes: Solutions, Treatments and Opportunities III*, pp. 313–318. CRC Press, 2019.

Dessie, Wubliker, Wenming Zhang, Fengxue Xin, Weiliang Dong, Min Zhang, Jiangfeng Ma, and Min Jiang. "Succinic acid production from fruit and vegetable wastes hydrolyzed by on-site enzyme mixtures through solid state fermentation." *Bioresource Technology* 247 (2018): 1177–1180.

Dimou, Charalampia, Haralabos C. Karantonis, Dimitrios Skalkos, and Antonios E. Koutelidakis. "Valorization of fruits by-products to unconventional sources of additives, oil, biomolecules and innovative functional foods." *Current Pharmaceutical Biotechnology* 20, no. 10 (2019): 776–786.

Dranca, Florina, and Mircea Oroian. "Optimization of pectin enzymatic extraction from Malus domestica 'fălticeni' apple pomace with celluclast 1.5 L." *Molecules* 24, no. 11 (2019): 2158.

El Barnossi, Azeddin, Fatimazhrae Moussaid, and Abdelilah Iraqi Housseini. "Tangerine, banana and pomegranate peels valorisation for sustainable environment: A review." *Biotechnology Reports* (2021): e00574.

Esparza, Irene, Nerea Jiménez-Moreno, Fernando Bimbela, Carmen Ancín-Azpilicueta, and Luis M. Gandía. "Fruit and vegetable waste management: Conventional and emerging approaches." *Journal of Environmental Management* 265 (2020): 110510.

EU. A Sustainable Bioeconomy for Europe: Strengthening the Connection between Economy, Society and the Environment; European Commission: Brussels, Belgium, 2018.

FAO. 2019. The State of Food and Agriculture 2019. Moving forward on food loss and waste reduction. Rome. Licence: CC BY-NC-SA 3.0 IGO

FAO. 2020. *The State of Food and Agriculture 2020. Overcoming water challenges in agriculture.* Rome.

Ferri, Maura, Anton Happel, Giulio Zanaroli, Marco Bertolini, Stefano Chiesa, Mauro Commisso, Flavia Guzzo, and Annalisa Tassoni. "Advances in combined enzymatic extraction of ferulic acid from wheat bran." *New Biotechnology* 56 (2020): 38–45.

Galanakis, Charis M., J. Cvejic, V. Verardo, and A. Segura-Carretero. "Food use for social innovation by optimizing food waste recovery strategies." In *Innovation Strategies in the Food Industry*, pp. 211–236. Academic Press, 2016.

Gu, Meidong, Huicheng Fang, Yuhang Gao, Tan Su, Yuge Niu, and Liangli Lucy Yu. "Characterization of enzymatic modified soluble dietary fiber from tomato peels with high release of lycopene." *Food Hydrocolloids* 99 (2020): 105321.

Hamid, Hekma Salem Hasan Ba, and Ku Syahidah Ku Ismail. "Optimization of enzymatic hydrolysis for acid pretreated date seeds into fermentable sugars." *Biocatalysis and Agricultural Biotechnology* 24 (2020): 101530.

Haque, Md Ariful, Vasiliki Kachrimanidou, Apostolis Koutinas, and Carol Sze Ki Lin. "Valorization of bakery waste for biocolorant and enzyme production by Monascus purpureus." *Journal of Biotechnology* 231 (2016): 55–64.

Hasanin, Mohamed S., and Amr H. Hashem. "Eco-friendly, economic fungal universal medium from watermelon peel waste." *Journal of Microbiological Methods* 168 (2020): 105802.

Hashemi, Seyed Sajad, Safoora Mirmohamadsadeghi, and Keikhosro Karimi. "Biorefinery development based on whole safflower plant." *Renewable Energy* 152 (2020): 399–408.

Hassan, Md Kamrul, Ranjana Chowdhury, Shiladitya Ghosh, Dinabandhu Manna, Ari Pappinen, and Suvi Kuittinen. "Energy and environmental impact assessment of Indian rice straw for the production of second-generation bioethanol." *Sustainable Energy Technologies and Assessments* 47 (2021): 101546.

Hennessey-Ramos, Licelander, Walter Murillo-Arango, Juliana Vasco-Correa, and Isabel Cristina Paz Astudillo. "Enzymatic Extraction and Characterization of Pectin from Cocoa Pod Husks (Theobroma cacao L.) Using Celluclast® 1.5 L." *Molecules* 26, no. 5 (2021): 1473.

Hoşgün, Emir Zafer, Suzan Biran Ay, and Berrin Bozan. "Effect of sequential pretreatment combinations on the composition and enzymatic hydrolysis of hazelnut shells." *Preparative Biochemistry & Biotechnology* 51, no. 6 (2021): 570–579.

Ioannidou, Sofia Maria, Katiana Filippi, Ioannis K. Kookos, Apostolis Koutinas, and Dimitrios Ladakis. "Techno-economic evaluation and life cycle assessment of a biorefinery using winery waste streams for the production of succinic acid and value-added co-products." *Bioresource Technology* 348 (2022): 126295.

Jain, Archana, Surendra Sarsaiya, Mukesh Kumar Awasthi, Ranjan Singh, Rishabh Rajput, Umesh C. Mishra, Jishuang Chen, and Jingshan Shi. "Bioenergy and bio-products from bio-waste and its associated modern circular economy: Current research trends, challenges, and future outlooks." *Fuel* 307 (2022): 121859.

Jegannathan, Kenthorai Raman, and Per Henning Nielsen. "Environmental assessment of enzyme use in industrial production–a literature review." *Journal of Cleaner Production* 42 (2013): 228–240.

Jiménez-Moreno, Nerea, Irene Esparza, Fernando Bimbela, Luis M. Gandía, and Carmen Ancín-Azpilicueta. "Valorization of selected fruit and vegetable wastes as bioactive compounds: Opportunities and challenges." *Critical Reviews in Environmental Science and Technology* 50, no. 20 (2020): 2061–2108.

Karimi, Reza, Mohammad Hossein Azizi, Mohammad Ali Sahari, and Ahmad Enosh Kazem. "In vitro fermentation profile of soluble dietary fibers obtained by different enzymatic extractions from barley bran." *Bioactive Carbohydrates and Dietary Fibre* 21 (2020): 100205.

Kavalopoulos, Michael, Vasileia Stoumpou, Andreas Christofi, Sofia Mai, Elli Maria Barampouti, Konstantinos Moustakas, Dimitris Malamis, and Maria Loizidou. "Sustainable valorisation pathways mitigating environmental pollution from brewers' spent grains." *Environmental Pollution* 270 (2021): 116069.

Kitrytė, Vaida, Vaida Kraujalienė, Vaida Šulniūtė, Audrius Pukalskas, and Petras Rimantas Venskutonis. "Chokeberry pomace valorization into food ingredients by enzyme-assisted extraction: Process optimization and product characterization." *Food and Bioproducts Processing* 105 (2017): 36–50.

Kochepka, Debora M., Lais P. Dill, Gustavo H. Couto, Nadia Krieger, and Luiz P. Ramos. "Production of fatty acid ethyl esters from waste cooking oil using Novozym 435 in a solvent-free system." *Energy & Fuels* 29, no. 12 (2015): 8074–8081.

Kover, Anna, Doris Kraljić, Rose Marinaro, and Eldon R. Rene. "Processes for the valorization of food and agricultural wastes to value-added products: recent practices and perspectives." *Systems Microbiology and Biomanufacturing* (2021): 1–17.

Kumar, Bikash, and Pradeep Verma. "Application of hydrolytic enzymes in biorefinery and its future prospects." In *Microbial Strategies for Techno-economic Biofuel Production*, pp. 59–83. Springer, 2020.

Ladole, Mayur R., Rajiv R. Nair, Yashomangalam D. Bhutada, Vinod D. Amritkar, and Aniruddha B. Pandit. "Synergistic effect of ultrasonication and co-immobilized enzymes on tomato peels for lycopene extraction." *Ultrasonics Sonochemistry* 48 (2018): 453–462.

Leong, Hui Yi, Chih-Kai Chang, Kuan Shiong Khoo, Kit Wayne Chew, Shir Reen Chia, Jun Wei Lim, Jo-Shu Chang, and Pau Loke Show. "Waste biorefinery towards a sustainable circular bioeconomy: a solution to global issues." *Biotechnology for Biofuels* 14, no. 1 (2021): 1–15.

Liu, Hong, Vinay Kumar, Linjing Jia, Surendra Sarsaiya, Deepak Kumar, Ankita Juneja, Zengqiang Zhang et al "Biopolymer poly-hydroxyalkanoates (PHA) production from apple industrial waste residues: A review." *Chemosphere* (2021): 131427.

Llimós, Jordi, Oscar Martínez-Avila, Elisabet Marti, Carlos Corchado-Lopo, Laia Llenas, Teresa Gea, and Sergio Ponsá. "Brewer's spent grain biotransformation to produce lignocellulolytic enzymes and polyhydroxyalkanoates in a two-stage valorization scheme." *Biomass Conversion and Biorefinery* (2020): 1–12.

Lu, Jiasheng, Yang Lv, Xiujuan Qian, Yujia Jiang, Min Wu, Wenming Zhang, Jie Zhou, Weiliang Dong, Fengxue Xin, and Min Jiang. "Current advances in organic acid production from organic wastes by using microbial co-cultivation systems." *Biofuels, Bioproducts and Biorefining* 14, no. 2 (2020): 481–492.

Madeira Jr, Jose Valdo, and Gabriela Alves Macedo. "Simultaneous extraction and biotransformation process to obtain high bioactivity phenolic compounds from Brazilian citrus residues." *Biotechnology Progress* 31, no. 5 (2015): 1273–1279.

Majumdar, Sayari, D. K. Bhattacharyya, and Jayati Bhowal. "Evaluation of nutraceutical application of xylooligosaccharide enzymatically produced from cauliflower stalk for its value addition through a sustainable approach." *Food & Function* 12, no. 12 (2021): 5501–5523.

Mak, Tiffany MW, Xinni Xiong, Daniel CW Tsang, K. M. Iris, and Chi Sun Poon. "Sustainable food waste management towards circular bioeconomy: Policy review, limitations and opportunities." *Bioresource Technology* 297 (2020): 122497.

Martínez-Avila, Oscar, Jordi Llimós, and Sergio Ponsá. "Integrated solid-state enzymatic hydrolysis and solid-state fermentation for producing sustainable polyhydroxyalkanoates from low-cost agro-industrial residues." *Food and Bioproducts Processing* 126 (2021): 334–344.

Martín-Lara, M. A., L. Chica-Redecillas, A. Pérez, G. Blázquez, G. Garcia-Garcia, and M. Calero. "Liquid hot water pretreatment and enzymatic hydrolysis as a valorization route of Italian green pepper waste to delivery free sugars." *Foods* 9, no. 11 (2020): 1640.

Martin-Rios, Carlos, Christine Demen-Meier, Stefan Gössling, and Clémence Cornuz. "Food waste management innovations in the foodservice industry." *Waste Management* 79 (2018): 196–206.

Mattila, Malla, Nina Mesiranta, Elina Närvänen, Outi Koskinen, and Ulla-Maija Sutinen. "Dances with potential food waste: Organising temporality in food waste reduction practices." *Time & Society* 28, no. 4 (2019): 1619–1644.

Meini, María-Rocío, Ignacio Cabezudo, Carlos E. Boschetti, and Diana Romanini. "Recovery of phenolic antioxidants from Syrah grape pomace through the optimization of an enzymatic extraction process." *Food Chemistry* 283 (2019): 257–264.

Melnichuk, Natasha, Mauricio J. Braia, Pablo A. Anselmi, María-Rocío Meini, and Diana Romanini. "Valorization of two agroindustrial wastes to produce alpha-amylase enzyme from Aspergillus oryzae by solid-state fermentation." *Waste Management* 106 (2020): 155-161.

Miranda, Isabel, Rita Simões, Barbara Medeiros, Kesavan Madhavan Nampoothiri, Rajeev K. Sukumaran, Devi Rajan, Helena Pereira, and Suzana Ferreira-Dias. "Valorization of lignocellulosic residues from the olive oil industry by production of lignin, glucose and functional sugars." *Bioresource technology* 292 (2019): 121936.

Mirpoor, Seyedeh Fatemeh, C. Valeria L. Giosafatto, and Raffaele Porta. "Biorefining of seed oil cakes as industrial co-streams for production of innovative bioplastics. A review." *Trends in Food Science & Technology* 109 (2021): 259–270.

Moccia, Federica, Adriana C. Flores-Gallegos, Mónica L. Chávez-González, Leonardo Sepúlveda, Stefania Marzorati, Luisella Verotta, Lucia Panzella, Juan A. Ascacio-Valdes, Cristobal N. Aguilar, and Alessandra Napolitano. "Ellagic acid recovery by solid state fermentation of pomegranate wastes by Aspergillus niger and Saccharomyces cerevisiae: A comparison." *Molecules* 24, no. 20 (2019): 3689.

Mora-Sandí, Anthony, Abigail Ramírez-González, Luis Castillo-Henríquez, Mary Lopretti-Correa, and José Roberto Vega-Baudrit. "Persea Americana agro-industrial waste biorefinery for sustainable high-value-added products." *Polymers* 13, no. 11 (2021): 1727.

Mou, Jinhua, Chong Li, Xiaofeng Yang, Guneet Kaur, and Carol Sze Ki Lin. "Overview of waste valorisation concepts from a circular economy perspective." *Waste Valorisation: Waste Streams in a Circular Economy* (2020): 1–11.

Nadar, Shamraja S., Priyanka Rao, and Virendra K. Rathod. "Enzyme assisted extraction of biomolecules as an approach to novel extraction technology: A review." *Food Research International* 108 (2018): 309–330.

Nair, Nimisha Rajendran, K. Madhavan Nampoothiri, Rintu Banarjee, and Gopal Reddy. "Simultaneous saccharification and fermentation (SSF) of jackfruit seed powder (JFSP) to L-lactic acid and to polylactide polymer." *Bioresource Technology* 213 (2016): 283–288.

NAAS National Academy of Agricultural Sciences. Saving the Harvest: Reducing the Food Loss and Waste; Policy Brief No. 5.; National Academy of Agricultural Sciences: New Delhi, India, 2019. http://naasindia.org/documents/Saving%20the%20Harvest.pdf

Ng, Hui Suan, Phei Er Kee, Hip Seng Yim, Po-Ting Chen, Yu-Hong Wei, and John Chi-Wei Lan. "Recent advances on the sustainable approaches for conversion and reutilization of food wastes to valuable bioproducts." *Bioresource technology* 302 (2020): 122889.

Nishad, Jyoti, Supradip Saha, and Charanjit Kaur. "Enzyme-and ultrasound-assisted extractions of polyphenols from Citrus sinensis (cv. Malta) peel: A comparative study." *Journal of Food Processing and Preservation* 43, no. 8 (2019): e14046.

Norrrahim, Mohd Nor Faiz, Muhammad Roslim Muhammad Huzaifah, Mohammed Abdillah Ahmad Farid, Siti Shazra Shazleen, Muhammad Syukri Mohamad Misenan, Tengku Arisyah Tengku Yasim-Anuar, Jesuarockiam Naveen et al "Greener pretreatment approaches for the valorisation of natural fibre biomass into bioproducts." *Polymers* 13, no. 17 (2021): 2971.

Nutongkaew, Tanawut, Poonsuk Prasertsan, Chonticha Leamdum, Supalak Sattayasamitsathit, and Pongsak Noparat. "Bioconversion of oil palm trunk residues hydrolyzed by enzymes from newly isolated fungi and use for ethanol and acetic acid production under two-stage and simultaneous fermentation." *Waste and Biomass Valorization* 11, no. 4 (2020): 1333–1347.

Okino-Delgado, Clarissa Hamaio, Débora Zanoni do Prado, Roselaine Facanali, Márcia Mayo Ortiz Marques, Augusto Santana Nascimento, Célio Junior da Costa Fernandes, William Fernando Zambuzzi, and Luciana Francisco Fleuri. "Bioremediation of cooking oil waste using lipases from wastes." *PLoS One* 12, no. 10 (2017): e0186246.

Onu, Peter, and Charles Mbohwa. *Agricultural Waste Diversity and Sustainability Issues: Sub-Saharan Africa as a Case Study*. Academic Press, 2021.

Ousaadi, Mouna Imene, Fateh Merouane, Mohammed Berkani, Fares Almomani, Yasser Vasseghian, and Mahmoud Kitouni. "Valorization and optimization of agro-industrial orange waste for the production of enzyme by halophilic Streptomyces sp." *Environmental Research* (2021): 111494.

Paiho, Satu, Elina Mäki, Nina Wessberg, Martta Paavola, Pekka Tuominen, Maria Antikainen, Jouko Heikkilä, Carmen Antuña Rozado, and Nusrat Jung. "Towards circular cities – Conceptualizing core aspects." *Sustainable Cities and Society* 59 (2020): 102143.

Papargyropoulou, Effie, Nigel Wright, Rodrigo Lozano, Julia Steinberger, Rory Padfield, and Zaini Ujang. "Conceptual framework for the study of food waste generation and prevention in the hospitality sector." *Waste Management* 49 (2016): 326–336.

Patel, Alok, Kateřina Hrůzová, Ulrika Rova, Paul Christakopoulos, and Leonidas Matsakas. "Sustainable biorefinery concept for biofuel production through holistic valorization of food waste." *Bioresource Technology* 294 (2019): 122247.

Peña-Barrientos, Alberto, María de Jesús Perea-Flores, Miguel Ángel Vega-Cuellar, Abelardo Flores-Vela, Mayra Beatriz Gómez-Patiño, Daniel Arrieta-Báez, and Gloria Davila-Ortiz. "Chemical and Microstructural Characterization of Vanilla Waste Compounds (Vanilla planifolia, Jackson) Using Eco-Friendly Technology." *Waste and Biomass Valorization* (2021): 1–16.

Pfaltzgraff, Lucie A., Emma C. Cooper, Vitaly Budarin, and James H. Clark. "Food waste biomass: a resource for high-value chemicals." *Green Chemistry* 15, no. 2 (2013): 307–314.

Pham, Thi Phuong Thuy, Rajni Kaushik, Ganesh K. Parshetti, Russell Mahmood, and Rajasekhar Balasubramanian. "Food waste-to-energy conversion technologies: Current status and future directions." *Waste Management* 38 (2015): 399–408.

Plazzotta Stella and Lara Manzocco Food waste valorization In: Galanakis, Charis M., ed. *Saving food: Production, supply chain, food waste and food consumption*. Academic Press, 2019. Pages 279–313.

Polprasert, Supawadee, Ornjira Choopakar, and Panagiotis Elefsiniotis. "Bioethanol production from pretreated palm empty fruit bunch (PEFB) using sequential enzymatic hydrolysis and yeast fermentation." *Biomass and Bioenergy* 149 (2021): 106088.

Ragauskas A. J., Williams C. K., Davison B. H., Britovsek G., Cairney J., Eckert C. A., Frederick W. J., Hallett J. P., Leak D. J., Liotta C. L., Mielenz J. R., Murphy R., Templer R., and T. Tschaplinski. The path forward for biofuels and biomaterials. *Science* (2006) 311: 484–489

Ratz-Łyko, Anna, and Jacek Arct. "Evaluation of antioxidant and antimicrobial properties of enzymatically hydrolysed Cucurbita pepo and Linum usitatissimum seedcakes." *Food Science and Biotechnology* 24, no. 5 (2015): 1789–1796.

Redondo-Gómez, Carlos, Maricruz Rodríguez Quesada, Silvia Vallejo Astúa, José Pablo Murillo Zamora, Mary Lopretti, and José Roberto Vega-Baudrit. "Biorefinery of biomass of agro-industrial banana waste to obtain high-value biopolymers." *Molecules* 25, no. 17 (2020): 3829.

Rifna, E. J., N. N. Misra, and Madhuresh Dwivedi. "Recent advances in extraction technologies for recovery of bioactive compounds derived from fruit and vegetable waste peels: A review." *Critical Reviews in Food Science and Nutrition* (2021): 1–34.

Rudraraju, Vaishnavi, Surekha Arasu, and Ashish Rawson. "Nutritional composition and utilization of pulse processing by-products." *Pulse Foods: Processing, Quality and Nutraceutical Applications* (2021): 461–486. In: Editor(s): Brijesh K. Tiwari, Aoife Gowen, Brian McKenna, Pulse Foods (Second Edition), Academic Press, 2021.

Ruviaro, Amanda Roggia, Paula de Paula Menezes Barbosa, and Gabriela Alves Macedo. "Enzyme-assisted biotransformation increases hesperetin content in citrus juice by-products." *Food Research International* 124 (2019): 213–221.

Ruviaro, Amanda Roggia, Paula de Paula Menezes Barbosa, Isabela Mateus Martins, Amanda Rejane Alves de Ávila, Vania Mayumi Nakajima, Aline Rodrigues Dos Prazeres, Juliana Alves Macedo, and Gabriela Alves Macedo. "Flavanones biotransformation of citrus by-products improves antioxidant and ACE inhibitory activities in vitro." *Food Bioscience* 38 (2020): 100787.

Saad, Naima, François Louvet, Stéphane Tarrade, Emmanuelle Meudec, Karine Grenier, Cornelia Landolt, Tan-Sothea Ouk, and Philippe Bressollier. "Enzyme-Assisted Extraction of Bioactive Compounds from Raspberry (Rubus idaeus L.) Pomace." *Journal of Food Science* 84, no. 6 (2019): 1371–1381.

Sabater, Carlos, Lorena Ruiz, Susana Delgado, Patricia Ruas-Madiedo, and Abelardo Margolles. "Valorization of Vegetable Food Waste and By-Products Through Fermentation Processes." *Frontiers in Microbiology* 11 (2020): 2604.

Šelo, Gordana, Mirela Planinić, Marina Tišma, Srećko Tomas, Daliborka Koceva Komlenić, and Ana Bucić-Kojić. "A comprehensive review on valorization of agro-food industrial residues by solid-state fermentation." *Foods* 10, no. 5 (2021): 927.

Sharma, Abha, Anamika Sharma, Jyoti Singh, Pushpendra Sharma, Govind Singh Tomar, Surender Singh, and Lata Nain. "A biorefinery approach for the production of ferulic acid from agroresidues through ferulic acid esterase of lactic acid bacteria." *3 Biotech* 10, no. 8 (2020): 1–10.

Sharma, Minaxi, Zeba Usmani, Vijai Kumar Gupta, and Rajeev Bhat. "Valorization of fruits and vegetable wastes and by-products to produce natural pigments." *Critical Reviews in Biotechnology* 41, no. 4 (2021): 535–563.

Sharma, Poonam, Vivek K. Gaur, Ranjna Sirohi, Sunita Varjani, Sang Hyon Kim, and Jonathan WC Wong. "Sustainable processing of food waste for production of bio-based products for circular bioeconomy." *Bioresource Technology* (2021): 124684.

Shehu, Isah, Taiwo O. Akanbi, Victor Wyatt, and Alberta NA Aryee. "Fruit, Nut, Cereal, and Vegetable Waste Valorization to Produce Biofuel." *By-products from Agriculture and Fisheries: Adding Value for Food, Feed, Pharma, and Fuels* (2019): 665–684.

Skendi, Adriana, Kyriaki G. Zinoviadou, Maria Papageorgiou, and João M. Rocha. "Advances on the valorisation and functionalization of by-products and wastes from cereal-based processing industry." *Foods* 9, no. 9 (2020): 1243.

Sodhi, A. S., Sharma, N., Bhatia, S., Verma, A., Soni, S., & Batra, N. (2021). Insights on sustainable approaches for production and applications of value added products. *Chemosphere*, 286(Pt 1), 131623. Advance online publication.

Sowbhagya, H. B., and V. N. Chitra. "Enzyme-assisted extraction of flavorings and colorants from plant materials." *Critical Reviews in Food Science and Nutrition* 50, no. 2 (2010): 146–161.

Strati, Irini F., and Vassiliki Oreopoulou. "Effect of extraction parameters on the carotenoid recovery from tomato waste." *International Journal of Food Science & Technology* 46, no. 1 (2011): 23–29.

Szymanowska, Urszula, and Barbara Baraniak. "Antioxidant and potentially anti-inflammatory activity of anthocyanin fractions from pomace obtained from enzymatically treated raspberries." *Antioxidants* 8, no. 8 (2019): 299.

Talekar, Sachin, Antonio F. Patti, R. Vijayraghavan, and Amit Arora. "Recyclable enzymatic recovery of pectin and punicalagin rich phenolics from waste pomegranate peels using magnetic nanobiocatalyst." *Food Hydrocolloids* 89 (2019): 468–480.

Tang, Wei, Xinxing Wu, Caoxing Huang, Zhe Ling, Chenhuan Lai, and Qiang Yong. "Natural surfactant-aided dilute sulfuric acid pretreatment of waste wheat straw to enhance enzymatic hydrolysis efficiency." *Bioresource Technology* 324 (2021): 124651.

Tanruean, Keerati, Watsana Penkhrue, Jaturong Kumla, Nakarin Suwannarach, and Saisamorn Lumyong. "Valorization of lignocellulosic wastes to produce phytase and cellulolytic enzymes from a-thermophilic fungus, Thermoascus aurantiacus SL16W, under semi-solid state fermentation." *Journal of Fungi* 7, no. 4 (2021): 286.

Teigiserova, Dominika Alexa, Joseph Bourgine, and Marianne Thomsen. "Closing the loop of cereal waste and residues with sustainable technologies: an overview of enzyme production via fungal solid-state fermentation." *Sustainable Production and Consumption* 27 (2021): 845–857.

Teigiserova, Dominika Alexa, Lorie Hamelin, and Marianne Thomsen. "Towards transparent valorization of food surplus, waste and loss: Clarifying definitions, food waste hierarchy, and role in the circular economy." *Science of the Total Environment* 706 (2020): 136033.

Trigo, João P., Elisabete MC Alexandre, Jorge A. Saraiva, and Manuela E. Pintado. "High value-added compounds from fruit and vegetable by-products–Characterization, bioactivities, and application in the development of novel food products." *Critical Reviews in Food Science and Nutrition* 60, no. 8 (2020): 1388–1416.

Truong, Khanh TP, and Pinthip Rumpagaporn. "Oligosaccharides preparation from rice bran arabinoxylan by two different commercial endoxylanase enzymes." *Journal of Nutritional Science and Vitaminology* 65, no. Supplement (2019): S171–S174.

Tsegaye, Bahiru, Swarna Jaiswal, and Amit K. Jaiswal. "Food waste biorefinery: pathway towards circular bioeconomy." *Foods* 10, no. 6 (2021): 1174.

Ubando, A. T., C. B. Felix, and W. H. Chen. "Bioresource technology biorefi neries in circular bioeconomy: A comprehensive review." *Bioresour. Technol. J* 299 (2020): 122585.

Ubando, Aristotle T., Aaron Jules R. Del Rosario, Wei-Hsin Chen, and Alvin B. Culaba. "A state-of-the-art review of biowaste biorefinery." Environmental Pollution 269 (2021): 116149.

UN. Transforming Our World: The 2030 Agenda for Sustainable Development Preamble; United Nations: New York, NY, USA, 2015; ISBN 9781138029415.

UNEP, United Nations Environment Programme. "Food waste index report 2021." (2021).

USEPA, United States Environmental Protection Agency. 2018 Wasted Food Report; EPA: Washington, DC, USA, 2018.

Usmani, Zeba, Minaxi Sharma, Abhishek Kumar Awasthi, Nallusamy Sivakumar, Tiit Lukk, Lorenzo Pecoraro, Vijay Kumar Thakur, Dave Roberts, John Newbold, and Vijai Kumar Gupta. "Bioprocessing of waste biomass for sustainable product development and minimizing environmental impact." *Bioresource Technology* (2020): 124548.

Vangsøe, Cecilie Toft, Jens Frisbæk Sørensen, and Knud Erik Bach Knudsen. "Aleurone cells are the primary contributor to arabinoxylan oligosaccharide production from wheat bran after treatment with cell wall-degrading enzymes." *International Journal of Food Science & Technology* 54, no. 10 (2019): 2847–2853.

Viayaraghavan, Ponnuswamy, Sujin Jeba Kumar, Mariadhas Valan Arasu, and Naif Abdullah Al-Dhabi. "Simultaneous production of commercial enzymes using agro industrial residues by statistical approach." *Journal of the Science of Food and Agriculture* 99, no. 6 (2019): 2685–2696.

Villacís-Chiriboga, José, Kathy Elst, John Van Camp, Edwin Vera, and Jenny Ruales. "Valorization of by-products from tropical fruits: Extraction methodologies, applications, environmental, and economic assessment: A review (Part 1: General overview of the by-products, traditional biorefinery practices, and possible applications)." *Comprehensive Reviews in Food Science and Food Safety* 19, no. 2 (2020): 405–447.

Wang, Lu, Yanan Wu, Yan Liu, and Zhenqiang Wu. "Complex enzyme-assisted extraction releases antioxidative phenolic compositions from guava leaves." *Molecules* 22, no. 10 (2017): 1648.

Wongkaew, Malaiporn, Bow Tinpovong, Korawan Sringarm, Noppol Leksawasdi, Kittisak Jantanasakulwong, Pornchai Rachtanapun, Prasert Hanmoungjai, and Sarana Rose Sommano. "Crude Pectic Oligosaccharide Recovery from Thai Chok Anan Mango Peel Using Pectinolytic Enzyme Hydrolysis." *Foods* 10, no. 3 (2021): 627.

Xavier, Janifer Raj, and Karna Venkata Ramana. "Optimization of levan production by cold-active Bacillus licheniformis ANT 179 and fructooligosaccharide synthesis by its levansucrase." *Applied Biochemistry and Biotechnology* 181, no. 3 (2017): 986–1006.

Xavier, Janifer Raj, Karna Venkata Ramana, and Rakesh Kumar Sharma. "Production of a thermostable and alkali resistant endoxylanase by Bacillus subtilis DFR40 and its application for preparation of prebiotic xylooligosaccharides." *Journal of Food Biochemistry* 42, no. 5 (2018): e12563.

Xavier, Janifer Raj, Mrithula Mahalakshmi Madhan Kumarr, Gopalan Natarajan, Karna Venkata Ramana, and Anil Dutt Semwal. "Optimized production of poly (γ-glutamic acid)(γ-PGA) using Bacillus licheniformis and its application as cryoprotectant for probiotics." *Biotechnology and Applied Biochemistry* 67, no. 6 (2020): 892–902.

Yahia, Elhadi M., M. E. Maldonado Celis, and Mette Svendsen. "The contribution of fruit and vegetable consumption to human health." *Fruit and Vegetable Phytochemicals.* Yahia, E. M., ed. Hoboken: John Wiley & Sons (2017): 3–52.

Zaharudin, Nor Athirah, Roslina Rashid, Lianash Azman, Siti Marsilawati Mohamed Esivan, Ani Idris, and Norasikin Othman. "Enzymatic hydrolysis of used cooking oil using immobilized lipase." In *Sustainable Technologies for the Management of Agricultural Wastes*, pp. 119–130. Springer, Singapore, 2018.

Zenevicz, Mara Cristina P., Artur Jacques, Agenor Furigo Furigo Jr, J. Vladimir Oliveira, and Debora de Oliveira. "Enzymatic hydrolysis of soybean and waste cooking oils under ultrasound system." *Industrial Crops and Products* 80 (2016): 235–241.

Zhu, Zhenzhou, Mohsen Gavahian, Francisco J. Barba, Elena Roselló-Soto, Danijela Bursać Kovačević, Predrag Putnik, and Gabriela I. Denoya. "Valorization of waste and by-products from food industries through the use of innovative technologies." In *Agri-food industry strategies for healthy diets and sustainability*, pp. 249–266. Academic Press, 2020.

Zuin, Vânia G., and Luize Z. Ramin. "Green and sustainable separation of natural products from agro-industrial waste: Challenges, potentialities, and perspectives on emerging approaches." *Chemistry and Chemical Technologies in Waste Valorization* (2018): 229–282.

14 The Treatment of Dairy Industry Waste

Shalagha Sharma,[1] Arup Giri,[2,] Neha Rani Bhagat,[3]*
Rajesh Kumar,[4] and Tilak Raj[3]

[1]School of Biological Engineering, Sobhit Deemed University, Meerut, U.P., India
[2]Baba Mastnath University, Asthal Bohar, Rohtak, Haryana, India
[3]DRDO-Defence Institute of High Altitude Research (DIHAR), Ladakh, India
[4]Department of Biosciences, Gyan Path, Himachal Pradesh University, Shimla, India
*Corresponding author:arupsatadal@gmail.com

CONTENTS

14.1   INTRODUCTION

The dairy industry is regarded as one of the most important in the food industry, producing a wide range of products such as milk, milk powder, butter, and cheese (Jaganmai and Jinka, 2017; Ahmad *et al.*, 2019). With rising demand for these products in a number of countries, the global dairy industry is rapidly expanding, having a significant impact on the country's economy (Chokshi, 2016; Ahmad *et al.*, 2019). Rapid industrial growth in the dairy industry, on the other hand, not only increases productivity but also increases the release of toxic solid and liquid wastes into the environment, posing serious health risks to living beings (Porwaland Velhal, 2015). Every year, approximately 4 to 11 million tonnes of dairy waste are released into the environment worldwide, and approximately 5 million tonnes of dairy waste per year in India are released into the environment, consequently causing environmental pollution and related health issues (Kushwaha *et al.*, 2011; Sinha *et al.*, 2018; Ahmad *et al.*, 2019).

These dairy wastes are characterized by high organic content, considerable variations in pH, increased suspended solids content, chemicals and waste water, where dairy waste water accounts for a larger proportion (Sinha *et al.*, 2018; Arvanitoyannis and Giakoundis, 2006). The open disposal of such waste water into the rivers, land, fields and other aquatic bodies without or with partial treatment will soon result in serious environmental and health issues (Tikariha and Sahu, 2014). As a result, the disposal of these wastes' water is quickly becoming a major social and economic issue confronting the dairy processing industry in a variety of ways, as all dairy factories face the problem of waste water treatment, disposal, and utilization (Tikariha and Sahu, 2014). Moreover, the cost of treatment and disposal of dairy waste appears to be in the order of a million dollars, which imposes a substantial financial burden on the industry, leading to either no or partial treatment of waste effluents. It was therefore crucial to develop efficient and economically profitable methods of waste management, thus resulting in a reduction of the levels of multiple severe contaminating factors (Arvanitoyannis and Giakoundis, 2006).

Furthermore, in order to discover solutions for environmental problems and to address the growing interest in the management of dairy waste to reduce pollution, it is critical to examine the waste composition, treatment, and possible ways of utilization (Daneshvar *et al.,* 2019; Ahmad *et al.,* 2019). Hence, this book chapter discusses the latest knowledge about the source of dairy waste and its characteristics, types of treatments being utilized, including physical, chemical, and biological treatments, and various challenges faced in dairy waste treatment in the following subheadings.

14.2   SOURCE OF DAIRY WASTE AND WASTE CHARACTERISTICS

Dairy waste is most complex in its biodegradation, as it contains mainly high levels of carbohydrates mixed with lipids, protein, and minerals (Janczukowicz *et al.*, 2008). Therefore, biological oxygen demand (BOD), chemical oxygen demand (COD), pH level varies significantly in dairy-waste product and makes them difficult to treat (Hur *et al.*, 2010; Slavov, 2017; Dhall *et al.*, 2012) (Figure 14.1(A–B)). For these properties, dairy waste management is a difficult process.

14.3   IMPACT OF DAIRY WASTE ON THE ENVIRONMENT

The growing global population with their increasing demand for dairy products leading to the fast development of the dairy industry, and so more generation of dairy-waste (Henchion *et al.*, 2017; Popkin *et al.*, 2012; Alae-Carew *et al.*, 2019), which is depicted in Figure 14.2. Therefore, a considerable amount of untreated dairy effluent and waste material are released into the environment without any treatment and impact our health. Ultimately, pollution is occurring, and the health of our mother nature is going to deteriorate (Boguniewicz-Zablocka *et al.*, 2019; Gil-Pulido *et al.*, 2018).

In the dairy sector, not only for milk production, but also for the production of dairy-industry-based various products, a huge amount of waste is produced. According to Vourch *et al.* (2008), about 0.2–10 liters of effluent are produced for every one liter of processed milk. The production of these wastes is increasing at a rate of 3.56% per year on average. Ultimately, most of these wastes are mixed with nearby water sources, and due to high BOD and COD, most of the water sources are becoming eutrophic (Dhall *et al.*, 2012; Gil-Pulido *et al.*, 2018).

Human activity on earth produces more greenhouse gases (GHG), with the associated climate-changing potential, than can be absorbed by natural carbon and nitrogen cycles (Cassia *et al.*, 2018; Moore, 2008). About 21% of all GHG from human activity is estimated to be from the growing, processing, transportation, consumption, and disposal of food (Solomon, 2007). As a result, an examination of food systems is warranted in order to understand where GHGs are produced in the

(a)

(b)

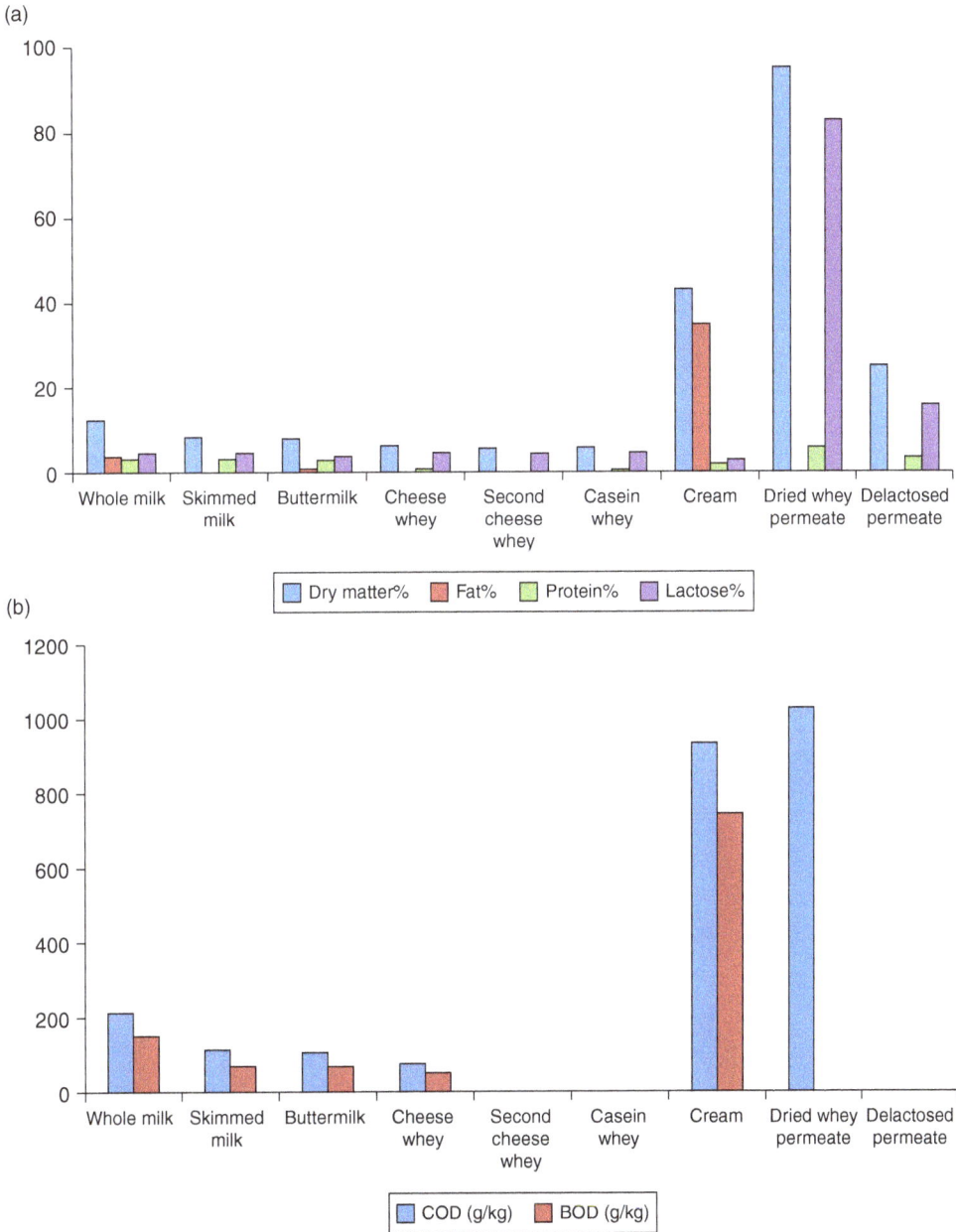

FIGURE 14.1(A–B)   Physico-chemical properties of different dairy waste products.

food chain and to aid in decision-making for their reduction. In a recent report from the United Nations Food and Agriculture Organization (FAO) division of Animal Production and Health titled "Greenhouse gas emissions from the dairy sector" (Gerber *et al.*, 2011), it concludes that the world dairy sector contributes 2.7%, with estimated ± 26% certainty, of the total world anthropogenic GHG. If meat is considered as a co-product of the dairy sector, then the contribution to total world anthropogenic GHG is 4.0%. These figures are significant when taken in context; given that dairy is a single food category. The largest world GHG-contributing sectors are electricity-heat generation, land-use change, transportation, and agriculture, with contributions to GHG of 24.6, 18.2, 13.5, and

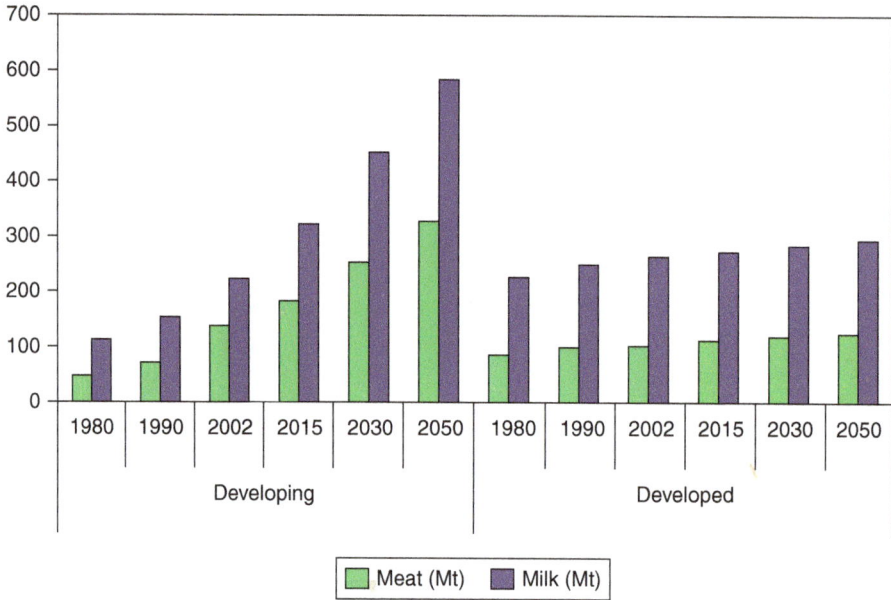

FIGURE 14.2 Production of milk and meat in developed and developing country.

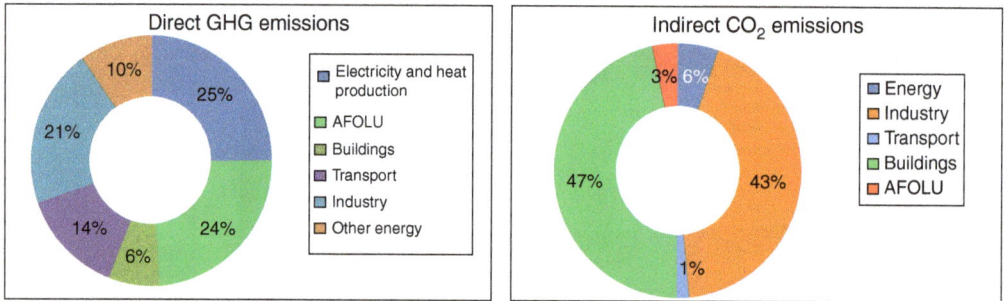

FIGURE 14.3 Global "greenhouse gas" (GHG) production by anthropogenic factors.

13.5%, respectively (Milani *et al.*, 2011). A direct comparison of world dairy GHG to the world GHG sectors is not totally correct, as world dairy GHG contribution also contributes to the world GHG sectors. A comparison of GHG associated with a dairy food as compared with a food with a similar nutritional composition of protein and fat would also illustrate the magnitude of the impact. The associated GHG of cheese is 5.9 kg of CO_2-Eq/kg versus the associated GHG of peanut butter, which is 0.17 kg of CO_2-Eq/kg (Kramer *et al.*, 1999) (Figure 14.3). The associated GHG in this comparison is 35 times greater for cheese compared with peanut butter.

14.4   METHOD OF TREATMENT FOR DAIRY WASTE

There are different methods for dairy waste management like anaerobic sludge blanket, trickling filters, anaerobic filters, aerated lagoons, aerated lagoons, etc (Slavov, 2017; Collivignarelli *et al.*, 2021; Joshiba *et al.*, 2019; Lateef *et al.*, 2013; Akansha *et al.*, 2020). Most of the process follows the anaerobic methodology followed by aerobic methods. Aerobic digestion has been used to treat municipal sewage; in aerobic fermentation, microorganisms grow rapidly and most of the energy is used for bacterial cell growth, not biogas production (Christy *et al.*, 2014; Hassan and Nelson, 2012). Only about half of the degradable organic compounds in wastewater can be stabilized by aerobic digestion, whereas up to 90% can be degraded in anaerobic digestion (Charles *et al.*, 2009; Van Lier *et al.*, 2008). Also, little or no dilution of high strength waste is required in the anaerobic process.

14.4.1 ANAEROBIC DIGESTION

The microbial composition of the initial dairy waste mixture exhibited a high concentration of viable aerobic and anaerobic bacteria in dairy waste (Kasmi *et al.*, 2017). Since LAB produces lactic acid by homolactic and heterolactic fermentation processes (Palomba *et al.*, 2012; Pradhan *et al.*, 2017), they are well-adapted to the acidic environment (Silva *et al.*, 2021). Major differences are found in the growth rates of various groups of microorganisms involved in anaerobic fermentation. For example, the minimum doubling time at 35°C is 30 min for sugar-fermenting acid-forming bacteria, 6 h for methanogens growing on hydrogen or formate, 1.4 d for acetogenic bacteria fermenting butyrate, 2.5 d for acetogenic bacteria fermenting propionate, and 2.6 d for methanogenic using acetate (Mosey and Fernandes, 1988).

14.4.2 COMMONLY INVOLVED BACTERIA AND MICROORGANISMS

Common fermentative bacteria *viz. Lactobacillus, Eubacterium, Clostridium, Escherichia coli, Fusobacterium, Bacteroides, Leuconostoc,* and *Klebsiella.* Examples of acetogens are *Acetobacterium, Clostridium,* and *Desulfovibrio,*which are used for dairy waste management (Ventorino *et al.*, 2018).

14.4.3 BACTERIAL DIGESTION

Hydrolytic bacteria, acetogenic bacteria, haemoacetogenic bacteria, methanogenic bacteria are able to degrade complex organic matter, produce acetate and hydrogen, and also covert methane to hydrogen (Angelidaki *et al.*, 2011; Guo *et al.*, 2010) (Table 14.1). *Bacillus* and *Cellulomonas, Bacteroides* and *Ruminococcus, Acetobacterium,* and *Desulfovibrio,* methylocaldum and methanobrevibacter have properties of hydrolysis, acetogenesis, haemoacetogenesis, and methanogenesis.

14.4.4 TREATMENT USING MICROALGAE

Microalgae are another wonder of nature. In the current demanding conditions, this is being used as a great source for biodiesel production (Ganesan *et al.*, 2020). Meanwhile, it has the capability to treat

---

TABLE 14.1
Bacterial Treatment for Dairy Waste Management

Sl. No.	Treatment	Effects	Reference
1.	Yeast strains, *Lactobacillus casei, Lactobacillus plantarum*	BOD, COD, chlorides, suspended solids↓	Keffala *et al.*, 2017
2.	*Bacillus Subtilis, Pseudomonas Aeruginosa*	BOD, COD, Turbidity↓	Gulhane and Shome, 2019
3.	*Enterococcus hira, Staphylococcus aureus, Bacillus subtilis, Pseudomonas aeruginosa, Lactobacillus delbrueckii*	Biological oxygen demand↓, enhancement of physicochemical quality	Al-Wasify *et al.*, 2017

TABLE 14.2
Microalgal Treatment for Dairy Waste Management

Sl. No.	Treatment	Effects	Reference
1.	*Chlorella vulgaris*	BOD, COD, suspended solids, total nitrogen, total phosphorus↓	Choi, 2016
2.	*Chlorella vulgaris, Botryococcus braunii*	Enhancement of physicochemical quality	Sreekanth *et al.*, 2014
3.	*Chlorella pyrenoidosa*	Nitrate and phosphate↓, eutrophication↓	Kothari *et al.*, 2002
4.	*Chlamydomonas reinhardtii*	Pollution load ↓	Gramegna *et al.*, 2020
5.	*Chlorella vulgaris*	Enhancement of physicochemical quality	Khalaji *et al.*, 2021

TABLE 14.3
Different Techniques for Dairy Waste Treatment

Sl. No.	Technique	Procedure	Reference
1.	Up-flow anaerobic sludge blanket reactor (USAB)	Organic material↓	Hwu *et al.*, 1998
2.	Cyanobacteria/microalgae-bacteria treatment	Organic material↓	Pires *et al.*, 2013
3.	Sequencing batch reactors (SBRs)	Microbial aggregation ↓, Biomass retention↑	Nicolella *et al.*, 2000
4.	Anaerobic digestion (AD)	Methane gas↓, COD↓	Murto *et al.*, 2004; Karakashev *et al.*, 2008
5.	Biological nitrogen removal (BNR)	Autotrophic nitrification↑, Heterotrophic denitrification↑	Henze *et al.*, 2002
6.	Chlorination	Disinfection rate↑	Hutton *et al.*, 2007

waste water by discharging oxygen-containing effluent, producing biomass, and trans-esterification reaction (Table 14.2). Therefore, this growing of microalgae will reduce environmental pollution and, in the meantime, will be helpful for the production of biodiesel (Yin *et al.*, 2020).

14.4.5 CHEMICAL TREATMENT

The various types of organic waste generated by the dairy-processing industry include whey (if not reused), fat streams, cheese curds, and sludge from treatment processes (Palmowski *et al.*, 2006; Ahmad *et al.*, 2019). After some chemical treatment, the produced organic waste is mainly used in the agriculture sector. Some of the findings are tabulated in the following Table 14.3.

14.4.6 MECHANICAL TREATMENT

Mechanical treatment is the preliminary stage for the treatment of dairy waste. Mechanical treatment removes suspended solids from wastewater, in this skimming tank, screens, primary sedimentation tank, and grit chamber or clarifiers are used. It has been found that sedimentation technology in water tank, clarifiers, etc. has the capability to reduce the amount of floating materials, sand, organic matter, and other heavy materials in the waste water (Rani *et al.*, 2012; Britz *et al.*, 2004).

Excessive variations in dairy effluent cause instability in the subsequent treatment facilities. A sufficient amount of equalization will resolve fluctuations in the flow, pH, and temperature organic loading, counterbalance residual cleaning agents, and completely destroy excess oxidizers.

14.4.7 ELECTROCOAGULATION (EC)

Dairy waste water has a higher level of organic waste, suspended solids, total dissolved solids, etc. EC is a process that uses a really appreciable technique to reduce all the parameters in water. In this process, there are electrodes that pass the electrical current, and finally, all the suspended materials will be settled down. Waste water will be cleaner and less contaminated. This mechanism is also applicable to any floating materials in waste water (Akansha *et al.*, 2020; Chezeau *et al.*, 2020). Some of the findings are tabulated in the following Table 14.4.

14.4.8 WATER HYACINTHS

Among all of the emerging techniques for dairy waste management, the aquatic treatment system is getting the most attention. Various aquatic plants are used for dairy waste treatment. Most of these aquatic plants have the capability to absorb or adsorb different types of pollutants, dissolve enzymes or chemicals, and recover nutrients from polluted aquatic systems (Omondi *et al.*, 2019; Arutselvy *et al.*, 2021) (Table 14.5).

The mechanism behind this treatment with hyacinth is unique. It has been found that during treatment, roots of this organism play an important role by increasing the population of both aerobic and anaerobic microorganisms grown. This microorganism then refines the waste water after absorbing and accumulating pollutants (Arutselvy *et al.*, 2021) (Table 14.5).

14.4.9 ULTRAFILTRATION (UF)

Ultrafiltration is a medium pressure-driven membrane filtration process. It is based on a membrane with a medium unbolted structure that allows the majority of dissolved components and some non-dissolved components to pass through it, while bigger components get rejected by the membrane (Bazrafshan *et al.*, 2021). The ultrafiltration process is widely used in the dairy and food industries for whey protein concentration and milk protein concentration and/or standardization (Habtu and

---

TABLE 14.4
Electrocoagulation Method for Dairy Waste Treatment

Sl. No.	Process	Result	Reference
1.	Coagulant combinations	FeClSO$_4$↓	Rusten *et al.*, 1993
2.	Coagulation-decantation (Iron chloride, aluminium sulfate and calcium hydroxide)	Organic matter ↓ 40%, TN↓, suspended matter↓, total phosphorus↓	Hamdani *et al.*, 2004
3.	Chitosan and alcohol precipitation treatment	BOD↓ to 87%	Mukhopadhyay *et al.*, 2003
4.	Dairy wastewater treatment	COD↓, Oil-grease↓	Sengil and Ozacar, 2006
5.	Treatment with electrical current	suspended colloidal particles↓	Sengil and Ozacar, 2006

TABLE 14.5
Different Types of Dairy Wastes with Various Treatments

Sl. No.	Type of wastewater	Result	Reference
1.	Domestic wastewater	Coliform ↓, Turbidity↓, pH↓, TDS↓	Alade and Ojoawo, 2009
2.	Domestic wastewater	pH↑, Turbidity↓, BOD and COD↓	Ting *et al.*, 2018
3.	Pulp and paper mill waste	Nitrogen↓, pH↓, TN ↓	FoxPaul *et al.*, 2008
4.	Crude oil pollution	After treatment by plants, pH, conductivity, temperature, conductivity and salinity↓	Ochekwu, 2013
5.	Textile wastewater effluent	Suspended solids↓, BOD↓, COD↓, TN↓	Gamage *et al.*, 2001
6.	Heavy-metal waste	Cr, Cu ↓	PN and Madhu, 2011

TABLE 14.6
Ultrafiltration and Nano-filtration Technology for Dairy Waste Management

Sl. No.	Treatment	Effects	Reference
1.	Ultrafiltration	Aflatoxin M1↓	Cattaneo *et al.*, 2013
2.	Ultrafiltration	Turbidity, BOD↓, proteins↓, suspended matter↓, conductivity and total dissolved salts ↓	Bennani *et al.*, 2015
3.	Ultrafiltration and nanofiltration	Proteins↓, Lactose↓, COD↓	Gong *et al.*, 2012
4.	Polyethersulfone (PES) ultrafiltration membrane	Enhancement of physicochemical quality	Anekar and Rao, 2009
5.	Ultrafiltration and nanofiltration	Enhancement of physicochemical quality	Luo *et al.*, 2011
6.	Vibratory membrane filtration	COD↓	Kertesz *et al.*, 2020

Zielińska, 2018) (Table 14.6). The pore size of the UF membrane is 0.001–0.01 μm, which is comparatively less than nanofiltration and reverse osmosis membranes. The pressure applied during the process is between 1–10 bars. The ultrafiltration process membranes are specifically distinguished on the basis of molecular weight cut off (MWCO) instead of a membrane material pore size. MWCO ranges from 1–200 kDa for ultrafiltration processes (Miller *et al.*, 2019).

14.4.10 NANO-FILTRATION (NF)

Nano-filtration (NF) is an intermediary process between ultrafiltration and reverse osmosis. It is a medium to high pressure-driven membrane filtration process. In this process, the membrane has a slightly more open structure, allowing predominantly monovalent ions to pass through the membrane (Table 14.7). It shows good performance in the removal of dissolved solutes, including multivalent ions and organic compounds, with a molecular weight ranging between 200 and 1,000 g/mol. In the dairy industry, nano-filtration is mainly used for special purposes, such as the production of lactose-free milk, partial demineralization of whey, and volume reduction of whey (Chen *et al.*, 2018; Kowalik-Klimczak and Stanisławek, 2018).

TABLE 14.7
Different Utra-filtration and Nano-filtration Application

Sl. No.	Ultra-filtration application	Nano-filtration application
1.	Protein concentration • Cheese milk • Milk protein concentrate • Whey protein concentrate	Concentration • Whey and permeate • Volume reduction • Lactose • Final concentration of WPC or WPI
2.	Protein standardization • Milk • Cheese milk • High-grade lactose by decalcification (calcium removal)	Partial demineralization • Demineralized whey • Demineralized whey powders
3.	Yield increase • White cheese • Fermented products	Lactose reduction • Lactose-free milk

14.4.11 PHYTOREMEDIATION

Phytoremediation is a biological waste water treatment method that uses diverse types of plants to remove, transfer, stabilize, and destroy contaminants in waste water as well as contaminated sites (Oh*et al.,* 2014). In this process, selective plant roots work by breaking up the contaminants that exist in the soil or by absorbing the contaminants and then storing them in the plant body. The significance of phytoremediation in the agricultural field offers removal of the contaminants, environmental restoration, wasteland site remediation, and preparation for cultivation. The process is also effectively used for the remediation of hydrocarbons, pesticides, heavy metals, explosives, chlorine compounds, hydrocarbons, excess nutrients, etc. (Rungwa, 2013).

The dairy industry involves the processing of different kinds of raw milk into products, for example, ghee milk, butter, skim milk powder, yogurt, and cheese. The major sources of BOD in waste water are cream, butter, cheese, and whey production. The wastewater contains carbohydrates such as lactose, fats and protein and other organic and inorganic materials. The waste water is also known to contain acidic washes like 3.5% nitric acid and alkaline washes like 1.5% sodium hydroxide, which are used for washing and cleaning (Munavalli and Saler, 2009).

It is known that dairy waste water has high BOD. The wastewater has an average BOD value of 360–1000 mg/L under Indian conditions. The COD to BOD ratio of dairy waste water is usually in the range of 1.4 to 1.6, which indicates that it is easily biodegradable. The other significant pollutants are total suspended solids (TSS) of about 1060 mg/L, oil and grease content of 290 mg/L, total nitrogen (TN) of 84 mg/L, and total phosphorous (TP) of 11.7 mg/L (Dipu *et al.*, 2011) (Table 14.8).

14.4.12 WETLAND AND INTERMITTENT SAND FILTERS

Wetland is now evolving as the new reactor for dairy waste management due to its low investment, very little maintenance, and operational technology required (Table 14.9). It acts as both a green space and a natural refinery. After a single construction, wetland may play an important role in acting as dairy waste refineries, where settled organic matter can be used for agriculture. Some studies show that waste stabilization ponds have a unique capacity for effluent refining (Parde *et al.*, 2020; Costello, 2020; Schierano *et al.*, 2020).

TABLE 14.8
Different Aquatic Plants for Dairy Waste Treatment

Sl. No.	Plant used for the process	Reduction in contaminants	References
1.	Azolla	BOD↓, Magnesium Hardness↓, Total Hardness↓ and Electrical Conductivity↓	Nair and Kani, 2016
2.	Water hyacinth	Neutral pH, COD↓, DO↑	Swati, 2017
3.	Cannas plant, Umbrella palm, Water hyacinths, Fern type vascular plants, Marigold, Pea plant	BOD, COD, turbidity, TSS, irons and chloride↓	Sheena and Harsha, 2018

TABLE 14.9
Ultrafiltration and Nano-filtration Technology for Dairy Waste Management

Sl. No.	Treatment	Effects	Reference
1.	Subsurface vertical flow constructed wetlands	Turbidity, BOD↓, Proteins↓, suspended matter↓, conductivity and total dissolved salts ↓	Dąbrowski et al., 2017
2.	Horizontal subsurface flow constructed wetlands	Nitrite, Nitrate, COD, total phosphorus, electrical conductivity, suspended solids↓	Schierano et al., 2020
3.	Vertical flow constructed wetlands	Turbidity, BOD↓, Proteins↓, suspended matter↓, conductivity and total dissolved salts ↓	Yazdania and Golestanib, 2019

14.5   CONCLUSION

All the findings showed that dairy waste management is mainly driven by anaerobic treatment followed by aerobic treatment. Some studies found that membrane methods, coagulation, and adsorption technology at some point are applicable for dairy waste management. Meanwhile, it has been found that anaerobic methodology has some constraints. A lot of studies should have been conducted on both the application of anaerobic and aerobic treatment along with physico-chemical treatment for dairy waste management.

REFERENCES

Ahmad, T., Aadil, R. M., Ahmed, H., ur Rahman, U., Soares, B. C., Souza, S. L.,& Cruz, A. G. (2019). Treatment and utilization of dairy industrial waste: A review. Trends in Food Science & Technology, 88, 361–372.https://doi.org/10.1016/j.tifs.2019.04.003

Akansha, J., Nidheesh, P. V., Gopinath, A., Anupama, K. V., & Kumar, M. S. (2020). Treatment of dairy industry wastewater by combined aerated electrocoagulation and phytoremediation process. Chemosphere, 253, 126652.https://doi.org/10.1016/j.chemosphere.2020.126652

Alade, G. A., & Ojoawo, S. O. (2009). Purification of domestic sewage by water-hyacinth (Eichhornia crassipes). International Journal of Environmental Technology and Management, 10(3–4), 286–294. https://doi.org/10.1504/ijetm.2009.023735

Alae-Carew, C., Bird, F. A., Choudhury, S., Harris, F., Aleksandrowicz, L., Milner, J., Joy, E. J., Agrawal, S., Dangour, A. D., & Green, R. (2019). Future diets in India: A systematic review of food consumption projection studies. Global Food Security, 23, 182–190. https://doi.org/10.1016/j.gfs.2019.05.006

Al-Wasify, R. S., Ali, M. N., & Hamed, S. R. (2017). Biodegradation of dairy wastewater using bacterial and fungal local isolates. Water science and technology: a Journalof the International Association on Water Pollution Research, 76(11–12), 3094–3100. https://doi.org/10.2166/wst.2017.481

Anekar, S., & Rao, C. R. (2009). Ultra filtration-tool to recover valuable constituent from dairy waste water. Journal of Applied Sciences in Environmental Sanitation, 4(2), 125–132.

Angelidaki, I., Karakashev, D., Batstone, D. J., Plugge, C. M., & Stams, A. J. (2011). Biomethanation and its potential. Methods in Enzymology, 494, 327–351. https://doi.org/10.1016/b978-0-12-385 112-3.00016-0

Arutselvy, B., Rajeswari, G., & Jacob, S. (2021). Sequential valorization strategies for dairy wastewater and water hyacinth to produce fuel and fertilizer. Journal of Food Process Engineering, 44(2), e13585. https://doi.org/10.1111/jfpe.13585

Arvanitoyannis, I. S., & Giakoundis, A. (2006). Current strategies for dairy waste management: a review. Critical Reviews in Food Science and Nutrition, 46(5), 379–390.https://doi.org/10.1080/1040839059 1000695

Bazrafshan, N., Firouzjaei, M. D., Elliott, M., Moradkhani, A., & Rahimpour, A. (2021). Preparation and modification of low-fouling ultrafiltration membranes for cheese whey treatment by membrane bio-reactor. Case Studies in Chemical and Environmental Engineering, 4, 100137.https://doi.org/10.1016/j.cscee.2021.100137

Bennani, C. F., Ousji, B., & Ennigrou, D. J. (2015). Reclamation of dairy wastewater using ultrafiltra-tion process. Desalination and Water Treatment, 55(2), 297–303. https://doi.org/10.1080/19443 994.2014.913996

Boguniewicz-Zablocka, J., Klosok-Bazan, I., & Naddeo, V. (2019). Water quality and resource management in the dairy industry. Environmental Science and Pollution Research International, 26(2), 1208–1216. https://doi.org/10.1007/s11356-017-0608-8

Britz, T. J., van Schalkwyk, C., & Hung, Y. T. (2004). Treatment of dairy processing wastewaters. In Handbook of Industrial and Hazardous Wastes Treatment (pp. 673–705). CRC Press.

Cassia, R., Nocioni, M., Correa-Aragunde, N., & Lamattina, L. (2018). Climate change and the impact of greenhouse gasses: CO2 and NO, friends and foes of plant oxidative stress. Frontiers in Plant Science, 9, 273.https://doi.org/10.3389/fpls.2018.00273

Cattaneo, T. M. P., Marinoni, L., Iametti, S., & Monti, L. (2013). Behavior of Aflatoxin M1 in dairy wastes subjected to different technological treatments: Ricotta cheese production, ultrafiltration and spray-drying. Food Control, 32(1), 77–82. https://doi.org/10.1016/j.foodcont.2012.11.007

Charles, W., Walker, L., & Cord-Ruwisch, R. (2009). Effect of pre-aeration and inoculum on the start-up of batch thermophilic anaerobic digestion of municipal solid waste. Bioresource Technology, 100(8), 2329–2335. https://doi.org/10.1016/j.biortech.2008.11.051

Chen, Z., Luo, J., Hang, X., & Wan, Y. (2018). Physicochemical characterization of tight nanofiltration membranes for dairy wastewater treatment. Journal of Membrane Science, 547, 51–63. https://doi.org/10.1016/j.memsci.2017.10.037

Chezeau, B., Boudriche, L., Vial, C., & Boudjemaa, A. (2020). Treatment of dairy wastewater by electrocoagu-lation process: Advantages of combined iron/aluminum electrodes. Separation Science and Technology, 55(14), 2510–2527. https://doi.org/10.1080/01496395.2019.1638935

Choi, H. J. (2016). Dairy wastewater treatment using microalgae for potential biodiesel application. Environmental Engineering Research, 21(4), 393–400. https://doi.org/10.4491/eer.2015.151

Chokshi, K., Pancha, I., Ghosh, A., & Mishra, S. (2016). Microalgal biomass generation by phycoremediation of dairy industry wastewater: an integrated approach towards sustainable biofuel production. Bioresource Technology, 221, 455–460. https://doi.org/10.1016/j.biortech.2016.09.070

Christy, P. M., Gopinath, L. R., & Divya, D. (2014). A review on anaerobic decomposition and enhancement of biogas production through enzymes and microorganisms. Renewable and Sustainable Energy Reviews, 34, 167–173. https://doi.org/10.1016/j.rser.2014.03.010

Collivignarelli, M. C., Abbà, A., Caccamo, F. M., Calatroni, S., Torretta, V., Katsoyiannis, I. A., Carnevale Miino, M., & Rada, E. C. (2021). Applications of up-flow anaerobic sludge blanket (UASB) and characteristics of its microbial community: areview of bibliometric trend and recent findings. International Journal of Environmental Research and Public Health, 18(19), 10326.https://doi.org/10.3390/ijerph181910326

Costello, C. J. (2020). Wetlands treatment of dairy animal wastes in Irish drumlin landscape. In Constructed Wetlands for Wastewater Treatment (pp. 702–709). CRC Press.

Dąbrowski, W., Karolinczak, B., Gajewska, M., & Wojciechowska, E. (2017). Application of subsurface vertical flow constructed wetlands to reject water treatment in dairy wastewater treatment plant. Environmental Technology, 38(2), 175–182.https://doi.org/10.1080/09593330.2016.1262459

Daneshvar, E., Zarrinmehr, M. J., Koutra, E., Kornaros, M., Farhadian, O., & Bhatnagar, A. (2019). Sequential cultivation of microalgae in raw and recycled dairy wastewater: microalgal growth, wastewater treatment and biochemical composition. Bioresource Technology, 273, 556–564. https://doi.org/10.1016/j.biortech.2018.11.059

Dhall, P., Siddiqi, T. O., Ahmad, A., Kumar, R., & Kumar, A. (2012). Restructuring BOD: COD ratio of dairy milk industrial wastewaters in BOD analysis by formulating a specific microbial seed. TheScientificWorldJournal, 2012, 105712. https://doi.org/10.1100/2012/105712

Dipu, S., Kumar, A. A., & Thanga, V. S. G. (2011). Phytoremediation of dairy effluent by constructed wetland technology. The Environmentalist, 31(3), 263–278. https://doi.org/10.1007/s10669-011-9331-z

Foladori, P., Ruaben, J., & Ortigara, A. R. (2013). Recirculation or artificial aeration in vertical flow constructed wetlands: a comparative study for treating high load wastewater. Bioresource Technology, 149, 398–405. https://doi.org/10.1016/j.biortech.2013.09.099

Fox, L. J., Struik, P. C., Appleton, B. L., & Rule, J. H. (2008). Nitrogen phytoremediation by water hyacinth (Eichhornia crassipes (Mart.)Solms). Water, Air, and Soil Pollution, 194(1), 199–207. https://doi.org/10.1007/s11270-008-9708-x

Gamage, N. S., & Yapa, P. A. J. (2001). Use of water Hyacinth (Eichhornia crassipes (Mart) Solms) in treatment systems for textile mill effluents-a case study. Journal of the National Science Foundation of Sri Lanka, 29(1–2). https://doi.org/10.4038/jnsfsr.v29i1-2.2615

Ganesan, R., Manigandan, S., Samuel, M. S., Shanmuganathan, R., Brindhadevi, K., Chi, N. T. L., ...& Pugazhendhi, A. (2020). A review on prospective production of biofuel from microalgae. Biotechnology Reports, e00509. https://doi.org/10.1016/j.btre.2020.e00509

Gerber, P., Vellinga, T., Opio, C., & Steinfeld, H. (2011). Productivity gains and greenhouse gas emissions intensity in dairy systems. Livestock Science, 139(1–2), 100–108. https://doi.org/10.1016/j.livsci.2011.03.012

Gil-Pulido, B., Tarpey, E., Almeida, E. L., Finnegan, W., Zhan, X., Dobson, A., & O'Leary, N. (2018). Evaluation of dairy processing wastewater biotreatment in an IASBR system: Aeration rate impacts on performance and microbial ecology. Biotechnology Reports (Amsterdam, Netherlands), 19, e00263.https://doi.org/10.1016/j.btre.2018.e00263

Gong, Y. W., Zhang, H. X., & Cheng, X. N. (2012). Treatment of dairy wastewater by two-stage membrane operation with ultrafiltration and nanofiltration. Water Science and Technology, 65(5), 915–919. https://doi.org/10.2166/wst.2012.937

Gramegna, G., Scortica, A., Scafati, V., Ferella, F., Gurrieri, L., Giovannoni, M., & Benedetti, M. (2020). Exploring the potential of microalgae in the recycling of dairy wastes. Bioresource Technology Reports, 12, 100604.https://doi.org/10.1016/j.biteb.2020.100604

Gulhane, V., & Shome, S. D. (2019). Treatment Efficiency Enhancement of Dairy Effluent by Bioaugmentation Using Bacterial Species. SSRN Electronic Journal.https://doi.org/10.2139/ssrn.3375406

Guo, X. M., Trably, E., Latrille, E., Carrere, H., & Steyer, J. P. (2010). Hydrogen production from agricultural waste by dark fermentation: a review. International Journal of Hydrogen Energy, 35(19), 10660–10673. https://doi.org/10.1016/j.ijhydene.2010.03.008

Habtu, T. S., & Zielińska, M. (2018). Post-treatment of dairy wastewater by activated sludge-ultrafiltration for water reuse. Desalination and Water Treatment, 115, 24–32.https://doi.org/10.5004/dwt.2018.22454

Hamdani, A., Chennaoui, M., Assobhei, O., & Mountadar, M. (2004). Dairy effluent characterizaton and treatment by coagulation decantation. Lait, 84(3), 317–328. https://doi.org/10.1051/lait:2004005

Harris, S. E., Reynolds, J. H., Hill, D. W., Filip, D. S., & Middlebrooks, E. J. (1977). Intermittent sand filtration for upgrading waste stabilization pond effluents. Journal (Water Pollution Control Federation), 83–102.

Hassan, A. N., & Nelson, B. K. (2012). Invited review: anaerobic fermentation of dairy food wastewater. Journal of Dairy Science, 95(11), 6188–6203. https://doi.org/10.3168/jds.2012-5732

Henchion, M., Hayes, M., Mullen, A. M., Fenelon, M., & Tiwari, B. (2017). Future Protein Supply and Demand: Strategies and Factors Influencing a Sustainable Equilibrium. Foods (Basel, Switzerland), 6(7), 53. https://doi.org/10.3390/foods6070053

Henze, M., Harremoes, P., la Cour Jansen, J., & Arvin, E. (1995). Wastewater treatment: biological and chemical processes:(2002). Springer. https://doi.org/10.1007/978-3-662-04806-1

Hur, J., Lee, B. M., Lee, T. H., & Park, D. H. (2010). Estimation of biological oxygen demand and chemical oxygen demand for combined sewer systems using synchronous fluorescence spectra. Sensors (Basel, Switzerland), 10(4), 2460–2471. https://doi.org/10.3390/s100402460

Hutton, G., Haller, L., & Bartram, J. (2007). Global cost-benefit analysis of water supply and sanitation interventions. Journal of Water and Health, 5(4), 481–502.

Hwu, C. S., Tseng, S. K., Yuan, C. Y., Kulik, Z., & Lettinga, G. (1998). Biosorption of long-chain fatty acids in UASB treatment process. Water Research, 32(5), 1571–1579. https://doi.org/10.1016/s0043-1354(97)00352-7

Jaganmai, G., & Jinka, R. (2017). Production of lipases from dairy industry wastes and its applications. Int J Curr Microbiol Appl Sci, 5, 67–73.

Janczukowicz, W., Zieliński, M., & Debowski, M. (2008). Biodegradability evaluation of dairy effluents originated in selected sections of dairy production. Bioresource Technology, 99(10), 4199–4205. https://doi.org/10.1016/j.biortech.2007.08.077

Joshiba, G. Janet, P. SenthilKumar, Carolin C. Femina, Eunice Jayashree, R. Racchana, & S. Sivanesan. Critical review on biological treatment strategies of dairy wastewater.Desalination and Water Treatment, 160 (2019): 94–109. https://doi.org/10.5004/dwt.2019.24194

Kadam, R. V., & Saxena, G. S. (1996). Managing a dairy effluent treatment plant. Indian Dairyman, 48,117–122.

Karakashev, D., Schmidt, J. E., & Angelidaki, I. (2008). Innovative process scheme for removal of organic matter, phosphorus and nitrogen from pig manure. Water Research, 42(15), 4083–4090. https://doi.org/10.1016/j.watres.2008.06.021

Kasmi, M., Hamdi, M., & Trabelsi, I. (2017). Processed milk waste recycling via thermal pretreatment and lactic acid bacteria fermentation. Environmental Science and Pollution Research, 24(15), 13604–13613. https://doi.org/10.1007/s11356-017-8932-6

Keffala, C., Zouhir, F., Abdallah, B. H., & Kammoun, S. (2017). Use of bacteria and yeast strains for dairy wastewater treatment. International Journal of Research in Engineering and Technology, 6(4), 108–113. https://doi.org/10.15623/ijret.2017.0603019

Kertesz, S., Szerencses, S. G., Vereb, G., Csanadi, J., Laszlo, Z., & Hodur, C. (2020). Single-and multi-stage dairy wastewater treatment by vibratory membrane separation processes. Membrane and Water Treatment, 11(6), 383–389.

Khalaji, M., Hosseini, S. A., Ghorbani, R., Agh, N., Rezaei, H., Kornaros, M., & Koutra, E. (2021). Treatment of dairy wastewater by microalgae Chlorella vulgaris for biofuels production. Biomass Conversion and Biorefinery, 1–7. https://doi.org/10.1007/s13399-021-01287-2

Kim, Y., Giokas, D. L., Chung, P. G., & Lee, D. R. (2004). Design of water hyacinth ponds for removing algal particles from waste stabilization ponds. Water Science and Technology, 48(11–12), 115–123. https://doi.org/10.2166/wst.2004.0818

Kosseva, M. R., Kent, C. A., & Lloyd, D. R. (2003). Thermophilic bioremediation strategies for a dairy engineering journal, waste. Biochemical, 15(2), 125–130. https://doi.org/10.1016/s1369-703x(02)00193-6

Kothari, R., Pathak, V. V., Kumar, V., & Singh, D. P. (2012). Experimental study for growth potential of unicellular alga Chlorella pyrenoidosa on dairy waste water: an integrated approach for treatment and biofuel production. Bioresource Technology, 116, 466–470. https://doi.org/10.1016/j.biortech.2012.03.121

Kowalik-Klimczak, A., & Stanisławek, E. (2018). Reclamation of water from dairy wastewater using polymeric nanofiltration membranes. Desalination and Water Treatment, 128, 364–371. https://doi.org/10.5004/dwt.2018.22981

Kramer, K. J., Moll, H. C., Nonhebel, S., & Wilting, H. C. (1999). Greenhouse gas emissions related to Dutch food consumption. Energy Policy, 27(4), 203–216.

Kushwaha, J. P., Srivastava, V. C., & Mall, I. D. (2011). An overview of various technologies for the treatment of dairy wastewaters. Critical Reviews in Food Science and Nutrition, 51(5), 442–452.https://doi.org/10.1080/10408391003663879

Lateef, A., Nawaz Chaudhry, M., & Ilyas, S. (2013). Biological treatment of dairy wastewater using activated sludge. ScienceAsia, 39(2), 179. https://doi.org/10.2306/scienceasia1513-1874.2013.39.179

Luo, J., Ding, L., Qi, B., Jaffrin, M. Y., & Wan, Y. (2011). A two-stage ultrafiltration and nanofiltration process for recycling dairy wastewater. Bioresource Technology, 102(16), 7437–7442. https://doi.org/10.1016/j.biortech.2011.05.012

Middlebrooks, E. J. (1995). Upgrading pond effluents: an overview. Water Science and Technology, 31(12), 353–368. https://doi.org/10.2166/wst.1995.0504

Milani, F. X., Nutter, D., & Thoma, G. (2011). Invited review: Environmental impacts of dairy processing and products: A review. Journal of Dairy Science, 94(9), 4243–4254. https://doi.org/10.3168/jds.2010-3955

Miller, N., Bosman, S. C., Malherbe, C. J., De Beer, D., & Joubert, E. (2019). Membrane selection and optimisation of tangential flow ultrafiltration of Cyclopia genistoides extract for benzophenone and xanthone enrichment. Food Chemistry, 292, 121–128. https://doi.org/10.1016/j.foodchem.2019.04.047

Moore T. G. (2008). Global warming. The good, the bad, the ugly and the efficient. EMBO reports, 9 Suppl, 1(Suppl 1), S41–S45. https://doi.org/10.1038/embor.2008.53

Mosey, F. E., & Fernandes, X. A. (1988). Patterns of hydrogen in biogas from the anaerobic digestion of milk-sugars. In Water Pollution Research and Control Brighton (pp. 187–196). Pergamon.

Mukhopadhyay, R., Talukdar, D., Chatterjee, B. P., & Guha, A. K. (2003). Whey processing with chitosan and isolation of lactose. Process Biochemistry, 39(3), 381–385. https://doi.org/10.1016/s0032-9592(03)00126-2

Munavalli, G. R., & Saler, P. S. (2009). Treatment of dairy wastewater by water hyacinth. Water Science & Technology, 59(4), 713–722. https://doi.org/10.2166/wst.2009.008

Murto, M., Björnsson, L., & Mattiasson, B. (2004). Impact of food industrial waste on anaerobic co-digestion of sewage sludge and pig manure. Journal of Environmental Management, 70(2), 101–107. https://doi.org/10.1016/j.jenvman.2003.11.001

Nair, C. S., & Kani, K. M. (2016). Phytoremediation of dairy effluent using aquatic macrophytes. International Journal of Scientific & Engineering Research, 7(4), 253–259.

Neder, K. D., Carneiro, G. A., Queiroz, T. R., & De Souza, M. A. A. (2002). Selection of natural treatment processes for algae removal from stabilisation ponds effluents in Brasilia, using multicriterion methods. Water Science and Technology, 46(4–5), 347–354. https://doi.org/10.2166/wst.2002.0622

Nicolella, C., Van Loosdrecht, M. C. M., & Heijnen, J. J. (2000). Wastewater treatment with particulate biofilm reactors. Journal of Biotechnology, 80(1), 1–33. https://doi.org/10.1002/chin.200036299

Ochekwu, E. B., & Madagwa, B. (2013). Phytoremediation potentials of water Hyacinth. Eichhornia crassipes (mart.) Solms in crude oil polluted water. Journal of Applied Sciences and Environmental Management, 17(4), 503–507.

Oh, K., Cao, T., Li, T., & Cheng, H. (2014). Study on application of phytoremediation technology in management and remediation of contaminated soils. Journal of Clean Energy Technologies, 2(3), 216–220. https://doi.org/10.7763/jocet.2014.v2.126

Omondi, E. A., Njuru, P. G., & Ndiba, P. K. (2019). Anaerobic Co-Digestion of Water Hyacinth (E. crassipes) with Ruminal Slaughterhouse Waste for Biogas Production. International Journal of Renewable Energy Development, 8(3), 253–259. https://doi.org/10.14710/ijred.8.3.253-259

Palmowski, L., Simons, L., & Brooks, R. (2006). Ultrasonic treatment to improve anaerobic digestibility of dairy waste streams. Water Science and Technology, 53(8), 281–288. https://doi.org/10.2166/wst.2006.259

Palomba, S., Cavella, S., Torrieri, E., Piccolo, A., Mazzei, P., Blaiotta, G.,& Pepe, O. (2012). Polyphasic screening, homopolysaccharide composition, and viscoelastic behavior of wheat sourdough from a Leuconostoc lactis and Lactobacillus curvatus exopolysaccharide-producing starter culture. Applied and Environmental Microbiology, 78(8), 2737–2747. https://doi.org/10.1128/aem.07302-11

Panico, A., d'Antonio, G., Esposito, G., Frunzo, L., Iodice, P., & Pirozzi, F. (2014). The effect of substrate-bulk interaction on hydrolysis modeling in anaerobic digestion process. Sustainability, 6(12), 8348–8363. https://doi.org/10.3390/su6128348

Parde, D., Patwa, A., Shukla, A., Vijay, R., Killedar, D. J., & Kumar, R. (2020). A review of constructed wetland on type, technology and treatment of wastewater. Environmental Technology & Innovation, 101261. https://doi.org/10.1016/j.eti.2020.101261

Pires, J. C. M., Alvim-Ferraz, M. C. M., Martins, F. G., & Simões, M. (2013). Wastewater treatment to enhance the economic viability of microalgae culture. Environmental Science and Pollution Research, 20(8), 5096–5105. https://doi.org/10.1007/s11356-013-1791-x

PN, A. M. L., & Madhu, G. (2011). Removal of heavy metals from waste water using water hyacinth. International Journal on Transportation and Urban Development, 1(1), 48.

Popkin, B. M., Adair, L. S., & Ng, S. W. (2012). Global nutrition transition and the pandemic of obesity in developing countries. Nutrition Reviews, 70(1), 3–21. https://doi.org/10.1111/j.1753-4887.2011.00456.x

Porwal, H. J., Mane, A. V., & Velhal, S. G. (2015). Biodegradation of dairy effluent by using microbial isolates obtained from activated sludge. Water Resources and Industry, 9, 1–15. https://doi.org/10.1016/j.wri.2014.11.002

Pradhan, N., Rene, E. R., Lens, P. N., Dipasquale, L., D'Ippolito, G., Fontana, A., & Esposito, G. (2017). Adsorption behaviour of lactic acid on granular activated carbon and anionic resins: thermodynamics, isotherms and kinetic studies. Energies, 10(5), 665. https://doi.org/10.3390/en10050665

Rani, R. U., Kaliappan, S., Kumar, S. A., & Banu, J. R. (2012). Combined treatment of alkaline and disperser for improving solubilization and anaerobic biodegradability of dairy waste activated sludge. Bioresource Technology, 126, 107–116. https://doi.org/10.1016/j.biortech.2012.09.027

Rinzema, A. (1993). Anaerobic Digestion of Long-Chain Fatty Acids in UASB and Expanded Granular Sludge Bed Reactors. Process Biochem. 28, 527–537. https://doi.org/10.1016/0032-9592(93)85014-7

Rungwa, S., Arpa, G., Sakulas, H., Harakuwe, A., & Timi, D. (2013). Phytoremediation–an eco-friendly and sustainable method of heavy metal removal from closed mine environments in Papua New Guinea. Procedia Earth and Planetary Science, 6, 269–277. https://doi.org/10.1016/j.proeps.2013.01.036

Rusten, B., Lundar, A., Eide, O., & Ødegaard, H. (1993). Chemical pretreatment of dairy wastewater. Water Science and Technology, 28(2), 67–76.https://doi.org/10.2166/wst.1993.0078

Saidam, M. Y., Ramadan, S. A., & Butler, D. (1995). Upgrading waste stabilization pond effluent by rock filters. Water Science and Technology, 31(12), 369–378.https://doi.org/10.2166/wst.1995.0505

Sayed, S., de Zeeuw, W., & Lettinga, G. (1984). Anaerobic treatment of slaughterhouse waste using a flocculant sludge UASB reactor. Agricultural Wastes, 11(3), 197–226. https://doi.org/10.1016/0141-4607(84)90045-3

Sayed, S., van Campen, L., & Lettinga, G. (1987). Anaerobic treatment of slaughterhouse waste using a granular sludge UASB reactor. Biological Wastes, 21(1), 11–28. https://doi.org/10.1016/0269-7483(87)90143-1

Schierano, M. C., Panigatti, M. C., Maine, M. A., Griffa, C. A., & Boglione, R. (2020). Horizontal subsurface flow constructed wetland for tertiary treatment of dairy wastewater: Removal efficiencies and plant uptake. Journal of Environmental Management, 272, 111094.https://doi.org/10.1016/j.jenvman.2020.111094

Sengil, I. A., & Ozacar, M. (2006). Treatment of dairy wastewaters by electrocoagulation using mild steel electrodes. Journal of Hazardous Materials, 137(2), 1197–1205. https://doi.org/10.1016/j.jhazmat.2006.04.009

Sheena, K. N., & Harsha, P. (2018) Feasibility Study of Phytoremediation in Wastewater Treatment. International Journal of Science and Research, www.ijsr.net/get_abstract.php?paper_id=ART20182544, 7(5), 1019–1026.

Silva, L. F., De Dea Lindner, J., Sunakozawa, T. N., Amaral, D. M. F., Casella, T., Nogueira, M. C. L., & Penna, A. L. B. (2021). Biodiversity and succession of lactic microbiota involved in Brazilian buffalo mozzarella cheese production. Brazilian Journal of Microbiology, 1–14. https://doi.org/10.1016/j.fm.2019.103383

Sinha, S., Srivastava, A., Mehrotra, T., & Singh, R. (2018). A Review on the Dairy Industry Waste Water Characteristics, Its Impact on Environment and Treatment Possibilities. SpringerBriefs in Environmental Science, 73–84. https://doi.org/10.1007/978-3-319-99398-0_6

Slavov, A. K. (2017). General characteristics and treatment possibilities of dairy wastewater - areview. Food Technology and Biotechnology, 55(1), 14–28. https://doi.org/10.17113/ftb.55.01.17.4520

Solomon, S., Manning, M., Marquis, M., & Qin, D. (2007). Climate change 2007-the physical science basis: Working group I contribution to the fourth assessment report of the IPCC (Vol. 4). Cambridge University Press.

Sreekanth, D., Pooja, K., Seeta, Y., Himabindu, V., & Reddy, P. M. (2014). Bioremediation of dairy wastewater using microalgae for the production of biodiesel. IJSEAT, 2(11), 783–791.

Swati A. Zingade, 2017An assessment of phytoremediation method for dairy effluent. International Journal of Applied and Pure Science and Agriculture Volume 03, Issue 1, 115

Thassitou, P. K., & Arvanitoyannis, I. S. (2001). Bioremediation: a novel approach to food waste management. Trends in Food Science & Technology, 12(5–6), 185–196. https://doi.org/10.1016/s0924-2244(01)00081-4

Tikariha, A., & Sahu, O. (2014). Study of characteristics and treatments of dairy industry waste water. Journal of Applied & Environmental Microbiology, 2(1), 16–22.

Ting, W. H. T., Tan, I. A. W., Salleh, S. F., & Wahab, N. A. (2018). Application of water hyacinth (*Eichhornia crassipes*) for phytoremediation of ammoniacal nitrogen: A review. Journal of Water Process Engineering, 22, 239–249. https://doi.org/10.1016/j.jwpe.2018.02.011

Van Lier, J. B., Mahmoud, N., & Zeeman, G. (2008). Anaerobic wastewater treatment. Biological wastewater treatment: principles, modelling and design, 415–456.https://doi.org/10.2166/9781789060362_0701

Ventorino, V., Romano, I., Pagliano, G., Robertiello, A., & Pepe, O. (2018). Pre-treatment and inoculum affect the microbial community structure and enhance the biogas reactor performance in a pilot-scale biodigestion of municipal solid waste. Waste Management, 73, 69–77. https://doi.org/10.1016/j.was man.2017.12.005

Vourch, M., Balannec, B., Chaufer, B., & Dorange, G. (2008). Treatment of dairy industry wastewater by reverse osmosis for water reuse. Desalination, 219(1–3), 190–202.https://doi.org/10.1016/j.desal.2007.05.013

Wang, X., Bai, X., Qiu, J., & Wang, B. (2005). Municipal wastewater treatment with pond–constructed wetland system: a case study. Water Science and Technology, 51(12), 325–329. https://doi.org/10.2166/wst.2005.0491

Yazdania, V., & Golestanib, H. A. (2019). Advanced treatment of dairy industrial wastewater using vertical flow constructed wetlands. Desalin Water Treat, 162,149–155. https://doi.org/10.1080/09593 330.2016.1262459

Yin, Z., Zhu, L., Li, S., Hu, T., Chu, R., Mo, F., & Li, B. (2020). A comprehensive review on cultivation and harvesting of microalgae for biodiesel production: Environmental pollution control and future directions. Bioresource Technology, 301, 122804.https://doi.org/10.1016/j.biortech.2020.122804

15 Aerobic Processes for the Treatment of Wheat Starch Effluents

Younis Ahmad Hajam,[1] *Dhiraj Singh Rawat,*[2]
and Rajesh Kumar[2,*]

[1]Department of Life Sciences and Allied Health Sciences, Sant Baba Bhag Singh University, Jalandhar, Punjab, India
[2]Department of Biosciences, Himachal Pradesh University, Shimla, Himachal Pradesh, India
*Corresponding author: drkumar83@rediffmail.com

CONTENTS

15.1 INTRODUCTION

There are so many species of wheat starch polysaccharides that help to form a viscous gel layer in wheat starch waste ultrafiltration. A test was performed with different enzymes on this wheat starch discharge to reduce its viscosity, in which Brew-n-Zyme Pentosanase shows its significant effect at minimum doses, i.e., 0.05 wt. % (dry solids' basis). At about 70 °C, it becomes stable and does not show any proteolytic activity. Permeate flux rate and concentration of wheat starch discharge increased up to 9=18 wt. % solids by hydrolysis of the enzyme. Both the wheat starch, which is hydrolyzed, and effluent flux-concentrations were also fitted by both models, i.e., pressure-driven and mass transfer models. In fresh wheat starch discharge, the main reason for resistance to permeate flux is adsorption of big or complex molecules that occurs on the membrane surface and further contributed total resistance of about 57% and 78% for wheat starch effluent and enzymatically hydrolyzed wheat starch discharge respectively. In this, enzymatic hydrolysis plays an important role to decrease the viscosity of the gel layer that helps to remove the layer easily through "shear stress". When polysaccharide thickness decreases then the wheat starch effluent gel layer exposes the membrane and leads to an increase in macromolecule adsorption and at last, an increase in the concentration of the reverse situation applied. About 93% of starch effluent is formed in the gel layer of the whole resistance and about 80% is formed for wheat starch effluent, which is hydrolyzed

DOI: 10.1201/9781003207689-15

enzymatically. To remove mass flux of macro-molecules, wall shear stress plays an important role that impacts the gel layer surface.

15.2 INDUSTRIAL PRODUCTION OF STARCH

The various researchers reported that starch was introduced commercially in different industries such as food industries and pharmaceutical markets that produce about 150 million tons of starch up to 2020 and it will further increase to more than 165 million tonnes in 2026 (Radley, 1976). Annually, about 11 million tonnes of starch and its derivatives is produced in the EU for which about 25 million tons of raw materials is required. Domestically about 9 million tons of starch and its derivatives are utilized per year and starch sweeteners are the major products produced annually (Fane, 1975). It is a polyhydroxylated natural compound found in a different part of the plant (roots, stem, and fruits). Corn, potato, and wheat are the major source of raw starch among all the crop plants. There is no general toxicity during the process of a starch formation, purification of soluble starch having some amount of protein shows zero effect or toxicity. Validation of this with real processing of starch enterprises means that sewage treatment and wastewater analysis produced from starch processing is combined. The starch factory is a big enterprise of "starch corn". Glucose crystallization and large-scale production of its by-products are the main processes from which huge waste is generated, which cannot be managed properly due to lack of sewage treatment stations.

15.3 TREATMENT PROCESSES

Transformation of the treatment of wastewater has a basic principle that is considered to decrease the cost of wastewater treatment and operation cost, whichencourages using the resources of wastewater as required. It also helps to increase economic benefits. On the other hand, the proper utilization of technology is required in an in-situ transformation process. Also, collaboration with civil engineering for the transformation process with low cost of investment should be formed. As the wastewater system has doubled, the aerobic and anaerobic system cannot fulfil the requirements of the system as well. The new anaerobic system has 3000 m^3/d ability to process gently with the use of UASB and IC named technology (Radley, 1976). In EU production, mainly corn, wheat, and potatoes are produced in the market. Meanwhile, its range of feedstocks is utilized across the whole world, including barley, oats, sweet potatoes, rice, cassava, sago palms, and milo (Waterschoot et al., 2015). The increase in the rate of wastewater production is one of the major challenges in starch-producing industries. There are various steps on the basis of feedstock and the different technologies such as feedstock washing, steeping, refinement, scarification, etc. (Waterschoot et al., 2015; Belfort, 1984). Among all the processes, some are thermal in state of the condensates accumulated and generally the production of starch is optimized.

Water pollution occurs in a huge amount due to industrial processes and in this, both starch and gluten processes from wheat flour have a significant role (Belfort, 1984). The mixture of proteins, pentosans, hexosans, suspended starch particles, and monosaccharides, having 0.35 wt.%, 0.35 wt.%, 0.35 wt.%, 0.13 wt.%, and 1.0 wt.%, is found in wheat starch effluent (WSE) (Matthiasson, 1983). Ultrafiltration of waste is one of the best methods to recover the protein-rich fraction in the market. Its economic cost-benefit depends upon WSE and high permeate flux rates having more than 16 wt% solids (Fane et al., 2000; Harris, 1985).

15.3.1 MARTIN PROCESS

In 1835, a process named Martin or dough ball was proposed in Paris (Rehwald, 1926). This process is very popular to date (Mittleider et al., 1978) and the raw material used was wheat flour rather than that of wheat grain. It has five steps to proceed, i.e.: (a) mixing of flour and preparing a dough;

(b) washing out the starch; (c) drying the gluten; (d) starch refining; and (e) drying the starch. In Martin, process variation occurs. The ratio of flour and water is 2:1, blended to form a uniform, lump-free, and smooth dough. Its ratio depends upon the type of flour used, e.g., if hard wheat flour is used then the dough will be strong elastic that requires more water than soft wheat flour. The water that is used in it should have mineral salts because the water has fewer mineral salts than gluten and becomes slimy (Knight, 1965). Before the washing step, dough should be fully hydrated and then washed out to separate starch from gluten without gluten breaking into small pieces. Various devices are used to continue this process such as ribbon blenders, twin-screw troughs, rotating drums, and agitated vessels (Knight, 1965)The ribbon blender is boat-shaped deep and narrow with twin open paddle rotors and in the rotor beds a groove is present that controls the paddle action with different speeds and rotating in opposite directions. Paddles are covered with the dough by adding the dough in a continuous manner. Both the fresh and treated water as well are put into the vessel and then the suspended water and starch get overflowed from it. Meanwhile, about 70% water and 70–80% proteins get discharged through the pipe with some amount of starch. Gluten may be dried through roller compression. About 8% moisture can be dried with spray, flash, or drum dryers, and among these mainly flash dryers are utilized to dry it. There is a need for spontaneous or controlled temperature to dry the gluten because it loses its vitality due to continuous exposure to extreme temperatures. The undried gluten is mixed with dried material. After that, it is exposed to a hot air stream. All the processes (mixing, dispersion, etc.) occur in a flash dryer to produce a fine powder. It is light brown in color. About 10% solids are found in the starch slurry on the sieve when washed in a dough washer. After separating it, all the macro-particles (like bran) are removed through some equipment like Dorr-Oliver D.S.M. screen. A wedge wire media is used having 100 μm openings "vibratory sieves or centrifugal conical filters". The starch slurry refinement occurs through washing it in nozzle discharge centrifuges and then the batch is dewatered in centrifuges with a basket-like shape and about 40% moisture in them. Some equipment such as solid-bowl, scroll-discharge, and decanter centrifuges are utilized for continuous dewatering of it. The supply of a flesh dryer with starch is mixed with the other dry products to maintain 36% moisture in it. Starch is dried continuously in hot air to reduce its ability of gelatinization. The minimum moisture content required for wheat starch gelatinization is 31% and at the end of the process it contains about 10–12% moisture and 0.3% protein content.

The secondary starch is produced when micro-granules of starch are dried and at the prime starch refining stage, other particulates get separated. Nozzle-type centrifuges play an important role in minimizing the insoluble material in the wastewater stream due to the huge requirement for water to wash the starch (15 parts by wt. of water/part of flour). The stream has about 10–13% and 0.85–1.2% dry flour substance and solids respectively. The solid effluents are either discarded through waste or are reutilized.

15.3.2 BIODEGRADATION OF STARCH STILLAGE

There are various methods that may not be able to use the complete starch stillage volume, so various hurdles are faced by distilleries. Mostly the stillage effluent is due to organic matter content. It cannot be discharged into the environment, i.e., water bodies and soil etc. A small portion of chemical oxygen demand (COD) should be removed at the source of origin. Nagano et al. (1992) reported that in sweet potato stillage the COD load is about 12.1 g O_2/lin and the feedstocks contain about 80% and 20% starch and potato starch respectively (Cibis, 2004). It completely shows the different chemical properties. The current finding shows the effect of COD content. Feedstock is not the only reason that affects the COD content, it is also affected by the technology of spirit production, its methods, and storage of stillage. Sour fermentation was shown by the distiller's stillage because of the formation of the acid such as organic acid (lactic acid) due to its dominance in stillage (Wilkie et al. 2000). All these chemicals' compositions considered that stillage is a biodegradable material and in both aerobic and anaerobic fermentation it was first reported about four decades ago (Fargey and Smith, 1965; Smith and Fargey, 1965). Various researchers have reported biodegradation of starch stillage in anaerobic processes (Weiland and Thomsen, 1990; Nagano et al. 1992; Goodwin and Stuart, 1994; Laubscher et al. 2001; Gao et al. 2007; Tang et al. 2007). Hutnan et al. (2003) reported that the initial COD range in the anaerobic treatment of wheat stillage was 91–107 g O_2/l. On the other hand, the COD of malt whisky distillery wastewater was in the range of 30.5 and 47.9 g O_2/l, which helps to reduce about 90% pollution (Goodwin and Stuart, 1994). The total COD in the grain distillation ranges from 20–30 g O_2/l with UASB (up-flow anaerobic sludge bed) system having the efficiency to remove about 80% COD reported by various researchers (Laubscher et al. 2001). Shin et al. (1992) reported that when both barley and sweet potato stillage of primary organic matter of about 29.5 g O_2/l processed continuously then about 80% COD gets reduced.

The different researchers studied that the starting pollution load is about 40 g O_2/L of wheat and sweet potato stillage, which helps to attain about 98% COD reduction (Nagano et al. 1992). Weiland and Thomsen (1990) reported that the organic content of potato stillage was between 20–55 g O_2/l and it shows about 80–95% reduction of COD.

With a potato and sugar beet stillage of an initial organic pollution load of 40 g O_2/l, the reduction in COD amounted to 90% (Wilkie et al. 2000). Biological treatment technologies have been utilized in wastewater reclamation for over a century. Out of the many different processes employed, the activated sludge system has proven to be the most popular. The implementation of membranes within the treatment sequence of a water pollution control facility was initially limited to tertiary treatment and polishing. Ultrafiltration, microfiltration, or reverse osmosis units were utilized in areas where discharge requirement was very stringent or direct reuse of the effluent was desired. High capital and operational costs as well as inadequate knowledge on membrane application in waste treatment were predominant factors in limiting the domain of this technology. However, with the emergence of less expensive and more effective membrane modules and the implementation of the ever-tightening water discharge standard, membrane systems regained interest.

In distilleries, stillage is a major by-product. The volume of distillery wastewater is more than that of ethanol produced and it has also become a serious problem throughout the world. Different methods were utilized to solve this problem but could not be resolved. About 90% of ethanol is produced in Poland mostly from grains and potatoes and the starch is also utilized for the production of spirits worldwide especially in European countries. The current study gives an overview of fuel production globally. Also, the methods are performed for biodegradation and the utilization of starch-based stillage. There are mainly two groups of different methods such as (a) the starch stillage utilization mode and (b) combined methods (aerobic as well as anaerobic) that help in the biodegradation of stillage. About 5% ethanol is produced synthetically and about 95% or more is produced from feedstock produced in the agriculture area in which about 42% sugar, 58% non-sugar-based feedstock respectively produces ethanol. Renewables (2018) reported that for the production of fuel about 67% ethanol is utilized in the whole world. Nowadays Poland produces about 90% of starch-based feedstock from the overall production of ethanol. Various researchers reported that mainly ethanol is produced from grain crops such as rye, triticale, and wheat, mainly potatoes in root crops, and different agriculture feedstock (Dzwonkowski et al., 2011). In Polish distilleries, maize grain is utilized as feedstock material and its demand increases day by day, as reported by different workers. The production of ethanol agriculture feedstock varies as per the situation occuring in the market however, about 70% of the production of ethanol is from potato-based feedstock for two decades from overall production of ethanol. In 2000, the potato-based feedstock contribution decreases up to less than 8% and in 2003 and 2004, the ethanol production from potato-based feedstock became 3.2% and 4.2% respectively but also dropped in 2005 again to 3.1%. Dzwonkowski et al. (2011) reported that ethanol production from potato-based feedstock decreased 9.1 million liters to 5.5 million liters in 2006. Generally, the most dominant feedstock is starch-based and is utilized in Poland, in which the main raw material for the production of ethanol is rye, i.e., about 90%. Starch-containing wastes, i.e., frozen potatoes, peelings, potato slops, waste flour, etc. are utilized in the production of spirits, i.e., up to 50–60% (Dzwonkowski et al., 2011). It was reported that about 95% of the production of ethanol was produced in Poland from rural distilleries and about 940 distilleries were operating in 1995, which decreased in 2006 and at last there are only 217 operating in the country (Kupczyk, 2007). In 1996, the ethyl alcohol production decreases due to the decrease in distilleries and in 1999–2000 it reduced from 278 million liters to 170 million liters, however, luckily a slight increase was observed in 2001, i.e., 181 million liters and then continuously increased day by day. In 2003 the production increased to 220 million liters and 230 million liters in 2004. As ethanol production increased day by day, in 2005 it reached 27.56% and its volume increased to 344 million liters in 2006 (Dzwonkowski et al., 2011) about 62.9% of production occurs in Poland by rural distilleries, which utilized it as a fuel additive and volume beame 161 million liters as biofuel (Kupczyk, 2007). The ethanol production again decreased in Q1–Q3 of 2007 (68 million liters only) (Licht, 2008). Orlen and Lotos become the source of production of ethanol (about 60%) due to less competition among these in Polish refineries. Licht (2008) reported that these domestic biofuel productions gave positive results and further the ethanol demand is increased and the "national biofuel program and as well as the Biofuel Directive 2003/30/EC" demanding more production from distilling industries.

15.3.3 METHODS OF STILLAGE UTILIZATION

Yeast, the starch-based feedstock fermented but contains some component of feedstock and also helps in yeast cell degradation (Sweeten et al., 1981–1982; Davis et al., 2005; Sanchez et al., 1985). All these substances have a high nutritive value, which contains vitamins in large amounts, proteins, and minerals (Mustafa et al. 1999). Mustafa et al. (1999) reported that there are particular mineral compounds present in the wheat and barley stillage. But in comparison to both, the maximum calcium (Ca), iron (Fe), and sodium (Na) content is found in barley. The various researchers reported that owing to the composition of the dry matter, the potato stillage has been regarded as valuable fodder (Maiorella et al., 1983; Larson et al., 1993; Ham et al., 1994; Fisher et al., 1999; Mustafa et al., 2000). Maiorella et al. (1983) reported that the potato stillage value is lower than grain stillage and among all the stillage of wheat, rye, and barley, barley has low nutritive value in liquid form and solid as well (Mustafa et al., 2000). The highest feeding value is found in unprocessed warm stillage. It has the disadvantage that it cannot be stored for a long time because of proneness to souring and mold growth, so it can be fed to animals as soon as possible and this process becomes troublesome due to this process. Raw stillage and feeding farm animals show positive results and are cost-effective only in one condition, i.e., users live in closed proximity of the distillery due to high water content (Ganesh and Mowat, 1985; Aines et al., 1986). The integration of rural distillery with the big animal farm is the best option to overcome the problem and they may be able to consume it but the opposite trend was observed in Poland and all the rural distilleries become independent (Carioca et al., 1981; Ganesh and Mowat, 1985) but big animal farms are lacking there, due to this the proper stillage utilization does not occur and animal fodder remains unsolved. Raw stillage should be processed as soon as possible because it loses its properties with time. To increase the storage life of starch-based stillage corn and hay are added into it, however, this type of fodder may be stored for a long time but its nutritive value may also become reduced (McCullough et al., 1963; Hunt et al., 1983; Muntifering et al., 1983). The various researchers reported that it also has another disadvantage: the maximum raw fiber content limits its value as fodder (for nonruminant animal farms) (Kienholz et al. 1979). Garcia and Kalscheur (2004) reported that lactic bacteria inoculants help in its preservation and also yeast plays an important role when utilized as fodder so the yeast can be cultivated for better use in starch stillage reported by various researchers (Murray and Marchant, 1986; Jamuna and Ramakrishna, 1989). The industrial scale of yeast cultivation has been investigated by various researchers (Tauk, 1982; Malnou et al., 1987; Moriya et al., 1990; Cibis et al., 1992). Due to high production, the COD level of effluent reduces pollutant about 70%. So ultimately the utilization method has been abandoned. This results in discontinuity in the cultivation of yeast on distillery stillage.

After the solid separation, the recirculation of liquid has been reported to the steamer and resulting from that when water (about 75%) was filled into the steamer having "grain sorghum-based thin stillage", the increase in COD and solid content in the huge amount wasreported by Egg et al. (1985). The thin stillage recirculation to the mash tub becomes very effective in some conditions. When about 40–50% of it is recycled then the recirculation is formed in its effective way (Sheehan and Greenfield, 1980; Wilkie et al., 2000), if value increases then decrease in alcohol yield formed and further it might accumulate these substances that have the ability to stop the yeast cells' activity (Egg et al., 1985; Kim et al., 1999; Wilkie et al., 2000). However, about a 25% decrease was observed in the volume of wastewater by recirculation of thin stillage, which resulted in increases in the COD level (Wilkie et al., 2000).

There are other utilization methods of starch-based stillage, i.e., concentration and drying, both require energy so they are used only in huge distilleries (Murphy and Power, 2008). In Poland the volume of production is more than that of present-day existing distilleries. Different researchers reported that in the US and Canada as well these two methods, i.e., drying and concentration used at its wide range have a high demand, which is utilized as fodder where the process occurs in different variants (The Mother Earth News, 1980; Aines et al., 1986; Wu, 1988). Solid substance separation occurs with the help of different types of equipment such as sieves, centrifuges, etc., and the product known to be wet distiller's grain. Distiller's dried grain was formed dry when liquid became 30–40% concentrated and the solid gets suspended and is called condensed distillers' soluble. Also, there was an investigated report on the method that is less energy consuming and increases the suspended solids' content more than that of other methods, however, ultrafiltration and high-pressure reverse osmosis get positive results. In reverse osmosis, the concentration of ash and solids as well as in permeate is less as compared with tap water (Wu, 1988). So due to this, permeate is utilized as a substitute in the ethanol production process and shows that it does not change its fermentation yield of permeate after eight-fold recirculation (Kim et al. 1999). In the dried form, the concentrated fraction is utilized as fodder and the presence of fouling and scaling phenomenon is the major reason for its disadvantage, which accounts for membrane clogging (Nguyen, 2003; Gryta, 2005). There is one more method that helps to increase suspended solids, the best example of it is soy hulls or dry beet pulp and this is also a way to form pellets (Garcia and Kalscheur, 2004). Jenkins et al. (1987) reported that characterization also creates starch-based feedstock utilization as a fertilizer and mainly for this vinasse is used (Jenkins et al. 1987; Monteiro, 1975; Maiorella et al. 1983). It was also reported by various workers that starch-based stillage was also utilized as a soil fertilizer directly to produce an organic fertilizer by composting in an easy way (Tanaka et al. 1995). Mostly cow feces are considered for this process. So organic fertilizer gives more benefit than chemical fertilizer by processing it at the right time and the crop grows without any disease or damage (Sheehan and Greenfield, 1980; Milewski et al. 2001; Czupryński et al. 2002). But for all this, the proper storage and management of stillage should be performed. Stillage does not dip into the soil directly, it may to the contamination due to its accumulation, it also causes odor, which is one of the major weaknesses of its dumping but in the case of vinasse and incineration of starch stillage it helps to increase soil fertility (Yamauchi et al. 1999; Maiorella et al. 1983). These methods are also used to produce rigid polyurethane-polyisocyanurate foams from this waste effluent (Czupryński et al. 2002). The products have properties that are identical to the standard foam, such as brittleness, and Leather (1998) reported that maize stillage fraction is also used as feedstock for some cosmetic products such as alternan and pullulan (West and Strohfus, 1996). Maize stillage was also utilized for astaxanthin carotenoid synthesis through Phaffiarhodozyma and other stillage like starch utilized for protease synthesis, chitosan, and plastic, which is bio-degradable in nature (Leathers, 2003; Morimura et al. 1994; Yang and Lin, 1998; Yokoi et al. 1998; Khardenavis et al. 2007).

15.4 BIODEGRADATION OF STARCH STILLAGE

The whole utilization of starch stillage volume is not considered through all of the methods discussed so all distilleries face a serious problem because the stillage is effluent having a high strength so it cannot be degraded easily and cannot be dumped anywhere (water bodies, soil). Some parts of COD are removed at its initial stage only but it is not a permanent solution for degradation. For potato stillage, the COD load varies as 12.1 g O_2/l–122.33 g O2/l from that stillage in which about 80% and 20% wheat starch and potato starch are found in waste feedstock (Nagano et al. 1992; Cibis, 2004), when feedstock are collected from the same place of the same feedstock matter their chemical properties are different even in that situation. So, it can be found due to COD content but it does not occur due to the feedstock. Also, the production of spirit and the methods or management affect this. For the sour fermentation distiller stillage shows proneness because of organic acids. In organic acid, lactic acid shows the dominating character in that stillage type.

As for the other carbon sources, reducing substances and glycerol are found to occur in large amounts. The presence of total and phosphate phosphorus, as well as the large amounts of total nitrogen, can be explained as being associated with the high protein proportion in the feedstock from which the stillage comes (Wilkie et al. 2000).

Such chemical composition suggests that starch stillage is biodegradable to a great extent. Aerobic and anaerobic fermentation of thin stillage was first reported four decades ago (Fargey and Smith, 1965; Smith and Fargey, 1965). In the past 15 years, several publications have dealt with the biodegradation of starch stillage, but they have described anaerobic processes only (Weiland and Thomsen, 1990; Nagano et al. 1992; Goodwin and Stuart, 1994; Laubscher et al. 2001; Gao et al. 2007; Tang et al. 2007). Laboratory investigations into the anaerobic treatment of the wheat stillage of an initial COD ranging from 91 to 107 g O_2/l (Hutnan et al. 2003) and of malt whisky distillery wastewater with an initial COD between 30.5 and 47.9 g O_2/l (Goodwin and Stuart, 1994) have revealed an approximately 90% reduction of this pollutant. The treatment of grain distillation wastewater whose initial COD level ranged from 20 to 30 g O_2/l with the aid of an up-flow anaerobic sludge bed (UASB) system has yielded an approximately 80% COD removal efficiency (Laubscher et al. 2001). When barley stillage and sweet potato stillage of an initial organic matter content of 29.5 g O_2/l were treated in a continuous process, COD reduction amounted to 80% (Shin et al. 1992). A 98% reduction in COD was attained with wheat stillage and sweet potato stillage of an initial pollution load of 40 g O_2/L (Nagano et al. 1992). When potato stillage with an initial content of organics ranging between 20 and 55 g O_2/l was treated, the extent of COD reduction varied from 80 to 95% (Weiland and Thomsen, 1990). With a potato and sugar beet stillage of an initial organic pollution load of 40 g O_2/l, the reduction in COD amounted to 90% (Wilkie et al. 2000). Stillage anaerobic biodegradation is regulated on an industrial scale and about 135 anaerobic bioreactors are operated worldwide. Among these, about nine bioreactors were utilized to treat starch stillage and others for the primary vinasse treatment (Wilkie et al. 2000). With the help of aerobic methods, the treatment of starch stillage was performed with equal efficiency and the publisher faced the problems that arise from 1965 reported by Smith and Fargey (1965). Different investigated reports showed the treatment of maize stillage at room temperature in STR (stirred tank reactor). In it, COD is reduced up to 60.7%. About a 98% decrease was obtained in the COD. The bioreactor used to have a fixed bed and then recirculation of stillage treatment has been started. There is no report found on starch aerobic biodegradation since in the last 35 years only stillage was reported but after some time again it should be paid attention (Cibis et al. 2002; Krzywonos et al. 2002; Cibis, 2004; Cibis et al. 2004; Ferzik et al. 2004; Cibis et al. 2006; Krzywonos et al. 2008). The potato stillage treatment was conducted batch-wise in STR with thermophilic and mesophilic bacteria (genus *Bacillus*) and about 89.8% COD reduction formed. Generally, temperature, pH, COD, and some nutrients like nitrogen (N), phosphorus (P), and potassium (K) play an important role in biodegradation (Cibis et al. 2002, Krzywonos et al. 2002; Cibis, 2004; Cibis et al. 2004; Cibis et al.

2006; Krzywonos et al. 2008). This biodegradation method with this bacterium is also performed by maize and rye stillage. The total COD reduction was shown in both maize and rye stillage, i.e., 82.6% and 84.6%, respectively (Cibis, 2004). When stillage of wheat and potato starch goes under treatment, about 80% wheat starch and 20% potato starch to produce "glucose syrup and chips" then reduction of COD becomes 94% (Cibis, 2004) also the same effect observed when continuous treatment of stillage occurs through the biodegradation process (Cibis, 2004).

Ferzik et al. (2004) reported the data availability on aerobic degradation with the bacterial culture of wheat grain, however, the total yielded COD reduction is about 64% through batch biodegradation system at 45 °C. On another hand, the continuous process gives more COD reduction, i.e., 90% at 55 °C. The aerobic biodegradation through thermophilic and mesophilic bacteria of starch stillage is not explored on the industrial scale in contrast to anaerobic methods. The different researchers reported that during lab work the aerobic biodegradation of distillery wastewater in thermophilic and mesophilic culture shows some effect of this for other micro-organisms as well in aerobic condition, and temperature (Anastassiadis and Rehm, 2006; Battestin and Macedo, 2007; Choorit and Wisarnwan, 2007; Cibis et al. 2002; Krzywonos et al. 2002; Krzywonos at al. 2008). The investigations of the laboratory seem to encourage the upgrading of the research scale as well.

15.5 CONCLUSION

Wheat starch effluents contain higher concentrations of toxic chemicals. There are various methods such as biochemical and flocculation, sedimentation methods available for the treatment of wheat starch effluents. However, aerobic processes are required to minimize the content in the wastes to recover the useful substances from starch effluents, and reuse them for other purposes. The available and commonly used methods are not fully efficient to remove the toxic substances and retain the useable ones, therefore, aerobic processes are very essential to retain all the useful substances as sources for future utilization.

REFERENCES

Aines, R. D., & Rossman, G. R., 1986. Relationships between radiation damage and trace water in zircon, quartz, and topaz. *American Mineralogist*, *71*(9–10), pp. 1186–1193.

Anastassiadis, S. and Rehm, H.J., 2006. Oxygen and temperature effect on continuous citric acid secretion in Candida oleophila. *Electronic Journal of Biotechnology*, *9*(4), pp.414–423.

Battestin, V. and Macedo, G.A., 2007. Effects of temperature, pH and additives on the activity of tannase produced by Paecilomycesvariotii. *Electronic Journal of Biotechnology*, *10*(2), pp.191–199.

Belfort, G., 1984. Membrane methods in water and wastewater treatment: an overview. *Synthetic membrane processes: fundamentals and water applications. G. Belfort, ed. Academic Press, Inc., Orlando, FL*, p.1.

Carioca, J.O.B., Arora, H.L. and Khan, A.S., 1981. Technological and socio-economic aspects of cassava-based autonomous minidistilleries in Brazil. *Biomass*, *1*(2), pp.99–114.

Choorit, W. and Wisarnwan, P., 2007. Effect of temperature on the anaerobic digestion of palm oil mill effluent. *Electronic Journal of Biotechnology*, *10*(3), pp.376–385.

Cibis, E., 2004. Aerobic biodegradation of starch stills using a mixed culture of thermophilic and mesophilic bacteria of the genus Bacillus. Scientific works of the Wrocław University of Economics. *Series: Monographs and Studies (nr 100)* (1028).

Cibis, E., Kent, C.A., Krzywonos, M., Garncarek, Z., Garncarek, B. and Miśkiewicz, T., 2002. Biodegradation of potato slops from a rural distillery by thermophilic aerobic bacteria. *Bioresource Technology*, *85*(1), pp.57–61.

Cibis, E., Krzywonos, M. and Miśkiewicz, T., 2006. Aerobic biodegradation of potato slops under moderate thermophilic conditions: effect of pollution load. *Bioresource Technology*, *97*(4), pp.679–685.

Cibis, E., Zmaczynski, K., Giryn, H. and Leupold, G., 1992. Nährstoffverwertungausmelasseangereicherter Rübenschlempedurch Candida utilis in einemzweistufigen, pHstatischenKultivierungssystem. *Chemie, Mikrobiologie, Technologie der Lebensmittel*, *14*(5–6), pp.149–156.

Czupryński, B., Klosowski, G., Kotarska, K. and Wolska, M., 2002. Unconventional directions of stillage management. Cz. 1. *Przemysł FermentacyjnyiOwocowo-Warzywny, 46*(01), pp.20–21.

Davis, L., Jeon, Y.J., Svenson, C., Rogers, P., Pearce, J. and Peiris, P., 2005. Evaluation of wheat stillage for ethanol production by recombinant Zymomonasmobilis. *Biomass and Bioenergy, 29*(1), pp.49–59.

Dzwonkowski, W., Szczepaniak, I., Rosiak, E., Chotkowski, J., Rembeza, J. and Bochińska, E., 2011. Rynekziemniaka–stan iperspektywy. *Wyd. IERiGŻ, ARR, MRiRW Warszawa.*

Egg, R.P., Sweeten, J.M. and Coble, C.G., 1985. Grain sorghum stillage recycling: Effect on ethanol yield and stillage quality. *Biotechnology and Bioengineering, 27*(12), pp.1735–1738.

Fane, A.G., 1975, January. Recovery of soluble protein from wheat starch factory effluents. In *AIChE Symp. Ser.; (United States)* (Vol. 73, No. 163).

Fargey, T.R. and Smith, R.E., 1965. Studies on the biological stabilization of thin stillage: II. Anaerobic fermentation. *Canadian Journal of Microbiology, 11*(5), pp.791–795.

Ferzik, S., Petrikova, K., Rychtera, M., Melzoch, K. and Lapisova, K., 2004, May. Aerobic degradation of distiller's slops by thermophilic bacteria. In *31 International Conference of SSCHE.*

Fisher, D.J., McKinnon, J.J., Mustafa, A.F., Christensen, D.A. and McCartney, D., 1999. Evaluation of wheat-based thin stillage as a water source for growing and finishing beef cattle. *Journal of Animal Science, 77*(10), pp.2810–2816.

Ganesh, D. and Mowat, D.N., 1985. Integrated systems of produci-ng beef and ethanol from fractionated maize silage. *Biomass, 7*(1), pp.13–25.

Gao, M., She, Z. and Jin, C., 2007. Performance evaluation of a mesophilic (37 C) upflow anaerobic sludge blanket reactor in treating distiller's grains wastewater. *Journal of Hazardous Materials, 141*(3), pp.808–813.

Garcia, A.D. and Kalscheur, K.F., 2004. *Ensiling wet distillers grains with other feeds.* SDSU, Cooperative Extension Service.

Goodwin, J.A.S. and Stuart, J.B., 1994. Anaerobic digestion of malt whisky distillery pot ale using upflow anaerobic sludge blanket reactors. *Bioresource Technology, 49*(1), pp.75–81.

Gryta, M., 2005. Long-term performance of membrane distillation process. *Journal of Membrane Science, 265*(1–2), pp.153–159.

Ham, G.A., Stock, R.A., Klopfenstein, T.J., Larson, E.M., Shain, D.H. and Huffman, R.P., 1994. Wet corn distillersby-products compared with dried corn distillers grains with solubles as a source of protein and energy for ruminants. *Journal of animal science, 72*(12), pp.3246–3257.

Harris, J.L., 1985. Protein recovery from wheat starch factory effluent by ultrafiltration: an economic appraisal. *Food technology in Australia.*37, pp. 564:567

Hunt, C.W., Paterson, J.A., Fischer, J.R. and Williams, J.E., 1983. The effect of sodium hydroxide treatment of fescue-corn stillage diets on intake, digestibility and performance with lambs. *Journal of Animal Science, 57*(4), pp.1013–1019.

Hutnan, M., Hornak, M., Bodik, I. and Hlavacka, V., 2003. Anaerobic treatment of wheat stillage. *Chemical and Biochemical Engineering Quarterly, 17*(3), pp.233–242.

Jamuna, R. and Ramakrishna, S.V., 1989. SCP production and removal of organic load from cassava starch industry waste by yeasts. *Journal of Fermentation and Bioengineering, 67*(2), pp.126–131.

Jenkins, J.W., Sweeten, J.M. and Reddell, D.L., 1987. Land application of thin stillage from a grain sorghum feedstock. *Biomass, 14*(4), pp.245–267.

Khardenavis, A.A., Kumar, M.S., Mudliar, S.N. and Chakrabarti, T., 2007. Biotechnological conversion of agro-industrial wastewaters into biodegradable plastic, poly β-hydroxybutyrate. *Bioresource Technology, 98*(18), pp.3579–3584.

Kienholz, E.W., Rossiter, D.L., Ward, G.M. and Matsushima, J.K., 1979. Grain Alcohol Fermentation By-Products for Feeding in Colorado. *Experiment Station General Services*, 983.

Kim, J.S., Kim, B.G. and Lee, C.H., 1999. Distillery waste recycle through membrane filtration in batch alcohol fermentation. *Biotechnology Letters, 21*(5), pp.401–405.

Knight, J.W. and Olson, R.M., 1984. Wheat starch: production, modification, and uses. In *Starch: Chemistry and Technology* (pp. 491–506). Academic Press.

Knight, J.W. and Olson, R.M., 1984. Wheat starch: production, modification, and uses. In *Starch: Chemistry and Technology* (pp. 491–506). Academic Press.

Knight, J.W., 1965. Wheat starch and gluten. *Leonard Hill, London.*

Krzywonos, M., Cibis, E. and Miskiewicz, T., 2002. Biodegradation of the potato slops with a mixed population of aerobic bacteria-optimisation of temperature and pH. *Polish Journal of Food and Nutrition Sciences*, 4(11/52).

Krzywonos, M., Cibis, E., Miśkiewicz, T. and Kent, C.A., 2008. Effect of temperature on the efficiency of the thermo-and mesophilic aerobic batch biodegradation of high-strength distillery wastewater (potato stillage). *Bioresource Technology*, 99(16), pp.7816–7824.

Krzywonos, M., Cibis, E., Miskiewicz, T. and Ryznar-Luty, A., 2009. Utilization and biodegradation of starch stillage (distillery wastewater). *Electronic Journal of Biotechnology*, 12(1), pp.6–7.

Kupczyk, A., 2007. The current state and prospects for the use of transport biofuels in Poland compared to the EU. *Part III, Instruments supporting the development of biofuels, Energy and Ecology, June–July*.

Larson, E.M., Stock, R.A., Klopfenstein, T.J., Sindt, M.H. and Huffman, R.P., 1993. Feeding value of wet distillersby-products for finishing ruminants. *Journal of Animal Science*, 71(8), pp.2228–2236.

Laubscher, A.C.J., Wentzel, M.C., Le Roux, J.M.W. and Ekama, G.A., 2001. Treatment of grain distillation wastewaters in an upflow anaerobic sludge bed (UASB) system. *Water SA*, 27(4), pp.433–444.

Leathers, T.D., 1998. Utilization of fuel ethanol residues in production of the biopolymer alternan. *Process Biochemistry*, 33(1), pp.15–19.

Leathers, T.D., 2003. Bioconversions of maize residues to value-added coproducts using yeast-like fungi. *FEMS Yeast Research*, 3(2), pp.133–140.

Licht, F.O., 2008. World Ethanol & Biofuels Report. Agra Informa Ltd.

Lonsdale, H.K., 1984. Synthetic membrane processes: Fundamentals and water applications: Georges Belfort (ed.), Academic Press, New York, 1984, xiii+ 552 pp.

Maiorella, B.L., Blanch, H.W. and Wilke, C.R., 1983. Distillery effluent treatment and by-product recovery. *Process Biochem.; (United Kingdom)*, 18(4).

Malnou, D., Huyard, A. and Faup, G.M., 1987. High load process using yeasts for vinasses of beet molasses treatment. *Water Science and Technology*, 19(1–2), p.11.

Matthiasson, E., 1983. The role of macromolecular adsorption in fouling of ultrafiltration membranes. *Journal of Membrane Science*, 16, pp.23–36.

McCullough, M.E., Sisk, L.R. and Sell, O.E., 1963. Influence of corn distillers dried grains with solubles on the feeding of wheat silage. *Journal of Dairy Science*, 46, pp.43–45.

Milewski, J., Sarnecka, J., Zalewska, T. and Labetowicz, J., 2001. Wywar z gorzelnirolniczej-wartosciowyproduktubocznyczyodpad?[2]. *Przemysł FermentacyjnyiOwocowo-Warzywny*, 7(45), pp.23–25.

Mittleider, J.F., 1978. Analysis of the economic feasibility of establishing wheat gluten processing plants in North Dakota.

Monteiro, C.E., 1975. Brazilian experience with the disposal of waste water from the cane sugar and alcohol industry. *Rocess Biochem*, 10(9), pp. 33–41.

Moore, S., 2018. Sustainable Energy Transformations, Power, and Politics: Morocco and the Mediterranean. Routledge. 276.

Morimura, S., Kida, K. and Sonoda, Y., 1994. Production of protease using wastewater from the manufacture of shochu. *Journal of fermentation and bioengineering*, 77(2), pp.183–187.

Moriya, K., Iefuji, H., Shimoi, H., Sato, S.I. and Tadenuma, M., 1990. Treatment of distillery wastewater discharged from beet molasses-spirits production using yeast. *Journal of Fermentation and Bioengineering*, 69(2), pp.138–140.

Muntifering, R.B., Burch, T.J., Miller, B.G. and Ely, D.G., 1983. Digestibility and metabolism of mature tall fescue hay reconstituted and ensiled with whole stillage. *Journal of Animal Science*, 57(5), pp.1286–1293.

Murphy, J.D. and Power, N.M., 2008. How can we improve the energy balance of ethanol production from wheat? *Fuel*, 87(10–11), pp.1799–1806.

Murray, A.P. and Marchant, R., 1986. Nitrogen utilization in rainbow trout fingerlings (Salmo gairdneri Richardson) fed mixed microbial biomass. *Aquaculture*, 54(4), pp.263–275.

Mustafa, A.F., McKinnon, J.J. and Christensen, D.A., 1999. Chemical characterization and in vitro crude protein degradability of thin stillage derived from barley-and wheat-based ethanol production. *Animal Feed Science and Technology*, 80(3–4), pp.247–256.

Mustafa, A.F., McKinnon, J.J., Ingledew, M.W. and Christensen, D.A., 2000. The nutritive value for ruminants of thin stillage and distillers' grains derived from wheat, rye, triticale and barley. *Journal of the Science of Food and Agriculture*, 80(5), pp.607–613.

Nagano, A., Arikawa, E. and Kobayashi, H., 1992. The treatment of liquor wastewater containing high-strength suspended solids by membrane bioreactor system. *Water Science and Technology*, 26(3–4), pp.887–895.

Nguyen, M.H., 2003. Alternatives to spray irrigation of starch waste based distillery effluent. *Journal of Food Engineering*, 60(4), pp.367–374.

Radley, J.A. ed., 1976. *Industrial uses of starch and its derivatives* (No. 664.22 R35). London: Applied Science Publishers.

Sanchez Riera, F., Cordoba, P. and Sineriz, F., 1985. Use of the UASB reactor for the anaerobic treatment of stillage from sugar cane molasses. *Biotechnology and Bioengineering*, 27(12), pp.1710–1716.

Sheehan, G.J. and Greenfield, P.F., 1980. Utilisation, treatment and disposal of distillery wastewater. *Water Research*, 14(3), pp.257–277.

Shin, H.S., Bae, B.U., Lee, J.J. and Paik, B.C., 1992. Anaerobic digestion of distillery wastewater in a two-phase UASB system. *Water Science and Technology*, 25(7), pp.361–371.

Smith, R.E. and Fargey, T.R., 1965. Studies on the biological stabilization of thin stillage: I. Aerobic fermentation. *Canadian Journal of Microbiology*, 11(3), pp.561–571.

Sweeten, J.M., Lawhon, J.T., Schelling, G.T., Gillespie, T.R. and Coble, C.G., 1981. Removal and utilization of ethanol stillage constituents. *Energy in Agriculture*, 1, pp.331–345.

Tanaka, Y., Murata, A. and Hayashida, S., 1995. Accelerated composting of cereal shochu-distillery wastes by actinomycetes. *Seibutsu-Kogaku Kaishi-Journal of The Society for Fermentation and Bioengineering*, 73(5), pp.365–372.

Tang, Y.Q., Fujimura, Y., Shigematsu, T., Morimura, S. and Kida, K., 2007. Anaerobic treatment performance and microbial population of thermophilic upflow anaerobic filter reactor treating awamori distillery wastewater. *Journal of Bioscience and Bioengineering*, 104(4), pp.281–287.

Tauk, S.M., 1982. Culture of Candida in vinasse and molasses: effect of acid and salt addition on biomass and raw protein production. *European Journal of applied Microbiology and Biotechnology*, 16(4), pp.223–227.

Weiland, P. and Thomsen, H., 1990. Operational behaviour of an industrial fixed bed reactor for biomethanation of alcohol slops from different crops. *Water Science and Technology*, 22(1–2), pp.385–394.

West, T.P. and Strohfus, B., 1996. Pullulan production by Aureobasidium pullulans grown on ethanol stillage as a nitrogen source. *Microbios*, 88, pp.7–18.

Wilkie, A.C., Riedesel, K.J. and Owens, J.M., 2000. Stillage characterization and anaerobic treatment of ethanol stillage from conventional and cellulosic feedstocks. *Biomass and Bioenergy*, 19(2), pp.63–102.

Wu, Y.V., 1988. Recovery of stillage soluble solids from corn and dry-milled corn fractions by high-pressure reverse osmosis and ultrafiltration. *Cereal Chemistry*, 65(4) pp. 345–348

Yang, F.C. and Lin, I.H., 1998. Production of acid protease using thin stillage from a rice-spirit distillery by Aspergillus niger. *Enzyme and Microbial Technology*, 23(6), pp.397–402.

Yokoi, H., Aratake, T., Nishio, S., Hirose, J., Hayashi, S. and Takasaki, Y., 1998. Chitosan production from shochu distillery wastewater by funguses. *Journal of Fermentation and Bioengineering*, 85(2), pp.246–249.

16 Anaerobic Treatment of Food Processing Wastes and Agricultural Effluents

Simmi Goel,[a] *Pankaj Kumar,*[b] *Mukul Sain,*[c] *and Ajay Singh*[d]

[a]Department of Biotechnology, Mata Gujri College Fatehgarh Sahib, Punjab, India

[b]Department of Microbiology, Dolphin (PG) Institute of Biomedical and Natural Sciences, Dehradun, India

[c]Dairy Engineering Division, National Dairy Research Institute, Karnal, Haryana, India

[d]Department of Food Technology, Mata Gujri College Fatehgarh Sahib, Punjab, India

CONTENTS

16.1 INTRODUCTION

A massive rate of industrialization and urbanization impacts everyone as a result of which, natural resources are on the forefront and are being exhausted at a very fast pace. A large number of effluents such as agricultural waste, food processing waste, municipal waste, fruit and vegetable waste have been generated annually. Among developing nations, a lack of proper transportation, storage, disposal, and treatment facilities for waste poses a great risk to the environment in terms of soil, water, and air pollution. Lots of vegetable and fruit crops get wasted to generate an enormous amount of agro-industrial waste. So, to counter this situation, various tactics of burning, pyrolysis, incineration, combustion, land filling, and animal feed have been currently in use for management of waste. Besides these suggested treatments, anaerobic digestion proved to be an economic and simple method for efficient treatment of agricultural and food processing waste. The large amounts of liquid and solid wastes generating from agriculture residues, manure, sewage sludge, food processing industries including fish, meat, poultry, fruits, vegetables, pulses, oil seeds, sugarcane bagasse, sheep and goat manure, cattle dung, dairy waste can be efficiently utilized for anaerobic decomposition. Biogas technology under anaerobic treatment is frequently utilized for the conversion of biomass into energy and sludge found its way as a fuel and biofertilizer respectively (Tamburini *et al.*, 2020).

16.2   CATEGORIES OF WASTE

16.2.1   MEAT PROCESSING INDUSTRIES

Slaughterhouse waste is the major contributor for meat, which consists of the remains of a slaughtered animal. The waste produced by the meat processing industries constitutes inedible parts like skin, bones, blood, gastro-intestinal tract, tendon, and visceral organs. Wastewater also contains several suspended solids of organic and inorganic nature resulting in an increase of turbidity and consumption of dissolved oxygen present in the water. Organic material constitutes fat, carbohydrates, and proteins. Protein degradation resulted in the production of ammonium ions, which eventually leads to the discharge of nitrites and nitrates that result in higher oxygen consumption. Viscera of animals and poultry birds have very high microbial load including several types of pathogenic bacteria like *Salmonella*, *Escherichia coli* and *Shigella*. The wastewater produced contains a variety of organic and inorganic pollutants, has a high concentration of etheric extract, suspended and biogenic matter as well as variable concentrations (Jayathilakan *et al.*, 2012). Effluent from slaughterhouses and meat packing corporations are having a high concentration of solids, fats, blood, manure, and a variety of organic compounds originating from proteins. The major sections that add contaminants are slaughtering, hide or hair removal, carcass washing, rendering, trimming, and clean-up activities. These contain a variety of readily degradable organic compounds like fats and proteins, present in both particulate and dissolved forms. The wastewater from these is heavily loaded with biochemical oxygen demand, chemical oxygen demand, suspended solids, nitrogen and phosphorus suitable for anaerobic treatment.

16.2.2   FRUIT AND VEGETABLE PROCESSING CORPORATIONS

Approximately 30 million tons of food waste is produced annually. Less than 3% is treated and the rest is disposed in land fillings. Food industrial waste including crop residues, fruits and vegetable peels, oil seeds, grains, meat and milk products, fishery waste, beverages, poultry waste, sugar and starch processing, corn milling waste, lignocellulosic waste, preservatives can be efficiently used for anaerobic digestion. At present, food waste either goes to animal farms as feedstock or to land filling. Serious health threats have been seen with animal feed stock and food waste landfills leachate resulting in ground water contamination. All these wastes need to be properly analyzed for their physico chemical composition before being used for anaerobic digestion. Food waste not only contains molecular organic matter, but also contains various trace elements. Currently, the anaerobic digestion process has become an intensive field of research, since the organic matter in the food waste is suited for anaerobic microbial growth (Zhang & Jahng, 2012).

Materials undergoing anaerobic decomposition can be classified based on their organic dry matter or total solids' content like low (0.2–1% dry organic matter), medium (1–5% dry organic matter), high (5–12% dry organic matter) and solid waste (20–40% dry organic matter).

16.2.3 Dairy Processing Enterprises

High load of COD, BOD, total nitrogen, total phosphorus, fat, oil and grease is found in dairy effluent due to the presence of dissolved and crystallized fats (glycerol, triglycerides), sugars (lactose) and protein (casein) in colloidal form. The major pollutant in milk processing wastewater is whey, which represents about 85–95% of the milk volume. Whey consists of carbohydrates (4–5%) in the form of lactose, proteins, and lactic acid less than 1%, fats to around 0.4–0.5%, while salts vary from 1 to 3%. During cheese manufacturing, cheese whey wastewater is produced whose volume and composition changes with respect to the type of cheese, technology, milk type, and the environmental conditions (Table 16.1). Anaerobic systems are more suitable for the direct utilization of high-strength dairy wastewater and are more cost-effective than aerobic processes. The addition of various additives like silica gel, bentonite, Tween 80, charcoal, pectin, gelatin in anaerobic digestion of cheese whey resulted in doubling the production of biogas (Aneja, 2003).

Milk processing wastewater is mainly treated in conventional one-phase systems i.e. an upflow anaerobic sludge blanket (UASB) reactor and an anaerobic filter (AF). If properly operated, these systems do not produce unpleasant odors. Some of the factors like long start-up periods due to complex substrate degradation, preliminary biomass adaptation prior to protein and fat utilization, inhibition of methane production due to lower pH, sludge disintegration by fats in the form of triglyceride emulsions and shock loadings need to be properly managed through regular monitoring of reactors. Cheese effluents are degraded in UASB reactors in laboratory tests and on an industrial scale. A laboratory-scale UASB reactor utilizing a cheese factory effluent eliminates around 90%

TABLE 16.1
Classification of Waste, Various Components, and Types of Anaerobic Reactors

SNo	Type of Waste	Components released	Anaerobic reactors
1.	Meat processing waste	skin, bones, blood, gastro-intestinal tract, tendon and visceral organs, high concentration of solids, fats, manure, and a variety of organic compounds originating from proteins	Upflow anaerobic sludge banket reactor (UASB)
2.	Food processing waste	Crop residues, sugarcane, sugarbeet, bagasse, potato. Vegetable, fruit processing, brewery, distillery waste, whey, oil seeds, corn milling waste, sugar and starch processing, fishery waste, fruit and vegetable peels	Anaerobic filter reactor, Anaerobic baffle reactor, Upflow anaerobic sludge banket reactor, Anaerobic contact reactor
3.	Dairy processing waste	Fats (glycerol, triglycerides), sugars (lactose) and protein (casein), cheese whey, salts, preservatives, oil, grease	Upflow anaerobic sludge banket reactor, Anaerobic packed bed reactor
4.	Cereal processing waste	Rice husk, rice bran, rice straw, wheat bran, wheat straw, maize stover, maize husk, skin, millet stover	Anaerobic contact reactor, Fixed film reactor
5.	Agro industrial waste	Straw, husk, bagasse, cow dung, woody waste, coir pith, animal sludge, poultry manure, milk whey, oil press residues, litter, vegetable, fruit and legume waste, cob, stalks, cattle dung, pig or chicken slurry, poultry, agricultural runoff nitrates, phosphates, herbicides, pesticides, farmyard slurry	Anaerobic contact reactor, Upflow anaerobic sludge banket reactor, Fixed film reactor

of effluents at an organic loading rate (OLR), expressed as chemical oxygen demand (COD), of 31 g/(L·day). Dairy effluents with a low TSS can be efficiently treated in anaerobic filter reactor. The COD decreased between 60 and 98% at hydraulic retention time (HRT) of 12–48 h and an OLR, expressed as COD, of 1.7–20 kg/(m³·day). A large specific surface of the filter media creates a precondition for higher biomass accumulation, which is less affected by shear stress. An anaerobic packed-bed bioreactor (PBB) can be successfully applied for dairy wastewater treatment of various organic loads. The cheese whey can be decomposed in a laboratory-scale PBB with a polyethylene carrier. The highest COD reduction will be achieved at a 3.5-day HRT with OLR, expressed as COD, of 3.8 kg/(m³·day) and biogas production of 0.42 m³ per kg of COD per day.

16.2.4 CEREAL PROCESSING INDUSTRIES

With the development of modern scientific techniques, the agriculture waste production has been increased at an exponential rate across the globe. Cereals are the staple food for humans as well as feed for cattle and is the major contributor to agricultural waste. Cereals occupy the maximum area and production in agriculture and become the largest source of lignocellulosic materials. About 75% of residues generated in the world is from cereals like corn, rice, wheat, millets, barley, rye. Amongst different crops, cereals, fibers, oil seeds, pulses, and sugarcane crops generate 352, 66, 29, 13, and 12 million tonne residues. Sustainable management of cereal waste seems to be the greatest challenge in today's era. Rice and wheat are the major contributors to the agri biowaste pool due to wider cultivation and large-scale production. There are two categories of crop residues: the first category includes materials like stalk and stubble in the case of cereals left over in agricultural fields after harvesting of crops and the second category is the processed residue comprising material left after the crop is processed like husk or hull, bagasse, molasses in sugarcane into a usable resource. The residue management in a cereal-cereal system like rice-rice and rice-maize, rice-wheat is the biggest challenge faced by the researchers (Hassan and Aadil, 2021). Most of the countries in Asia, Africa, and America have failed to cope with the large volume of crop residues although a majority of these are used as fodder and fuel. In agricultural fields, crop residues are burnt every year releasing toxic fumes into the atmosphere that can cause severe health infections. Small farmers usually practice burning of crop residues as it is the inexpensive alternative in the absence of any technical knowledge. The lignin and polyphenol content present in cereals reduces the rate of anaerobic degradation. Various environmental factors like low temperature decreases the rate of degradation while moisture content affects the rate of degradation by influencing the growth of microbes (Shahane and Shivay, 2016). The size of crop residue also affects the rate of decomposition, smaller size increases the area exposed for microbial activity thus the rate of degradation is faster. Leguminous plants' waste decomposes faster than cereal residues because of higher nitrogen content. Generally, cereal processing waste has high BOD (1050–7950mg/l), COD (1400–29000mg/l) and pH in the range of 3.5–8.5 and is also the major contributors of sulphur and phosphorous. Cereal industry waste can be utilized for the generation of biohydrogen, bioethanol, biobutanol, biogas, biocoal, biomass, biofertilizer, proteins, organic acids, and polysaccharides.

16.2.5 AGRO INDUSTRIAL EFFLUENT

Agro industries generate a large number of lignocellulosic wastes in the form of coir pith waste, coffee husk, and bagasse that accumulate in the environment. In addition, animal sludge, poultry manure, milk whey, oil press residues, litter, vegetable, fruit and legume waste, straw, cob, stalks are also the major contributors. The liquid agricultural waste includes cattle dung, pig or chicken slurry, poultry, agricultural runoff nitrates, phosphates, herbicides, pesticides, farmyard slurry. Agro wastes and energy crops are an important source of biomass that can be efficiently utilized as a substrate in anaerobic digestion, resulting in the production of renewable energy. The efficiency of all these agro-wastes in combination with animal dung for methane production by anaerobic digestion have been investigated by various researchers. The major components of lignocellulosic agricultural biomass are cellulose, hemicellulose, and lignin and minor quantity of oils, proteins, and ash. These can in turn be utilized for generation of biofuels by thermochemical and biochemical conversion. The different forms of bioenergy that can be generated from agricultural residues are bioethanol, biogas, biohydrogen, and biodiesel. The process of anaerobic decomposition is a biological approach in which anaerobic microorganisms decompose biodegradable substrate into biogas, nutrient-rich fertilizer in the form of sludge and additional cell matter. The microbial biomass breaks down recalcitrant biomass lignocellulose into a simpler form of substrate. Complex and unsterile raw material can be transformed into energy-rich biogas through anaerobic decomposition. Biogas is a mixture of methane (CH_4, 50–70%), carbon dioxide (CO_2, 30–40%), hydrogen (H_2, 1–5%), nitrogen (N_2, 0.5%), carbon monoxide (CO), hydrogen sulphide (H_2S), and water vapors in traces (Joshi & Pandey, 1999).

16.3 MECHANISM OF ANAEROBIC DECOMPOSITION

Anaerobic decomposition is a biological process in which food waste and organic materials are broken down by microorganisms in the absence of light and oxygen to produce renewable energy (biogas) and a nutrient-rich biofertilizer. Anaerobic decomposition is suitable for agricultural and industrial waste containing high concentration of organic compounds.

It is the microbial decomposition of organic waste under optimized conditions into CO_2 and CH_4 strictly in the absence of oxygen. Organic matter, suspended solids, acids, inorganic salts, fats, oils, chlorides, inorganic matter are readily degradable and are free from any toxicity (Singh, 2012). Efficiency of decomposition can be increased by increasing the density of anaerobes using a solid matrix to retain biomass or by reuse of activated biomass or by minimizing biomass washout (Figure 16.1).

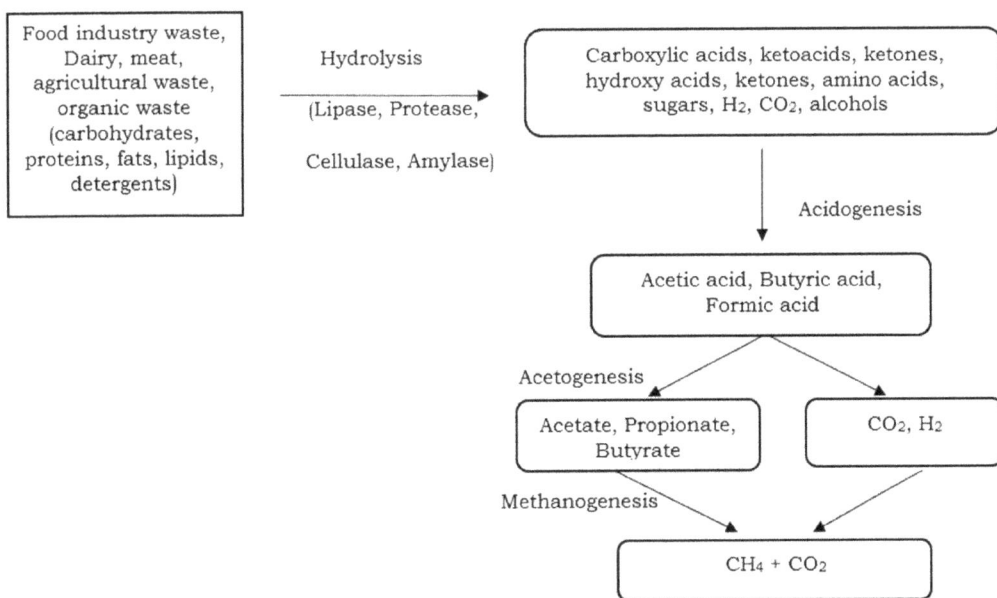

FIGURE 16.1 Steps of anaerobic decomposition of organic waste into biogas.

Source: Scragg, 2010.

The whole anaerobic process can be divided into various steps.

1. **Hydrolysis:** in this step insoluble complex organic matter like proteins, fats, and polysaccharides are enzymatically hydrolyzed into soluble monomers like peptides, fatty acids, and saccharides using anaerobic cellulolytic bacteria. A variety of exoenzymes of obligatory and facultative anaerobic bacteria like *Clostridium* sp., *Bacteroides, Fusobacterium, Micrococci, Eubacterium, Streptococcus* and *Selenomonas* are involved in hydrolysis. Production of carboxylic acids, keto acids, hydroxy acids, ketones and alcohols, H_2 and CO_2 by the action of anaerobic microbes. The capacity of hydrolytic microbes to degrade the raw solid waste and its biomethanation is the crucial or rate-limiting step. So, to make the substrate available to hydrolytic microbes, pre-treatment is necessary. This is a slow and rate limiting step in anaerobic decomposition.

2. **Acidogenesis:** these monomers are further fermented into intermediate products of low molecular weight like acetate, propionate, butyrate using acetogenic bacteria. Lipids can be fermented at a faster rate than carbohydrates and proteins. The higher concentration of acetic acid, propionic acid, butyric acid, and other fatty acids inhibit anaerobic digestion. Acidogenesis is generally an expeditious phase in the anaerobic transformation of the substance produced in the hydrolysis stage into simpler organic acids (formic acid, acetic acid, propionic acid, and butyric acid), alcohols, H_2, and CO_2. The pH of the system decreases on biodegradation of organic material into organic acids thus creating a suitable environment for acidogenic and acetogenic bacteria as these microorganisms prefer moderately acidic conditions, with a pH range of 4.5–5.5.

3. **Acetogenesis:** in this stage, the products obtained from the previous (acidogenic) stage are utilized as a substrate for other microbes. In this stage, homoacetogenic microbes continuously reduce H_2 and CO_2 to acetic acid (CH_3COOH). Acetogenic bacteria are strict anaerobes and grow symbiotically with methane-producing bacteria. During this step, acetate (substrate for methane-producing bacteria) is formed from organic acids and alcohols (Scragg, 2005).

4. **Methanogenesis:** then acetic acid is converted to methane and CO_2 using methanogenic bacteria (Dar *et al.*, 2021). During this stage, methane production occurs under stringent anaerobic environment. Not all the methanogenic species can degrade all substrates. Two biochemical characteristics that are peculiar to methanogens are the mechanism to oxidize hydrogen and reduce CO_2 (Table 16.2). Methanogenic bacteria use hydrogen along with carbon dioxide, formate, methanol, and acetate as substrates for methane production. Carbon dioxide is reduced to methane as it acts as a terminal electron acceptor.

TABLE 16.2
Classification of Microbes Used in Anaerobic Decomposition

Stages	Type of Process	Category of microbes	Examples
1	Hydrolysis	Hydrolytic anaerobes	*Clostridium sp., Bacteroides, Fusobacterium, Micrococci, Eubacterium, Streptococcus and Selenomonas*
2	Acidogenesis	Acidogenic bacteria	*Peptococcus, Propionibacterium*
3	Acetogenesis	Acetogenic bacteria	*Syntrophobacter, Desulphovibrio, Syntrophomonas*
4	Methanogenesis	Methanogenic bacteria	*Methanococcusvanielii, Methanococcusvoltae, Methanobacteriumformicicum, Methanobacteriumsmithii, Methanobacillus, Methanosarcinabarkeri*

Hydrolytic and acidogenic anaerobes are heterogenous, fast growing, less susceptible to changes in nutritional and environmental conditions. On the other hand, methanogens are more fragile, obligate anaerobes, have a slow growth rate, and are fastidious in their nutritional requirements for growth and methane production. Morphologically, methanogens can be of different shapes like cocci, bacilli, spirilli, and sarcina.

The efficiency of treatment totally depends upon the active growth of microbes and decomposition of raw material so various factors like composition of waste, temperature, concentration of solids, toxic or inhibitory compound, salinity, ions' concentration, nutrient concentration, and pH needs to be optimized. Excess of acids retards the growth of microbes and hence lowers the methane production. Highly saline waste and high temperature can retard the rate of methane production. Concentration of nitrogen, phosphorous, and trace elements like calcium, magnesium, barium, iron, and nickel need to be optimized as they serve as nutrients for microbial growth. Fats give maximum yield of biogas whereas proteins yield the least. Carbohydrates are poor sources for biogas generation (Dubey, 2002).

Various anaerobic reactors have been designed to develop continuous treatment technology on an industrial scale.

16.3.1 Design of Anaerobic System Reactor

Anaerobic decomposition takes place in a sealed vessel known as a reactor, which is designed and constructed in various shapes and sizes specific to the site, environmental conditions, and feedstock material. The reactor is designed for complete degradation of organic waste into biogas by providing optimal conditions for proper growth of anaerobes. The anaerobic reactor needs to be monitored regularly during the process of degradation. It is a controlled process that effectively produces 60% methane, 40% carbon dioxide, and highly nutritive sludge. It can be broadly classified into various steps:

1. **Organic waste collection system:** waste is collected and placed in a centralized location. Food waste, food scraps, agricultural waste etc. are delivered to a facility where it is prepared for further processing. It can be collected in the form of liquid, slurry, or semi solid at a single point (lagoon, pits, tank). The waste should be free of bedding materials like rocks, stones, straw, or sand etc., which can clog the pipes of the digester. Materials like detergents, acids, antibiotics or halogenated compounds, which can retard the anaerobic decomposition, must be avoided in the system.

2. **Organic waste handling system:** organic waste needs to be pre-treated before it is fed into the anaerobic digestor. The various steps of pre-treatment include the following:
   i. **Size reduction:** the raw material that is to be fed into the system must be shredded or in the case of animal waste slurry should be prepared with water in the ratio of 1:1 so as to fasten the rate of degradation. The smaller the grain size, the faster will be the rate of methane production since more surface area is available for bacterial attack. So, shredding the feedstock can increase the rate of decomposition especially in waste with high concentration of cellulose and lignin whose presence retards the microbes to access and degrade the waste. Shredding by screw mills is preferable in large lumps, green cuttings, straw and other agricultural residues. The process of shredding can increase the biogas yield for hay and foliage by 20% (Deublein & Steinhauser, 2008).
   ii. **Contaminant removal:** by using sand separators, screens, hydrocyclones, plastic bags, sand settling lanes, various objectionable material can be separated so as to avoid any inhibition by these contaminants.
   iii. **Equilibration and storage:** the diluted feedstock material can be stored temporarily for its proper homogenization and settling and minimizing any fluctuations in the amount or

physico chemical or biological other characteristics. Plant waste can be left outside to rot before being put into the reactor (FAO, 1984). Long term storage should be avoided in the case of animal waste so that flies do not breed. The covered storage system can be built of bamboo, woody cuttings, bricks, tiles, and mortar to avoid the exposure to rain (NAS, 1977). The process of hydrolysis and acidogenesis can start in an equalization tank depending on its size or other environmental conditions.

16.3.2 ANAEROBIC DECOMPOSITION: BIOGAS PRODUCTION

Once the feedstock material is collected and prepared it is then loaded into the reactor vessel. The vessel used may be made of bricks, cement, stainless steel, or fiber and must be air tight. The upper part of this vessel is fitted with another container to collect the biogas. The vessel is sealed for about 40–45 days for complete decomposition. The various conditions need to be optimized for maximum biogas generation. Various factors that need to be controlled during anaerobic decomposition for maximum yield of biogas are as follows:

16.3.3 FACTORS THAT NEED TO BE OPTIMIZED FOR DESIGNING AN EFFICIENT ANAEROBIC BIOREACTOR

1. Organic loading rate (OLR) is the mass of volatile solids per unit volume of reactor and time. Its unit is Kg VS/m³day. It should be less than 2Kg VS/m³day for unstirred reactors and 4–8 kg VS/m³day for stirred reactors. Too high OLR can cause risk of acidification by changing digester conditions inhibiting microbial activity resulting in accumulation of fatty acids.

2. pH range should be between 6.5–7.5. Beyond this methanogenic bacterium will be inhibited. Variations in pH may occur due to bicarbonate concentration, volatile fatty acids, CO_2 produced.

3. Mixing equipment or manual mixing is necessary to avoid any scum formation and to have maximum biogas generation.

4. Temperature range 45–60°C thermophilic fast growth, 30–40°C mesophilic slow growth, and at less than 15°C no activity will be shown by anaerobes.

5. Composition of feedstock i.e. the type of raw material fed into the anaerobic reactor also affects the rate of degradation. Lignin and polyphenols present in cereals reduces the rate of degradation (Shahane and Shivay, 2016).

6. The anaerobic reactor should be completely sealed and strict anaerobic conditions are required for maximum biogas production.

7. Various nutrients like urea and ammonium dihydrogen orthophosphate can be added to increase the rate of growth of microbes and to have maximum biogas generation.

8. Hydraulic retention time (HRT) should be between 10–40 days depending upon composition of waste for an efficient anaerobic digestion. HRT is the time for which waste resides in the reactor for complete degradation. For a given HRT, the size of the reactor can be calculated.

$$HRT \ (days) = volume \ (m^3)/ \ flow \ rate(m^3.day)$$

$$Reactor \ volume, \ V = flow \ rate \times HRT$$

9. C/N ratio must be between 16–25. A higher ratio reduces the gas production while a lower ratio resulted in ammonia accumulation, which will inhibit the activity of anaerobes.

10. Lipid containing waste are fermented at a quicker rate than protein or carbohydrate containing feeds. Carbohydrates are a poor source of biogas and methane generation and fats gave the maximum yield of biogas.

11. High concentration of acetic, propionic, butyric, valeric and other short chain fatty acids inhibits the rate of biogas generation.
12. Particle size of input waste should be less than 5cm. For this purpose, choppers or mechanical shredders can be opted. Higher surface area of material resulted in faster degradation and rate of hydrolysis by microbes.
13. The size of anaerobic reactor can be determined using volatile solids' loading rate, VSLR (Kg VS/m^3.day).

Volatile solids loading rate (Kg VS/m^3.day) = influent volatile solids (Kg/day)/reactor volume Reactor volume (m^3) = influent VS/VSLR

16.4 VARIOUS REACTOR CONFIGURATIONS FOR ANAEROBIC DECOMPOSITION

Several modifications have been incorporated in traditional designs so as to inculcate tolerance towards high biochemical oxygen demand loading and exposure to toxic compounds. Various types of designs like anaerobic lagoons, plug flow, completely mix, batch reactor, blanket reactor, fixed film reactors have been used for biogas production.

16.4.1 ANAEROBIC CONTACT REACTOR (ACR)

The mechanism of ACR is based on the concept of activated sludge treatment where sludge is recycled back to the main tank for good microbial growth. To maximize the contact between feed and activated biomass good mixing must be provided by stirring (Figure 16.2). Organics and pathogen loaded domestic, sugar and starch processing, distilleries, vegetable, meat processing, agricultural and municipal wastewater can be efficiently treated. The presence of degassifier allows the removal of biogas bubbles attached to sludge, which may float to the surface. The efficiency of process depends on favorable settling properties of anaerobic sludge. In the anaerobic contact process the rate limiting steps are the hydrolysis conversion rate and the soluble sub utilization rate of methanogenesis. The anaerobic process does not require any aeration energy and ACR is easy to operate and requires low maintenance and helps in the recovery of fuel in the form of biogas (Fulekar, 2005).

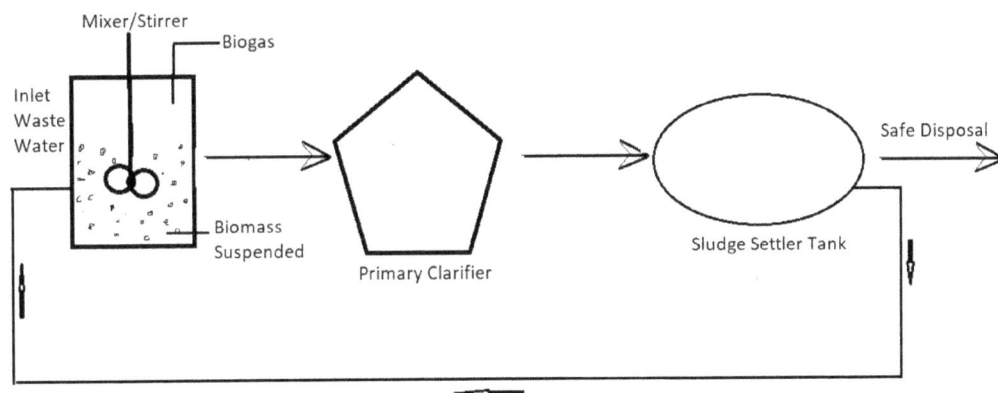

FIGURE 16.2 Anaerobic contact reactor.

16.4.2 ANAEROBIC FILTER REACTOR (AFR)

It is an attached form of biological treatment in which biomass is supported on some suitable matrix. The selected matrix must have high surface area to volume ratio for maximum attachment of biomass in the form of biofilm and low void volumes to restrict biomass washout. The different matrices used so far by various researchers are sand, plastic film, polymers, granite, quartz, stones, activated carbon. AFR is operated in an upflow mode in which wastewater enters from the base of the reactor and flows in upward direction (Figure 16.3). As it comes in contact with biofilm, microbes remove the contaminants through biosorption. The matrix retains or holds the active biomass within the filter. The wastewater with less solids' concentration is more preferred for treatment to prevent clogging of the matrix by gradual deposition of solids in voids. The anaerobic filter reactor is simple in design, easy to operate and has less maintenance, and high biofilm growth.

16.4.3 FLUIDIZED BED REACTOR (FBR)

It is an attached form of anaerobic treatment in which microbes attach and grow on the surface of a small sized solid matrix in the form of biofilm. The adsorption of biomass occurs on an inert support material by means of slime growth. This solid matrix can be spherical silica, basalt, sand, coal particles, granular activated carbon, which are usually inert in nature. The choice of matrix depends on its inertness, not reacting with components of wastewater or to interfere or inhibit the growth of anaerobes and also having more surface area for maximum growth of biofilm. FBR utilizes small sized particles like sand on the surface of which biomass attaches and grows so as to enable the matrix in continuous motion due to upward flow of wastewater. Some of the salient features of FBR include large surface area, high degree of mixing and high concentration of biomass, efficiently treating wastewater with high organic loads. Due to small sized solid particles, problem of clogging can be resolved, better substrate diffusion within the biomass, gas, and treated water can easily pass through the suspension of particles (Figure 16.4). These particles can be reused and recycled back for fresh wastewater treatment after proper pre-treatment.

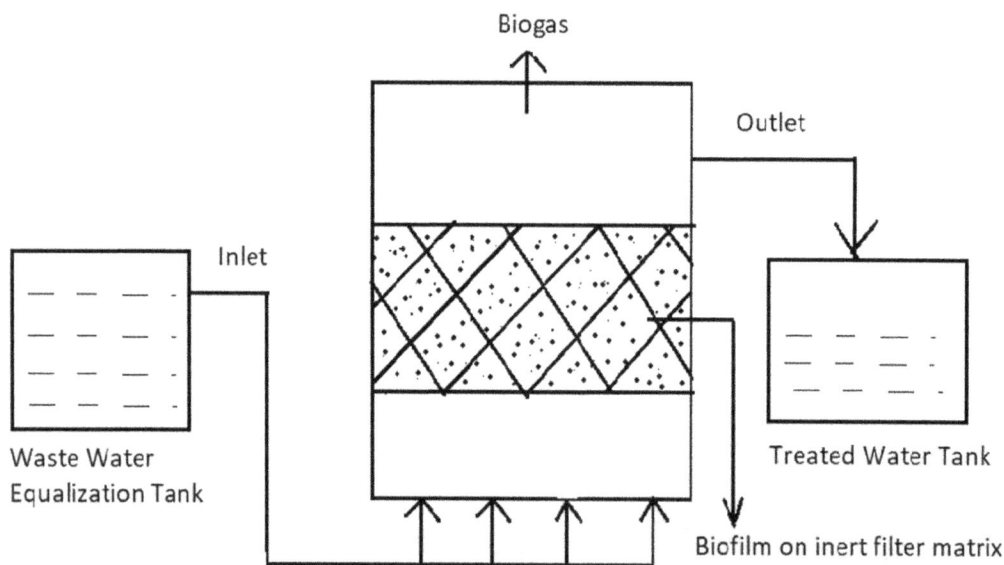

FIGURE 16.3   Anaerobic filter reactor.

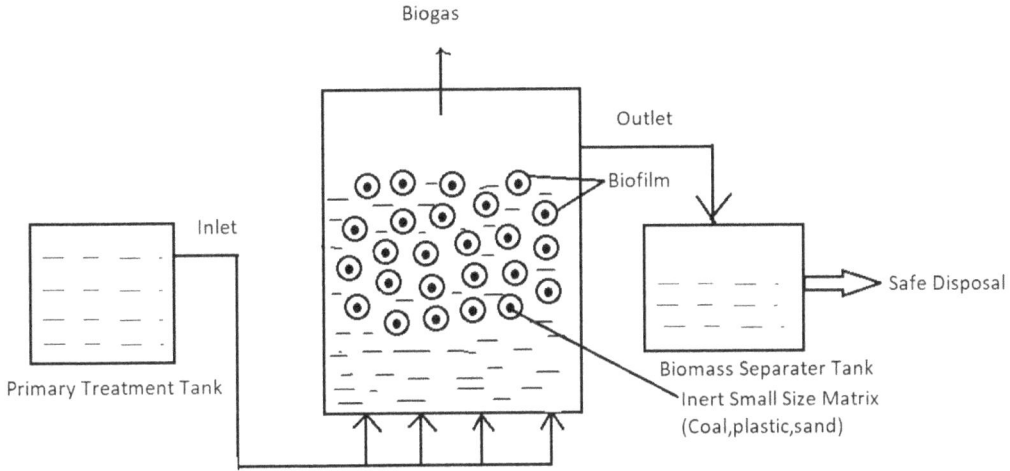

FIGURE 16.4 Fluidized bed reactor.

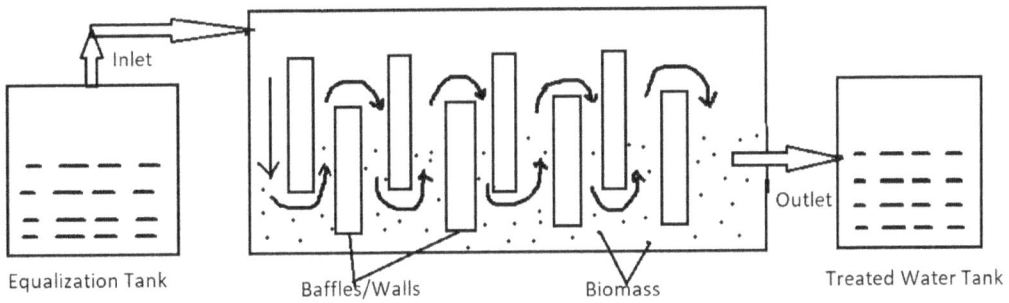

FIGURE 16.5 Anaerobic baffled reactor.

16.4.4 ANAEROBIC BAFFLED REACTOR (ABR)

It consists of a rectangular tank in which partitions are made through a series of baffles or walls, which are arranged vertically throughout the reactor. Baffles allow horizontal movement of biomass and maintains high concentration of biomass within the reactor. These baffles ensure complete mixing and uniform distribution of wastewater and biomass. The biomass gets accumulated in between the baffles resulting in the formation of granules with the passage of time. The wastewater comes from one end and flows over and under the baffles till it reaches the outlet. The biomass is present in the suspended form within the reactor. The baffles enable the system to retain the biomass and also maximize the hydraulic retention time for complete degradation and minimizes the biomass washout (Figure 16.5). All the three phases of anaerobic digestion are well separated for the complete degradation of effluent. ABR is simple in design and easy to operate (Murugesan & Rajakumari, 2006).

16.4.5 UPFLOW ANAEROBIC SLUDGE BLANKET (UASB)

It is a suspended form of anaerobic treatment in which biomass remains in suspension by the upward flow of effluent. Wastewater enters through the bottom of the reactor and passes through a high concentration of the sludge bed where maximum degradation of contaminants occurs. The remaining solids are degraded in the sludge blanket zone. Addition of calcium promotes granulation and increases the efficiency of treatment. UASB maintains hydraulic and organic loading rate, which ultimately helps in more biomass aggregation or granulation. The size of these biomass granules is 1–3mm diameter making them heavier and settled and retained at the bottom of the reactor. The whole UASB reactor can be divided into three major zones:

   i. Sludge bed: it has the highest biomass concentration in which maximum degradation occurs when the effluent enters in the reactor in the upflow mode.
   ii. Sludge blanket zone: the solids, which escape degradation in the sludge bed, are degraded in the sludge blanket zone.
   iii. Gas liquid solid separator (GLSS) zone: it is also known as the three phase separator zone. GLSS helps to separate biogas from the reactor and also prevents activated biomass washout. It helps to enable sludge to be retained in the reactor and also prevents washout of floating granular sludge.

The wastewater from alcohol manufacturing, bakery, dairy and cheese, distilleries, domestic sewage, pulp and paper, pharmaceutical, slaughter house, sugar processing, vegetable and fruit processing, starch industries are commonly treated by the UASB reactor (Figure 16.6).

Biogas processing, transportation and uses: the prepared biogas is transported through pipelines from the reactor to the gas using devices or if required to gas treatment stations. Excess of moisture must be removed from the biogas before using it for combustion or high concentrations of sulphur in the form of hydrogen sulphide gas should be removed to avoid any corrosion problems. Biogas storage tanks can be maintained from where supply can be regularized according to the needs of consumers.

   i. Combustion purposes in case of boilers, furnaces, cement kilns, brick kilns, cooking purposes, heat up of green houses.

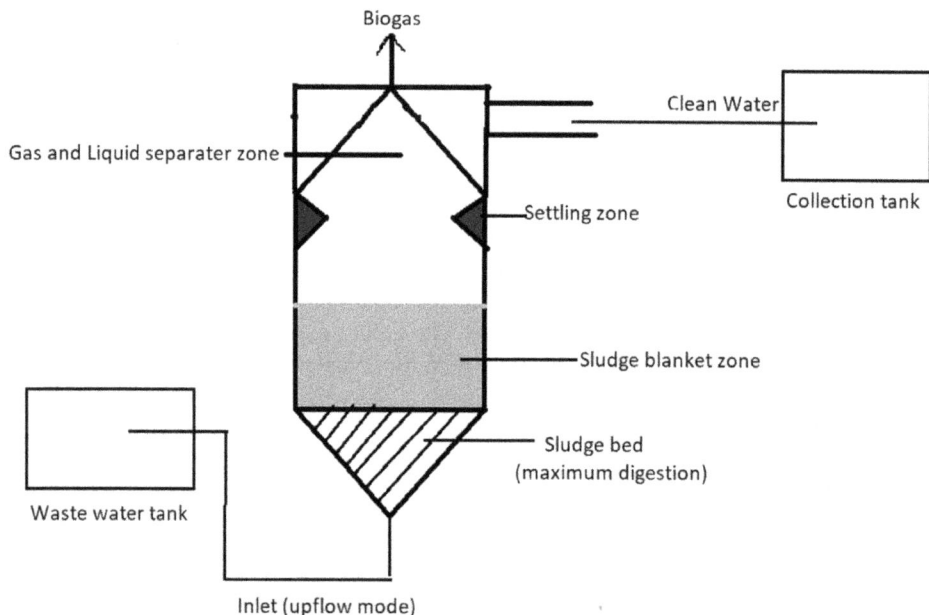

FIGURE 16.6   Upflow anaerobic sludge blanket reactor.

ii. Electricity production using internal combustion engines or gas turbines. Thermal energy in the form of waste heat produced during electricity generation can be utilized to heat reactors or supplied to maintain heating systems in buildings.

iii. Major component of biogas i.e. pure methane gas can be used as compressed gas and exported on large scale.

iv. Biogas can be processed and converted to compressed natural gas (bioCNG), which can be further used as transportation fuel.

16.5 SIGNIFICANCE OF ANAEROBIC DECOMPOSITION

- Low quantity of sludge production.
- Low energy consumption due to anoxic conditions.
- Less space required (underground).
- Biogas production.
- Better treatment efficiency at high organic loading rate.
- Low nutrient requirement.
- Short retention time.
- Low cost of operation.
- Activated microbes can be reused.
- No environmental pollution.
- High degree of purification.
- Source of renewable energy.
- Reduction of greenhouse gas emissions.
- Air proof reactor tanks (below ground).
- Reduces solid waste volume.
- Production of slurry utilized as manure in fields.
- Lesser dependence on fossil fuels.
- No odor problem due to enclosed anaerobic reactor.

REFERENCES

Agyeman, F. O., & Tao, W. (2014). Anaerobic co-digestion of food waste and dairy manure: Effects of food waste particle size and organic loading rate. Journal of Environmental Management, 133, 268–274.

Ahamed, A., Chen, C. L., Rajagopal, R., Wu, D., Mao, Y., Ho, I. J. R., ... & Wang, J. Y. (2015). Multi-phased anaerobic baffled reactor treating food waste. Bioresource Technology, 182, 239–244.

Ajila, C. M., Brar, S. K., Verma, M., & Rao, U. P. (2012). Sustainable solutions for agro processing waste management: an overview. Environmental Protection Strategies for Sustainable Development, 65–109.

Amenu, D. (2014). Characterization of wastewater and evaluation of the effectiveness of the wastewater treatement systems. World Journal of Life Sciences Research, 1(1), 1–11.

Aneja, K R. (2003). Experiments in microbiology, plant pathology and biotechnology. Fourth revised Ed., New Age International (p). Ltd., Publishers.

Angelidaki, I., & Ahring, B. K. (1993). Thermophilic anaerobic digestion of livestock waste: the effect of ammonia. Applied Microbiology and Biotechnology, 38(4), 560–564.

Argelier, S., Delgenes, J. P., & Moletta, R. (1998). Design of acidogenic reactors for the anaerobic treatment of the organic fraction of solid food waste. Bioprocess Engineering, 18(4), 309–315.

Arvanitoyannis, I. S., & Ladas, D. (2008). Meat waste treatment methods and potential uses. International Journal of Food Science & Technology, 43(3), 543–559.

Barber, W. P., & Stuckey, D. C. (2000). Nitrogen removal in a modified anaerobic baffled reactor (ABR): 1, denitrification. Water Research, 34(9), 2413–2422.

Biswas, J., Chowdhury, R., & Bhattacharya, P. (2007). Mathematical modeling for the prediction of biogas generation characteristics of an anaerobic digester based on food/vegetable residues. Biomass and Bioenergy, 31(1), 80–86.

Bodkhe, S. Y. (2009). A modified anaerobic baffled reactor for municipal wastewater treatment. Journal of Environmental Management, 90(8), 2488–2493.

Bres, P., Beily, M. E., Young, B. J., Gasulla, J., Butti, M., Crespo, D., & Komilis, D. (2018). Performance of semi-continuous anaerobic co-digestion of poultry manure with fruit and vegetable waste and analysis of digestate quality: A bench scale study. Waste Management, 82, 276–284.

Browne, J. D., & Murphy, J. D. (2013). Assessment of the resource associated with biomethane from food waste. Applied Energy, 104, 170–177.

Buncic, S., & Sofos, J. (2012). Interventions to control Salmonella contamination during poultry, cattle and pig slaughter. Food Research International, 45(2), 641–655.

Christofoletti, C. A., Escher, J. P., Correia, J. E., Marinho, J. F. U., & Fontanetti, C. S. (2013). Sugarcane vinasse: environmental implications of its use. Waste Management, 33(12), 2752–2761.

Dalemo, M., Sonesson, U., Björklund, A., Mingarini, K., Frostell, B., Jönsson, H., & Thyselius, L. (1997). ORWARE–A simulation model for organic waste handling systems. Part 1: Model description. Resources, Conservation and recycling, 21(1), 17–37.

Dar, R.A., Parmar, M., Dar E.A., Sani, R.K., Phutela, U.G. (2021) Biomethanation of agricultural residues: Potential, limitations and possible solutions, Renewable and sustainable energy reviews, 135, 110217

De, A.K. (2010). Environmental Chemistry, New Age International Publishers, 7th Ed.

Del Nery, V., De Nardi, I. R., Damianovic, M. H. R. Z., Pozzi, E., Amorim, A. K. B., & Zaiat, M. (2007). Long-term operating performance of a poultry slaughterhouse wastewater treatment plant. Resources, conservation and Recycling, 50(1), 102–114.

Deublein, D. and Steinhauser, A. (2008). Biogas from waste and renewable resources; An Introduction, Willey-VCH Verlag Gmb H & Co.

Dubey, R.C. (2002) A Text book of Biotechnology, S. Chand & company Ltd. New Delhi.

Fujihira, T., Seo, S., Yamaguchi, T., Hatamoto, M., & Tanikawa, D. (2018). High-rate anaerobic treatment system for solid/lipid-rich wastewater using anaerobic baffled reactor with scum recovery. Bioresource Technology, 263, 145–152.

Fulekar, M.H. (2005). Environmental Biotechnology, Oxford and IBH Publishing Co. Pvt. ltd.

Gannoun, H., Khelifi, E., Bouallagui, H., Touhami, Y., & Hamdi, M. (2008). Ecological clarification of cheese whey prior to anaerobic digestion in upflow anaerobic filter. Bioresource Technology, 99(14), 6105–6111.

Garrido, J. M., Omil, F., Arrojo, B., Mendez, R., & Lema, J. M. (2001). Carbon and nitrogen removal from a wastewater of an industrial dairy laboratory with a coupled anaerobic filter-sequencing batch reactor system. Water Science and Technology, 43(3), 249–256.

Gavala, H. N., Kopsinis, H., Skiadas, I. V., Stamatelatou, K., & Lyberatos, G. (1999). Treatment of dairy wastewater using an upflow anaerobic sludge blanket reactor. Journal of Agricultural Engineering Research, 73(1), 59–63.

Gelegenis, J., Georgakakis, D., Angelidaki, I., & Mavris, V. (2007). Optimization of biogas production by co-digesting whey with diluted poultry manure. Renewable Energy, 32(13), 2147–2160.

Girotto, F., Alibardi, L., & Cossu, R. (2015). Food waste generation and industrial uses: a review. Waste Management, 45, 32–41.

Goel, P.K. (1999) Advances in Industrial wastewater treatment, Technoscience publications, First Ed.

Harris, P. W., & McCabe, B. K. (2015). Review of pre-treatments used in anaerobic digestion and their potential application in high-fat cattle slaughterhouse wastewater. Applied Energy, 155, 560–575.

Hassan, Gul and Aadil, R. M. (2021) Cereal processing waste, an environmental impact and value addition perspectives: A comprehensive treatise, Food Chemistry, vol. 363.

Herndon, M. (2018). Food waste utilization as a viable, alternative energy generating feedstock. Journal of Clean Energy Technologies, 6(3), 197–203.

Jayathilakan, K., Sultana, K., Radhakrishna, K., & Bawa, A. S. (2012). Utilization of by-products and waste materials from meat, poultry and fish processing industries: a review. Journal of Food Science and Technology, 49(3), 278–293.

Jogdand, S.N. (2010). Environmental Biotechnology (Industrial pollution management) Himalaya Publishing House.

Joshi, V.K. and Pandey, Ashok (1999) Biotechnology: Food Fermentation (Microbiology, Biochemistry and technology) vol. II: Applied, Educational Publishers & Distributors.

Khalid, A., Arshad, M., Anjum, M., Mahmood, T., & Dawson, L. (2011). The anaerobic digestion of solid organic waste. Waste Management, 31(8), 1737–1744.

Kizhekkedath, J., Sultans, K., RadhaKrishna, K., Bawa, A.S. (2012) Utilization of by-products and waste materials from meat, poultry and fish processing industries: A review, J. of Food Science and Technology, 49(3), 78–93.

Kolev Slavov, A. (2017). General characteristics and treatment possibilities of dairy wastewater–a review. Food Technology and Biotechnology, 55(1), 14–28.

Kolev Slavov, A. (2017). General characteristics and treatment possibilities of dairy wastewater–a review. Food Technology and Biotechnology, 55(1), 14–28.

Kondusamy, D., & Kalamdhad, A. S. (2014). Pre-treatment and anaerobic digestion of food waste for high rate methane production–A review. Journal of Environmental Chemical Engineering, 2(3), 1821–1830.

Kosseva, M. R. (2009). Processing of food wastes. Advances in Food and Nutrition Research, 58, 57–136.

Laufenberg, G., Kunz, B., & Nystroem, M. (2003). Transformation of vegetable waste into value added products: (A) the upgrading concept; (B) practical implementations. Bioresource Technology, 87(2), 167–198.

Li, Y. Y., Sasaki, H., Yamashita, K., Seki, K., & Kamigochi, I. (2002). High-rate methane fermentation of lipid-rich food wastes by a high-solids co-digestion process. Water Science and Technology, 45(12), 143–150.

Melikoglu, M., Lin, C. S. K., & Webb, C. (2013). Analysing global food waste problem: pinpointing the facts and estimating the energy content. Central European Journal of Engineering, 3(2), 157–164.

Murugesan A.G. & Rajakumari, C. (2006), Environmental Science and Biotechnology, MJP Publishers.

NAS (1977) Methane generation from human, animal and agricultural waste, National Academy of Sciences, Washington DC.

Novaes, R. F. (1986). Microbiology of anaerobic digestion. Water Science and Technology, 18(12), 1–14.

Oliveira, D., Fox, P., & O'Mahony, J. A. (2019). By-products from dairy processing. by-products from agriculture and fisheries: Adding value for food, Feed, Pharma, and Fuels, 57–106. https://doi.org/10.1002/9781119383956.ch4

Omil, F., Garrido, J. M., Arrojo, B., & Méndez, R. (2003). Anaerobic filter reactor performance for the treatment of complex dairy wastewater at industrial scale. Water Research, 37(17), 4099–4108.

Peavy, H.S., Rowe, D.R., Tchobanoglous, G. (1985). Wastewater treatment and disposal, Environmental Engineering. McGraw Hill Education Pvt. Ltd.

Pramanik, S. K., Suja, F. B., Zain, S. M., & Pramanik, B. K. (2019). The anaerobic digestion process of biogas production from food waste: Prospects and constraints. Bioresource Technology Reports, 8, 100310.

Rajeshwari, K. V., Balakrishnan, M., Kansal, A., Lata, K., & Kishore, V. V. N. (2000). State-of-the-art of anaerobic digestion technology for industrial wastewater treatment. Renewable and sustainable energy, Reviews, 4(2), 135–156.

Rizvi, H., Ahmad, N., Abbas, F., Bukhari, I. H., Yasar, A., Ali, S., ... & Riaz, M. (2015). Start-up of UASB reactors treating municipal wastewater and effect of temperature/sludge age and hydraulic retention time (HRT) on its performance. Arabian Journal of Chemistry, 8(6), 780–786.

Schiopu, A. M., & Gavrilescu, M. (2010). Municipal solid waste landfilling and treatment of resulting liquid effluents. Environmental Engineering & Management Journal (EEMJ), 9(7), 993–1019.

Scragg, Alan (2005). Environmental Biotechnology, 2nd Edition, Oxford University Press.

Shahane, A.A. and Shivay, Y.S. (2016) Int. J. of Bioresource and Stress Mgt., 7(1): 162–173.

Shamurad, B., Sallis, P., Petropoulos, E., Tabraiz, S., Ospina, C., Leary, P., & Gray, N. (2020). Stable biogas production from single-stage anaerobic digestion of food waste. Applied Energy, 263, 114609.

Shete, B. S., & Shinkar, N. P. (2013). Dairy industry wastewater sources, characteristics & its effects on environment. International Journal of Current Engineering and Technology, 3(5), 1611–1615.

Singh, B.D. (2012). Biotechnology Expanding Horizons, Kalyani Publishers, Fourth Edition.

Speece, R. E. (1983). Anaerobic biotechnology for industrial wastewater treatment. Environmental Science & Technology, 17(9), 416A-427A.

Sponza, D. T., & Işik, M. (2002). Decolorization and azo dye degradation by anaerobic/aerobic sequential process. Enzyme and Microbial Technology, 31(1–2), 102–110.

Tamburini, E., Gaglio, M., Castaldelli, G., Fano, E.A. (2020) Biogas from agricultural food and agricultural waste can appreciate agro ecosystem services: The case study of Envilia Romagna Region, Sustainability, 12(20):8392

Tanner, C. C., Clayton, J. S., & Upsdell, M. P. (1995). Effect of loading rate and planting on treatment of dairy farm wastewaters in constructed wetlands—I. Removal of oxygen demand, suspended solids and faecal coliforms. Water Research, 29(1), 17–26.

Torkian, A., Eqbali, A., & Hashemian, S. J. (2003). The effect of organic loading rate on the performance of UASB reactor treating slaughterhouse effluent. Resources, Conservation and Recycling, 40(1), 1–11.

ur Rahman, U., Sahar, A., & Khan, M. A. (2014). Recovery and utilization of effluents from meat processing industries. Food Research International, 65, 322–328.

Van Der Merwe, M., & Britz, T. J. (1993). Anaerobic digestion of baker's yeast factory effluent using an anaerobic filter and a hybrid digester. Bioresource Technology, 43(2), 169–174.

Wesley, W. Elbenfields Jr. (1998) Industrial water pollution Control, Mc Graw Hill International Editions.

Wulf, Crueger and Anneliese, Crueger (2005). Textbook of Industrial Microbiology, Second Ed. Panima Publishing Corporation.

Yates, J. G., & Newton, D. (1986). Fine particle effects in a fluidized-bed reactor. Chemical Engineering Science, 41(4), 801–806.

Zeikus, J. G. (1977). The biology of methanogenic bacteria. Bacteriological reviews, 41(2), 514–541.

Zhang, L. and Jahng, D. (2012) Long term anaerobic digestion of food waste stabilized by trace elements, Waste Management, 32(8): 1509–1515.

Zhang, Y., Banks, C. J., & Heaven, S. (2012). Anaerobic digestion of two biodegradable municipal waste streams. Journal of Environmental Management, 104, 166–174.

Ziemiński, K., & Frąc, M. (2012). Methane fermentation process as anaerobic digestion of biomass: Transformations, stages and microorganisms. African Journal of Biotechnology, 11(18), 4127–4139.

Zinder, S. H. (1990). Conversion of acetic acid to methane by thermophiles. FEMS Microbiology Reviews, 6(2–3), 125–137.

17 Fishery By-product Valorization

Shreya Panwar,[1] Ritu Sindhu,[1,] and Shalini Arora[2]*

[1]Centre of Food Science and Technology, ChaudharyCharan Singh Haryana Agricultural University, Hisar, Haryana, India
[2]College of Dairy Science and Technology, Lala Lajpat Rai University of Veterinary and Animal Sciences, Hisar, Haryana, India
*Corresponding author: ritu3400@gmail.com

CONTENTS

17.1 INTRODUCTION

The fisheries industry is an avenue of economic significance as it is now being perceived as a potential for many resources instead of mere waste. The fisheries industry globally has observed a many fold rise in production from 37 to 200 million tonnes and the per capita consumption has increased from 9 to 20 kg, in the years 1961 to 2016 respectively (FAO,2018). According to the FAO, the statistics also show that there is a 14% rise in global fisheries catches from 1960 to 2018. There is a 527% rise in global aquaculture production from 1990 to 2018 and global fish consumption has increased 122% from 1990 to 2018 as illustrated in Figure 17.1 (FAO, 2020).

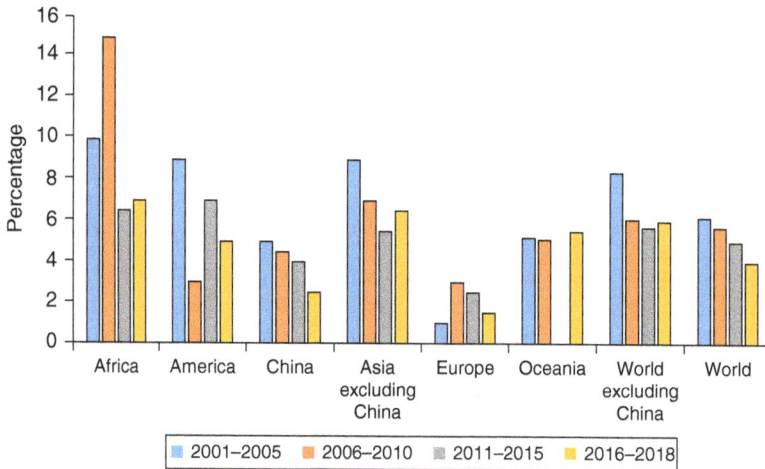

FIGURE 17.1 Annual growth rate of aquaculture fish.

Source: FAO, 2020.

TABLE 17.1
World Fisheries and Aquaculture Production

Production	1986–1995	1996–2005	2006–2015	2016	2017	2018
Capture	(million tonnes, live weight)					
Inland	6.4	8.3	10.6	11.4	11.9	12.0
Marine	80.5	83.0	79.3	78.3	81.2	84.4
Total capture	**89.9**	**91.4**	**89.8**	**89.6**	**93.1**	**96.4**
Aquaculture						
Inland	8.6	19.8	36.8	48	49.6	51.3
Marine	6.3	14.4	22.8	28.5	30.0	30.8
Total aquaculture	**14.9**	**34.2**	**59.7**	**76.5**	**79.5**	**82.1**

Source: Agriculture Organization of the United Nations, Fisheries Department (2000).

This indicates the ever-growing consumption of fish globally. As a result in 2018 the total capture fisheries production globally touched an all-time high mark of 96.4 million tonnes (FAO, 2020). The world fisheries and aquaculture production is given in Table 17.1. China was the largest producer of captures in the year 2018 (Figure 17.2). However, a large chunk of global harvest as high as 35% gets lost or wasted that points toward the significant amount of waste generated from the marine industry. Global food wastage is a serious problem. Therefore, it is crucial to utilize the by-product obtained from fish processing to not only valorize the waste but also to reduce the burden on the environment. The fish falls under the category of a very perishable commodity and thus is considered for post-harvest processing, which typically includes washing, beheading, de-gutting, de-scaling, and peeling. The processing operations are done to prevent spoilage of fish, which occurs due to invariably high percentage of PUFA and moisture (Nahid et al., 2016). The total production of fish accounts for 133,000 tons taking into consideration both the catching sector as well as fish procured by the processors and retailers (Waste and Resources Action Programme, 2012).

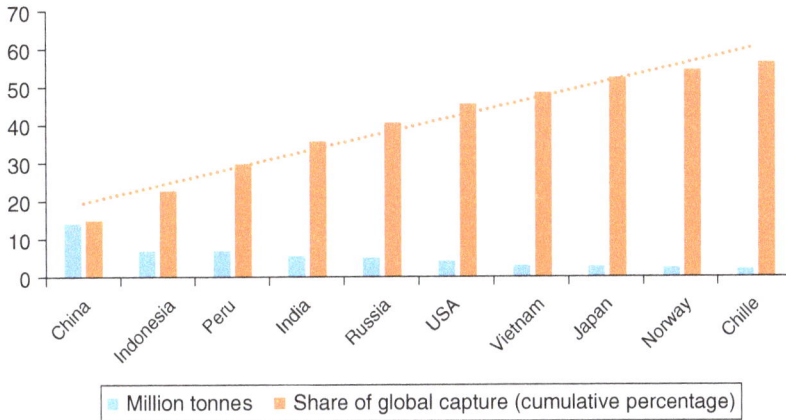

FIGURE 17.2 Top ten global capture producers in 2018.

Source: FAO, 2020.

As the fish is subjected to processing, the waste generated during these unit operations can be as high as 60% depending on the fish type and the degree of processing. This waste has the potential to be valorized for the production of different value-added products like bioactive compounds, fish silage, fish oil, fish meal, fish manure, is in glass, chitin–chitosan, gelatin, fish sauce, pigments etc.

17.1.1 CONSUMPTION PATTERN AND NUTRITION

Any aquatic organism that has been harvested either from the fish farming that is aquaculture or from the wild with an intention for the utilization for the commercial purpose falls under the umbrella term fish (Kumar et al., 2018). The class of organisms that are edible and biologically divergent comprised of finfish and shellfish constitute the group of seafood organisms (Venugopal, 2006; FAO, 2010). The fisheries industry is very dynamic and the consumption pattern varies depending on region, economy, culture etc. In 2018 the major fraction (88%) of the total catch fish was utilized for direct consumption and the rest 12 % was utilized for the production of fish meal, ornamental fish, and fishoil for non-food purposes. The report suggests that the direct usage of fishes have however seen huge growth from the last few decades (FAO, 2020). The fisheries industry is blooming due to the synergistic effect of various reasons as the nutritional significance of fish is now more in the public eye due to consumer awareness, fish is considered as a crucial source of protein as it suffices almost 20% of the meat protein requirement of 3.3 billion people globally, landed marine fishes account for 14–16% of the animal proteins consumed by the humans. Apart from proteins, they are also a good source of essential fatty acids, iodine, essential oil etc.

17.2 NEED FOR BY-PRODUCT UTILIZATION

Fish discarding is a frequent and fatal phenomenon where the unwanted catches are returned to the sea either in the dead or in the alive form. This in turn affects not only the fish as a species but also the entire marine environment. As per the stats by the Food and Agricultural Organization (FAO), it is estimated that almost 50% of the more than 90 million tons of fish and shellfish caught across the globe are discarded each year (Roda, 2019). In seafood processing, the edible portion recovery is around 20–50% and the remaining 50% is rendered as inedible (Guerard, 2007). Technological development and progress have led to massive production and at the same time massive wastage. Around 20–80% of landed weight often contributes to the by-product are illustrated in Figure 17.3.

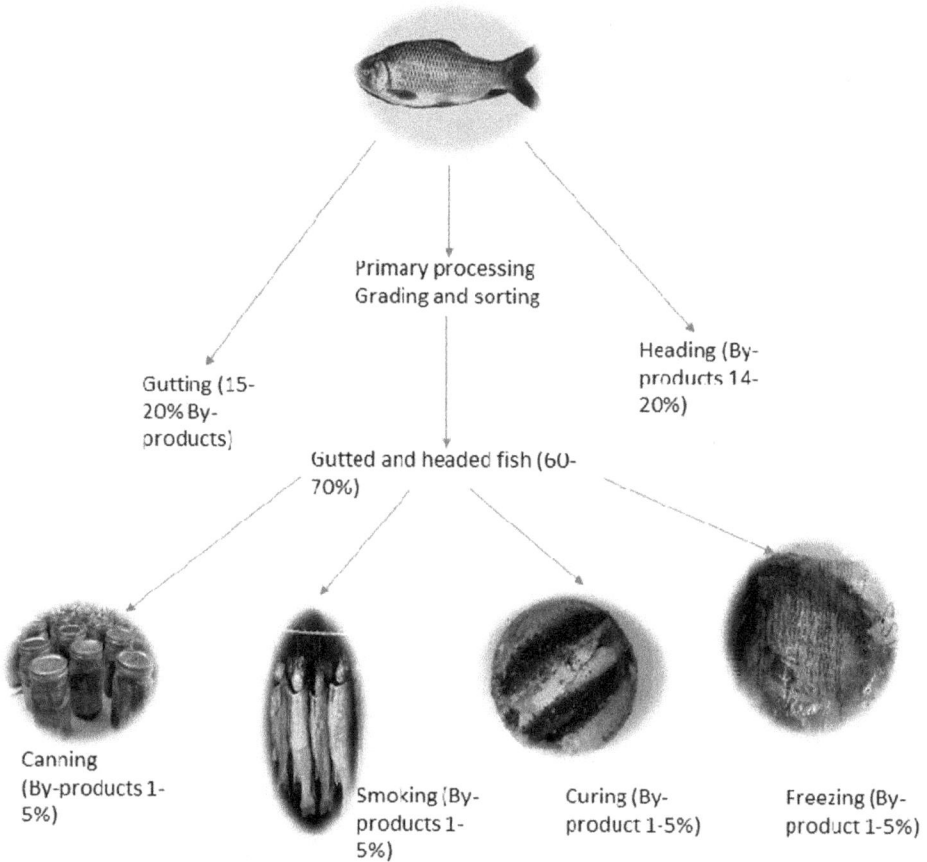

FIGURE 17.3 Finfish processing and possible by-products.

By-product is the terminology used for the commodity that is not fit for direct sale, however, can be used after subjecting it to the appropriate treatments. Waste is a commodity that can neither be used for food purposes nor for feed purposes and is usually discarded, decomposed, or destroyed (Suresh & Prabhu, 2012). By-product comprises all the products that are either inedible or waste left behind during the processing of the primary product (Gildberg, 2002). The by-products of the fisheries industry can be treated as raw material for further processing. These inedible or waste fractions comprise nutritionally significant portions of proteins, fats, vitamins etc. However, these fractions vary depending on the type of processing and portion of the fish. Aquatic by-products being rich in nutritional value and bioactive compounds are rightfully considered as the potential source for the production of valuable food ingredients (Nawaz et al., 2020). Finfish processing by-products are given in Table 17.2.

17.3 CLASSIFICATION OF WASTE IN THE FISH INDUSTRY

Fish is a very perishable commodity and requires care during harvest (catch) and therefore it is challenging to valorize the by-products of such a perishable commodity in a very short period. The waste or the leftover fraction during processing to produce a primary product in the fishery industry can be classified based on its organic or inorganic nature. The organic fraction is the most abundant out of the two fractions. The organic waste comprises the viscera, fish muscle, scales

TABLE 17.2
Finfish Processing By-products

By-products	% of by-products	Valuable bioactive components
Head	14-20	Collagen, Gelatin, oil, lipid, protein hydrolysates, bioactive, Peptides, flavour
Gut	15-20	Oil, lipid, Protein hydrolysates, bioactive peptides, Squalenes, enzyme, flavour
Skin	1-3	Collagen, Gelatin, elastin, cartilage
Bones	10-16	Collagen, Gelatin, elastin, cartilage
Trimmings	1-5	Protein hydrolysates, Peptides

Source: Suresh& Prabhu (2012).

and fins that fall under the category of organic waste. They are potential sources of various functional ingredients like collagen, gelatin, bioactive peptides, chitin, fish protein isolate, pigments, vitamins, and minerals. The fish bones fall under the category of inorganic fraction, and are a fairly rich source of calcium and phosphorus like minerals and are invariably used to prepare fish bone powder and calcium supplements (Nawaz et al., 2020). Another way of classifying the fishery waste or leftover is by dividing it into two categories based on the purpose of usage, which is for food and non-food use. The food-use includes protein, protein hydrolysate, fish oil, collagen, gelatin, flavor, calcium, pigments, enzymes etc. and the non-food use includes by-product utilization as animal feed, as an ingredient in the pharmaceutical and cosmetic industry, as biogas and biodiesel, and as compost etc.

17.4   BY-PRODUCT UTILIZATION

17.4.1   PROTEIN

The protein requirement across the globe is increasing owing to an increase in population and consumer awareness along with various other factors and on the other hand, the conventional sources to cater to the protein requirements are falling short. The seafood processing by-products globally have been estimated to be around 32 million tonnes (Nguyen et al., 2020). Thus, marine by-products have the potential to be explored as reliable sources of dietary proteins as they are inexpensive and of excellent quality. Additionally, these proteins have great functionality as well as bioactive properties like antihypertensive, antioxidant, antidiabetic, anticancer, and antimicrobial properties. The beneficial bioactive peptides get activated upon hydrolysis in the gastro-intestinal tract after consumption or they can be prepared by simulating conditions of hydrolysis using heat or chemicals. It provides an opportunity to contribute to the concept of the circular economy of the UN where not only the bio-resources are being saved from excess exploitation at the same time the bio-waste burden of the environment is also being taken care of. Owing to the unique quality of the marine proteins the popular products that are manufactured using these proteins are protein concentrate and protein hydrolysates (Chen et al., 2020). These forms of proteins are extracted employing chemical treatments or enzymatic processes (Kristinsson & Rasco, 2000). Various derived proteins from finfish by-products are mentioned in Table 17.3. The functional properties of the fish proteins are directly influenced by the size, shape, and charge distribution on the structure of proteins. It has been reported that protein in the skin is more than protein in the bone, e.g. gilthead seabream, which has approximately 24.8% protein in the skin but its bone accounts for only 16.4% protein (Pateiro et al., 2020).

TABLE 17.3
Derived Proteins from Fish By-products

By-product type	% By-product of total fish volume	Derived protein	Reference
Skin, scale, Bones	30%	Collagen, Gelatin	Pateiro et al. (2020)
Viscera	12-18%	Intestinal enzymes and biopeptides	Villamil et al. (2017)
Fish blood	2-7%	Protein Hydrolysates, Biopeptides, Bioactive Amino Acids	Hayes & Gallagher (2019)

TABLE 17.4
Method of Extraction of Collagen from Fish By-products

Pre-treatment	Method	Purpose	Reference
Alkaline pre-treatment	Dilute NaOH is used	To remove contaminants	Nagai et al. (2002)
Alcohol pre-treatment	10 % Butyl alcohol is used	To remove lipids	Nagai & Suzuki (2000)
De-mineralisation treatment	0.05M EDTA in neutral buffer or HCl solution	To remove the calcium content	Nagai & Suzuki (2000); Nomura et al. (1996)

17.4.1.1 Collagen and Gelatin

Marine animals are considered as treasures of collagen, which is extracted from both the under-utilized as well as discarded fish waste. The aquatic animals are a unique source of collagen as the marine animals unlike land animals (bovines) do not harbor transmissible diseases, which is one of the major concerns behind the usage of land animals apart from some religious restrictions in the usage of the bovine animals (Coppola et al., 2020). Various methods of extraction of collagen from fish by-products are given in Table 17.4. Collagen is associated with structural function in aquatic animals (Regenstein & Zhou, 2007). Collagen finds its use owing to its excellent water adsorption capacity, gelling capacity, and texture modifying capacity in the food processing sector, cosmetics sector, and pharmaceutical sector. Collagen is one of the most abundant proteins and accounts for almost 20 to 30 % of total proteins in the living organism (Muller, 2003); although gelatin is not found naturally but is derived from the parent component collagen using denaturation. This procedure encompasses the breakage of cross-linking present between polypeptide chains. Out of these two proteins, usage of gelatin is subsequently higher in the food industry and most of the collagen is therefore converted into gelatin. The conversion of collagen into gelatin is done using acid or alkali and the product formed by acid treatment is called type 'A' gelatin and alkali-treated gelatin is called type 'B' gelatin (Regenstein & Zhou, 2007).

Collagen is prepared by using the waste or under-utilized components from the marine industry, by subjecting the raw material to pre-treatments first to yield the pure final product as non-collagenous compounds are usually considered as contaminants during collagen extraction. Different pre-treatments that are given to the raw material have been summarized in the table below. After subjecting it to the different pre-treatments the raw material is processed by different methods like acidic hydrolysis or enzymatic hydrolysis to facilitate soluble collagen production. Depending on the solubility of collagen fraction the collagen can be pepsin soluble, salt soluble, or acid-soluble collagen (Shahidi, 2006). Collagen produced in turn acts as the raw material for the production of

TABLE 17.5
Applications of Marine Collagen, Gelatin, and Collagen Peptides

Functional ingredient	By-product	Applications
Omega -3	Salmon and cod liver oil	Health benefits
Carotenoid	Astaxanthin, fucoxanthin	Dietary supplements, Food colour, Cosmetics
Vitamin and mineral	Marine fish	Health benefits
Collagen	Fishbone, scale, fin	Healthy joints and skin elasticity

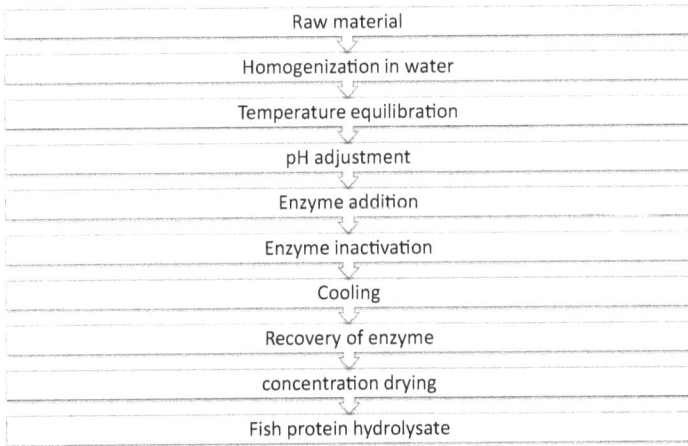

```
Raw material
    ↓
Homogenization in water
    ↓
Temperature equilibration
    ↓
pH adjustment
    ↓
Enzyme addition
    ↓
Enzyme inactivation
    ↓
Cooling
    ↓
Recovery of enzyme
    ↓
concentration drying
    ↓
Fish protein hydrolysate
```

FIGURE 17.4 Fish protein hydrolysate preparation.

the gelatin. The quantity and quality of the product are dependent on the species used for its extraction and the process used for its manufacturing. Applications of various functional ingredients from marine By-products are mentioned in Table 17.5.

17.4.1.2 Protein Hydrolysates and Peptides

The marine waste can also be utilized by processing it after using protease enzyme, which can hydrolyze the fish protein into protein hydrolysates and peptides (Kristinsson, 2007). Fish protein hydrolysate preparation is shown in Figure 17.4.

These fish protein hydrolysates (FPH) have nutritional as well as techno-functional properties. The fish protein hydrolysates and peptides can be predominantly produced by three methods – fermentation, chemical synthesis, and enzymatic hydrolysis. Fermentation is the process that involves incubating substrate with the desirable specific types of microorganisms while chemical synthesis is used to extract bioactive peptides with predetermined amino acid sequence and enzymatic hydrolysis is the usage of the protein hydrolyzing enzymes to produce desirable peptides. Enzymatic hydrolysis is the safest, cheapest, most controllable method without any drawbacks to produce the hydrolysates and peptides. The standard process to produce fish protein hydrolysates (FPH) using enzymatic hydrolysis has been depicted. The FPH however has gained a lot of popularity in recent times due to its supporting activity in digestion, absorption, and utilization of nutrients in animals when added as an additive in fish meal. It also acts as a nutritional supplement due to the dynamic amino acid profile. Moreover, it also increases techno-functional properties in food applications like food emulsification, foaming, or dispersion activities (Gao et al., 2021). Various properties of fish protein hydrolysates and their applications are given in Table 17.6.

TABLE 17.6
Properties of Fish Protein Hydrolysates and Their Applications

FPH components/properties	Application
Essential amino acid, small peptides	Animal feed
Emulsifying properties	Food ingredients and additives
Antioxidants, anti-inflammatory, anticancer, antihypertensive activity	Functional food

Source: Gaoet al. (2021).

17.4.2 Lipids

Marine oil is derived from not only fatty fishes like salmon, tuna, sardines, but also from the offal of different kinds of fishes as well as the processing operations like gutting, filleting provides the good raw material for extraction of marine lipid-derived products. The important lipid-based products that can be extracted from marine lipids are fish oils, omega-3 fatty acids, phospholipids, squalene, vitamins, and cholesterol (Amit et al., 2010). The fat-rich flesh is primarily associated with the extraction of fat and fish meal. The essential fatty acids, which are required to be consumed through diet like alpha linoleic acid (ALA), docosapentaenoic acid (DPA), eicosapentaenoic acid (EPA), and docosahexaenoic acid (DHA) are a very crucial part of the human diet and are present in fishes in abundance. Fish meal is a very popular fish-based product and the by-product of fish meal is the fish oil and is therefore produced from both the fishes and the by-catch fishes in considerable amounts (Shahidi, 2007). The deviation in the fat obtained from marine fishes varies invariably with the composition of the part it is derived from (Aidos et al., 2002). The fish oil is extracted by the protease catalyzed hydrolysis process or through the lipase catalyzed hydrolysis process (Wanasundara, 2011). The fishes are classified as finfish and shellfish, under shellfish, fishes are further categorized as molluscs and crustaceans. Crustacean farming across the globe alone accounts for 23% of the total global aquaculture market with an estimated value of over 57 billion dollars (UN FAO Fisheries and Aquaculture Department, 2018). The lipids associated with the crustaceans for example in shrimp are rich in omega-3 fatty acids and carotene as well (Amiguet et al., 2012). Fat liquor production from fish waste is a recent development in fish waste valorization (Saranya, 2020).

17.4.2.1 Fish Oil

The oil content varies in the fish by-products from 1.4% to 40.10% depending on the species and the tissue being used for oil extraction. Fish oil is considered a brilliant source of omega-3 fatty acid, which is crucial as a source for EPA and DHA that play a very important role in human health. The advantages of omega-3 fatty acids in human health avenue have been one of the most important developments in nutritional sciences as n-3 fatty acid is effective against coronary heart disease, cancer, hypertension, and many other non-communicable diseases. The key elements in good quality fish oil production are the source being used for oil production and the process being followed. Recently it has been seen that the most efficient method of fish oil extraction is enzymatic hydrolysis in combination with high-pressure processing and ultra-high pressure processing (Zhang et al., 2021).

17.4.3 Flavor

Flavor is considered a premium entity and seafood flavors are very popular across the globe. It finds its application in the food sector as it is incorporated in different seafood like sauces, powders, gravies, soups, instant noodles, snacks and surimi. Flavors can be produced from the marine industry by aqueous extraction of flavor from by-products, by enzymatic hydrolysis, and through fermentation (Lee, 2007). Seafood flavor comprises both the volatile components responsible for the aroma and the non-volatile components responsible for the taste (Suresh and Prabhu, 2012). Due to the proximity between the health and the diet of the aware consumers, the clean label movement has picked up and therefore it becomes crucial for the industries to choose bio-ingredients from natural origin to meet the consumer demand and consumer satisfaction (Siewe et al., 2021).

17.4.4 CALCIUM

Calcium is essential for many physiological functions in the human body; it is crucial for the structure of the human body and plays an important role in the formation of the teeth and bones. Calcium is utilized in various industries like electronics, food, leather etc. It has nutritional significance as well as being required for the many enzyme-based reactions that require calcium as a cofactor to occur (Suresh and Prabhu, 2012). The inorganic kind of waste in the fish industry is used for the preparation of bone powder, which is fairly rich in calcium, the bone powder products are found prevalently in the form of hydroxyapatite and calcium carbonate (Narayandas et al., 2021).

17.4.5 PIGMENTS

Pigments are an integral part of human life and are vividly used in many avenues like cosmetics, foods, beverages, textiles etc. Marine pigments comprise different molecules that contain chromophore, which is responsible for imparting the color (Pereira et al., 2014). The aquatic animals are brilliantly colored and therefore they can be used as a source of pigments. The aquatic species usually encompasses carotenoids, carotenoproteins, tetrapyrroles, quinones, indigoids, azulenes, and melanins (Ye et al., 2019). These pigments are the result of several processes and serve different purposes (Bandaranayake, 2006). These marine pigments vary with the depth and geographical position of the associated marine animals. These pigments are not only crucial because of their economic importance, but also play a significant role due to their bioactive properties and therefore applications. Few bioactive pigments have been known to show antimicrobial, anticancer, antioxidant, and bioluminance like properties.

17.4.5.1 Carotenoid

Carotenoids encompass phytochemicals responsible for the bright red, orange, and yellow coloration of the flesh and skins of fish. The common examples of fishes that contain carotenoids are salmonids and the crustacean exoskeleton (Suresh & Prabhu, 2012). As per the reports cited by Sachindra et al. (2007) total carotenoid content (μg/g) in marine shrimps from shallow Indian waters ranged from 35.8 to 185.3 in head and from 59.8 to 117.4 in the carapace. Carotenoids are crucial as apart from imparting color it also participates in the process of photosynthesis for energy transfer reactions (Rao & Rao, 2007) reported that carotenoids are a dietary source of vitamin A and potent antioxidant as well. The extraction of carotene from the marine by-products includes the fermentation or enzymatic approach to obtain good quality and a good quantity of the carotene pigment.

17.4.5.2 Carotenoproteins

Carotenoproteins are molecules that are formed by the association of carotenoids and proteins. There is a mutual advantage due to this association as it can assist in preventing the photo-oxidation of the carotenoids and also molecular changes that occur in the tertiary structure of the proteins that may induce denaturation is also checked (Pereira et al., 2014). Carotenoproteins perform protective coloration; additionally, they help in a variety of activities like photosensitivity, electron transport and enzymatic activity (Heras et al., 2007). α-Crustacyanin is a carotenoprotein that is present in lobsters

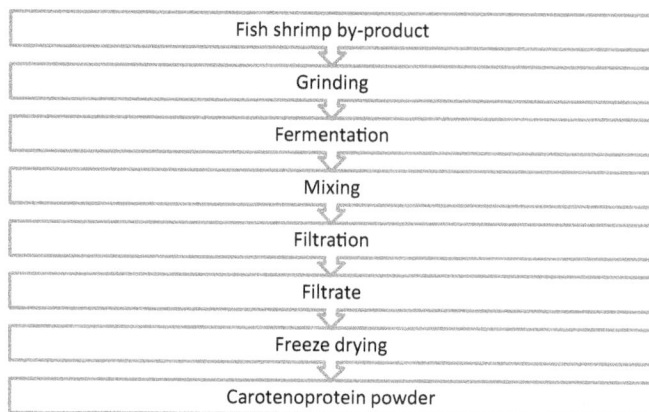

FIGURE 17.5 General fermentation isolation method for carotenoid as carotenoprotein from shrimp processing.

while blue carotenoprotein linkiacyanin, is present in the starfish Linckia laevigate L. (Zagalsky et al., 1989).

17.4.5.3 Tetrapyrroles

These are a group of chemical compounds having four pyrrole rings whose derivatives are linked by (=(CH)- or -CH. - units), the linkage can exist both linearly and cyclically. Tetrapyrroles are present in plentiful amounts after carotenoids when natural pigments are considered (Bandaranayake, 2006). Purpleink produced by sea hares also contains these pigments. The ink released by these organisms varies across different species and is predominantly constituted of the red, purple, and blue molecules (MacColl, 1990). However, color impartation is the secondary function of the tetrapyrrole molecule and the primary function is associated with the process of biological oxidation in these organisms.

17.4.5.4 Quinones

The quinones are the category of organic molecules that are produced from aromatic compounds by conversion of an even number of –CH= groups into –C– groups with any necessary reshuffling of double bonds, producing a cyclic dione structure. The main natural pigments that fall under the quinones are echinochromes and spinochromes from marine sources. Two spinochromes, echinamines A and echinamines B, were reportedly extracted from the sea urchin Scaphechinus mirabilis. The quinones however have several protective and biological functions. It includes activities like antioxidant and scavenging reactive oxygen species (ROS).

17.4.5.5 Indigoids

Indigoids are a well-known class of natural pigments that are not only found in several plants but also in marine organisms. There are around 13 compounds that fall under this class, with indigo being the main compound of Indigoids dyes. The colors imparted by indigoids are, green, blue, or purple, generally found in several species of indigo hamlet fishes from the family of ray finfishes.

17.4.5.6 Azulenes

Azulene is an organic compound and an isomer of naphthalene. Naphthalenes are colorless but in contrast, azulene imparts dark blue color. The terpenoids vetivazulene and guaiazulene are responsible to constitute the azulene structure. These pigments are found in association with the natural pigments of different plants and invertebrates like in mushrooms, guaiac wood oil, and some marine invertebrates. Various verities of Gaurami fishes have azulenes pigment in them.

17.4.5.7 Melanins

Melanins are the category of brown pigments, which are formed as the result of oxidation and polymerization of the phenolic compounds. However, its poor solubility in particular solvents and its complex polymeric nature constrain its study and usages (Glagoleva, 2020). It has been reported that melanin pigments are invariably present in the peritoneal lining as well as in different organs of certain species of fish (Simpson, 2007). Melanin is found in mammals including humans, especially in the hair and the eyes. In the case of birds, it is found in the feathers and the reptiles; they are vividly present in skin or scales. They are often found in the ink of cephalopods like octopuses and squids.

17.5 ENZYMES

Enzymes, because of their excellent catalytic nature, are very crucial in biological sciences. The global enzymes' market is expected to rise at a compound annual growth rate of 7.1% from 2020 to 2027 to reach USD 14.9 billion by 2027, as predicted in the Enzymes Market Size and Share Industry Report (2020–2027). In this regard, the under-utilized or surplus catch of the marine industry can be an excellent source of enzymes. These have several applications in the food as well as in the non-food industry (Likhar & Chudasama, 2021). Different kinds of protease enzymes are given in Table 17.7. These enzymes based on their functions can be classified into four groups including enzymes responsible for the degradation of carbohydrates, lipids and proteins and other miscellaneous enzymes (Hathwar et al., 2010). However, the fifth group of enzymes nucleotide-degrading enzymes have been recently cited by Likhar & Chudasama (2021). Protein-degrading enzymes, lipid-degrading enzymes (lipase, phospholipase), carbohydrate-degrading enzymes (alginate lyase, chitinase), and nucleotide-degrading enzymes (ATPase, AMP deaminase, Nucleotidase, Xanthine oxidase). These enzymes have the potential to be used owing to the characteristic properties like low-temperature reactivity, stability in diverse applications such as food processing, biotechnology, clinical diagnosis, detergents, leather and fabric upgrading, organic synthesis, therapeutics, biosensors, among others (Likhar & Chudasama, 2021).

17.6 CONCLUSION

The amount of waste generated due to fish by-products is huge. Thus, there is immense scope in utilizing this waste into several value-added products at the same time it is required to understand that fish and other marine products are finite and thus must be taken utmost care of. Fish by-products are neither produced nor utilized for their full potential. There is still large potential in

TABLE 17.7
Types of Protease Enzymes

Proteases	Body parts	Molecular weight (kDa)	Optimum activity
Trypsin	Pancreatic tissue	22.5-24	pH 7.5-10
Chymotrypsin	Pyloric caeca	25-28	pH 9
Pepsin	Digestive glands, stomach tissue	27-42	pH 2-4
Chymosin	Gastric mucosa	33.8	pH 2.2-3.5
Gastric	Gastric juices	32.3-33.9	PH 3

Source: Hathwar et al. (2010).

the by-product valorization of marine products. To transform these goals into reality, it is important to innovate and develop a technological intervention for cautious utilization of the waste with the help of by-products' valorization and value addition. It will not only enhance the circular economy but also will release the bio-burden due to waste generated in the environment. This can open new avenues in novel food ingredient development from blue industry organizations along with efficient waste management. In contrast to several other meat products, fish is acceptable regardless of any religion and irrespective of its handling and processing type. The only intervention that must be done is the optimum regulation as this by-product valorization also comprises various food safety-associated risks.

REFERENCES

Aidos, I., Masbernat-Martinez, S., Luten, J. B., Boom, R. M., and Van der, P. A. (2002). Composition and stability of herring oil recovered from sorted by-products as compared to oil from mixed by-products. Journal of Agriculture Food Chemistry, 50: 2818–2824

Agriculture Organization of the United Nations. Fisheries Department. (2000). The State of World Fisheries and Aquaculture, 2000 (Vol. 3). Rome: Food & Agriculture Org.

Amiguet, T. V., Kramp, K. L., Mao, J., McRae, C., Goulah, A., Kimpe, L. E., and Arnason, J. T. (2012). Supercritical carbon dioxide extraction of polyunsaturated fatty acids from Northern shrimp (Pandalus borealis Kreyer) processing by-products. Food Chemistry, 130(4), 853–858

Amit, K. R., Swapna, H. C., Bhaskar, N., Halami, P. M., and Sachindra, N. M. (2010). Effect of fermentation ensilaging on recovery of oil from fresh water fish viscera. Enzyme and Microbial Technology, 46, 9–13

Bandaranayake, W. M. (2006). The nature and role of pigments of marine invertebrates. Natural Product Reports, 23(2), 223–255

Bandaranayake, W. M. (2006). The nature and role of pigments of marine invertebrates. Natural Product Reports, 23(2), 223–255

Chen, Y., Wang, C., and Xu, C. (2020). Nutritional evaluation of two marine microalgae as feedstock for aquafeed. Aquac. Res., 51, 946–956

Coppola, D., Oliviero, M., Vitale, G. A., Lauritano, C., D'Ambra, I., Iannace, S., and de Pascale, D. (2020). Marine collagen from alternative and sustainable sources: Extraction, processing and applications. Marine Drugs, 18(4), 214

FAO. (2010). Yearbook of Fishery Statistics. Statistics and Information Service, FAO Fisheries and Aquaculture Department. Viale delle Terme di Caracalla 00153, Rome: Food and Agriculture Organization of the United Nations. www.fao.org/fishery/statistics/en

FAO. (2020). The State of World Fisheries and Aquaculture 2020. Sustainability in action. Rome. https://doi.org/10.4060/ca9229en

Gao, R., Yu, Q., Shen, Y., Chu, Q., Ge, C., Fen, S., and Sun, Q. (2021). Production, bioactive properties, and potential applications of fish protein hydrolysates: Developments and challenges. Trends in Food Science & Technology, 110, 687–699

Gildberg, A. (2002). Enhancing returns from greater utilization. In: H. A. Bremner (Ed.) Safety and quality issues in fish processing. Woodhead Publishing Limited and CRC Press LLC, Cambridge, pp. 425–449

Glagoleva, A. Y., Shoeva, O. Y., and Khlestkina, E. K. (2020). Melanin pigment in plants: Current knowledge and future perspectives. Frontiers in Plant Science, 11, 770

Guerard, F. (2007). Enzymatic methods for marine by-products recovery. In Maximizing the Value of Marine By-Products, ed. F. Shahidi, pp. 107–136. Cambridge: Woodhead Publishing Ltd., CRC Press

Hathwar, S. C., Rai, A. K., Nakkarike, S. M., and Bhaskar, N. (2010). Seafood enzymes and their potential industrial application. Handbook of seafood quality, safety and health applications, ed. Cesarettin Alasalvar, Fereidoon Shahidi, Kazuo Miyashita and Udaya Wanasundara, pp. 522–535. Blackwell Publishing Ltd.

Hayes, M., and Gallagher, M. Processing and recovery of valuable components from pelagic blood-water waste streams: A review and recommendations. J. Clean. Prod., 215, 410–422

Heras, H., Dreon, M. S., Ituarte, S., and Pollero, R. J. (2007). Egg carotenoproteins in neotropical Ampullariidae (Gastropoda: arquitaenioglossa). Comparative Biochemistry and Physiology Part C: Toxicology & Pharmacology, 146(1–2), 158–167

Kristinsson, H. G. (2007). Aquatic food protein hydrolysates. In *Maximising the value of marine by-products* (pp. 229–248). Cambridge: Woodhead Publishing

Kristinsson, H. G. and Rasco, B. A. (2000). Fish protein hydrolysates: Production, biochemical and functional properties. CRC Critical Reviews in Food Science & Nutrition, 32, 1–39

Kumar, V., Muzaddadi, A. U., Mann, S., Balakrishnan, R., Bembem, K., and Kalnar, Y. (2018). Utilization of Fish Processing Waste: A Waste to Wealth Approach. Ludhiana: ICAR-CIPHET.

Lee, C. M. (2007). Seafood flavor from processing by-products. In *Maximising the value of marine by-products* (pp. 304–327). Cambridge: Woodhead Publishing

Likhar, V., and Chudasama, B. J. (2021). Seafood enzymes and their potential industrial applications. J Entomol Zool Stud, 9(1), 1410–1417

MacColl, R., Galivan, J., Berns, D. S., Nimec, Z., Guard-Friar, D., and Wagoner, D. (1990). The chromophore and polypeptide composition of Aplysia ink. The Biological Bulletin, 179(3), 326–331

Muller, W.E. The origin of metazoan complexity: Porifera as integrated animals. Integr. Comp. Biol. 2003, 43, 3–10

NAGAI T and SUZUKI N (2000). `Isolation of collagen from fish waste material -skin, bone and fins', Food Chem, 68, 277–281

Nagai, T., Araki, Y. and Suzuki, N. (2002). Collagen of the skin of ocellate puffer fish (Takifugu rubripes), Food Chem, 78, 173–177

Nahid, M.N., Latifa, G.A., Chandra, S., Farid, F.B., and Begum, M. (2016). Shelf-life quality of smoke-dried freshwater SIS fish; chapila (Gudusia chapra, Hamilton-Buchanan) Kaika (Xenentodon cancila, Hamilton-Buchanan;) and Baim (Mastacembelus pancalus, Hamilton-Buchanan;) stored at laboratory. J. Agric. Vet. Sci., 9, 23–32

Narayandas, A., Baig, M. S., Gayatri, D., and Penchalaraju, M. (2021). Valorisation of organic and inorganic waste and by-products from the agricultural food processing industry. The Pharma Innovation Journal, 10(3), 893-895

Nawaz, A., Li, E., Irshad, S., Xiong, Z., Xiong, H., Shahbaz, H. M., and Siddique, F. (2020). Valorization of fisheries by-products: Challenges and technical concerns to food industry. Trends in Food Science & Technology, 99, 34–43

Nguyen, T. T., Heimann, K., and Zhang, W. (2020). Protein recovery from underutilised marine bioresources for product development with nutraceutical and pharmaceutical bioactivities. Marine Drugs, 18(8), 391.

Nomura, Y., Sakai, H., Ishii, Y. and Shirai, K. (1996). Preparation and some properties of type I collagen from fish scales. Biosci Biotech and Biochem, 60, 2092–2094

Pateiro, M., Munekata, P.E., Domínguez, R., Wang, M., Barba, F.J., Bermúdez, R., and Lorenzo, J.M. (2020) Nutritional profiling and the value of processing by-products from gilthead sea bream (Sparus aurata). Mar. Drugs

Pereira, D. M., Valentão, P., and Andrade, P. B. (2014). Marine natural pigments: Chemistry, distribution and analysis. Dyes and Pigments, 111, 124–134

Rao, A. V., & Rao, L. G. (2007). Carotenoids and human health. Pharmacological Research, 55(3), 207–216

Regenstein, J. M. and Zhou, P. (2007). Collagen and gelatin from marine by-products. In *Maximizing the Value of Marine By-Products*, ed. F. Shahidi, pp. 279–299. Cambridge: Woodhead Publishing Ltd., CRC Press

Roda, M.A.P., Gilman, E., Huntington, T., Kennelly, S.J., Suuronen, P., Chaloupka, M. and Medley, P.A. (2019). A Third Assessment of Global Marine Fisheries Discards; FAO Fisheries and Aquaculture Technical Paper No. 633; FAO: Rome, Italy, p. 78

Sachindra, N. M., Bhaskar, N., Siddegowda, G. S., Sathisha, A. D., and Suresh, P. V. (2007). Recovery of carotenoids from ensilaged shrimp waste. Bioresource Technology 98: 1642–1646

Saranya, R., Selvi, A. T., Jayapriya, J., and Aravindhan, R. (2020). Synthesis of fat liquor through fish waste valorization, characterization and applications in tannery industry. Waste and Biomass Valorization, 11(12), 6637–6647.

Shahidi, F. (2007). Marine oils from seafood waste. In *Maximising the value of marine by-products* (pp. 258–278). Cambridge: Woodhead Publishing

Shahidi, F. (Ed.). (2006). *Maximising the value of marine by-products*. Cambridge: Woodhead Publishing

Siewe, F. B., Kudre, T. G., and Narayan, B. (2021). Optimisation of ultrasound-assisted enzymatic extraction conditions of umami compounds from fish by-products using the combination of fractional factorial design and central composite design. Food Chemistry, 334, 127498

Simpson, B. K. (2007). Pigments from by-products of seafood processing. In Maximizing the Value of Marine By-Products, ed. F. Shahidi, pp. 413–432. Cambridge: Woodhead Publishing Ltd., CRC Press

Suresh, P.V. and Prabhu, G.N. (2012). Seafood. In: Valorization of food processing by-products (Chandrasekaran, M., Ed.). CRC Press. Chapter-23, pp. 685–736

UN FAO Fisheries and Aquaculture Department. (2018). Meeting the sustainable development goals. The state of world fisheries and aquaculture. https://doi.org/10.1093/japr/3.1.101

Venugopal, V. (2006). Seafood Processing: Adding Value through Quick Freezing Retortable Packing and Cook-Chilling. Boca Raton, FL: Taylor & Francis, CRC Press

Villamil, O., Váquiro, H., and Solanilla, J.F. (2017). Fish viscera protein hydrolysates: Production, potential applications and functional and bioactive properties. Food Chem., 224, 160–171

Wanasundara, U. (2011). Preparative and industrial-scale isolation and purification of omega-3 polyunsaturated fatty acids from marine sources. In Handbook of Seafood Quality, Safety and Health Application, eds. C. Alasalvar, F. Shahidi, K. Miyashita, and U. Wanasundara, pp. 464–475. UK: Wiley-Blackwell

Waste and Resources Action Programme (2012). Sector Guidance Note: Preventing Waste in the Fish Processing Supply Chain Sector Guidance Note www.greenindustryplatform.org

Ye, K. X., Fan, T. T., Keen, L. J., and Han, B. N. (2019). A review of pigments derived from marine natural products. Israel Journal of Chemistry, 59(5), 327–33

Zagalsky, P. F., Haxo, F., Hertzberg, S., and Liaaen-Jensen, S. (1989). Studies on a blue carotenoprotein, linckiacyanin, isolated from the starfish Linckia laevigata (Echinodermata: Asteroidea). Comparative Biochemistry and Physiology Part B: Comparative Biochemistry, 93(2), 339–353

Zhang, Y., Sun, Q., Liu, S., Wei, S., Xia, Q., Ji, H., & Hao, J. (2021). Extraction of fish oil from fish heads using ultra-high pressure pre-treatment prior to enzymatic hydrolysis. Innovative Food Science & Emerging Technologies, 70, 102670

18 Food Industry Waste
Potential Pollutants and Their Bioremediation Strategies

Pawan Kumar Rose,[1] Mohd. Kashif Kidwai,[1] and Sanju Bala Dhull[2]*

[1]Department of Energy and Environmental Sciences, Chaudhary Devi Lal University, Sirsa, Haryana, India
[2]Department of Food Science and Technology, Chaudhary Devi Lal University, Sirsa, Haryana, India
*Corresponding author: kashif357313@yahoo.co.in

CONTENTS

18.1 INTRODUCTION

Sustainable production of different commodities is attaining momentum as various environmental issues are associated with different industrial processes all over the globe. Both food security and environmental security are considered big challenges for sustainable development recognized at the global level. Since the global population is gradually increasing with time and to overcome hunger from a food security point of view is a mammoth challenge, especially for developing and poor nations. According to Osorio et al. (2021), about 250 million people experience acute hunger globally, which makes it a difficult task for achievement of zero hunger by 2030. The issue of the food crisis is massive as in the United States, over 50 million people including more than 15 million children suffer from insufficiency of food (Truong et al. 2019). As a strategy to combat food and nutritional security issues various types of food processing industries are established, which produce a large volume of different types of waste, particularly solid and liquid waste, which negatively influences the biotic and abiotic components of different ecosystems. The agri-food industrial sector contributes to the production of different greenhouse gases in different production and processing stages with an annual contribution of 19–29 percent of the total emitted greenhouse gases, which are one of the prime responsible factors for global climate change (Paini et al. 2021). The agro-food industrial sector also contributes to the depletion of various natural and anthropogenic resources, such as the use of fresh water, fossil fuels, xenobiotic chemicals, etc., involved in the production of various value-added food products. Industrial food processing consumes a large amount of energy in comparison to the cultivation of raw materials. Milk products are a very good example as milk consumes less energy, but cheese production consumes five- to seven-fold energy resources (Paini et al. 2021). Moreover, diverse food industry waste contains heavy metals, which are reported to adversely affect the overall biodiversity, thereby disturbing the ecosystem through the process of biomagnification, bioaccumulation, etc. The disposal and management of waste generated in the food processing industry create issues in environmental quality and sustainability (Jayathilakan et al. 2012). Improper disposal methods of industrial waste generated from different industries, including the food industry is an alarming issue affecting the overall environment (Jiang et al. 2021). The increasing trend of the generation of waste from the food industry is one of the serious challenges throughout the world (Dora et al. 2020, Salmani et al. 2022).

Both environmental sustainability along with food security are burning issues that require extensive research. Approximately 70 percent of the global human population will be living in urban area and big cities by 2050, with the annual population expected to be more than 9 billion. The generation of huge waste from food-based sectors is recorded high in developed countries (Figure 18.1).

At the global level, enough food is available to feed the mammoth global population. In reality, however, this total hides a broad range of food consumption, from acute hunger to excess consumption as well as excessive food waste generation. One billion people suffer from food insecurity; however, more than a billion people consume too much food or waste too much food. From agricultural production to household consumption, food is wasted throughout the food supply chain (FSC). In the lower-middle income and high-income countries, food has been found to be thrown away even if it still qualifies for human consumption.

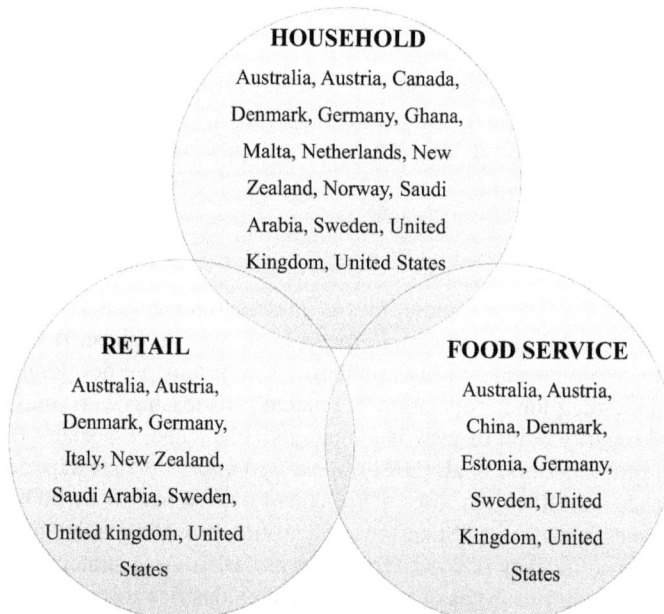

HOUSEHOLD
Australia, Austria, Canada, Denmark, Germany, Ghana, Malta, Netherlands, New Zealand, Norway, Saudi Arabia, Sweden, United Kingdom, United States

RETAIL
Australia, Austria, Denmark, Germany, Italy, New Zealand, Saudi Arabia, Sweden, United kingdom, United States

FOOD SERVICE
Australia, Austria, China, Denmark, Estonia, Germany, Sweden, United Kingdom, United States

FIGURE 18.1   Countries with high confidence of food waste estimates, by sector.

According to the Food and Agriculture Organization of the United Nations (FAO) report 2021, around 931 million tonnes of food waste was produced in 2019, shared among households (61 percent), food service (26 percent), and retail (13 percent). The total global food production wastage at consumer level was about 17 percent (11 percent in households, 5 percent in food service, and 2 percent in retail). Further, the global average per capita of food wasted each year, i.e., 74 kg shows comparatively similar data existence between low-middle income and high-income countries, indicating that a number of countries still have room for improvement and can meet the Sustainable Development Goal 12.3: halving food waste and reducing food loss by 2030. The household food waste production of India was 50 kg/capita/yr (UNEP 2021). According to Grover and Singh (2014), the food waste (kg/capita/yr) generated was 63, 68, and 90 for the low-income group, medium-income group, and high-income group, respectively in India. The average value of food waste generation of 73.66 kg/capita/yr exhibited closeness to the global average value (i.e., 74 kg/capita/yr).

Wastage of food-based commodities has surfaced as a global challenge as various food crops, fruits, vegetables, etc., are damaged before their use for processing. The food processing industry played a vital role by using different food materials as raw materials and producing products with extended shelf life (Singh et al. 2021).

An integrated approach to reducing hunger should include (i) investments from public and private sectors to raise agricultural productivity; (ii) well access to inputs, land, services, technologies, and markets; (iii) measures to enhance rural development; (iv) social protection program for the most vulnerable population, involving strengthening resilience to conflicts and natural disasters; and (v) specific nutrition programs for mothers and children under the age of 5. A variety of factors contribute to maintaining food security, including shortage of natural resources, population growth, food price fluctuations, the disparity in food consumption habits, climate change, and food loss and waste. For instance, global food production must increase 60–70 percent by 2050 to meet the needs of a population of 9 billion (Truong et al. 2019).

India is claimed to be the top producer of various agricultural commodities, i.e., milk, fruits such as banana, papaya, mango, cereals, millets, pulses, meat products etc., at the global level (Singh et al. 2021) due to various successful agriculture-based revolutions adopted by the farming community in different parts of India such as green revolution, white revolution, pink revolution, etc. Over the past fifty years, dairy farming has become more and more intensive, especially in India, which contributes a 35 percent share of total Asian milk generation (Qasim and Mane 2013). From 1950–1951, milk production in India is increased from 17 million tons to 100 million tons (2006–2007), with a per capita availability rate increase from 124 g/d to 245 g/d during the same period credited to operation flood, a dairy development program (Gautam et al. 2010). Various types of waste by-products produced from the dairy industry pose significant environmental impacts, so they need various treatment and disposal practices. Some of the environmental impacts include greenhouse gas emissions, soil and water pollution, loss of biodiversity and wildlife, health issues, alteration in nutrition cycle, etc. (Clay et al. 2020).

18.2   WASTE FROM DIFFERENT FOOD INDUSTRIES

18.2.1   FRUIT AND VEGETABLE WASTE

Vegetables and fruits are important food commodities which are essentially needed for the nutritional requirements of humans on a day to day basis. Waste from fruit and vegetable-based industries include pomace, wet and dry citrus peels, fleshy scales and roots of tuberous vegetables, peels, screen solids, kernels, seeds, rind, rags, stones, vine, shell, skin plant residue, etc., sludge from wastewater treatment plants. The wastewater from fruits and vegetable processing industrial units is reported to have high BOD, COD, TSS, and pH values (Valta et al. 2016).

18.2.2 Distillery Waste

Various types of solid and liquid waste and wastewater are generated in the industrial process for production of alcohol, ethanol, etc. Vinasse, betaine, stillage are the specific wastes generated from distilleries having environmental concerns, vinasse possesses high organic value because of the presence of polyphenols, organic acids, etc., and the wastewater have high biological oxygen demand (BOD), chemical oxygen demand (COD), phosphorus, nitrogen, magnesium, potassium, sulphates, etc. Stillage is reported to reduce the intensity of solar light to water bodies, which inhibits the process of photosynthesis, adversely affecting the aquatic life and leading to eutrophication of water bodies (Mikucka and Zielinska 2020).

18.2.3 Milk Industry Waste

The dairy industry wastewater is produced from various cleaning and washing procedures carried out in the milk processing industrial setup. On an estimate, about 2 percent of the processed milk gets washed out in drains. Dairy wastewaters have high BOD value upto 0.50g/L to 40g/L, high COD upto 60,000mg/L, and high volatile solids (Bella and Rao 2021). The untreated dairy wastewater has a high organic load with dissolved and suspended solids containing proteins, salts, fats, carbohydrates, residues of detergents, and cleaning agents such as sodium hypochorite, alkyl sulphonate, nitric acid, phosphoric acid, etc. (Arvanitoyannis and Giakoundis 2006, Kushwaha et al. 2011) causing eutrophication in receiving water bodies, which pose environmental risks to the biodiversity of aquatic ecosystems and alter the physico-chemical features of soil affecting its productivity and microbiological diversity. Along with wastewater, other wastes are also generated, i.e., whey permeate clarified butter sediment waste, black heavy sludge, etc. (Bella and Rao 2021).

18.2.4 Meat and Poultry Industry

The generation of waste from the meat industry is mainly slaughter wastewater considered one of the most hazardous wastewaters along with other solid waste produced from various processes. The meat industry is reported to consume approximately 24 percent of the freshwater. Slaughter wastewater is composed of fats, fibers, disinfectants, cleaning agents, total suspended solids, total phosphorus, total nitrogen, high COD, high BOD, etc. Because manure material is present in the gut of animals, wastewater contains a high concentration of total coliforms, fecal coliforms, and fecal *Streptococcus* species (Bustillo-Lecompte and Mehrab 2015). Solid waste is composed of bones, ligaments, blood, blood plasma, etc., with a strong bad smell. In poultry processing, 26.5 liters of water are used per bird and the wastewater contains proteins, fats, and carbohydrates from the meat, blood, skin, and feathers, resulting in high BOD and COD values.

18.2.5 Oil Industry

Among diverse types of industries in the food sector, edible oil-based industries generate different types of waste in different industrial processes as exhibited in Table 18.1. Mostly solid waste is produced as industrial by-products such as meal, oil cakes, hull, vines, etc., whereas the industrial wastewater such as the olive oil industry appears as dark liquid enriched with various organic substances, i.e., organic acids, polyalcohols, pectin, colloids, tannins, lipids, etc.

18.3 BIOREMEDIATION OF FOOD INDUSTRY WASTE

The waste generated by different food industries is further processed by applying methods via. fermentation, composting, pyrolysis, gasification, briquetting, synthesis of biocomposites, anaerobic digestion, etc., collectively known as valorization, and the waste gets transformed into various

TABLE 18.1

Waste Generated from the Different Edible Oil Industries

Sr.No.	Industry	Waste type	Reference
1.	Soybean oil industry	Soybean cakes, Soybean straw, Soybean oil residue, Refined soybean oil wastewater	Ramachandran et al. 2007, Zhu et al. 2014, Yu et al. 2018, Mohanty et al. 2021
2.	Rapeseed oil industry	Rapeseed cakes, Rapeseed meal	Ramachandran et al. 2007, Lomascolo et al. 2012, Mohanty et al. 2021
3.	Groundnut oil industry	Peanut meal, Peanut hull, Peanut skin, Peanut vine	Ramachandran et al. 2007, Zhao et al. 2012, Mohanty et al. 2021
4.	Sunflower oil industry	Sunflower cakes, Sunflower meal	Ramachandran et al. 2007, Raposo et al. 2008, De la rubia et al. 2012, Lomascolo et al. 2012, Mohanty et al. 2021
5.	Mustard oil industry	Mustard cakes	Ramachandran et al. 2007, Mohanty et al. 2021
6.	Palm oil industry	Palm press fiber, palm cakes, palm oil empty fruit bunch	Ramachandran et al. 2007, Chaikitkaew et al. 2015, Anwar et al. 2021, Mohanty et al. 2021
7.	Safflower oil industry	Safflower cake	Hashemi et al. 2020
8.	Olive oil industry	Olive mill wastewater	Hamza and Sayyadi 2015, Caporaso et al. 2018, Dias et al. 2018, Paulo and Santos 2021, Sounni et al. 2021

value-added products, i.e., biopolymer, biofuels, food supplement, animal feed, biochemicals, etc. (Lomascolo et al. 2012, Arancon et al. 2013, Mohanty et al. 2021). The food industries generate a vast amount of waste, which is valorized into various beneficial products, which increases the economical value of waste itself. However, various key aspects must be considered before considering food waste as feedstock for any valorization processes, which include the amount of food waste available, production pattern, quality of food waste, irregularity in generation of food waste, etc. These aspects must be considered and analyzed for every food waste from each industrial site to find optimal food waste valorization opportunities (Garcia-Garcia et al. 2019). Animal feed, fertilizer supply, composting, and anaerobic digestion are the common and current traditional food waste valorization practices (Lin et al. 2013). The traditional valorization strategies are globally accepted techniques with a broad spectrum of general applications however their effective conversion of wastes into the valuable product remains quite limited. Nowadays, extraction, fermentation, and enzyme technologies are the major paths used for food waste valorization. Food waste (except the animal-derived by-products) consists of lignocellulosic material, which has shown its use in biofuel production. Similarly, the extraction technique shows its large-scale applicability in the production of bioactive compounds, nutraceuticals, and phytochemicals (Otles and Kartal 2018). They are considered as the non-conventional strategies for food waste valorization. According to the well-known EPA hierarchy for food waste management, the transformation of food waste into energy such as biofuel, and composting are the most preferred strategies after maximizing the source reduction, feeding the hungry people, and feeding animals (Figure 18.2). The top three levels of the hierarchy that prevent and divert food waste into direct use are the best strategies for preventing food from being wasted. The next two levels are environmentally, socially, and economically beneficial

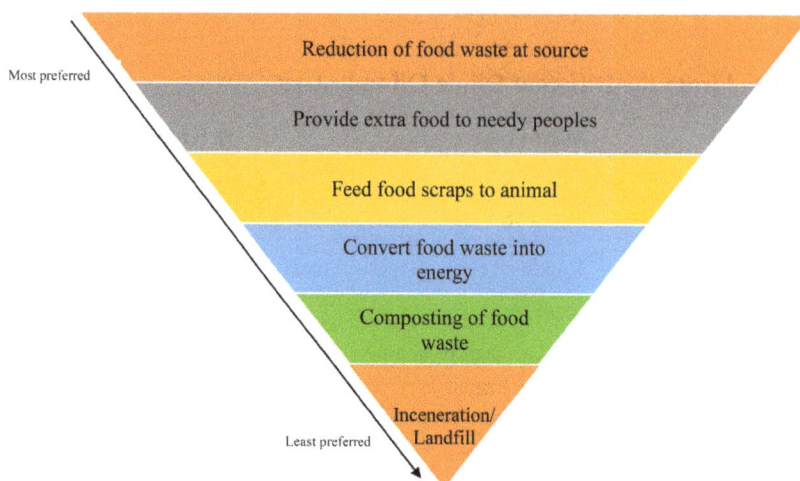

FIGURE 18.2 Food waste hierarchy of conventional management options.

Source: EPA 2021.

ways to manage waste from the food industry. This chapter focuses on these two levels of waste management strategies for food industry waste.

18.3.1 BIOFUELS PRODUCTION

Biofuels are a sustainable and renewable form of the fuel present in solid, liquid, and gaseous forms (based on the physical state of the application) such as bioethanol, biobutanol, biodiesel, biohydrogen, bio-oil, biogas, biochar, syngas, etc. (Dhiman and Mukherjee 2021). The most significant step in the production of biofuels is the selection of adequate feedstock, e.g., one-third cost of per gallon of ethanol production has been shared by the pre-treatment process (Edeh 2020). The search for the identification of the best feedstock for biofuels production increased over the years. In this regard, the use of food waste offers various social, economic, and environmental benefits. The organic portion of food waste generally consists of carbohydrates (35.5–69 percent), proteins (3.9–21.9 percent), oils, fats, and organic acids (Kiran et al. 2014, Dhull et al. 2021). These complex structures can be disintegrated to their simple form, which acts as feedstock for microorganism to produce various biofuels (Rose 2022) having more worth compared to other utilization of food waste such as electricity generation and animal feed (Lin et al. 2013). Production of biomass-based biofuels is still at the proof-of-concept stage, however, in the present scenario food waste can be considered as sustainable and environmentally friendly feedstock that can be converted to bioenergy such as bioethanol, biobutanol, biodiesel (Dong et al. 2016, Patel et al. 2019). Food industry waste based bioethanol production involves various steps depending on the nature of the feedstock, such as direct sugar as feedstock involves fermentation followed by distillation, starchy feedstock involves the addition of a hydrolysis step to transform starch into sugar, followed by fermentation and distillation, lignocellulosic biomass involves one more additional step pre-treatment followed by hydrolysis, fermentation, and distillation to get fuel-grade bioethanol. However, a major concern with utilizing food processing waste as a sole carbon source for bioethanol production is low carbohydrate content, which can be improved by mixing with another lignocellulosic biomass. Another biofuel that can be produced from food waste is n-butanol having the superior quality to bioethanol and has other associated benefits such as an encouraging cost effective alternative with commercial application. The conventional production of butanol involves acetone-butanol-ethanol

TABLE 18.2
Various Food Waste Fractions and Possible Production of Biofuels

Food type	Food waste type	Waste fraction type	Biofuel production
Oil (palm)	Palm empty fruit bunch	Lignocellulose	Bioethanol, Biobutanol, Biodiesel
Oil (soybean)	Soybean meal	Protein, carbohydrate	Bioethanol, Biobutanol, Biodiesel
Oil (rapeseed)	Rapeseed meal	protein, carbohydrate	Bioethanol, Biobutanol, Biodiesel
Rice	Rice hull	lignocellulose, ash	Bioethanol, Biobutanol
Wheat	Bran	arabinoxylan, cellulose, protein	Bioethanol, Biobutanol
Potatoes	Potato peel and other processing waste	lignocellulose, starch	Bioethanol, Biobutanol
Banana	Rejected banana	lignocellulose, pectin, starch	Bioethanol, Biobutanol
Apple	Rejected apples	fucogalactoxyloglucan, lignocellulose, glucose, fructose	Bioethanol, Biobutanol
Raw sugar	Molasses	sucrose, glucose and fructose	Bioethanol, Biobutanol
Beer (barley)	Brewery waste	carbohydrate, protein, organic acids	Bioethanol, Biobutanol
Wine	Brewery waste	carbohydrate, organic acids	Bioethanol, Biobutanol

Source: Zhang et al. 2016

(ABE) fermentation of sugar using solventogenic *Clostridia* spp. with less than 15 g/L concentration of butanol. This concentration can be improved by using genetic and metabolic engineering species of *Clostridia* spp., *Escherichia coli*, *S. cerevisiae* strains, etc. (Rose 2022) and integrating the gas stripping-pervaporation process with ABE fermentation further increases the biobutanol concentration upto 75.5 g/L (Xue et al. 2016, Zhang et al. 2016). Biodiesel production is a chemical transformation called transesterification of the tri-, di-, and mono-glycerides in the presence of alcohol, base, acid, enzyme, or solid catalyst into biodiesel. Biodiesel can be produced from discarded oils derived from soybean, rapeseed, and canola plants. Food waste is a non-edible source of carbohydrates, and lipids can be used to produce bioethanol and biodiesel, respectively. The commercialization of biofuel (bioethanol, biobutanol, biodiesel) produced from food waste is mostly dependent on (i) accessibility of food waste throughout the year; (ii) the carbohydrate and lipid fraction in food waste; and (iii) efficiency of the bioconversion process (Karmee and Lin 2014). The non-edible and zero-cost food waste can be a resource for low-cost biofuel production, thereby reducing direct competition between food and fuel. The possible utility of various food waste for various biofuels production is summarized in Table 18.2.

18.3.2 COMPOSTING

Composting is the aerobic biological decomposition of organic matter. Composting is an anthropogenic process performed under a controlled environmental condition in contrast to decomposition, which is a natural and uncontrolled process (Hoitink and Fahy 1986, Ayilara et al. 2020). Industrial-scale composting requires proper control on optimum operation conditions such as temperature, oxygen level, moisture level, aeration rate, turning rate, C/N ratio, and particle size of compost material. The decomposition of food waste, which consists of carbohydrates, proteins, and fats are easily metabolized by varieties of microorganism at both mesophilic and thermophilic temperatures. However, mesophilic conditions promote an initial rapid decomposition rate, which provides at the end a more stable and mature product. The properties of food waste as a raw composting agent are unique. Because fresh food waste contains high moisture content and is low in structure, it is important to combine it with a bulking agent to absorb the surplus moisture, which adds structural strength to the mixture. Bulking agents with a high C/N ratio, such as agricultural waste is a good choice. Composting process aerobic degrades and transforms the complex degradable food materials by microorganism into 'humic-like' compounds that can be used safely and beneficially as biofertilizers and soil amendments to increase agricultural productivity (Yu et al. 2019). Along with this, compost has also shown application in bioremediation, weed and plant disease control, pollution prevention, erosion control, soil biodiversity improvement, wetland restoration, decreased environmental risk of the use of synthetic fertilizer, support, and promoting plant growth-promoting organisms, etc. (Ayilara et al. 2020). Aerobic bacteria, anaerobic bacteria, fungi, actinobacteria, pseudomonads, and nitrogen-fixing bacteria are six different groups of microorganism that tangle in the composting of food waste. Starch, cellulose, lignin, and pectin portion of food waste is degraded by mesophilic and thermophilic bacteria and fungi or actinomycetes degrade the natural polymers portion of food waste (Patwa et al. 2020). The composting of food waste is accomplished in two temperature regimes (mesophilic, 15°C to 45°C and thermophilic, 40°C to 80°C) by microorganism (bacteria, fungi, and actinomycetes) in three phases (mesophilic, thermophilic, and cooling or maturation phase) (Moreno et al. 2011). The duration of each phase is a function of the quantity and nature of microbial community, nature of the waste (organic content), moisture content, etc. (Fischer and Glaser 2012). The first phase of food waste composting is mesophilic, and carbon-rich biomass is decomposed by mesophilic fungi (mold and yeast) commonly belonging to the genus *Aspergillus*, *Penicillium*, *Fusarium*, *Rhizopus*, *Streptomyces*, and by mesophilic bacteria belong to genus *Bacillus*, *Staphylococcus*, *Mycobacterium*, etc. The complex structure of food waste is initially broken down by fungi into amino acids, sugars, and other simple components, which are later degraded by bacteria. This conversion lowers the pH and increases the temperature of the compost and composting process progresses into the second thermophilic phase. In the thermophilic phase, thermophiles replace the mesophiles, and elevated temperatures between 40°C to 80°C favor the growth of thermophilic fungal species belonging to genus *Aspergillus*, *Talaromyces*, *Thermocyces*, *Thermatinomyces*, *Thermop*, etc. and thermophilic bacterial species belong to genus *Bacillus*, *Clostridium*, *Geobacillus*, *Pseudomonas*, *Thermobacillus*, etc. to degrade the holocellulose, fats, and some lignin. The elevated temperature accelerates the breakdown of the organic matter, which allows maximum decomposition of the available carbon source in the compost. The exhaustion of carbon sources causes a decline in the temperature in the compost and compositing process to enter its last stage, i.e., maturation stage (Palaniveloo et al. 2020). The final maturation phase exhibits lowering in temperature (below 25°C), cease in thermophiles activity, re-appearance of mesophiles, reduction in oxygen uptake, and degradation of leftover organic matter by fungi. The product after the maturation stage is called matured compost and has a high pH value, lower C/N ratio (15–20/1), high NO_3, and lower NH^{4+} content (Palaniveloo et al. 2020). The optimum C/N ratio for the compositing of food waste composition must be in the range of 25–33/1 because below 25 value, ammonia start releasing which ultimately causes loss in nitrogen content, whereas above 35 value the degradation process reduces (Kapetanios et al. 1993, Patwa et al. 2020). Food waste can be composted in various ways, including windrow composting, vermicomposting, anaerobic bio-digester, etc. Aerobic composting is a fast process of decomposition of food waste by aerobic microorganisms and produces CO_2, NO_2, and NO_3. However, anaerobic composting is a slow reduction process of food waste biodegradation by anaerobic microorganisms and produces CH_4 and H_2S (Patwa et al. 2020).

18.3.2.1 Windrow Composting

Windrow composting is a traditional widely performed technique that can treat large amounts of waste. Windrow is a passive aeration technique based on both natural aerations known as passive windrow composting and mechanical aeration known as turned windrow composting. Both techniques have their advantages and limitation however homogeneous compost is observed with turned composting (Ab Muttalib et al. 2016). A windrow composting process begins with the blending of food waste with agriculture or wood residue to obtain the required initial condition for composting, i.e., moisture content (60 percent), C/N ratio (27:1), bulk density (400–600 kg/m³), free air space (50–60 percent) and particle size (1/8–2 inches) (Saer et al. 2013). After that waste blend is covered for a few weeks for the intensive composting process, followed by an additional two weeks of uncovered aeration on an aerated pad. After that, compost is stored in a large windrow for final curing, which is later screened for the appropriate particle size and also undergoes quality check as per necessary regulations such as testing of heavy metals, fecal coliform, *salmonella* presence, and nutrient content counting NPK, which ensure a suitable compost quality for the consumers (Sánchez et al. 2017, Al-Rumaihi et al. 2020).

18.3.2.2 Vermicomposting

Vermicomposting is an ecologically sound and eco-friendly successful technology for bioconversion of food industry waste into organic-rich manure having high economic value (Ali et al. 2015). Vermicomposting is a waste management technique that involves biotransformation and stabilization of the organic section of food industry waste involving earthworms, microorganisms, and other degradable communities into a granular humus-like material called vermicompost, which can be easily stored, handled, and used in agricultural fields without any negative impacts. During vermicomposting, the organic matter is transformed by physical (aeration, mixing, and grinding of organic waste) as well as biochemical action (microbial degradation of organic waste) (Garg et al. 2012). Microbial action on the organic matter occurs inside as well as outside of the gut of the earthworm. The major transformation of the inorganic matter occurs inside the gut of the earthworm, which causes the low temperature in the waste during the vermicomposting (Figure 18.3). The degradation of the organic matter by earthworms causes a rapid decline in the C/N ratio and an increase in mineral oxygen with time (Patwa et al. 2020). The earthworm performance ability is a function of their tolerance toward environmental fluctuations (pH, moisture, aeration, temperature, etc.), handling and disruption to the worm bed, and growth and reproduction rate. Better degradation of the organic portion of food waste can be achieved by using earthworm species having a high reproduction rate, fast growth rate, and short life span as these maintain a high number of the juvenile worm as they are insatiable for feeding and keep the vermicomposting process on high speed (Othman et al. 2012). Non-burrowing type earthworms are considered better for food waste vermicomposting, such as *Eisenia fetida* (Munnoli and Bhosle 2009, Yadav and Garg 2011). The abiotic and biotic conditions for suitable agro-food industry waste (organic matter) transformation into vermicompost as shown in Table 18.3. Management of food industry waste through vermicomposting can be cost-effective and environmentally friendly technique to address various environmental issues related to conventional treatment methods of food industrial wastes. The vermicomposting technique requires less energy consumption and exhibits economical feasibility over conventional methods, especially in terms of nutrient recovery. However, no comprehensive and concise information is not available on the sole and direct utilization of food industry waste in vermicomposting (Ali et al. 2015, Mago et al. 2021).

18.3.3 Anaerobic Digestion

Biogas generation using anaerobic digestion of food waste is a clean, efficient, and renewable energy source. The major fraction of biogas is methane (CH_4) and carbon dioxide (CO_2) and the minor fraction is contributed by water vapor (H_2O), hydrogen sulfide (H_2S), hydrogen (H_2), and siloxanes (Mirmohamadsadeghi et al. 2019, Yang et al. 2021). Generally, biogas consists of 40–70 percent methane content, which can be enhanced to 75–99 percent (natural gas quality) (Mittal

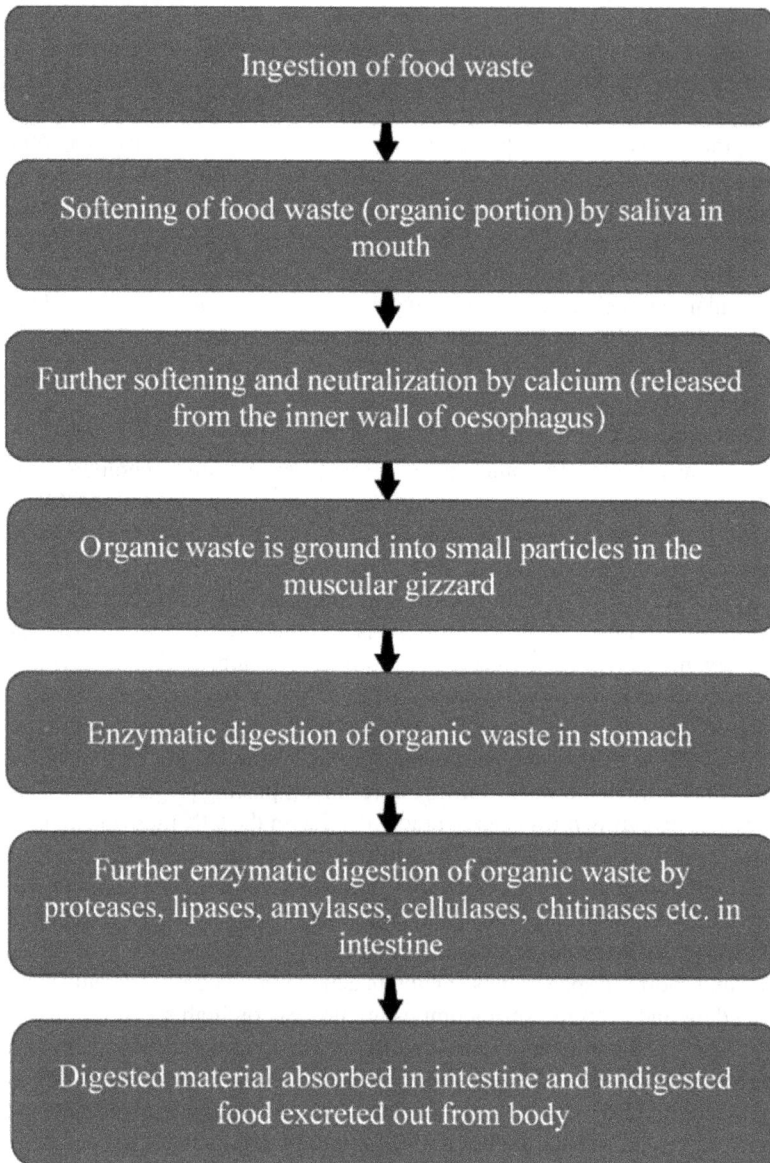

FIGURE 18.3   Food waste (organic matter) digestion steps in earthworms.

Source: Yadav and Garg 2011.

et al. 2018). The products of anaerobic digestion include biogas and digestate. Biogas can be used in a variety of ways, including for cooking, heating, electricity generation, as fuel for transportation, etc. Digestate is the nutrient-rich solid or liquid material by-product digestion process that can be used as fertilizers, soil amendments, livestock bedding, etc. The common utility of biogas is in household applications due to the high cost of its distribution to end-user (Sorathiya et al. 2014). This problem has been investigated from the research outcomes of IIT, New Delhi (Vijay 2011). They developed a unique technique of biogas purification and bottling, which allowed storage of pure biogas (biomethane) in cylinders and its availability anywhere as an LPG cylinder. Pure biomethane also known as renewable natural gas (RNG), has shown similarity with CNG that

TABLE 18.3
Conditions Required for Vermicomposting of Waste from Food-based Sectors

Components	Optimum condition
Abiotic	
pH	5.5–8.5
Temperature	12–28°C
Moisture level	60–80%
Salt level	< 0.5%
C/N ratio	25–30
Light	kept away from light
Biotic	
Earthworms stocking density	approx 1.60 kg-worms/m^2
Feeding rate	1.25 kg feed/kg worm/d

Source: Yadav and Garg 2011

allows its utility where CNG is being applicable (Sorathiya et al. 2014). Presently, 21 BGFP (Biogas Filling Plant) projects with 37,016 cu^m/d of aggregate capacity have been sanctioned in ten states of India, i.e., Chhattisgarh, Gujarat, Haryana, Karnataka, Maharashtra, Punjab, Madhya Pradesh, Andhra Pradesh, Uttar Pradesh, and Rajasthan (www.terienvis.nic.in). Anaerobic digestion of food waste offers numerous social and environmental benefits along with major outcomes in the form of energy and manure such as lowering in pollution (groundwater pollution, soil pollution, air pollution through the release of dioxins, furans, methane, etc.), reducing the burden on non-renewable fossil fuels, the slurry is rich in nitrogen, which can be used as organic fertilizer, etc. (Lewis et al. 2017, Mittal et al. 2018).

Food processing industries and retailers are the major sources of food waste generation. The simple and homogeneous composition of food waste in food processing industries allows direct application in anaerobic digestion. However, retailer-based food wastes are commonly disposed of as municipal solid waste (MSW) and share 20 to 54 percent of the waste (Yasin et al. 2013). Due to heterogeneous composition, sorting of the MSW before anaerobic digestion becomes an important step. The anaerobic digestion allows simultaneous resource recovery and treatment of food waste (Xu et al. 2018). Solid food waste consists of about 90 percent (weight basis) of total solids (organic). Generally, food waste exhibits a high VS/TS (volatile solid/total solid) ratio, which indicates its convertible suitability into biogas (Yang et al. 2021). The utility of the food waste as the sole substrate in an anaerobic digester was found unstable (Zhang et al. 2011, Banks et al. 2012). Various studies indicated the co-digestion of food waste with other suitable waste exhibits improvement in degradation rate and biogas production (Edelmann et al. 2000, Labatut et al. 2014, Zamanzadeh et al. 2017). Biogas production is an anaerobic degradation of the carbohydrate, protein, and lipid portion of food waste via four diverse groups of microorganisms, i.e., fermentative, syntrophic, acetogenic, and methanogenic bacteria (Figure 18.4) (Al-Wahaibi et al. 2020). Carbohydrate digestion results in higher biogas production in a short residence time due to its high degradability. However, biogas produced from lipids is of superior quality to carbohydrates but requires longer residence time, indicating their lower biodegradability rate (Yang et al. 2015). The efficiency of anaerobic digestion is dependent on the characteristics of the substrate and process parameters (Almasi et al. 2018). Animal by-products category 1 and packaged wastes (i.e., non-separable from packaging) in nonbiodegradable packaging are excluded from food types that cannot be digested by anaerobic digestion (Garcia-Garcia et al. 2017). Nowadays, biochemical

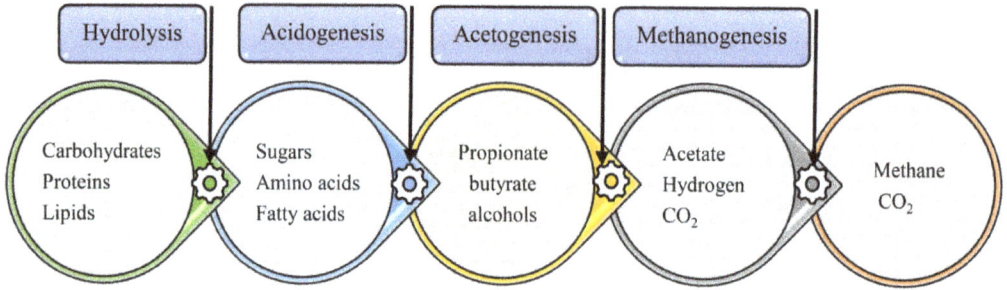

FIGURE 18.4   Biogas production process via anaerobic digestion of food waste.

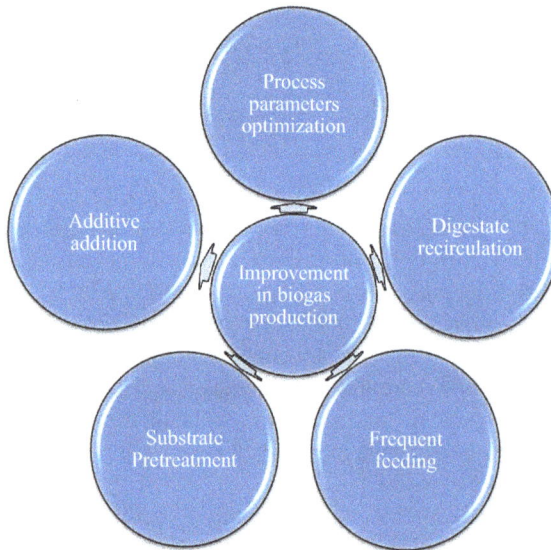

FIGURE 18.5   Various strategies for the enhancement in biogas production.

methane potential (BMP), which can be determined through batch mode at laboratory conditions is considered a reliable analysis in the selection of adequate substrate and optimization of the influential process parameters (Raposo et al. 2012, Kondusamy and Kalamdhad 2014). The adequate substrate should be balanced in nutrient composition, which includes macronutrients (carbon, nitrogen, phosphor, and sulfur) and micronutrients (iron, cobalt, nickel, zinc, selenium, tungsten, magnesium, chromium, and molybdenum). Food waste is the major contributor of carbon sources for growing microorganism. Generally, food waste comprises a 3–55/1 C/N ratio (Yang et al. 2021). The recommended C/N ratio value for efficient anaerobic digestion of food waste is 16–25/1. Micronutrients and metal elements are necessary for microorganism survival but are required in very low concentrations as their high concentration can hinder the anaerobic digestion process (Choong et al. 2016). The shredding of food waste is another essential step, which reduces the waste particle size ultimately increasing the available surface area for better action of the exo-enzymes (Agyeman and Tao 2014) (Figure 18.5). Several process operating conditions have a significant effect on biogas production, including pH, temperature, hydraulic retention time, organic loading rate, inoculum, pre-treatment, feeding pattern, etc. The ideal pH range for efficient methanogenesis of food waste is 6.8 to 8.0 and for acidogenesis is 5.0 to 6.5 (Al-Wahaibi et al. 2020). The anaerobic digestion of food waste is executed at both mesophilic (32–45°C) and thermophilic (55–70°C) temperatures. However, due to the high organic load of food waste, mesophilic anaerobic digestion exhibits better performance and stability than thermophilic anaerobic digestion (Guo et al. 2014). Further, thermophilic digesters are more prone to process inhibition because, at high temperatures, the toxicity of intermediate compounds also increases (Angelidaki and Ahring 1993, Zamanzadeh et al. 2017). Mesophilic anaerobic digestion required less process heat, which ultimately reduced the operating cost (Al-Wahaibi et al. 2020). Another significant parament process is the retention time for bioconversion of food waste into biogas and optimum retention time depends on temperature, organic loading rate, and substrate composition. Morken et al. (2018) increased the organic loading rate from 1.8 to 5.0 kg of volatile solid per m³/d and observed a 479 percent enhancement in the methane production per liter of digester fed with food waste. The addition of volatile solids increased methane production and is within the range of 0.31–1.1 m³ methane per kg (Xu et al. 2018, Yang et al. 2021).

In order to ensure the efficient digestion of food waste, proper mixing of food waste in the digester at regular intervals, maintaining an optimum moisture content throughout the digestive process, adequate quantity and quality of inoculum and carefully optimizing the process configurations are all important. Various types of anaerobic digester configurations have been developed for food waste based biogas production such as anaerobic sequencing batch reactor (ASBR), continuous stirred tank reactor (CSTR), anaerobic plug-flow reactor (APFR), anaerobic contact reactor (ACR), up-flow anaerobic sludge bed reactor (UASB), up-flow anaerobic solid-state reactor (UASS), anaerobic baffled reactor (ABR), internal circulation reactor (ICR), leach bed reactor (LBR), hollow fiber type anaerobic membrane bioreactor (HF-AnMBR) (Mirmohamdsadeghi et al. 2019).

18.4 CONCLUSION

The generation of agro-industrial waste is a big environmental issue as it adversely affects the environmental quality affecting the different life forms. Several studies and investigations conducted earlier identified the agro-industrial wastes and their utilization as a strategic raw material for economic gains by transforming waste into value added products such as compost and bioplastics along with cleaner fuels such as biofuels and recovery of other important chemicals of industrial use. Intensive investigations are required for the development of novel and efficient extraction methods for the recovery of bioactive compounds from various food industry wastes, which may be utilized further in diverse industrial processes such as the development of natural coloring agents, synthetic antioxidants, preservatives, anti-microbial agents, etc. Modern delivery processes such as nano-encapsulation, nano-emulsions etc., must be developed, which are economically effective methods.

So in order to maintain sustainability, some factors such as economic costs, social priorities, and environmental concerns are considered. Key aspects of sustainable food industry waste strategies include its treatment with the most environmentally sound technologies and also to recover the industrially important substances. The conventional agro-industry waste management methods (wastewater treatment, incineration, and landfilling) are considered less effective from an environmental viewpoint as several open waste dumps in many parts of the developing world are creating strong disagreements. However, if modern biotechnological strategies are used and various system outputs (including methane-rich biogas, syngas, waste heat, etc.) are managed and utilized, the adverse environmental impacts of these methods can be minimized. The concept of industrial ecology must be thoroughly explored for sustainable and holistic solutions address various environmental issues pertaining to the generation and disposal of different wastes. U.N. Sustainable Development Goal 12:3 focuses on sustainable consumption and production of food waste and aims for its reduction by 2030.

REFERENCES

Ab Muttalib, S.A., Ismail, S.N.S., & Praveena, S.M., 2016. Application of effective microorganism (EM) in food waste composting: A review. *Asia Pacific Environmental and Occupational Health Journal* 2(2): 37–47.

Agyeman, F.O., & Tao, W. 2014. Anaerobic co-digestion of food waste and dairy manure: Effects of food waste particle size and organic loading rate. *Journal of Environmental Management* 133: 268–274.

Ali, U., Sajid, N., Khalid, A., Riaz, L., Rabbani, M.M., Syed, J.H., & Malik, R.N. 2015. A review on vermicomposting of organic wastes. *Environmental Progress & Sustainable Energy* 34(4):1050–1062.

Almasi F., Soltanian S., Hosseinpour S., Aghbashlo M., & Tabatabaei M., 2018. Advanced Soft Computing Techniques in Biogas Production Technology. In: Tabatabaei M., Ghanavati H. (eds), Biogas. Biofuel and Biorefinery Technologies 6: 387–417. Springer, Cham.

Al-Rumaihi, A., McKay, G., Mackey, H.R., & Al-Ansari, T. 2020. Environmental impact assessment of food waste management using two composting techniques. *Sustainability* 12(4):1–23.

Al-Wahaibi, A., Osman, A.I., Ala'a, H., Alqaisi, O., Baawain, M., Fawzy, S., & Rooney, D.W. 2020. Techno-economic evaluation of biogas production from food waste via anaerobic digestion. *Scientific Reports* 10(1):1–16.

Angelidaki, I., & Ahring, B.K. 1993. Thermophilic anaerobic digestion of livestock waste: the effect of ammonia. *Applied Microbiology and Biotechnology* 38(4): 560–564.

Anwar, N.A.K., Hussain, N., Idris, A., Ramli, S., & Malek, R.A. 2021. Large scale production of succinic acid by fermentation of sequential inorganic salt pretreated oil palm empty fruit bunch. *Biomass and Bioenergy* 155: 106307.

Arancon, R.A.D., Lin, C.S.K., Chan, K.M., Kwan, T.H., & Luque, R. 2013. Advances on waste valorization: new horizons for a more sustainable society. *Energy Science & Engineering* 1(2): 53–71.

Arvanitoyannis, I.S., & Giakoundis. A., 2006. Current strategies for dairy waste management: A review. *Critical Reviews in Food Science and Nutrition* 46(5): 379–390.

Ayilara, M.S., Olanrewaju, O.S., Babalola, O.O., & Odeyemi, O. 2020. Waste management through composting: Challenges and potentials. *Sustainability* 12(11): 1–23.

Baisali S., Chakrabarti, P.P., Vijaykumar, A., & Kale, V. 2006. Wastewater treatment in dairy industries-possibility of reuse. *Desalination* 195: 141–152.

Banks, C.J., Zhang, Y., Jiang, Y. & Heaven, S. 2012. Trace element requirements for stable food waste digestion at elevated ammonia concentrations. *Bioresource Technology* 104: 127–135.

Bella, K. & Rao, P.V., 2021. Anaerobic digestion of dairy wastewater: effect of different parameters and co-digestion options-a review. *Biomass Conversion and Biorefinery* 1–26.

Bohdziewicz, J., & Sroka, E. 2005. Integrated system of activated sludge–reverse osmosis in the treatment of the wastewater from the meat industry. *Process Biochemistry* 40:1517–1523.

Bustillo-Lecompte, C.F., & Mehrab, M. 2015. Slaughterhouse wastewater characteristics, treatment, and management in the meat processing industry: A review on trends and advances. *Journal of Environmental Management* 161: 287–302.

Caporaso, N., Formisano, D., & Genovese, A. 2018. Use of phenolic compounds from olive mill wastewater as valuable ingredients for functional foods. *Critical Reviews in Food Science and Nutrition* 58(16): 2829–2841.

Chaikitkaew, S., Kongjan, P., & Sompong, O. 2015. Biogas production from biomass residues of palm oil mill by solid state anaerobic digestion. *Energy Procedia* 79: 838–844

Choong, Y.Y., Norli, I., Abdullah, A.Z., & Yhaya, M.F. 2016. Impacts of trace element supplementation on the performance of anaerobic digestion process: A critical review. *Bioresource Technology* 209: 369–379.

Clay, N., Garnett, T., & Lorimer, J. 2020. Dairy intensification: Drivers, impacts and alternatives. *Ambio* 49(1): 35–48.

De la Rubia, M. A., Fernández-Cegrí, V., Raposo, F., & Borja, R. 2012. Anaerobic digestion of sunflower oil cake: a current overview. *Water Science & Technology* 67(2): 410–417.

Dhiman, S., & Mukherjee, G. 2021. Present scenario and future scope of food waste to biofuel production. *Journal of Food Process Engineering* 44(2): e13594.

Dhull, S.B., Kidwai, M.K., Noor, R., Chawla, P., & Rose, P.K. 2021. A review of nutritional profile and processing of faba bean (*Vicia faba* L.). *Legume Science* e129.

Dias, B., Lopes, M., Ramoa, R., Pareira, A.S., & Belo, I. 2018. *Candida tropicalis* as a promising oleaginous yeast for olive mill wastewater bioconversion. *Energies* 14: 640.

Dong, T., Knoshaug, E.P., Pienkos, P.T., & Laurens, L.M. 2016. Lipid recovery from wet oleaginous microbial biomass for biofuel production: a critical review. *Applied Energy* 177: 879–895.

Dora, M., Wesana, J., Gellynck, X., Seth, N., Dey, B., & De Steur, H., 2020. Importance of sustainable operations in food loss: Evidence from the Belgian food processing industry. *Annals of Operations Research* 290(1): 47–72.

Edeh, I., 2020. Bioethanol Production: An Overview. Intechopen, United Kingdom.

Edelmann, W., Engeli, H., & Gradenecker, M. 2000. Co-digestion of organic solid waste and sludge from sewage treatment. *Water Science and Technology* 41(3): 213–221.

EPA, 2021. U.S. Environmental Protection Agency, viewed 25 January 2022, www.epa.gov/Sustainable-Mana gements-Food/Food-Recovery-Hierarchy.

Fischer, D., & Glaser, B. 2012. Synergisms between compost and biochar for sustainable soil amelioration. In: Kumar, S., Bharti, A. (eds), Management of organic waste, 167–198. Intechopen, United Kingdom.

Garcia-Garcia, G., Stone, J., & Rahimifard, S. 2019. Opportunities for waste valorisation in the food industry-A case study with four UK food manufacturers. *Journal of Cleaner Production* 211: 1339–1356.

Garcia-Garcia, G., Woolley, E., Rahimifard, S., Colwill, J., White, R., & Needham, L. 2017. A methodology for sustainable management of food waste. *Waste and Biomass Valorization* 8(6): 2209–2227.

Garg, V.K., Suthar, S., & Yadav, A. 2012. Management of food industry waste employing vermicomposting technology. *Bioresource Technology* 126: 437–443.

Gautam, Dalal, R.S., & Pathak, V., 2010. Indian dairy sector: time to revisit operation flood. *Livestock Science* 127(2–3): 164–175.

Grover, P., and Singh, P. 2014. An analytical study of effect of family income and size on per capita household solid waste generation in developing countries. *Review of Arts and Humanities* 3(1): 127–143.

Guo, X., Wang, C., Sun, F., Zhu, W., & Wu, W. 2014. A comparison of microbial characteristics between the thermophilic and mesophilic anaerobic digesters exposed to elevated food waste loadings. *Bioresource Technology* 152: 420–428.

Hamza, M., & S. Sayadi. 2015. Valorisation of olive mill wastewater by enhancement of natural hydroxytyrosol recovery. *International Journal of Food Science & Technology* 50(3): 826–833.

Hashemi, S.S., Mirmohamadsadhegi, S., & Karimi, K. 2020. Biorefinery development based on whole safflower plant. *Renewable Energy* 152: 399–408.

Hoitink, H.A., & Fahy, P.C. 1986. Basis for the control of soilborne plant pathogens with composts. *Annual Review of Phytopathology* 24(1): 93–114.

Jayathilakan, K., Sultana, K., Radhakrishna, K., & Bawa, A.S. 2012. Utilization of by-products and waste materials from meat, poultry and fish processing industries: a review. *Journal of Food Science and Technology* 49(3): 278–293.

Jiang, S., Wang, F., Li, Q., Sun, H., Wang, H., & Yao, Z. 2021. Environment and food safety: a novel integrative review. *Environmental Science and Pollution Research* 28(39): 54511–54530.

Kapetanios, E.G., Loizidou, M., & Valkanas, G. 1993. Compost production from Greek domestic refuse. *Bioresource Technology* 44(1): 13–16.

Karmee, S.K., & Lin, C.S.K. 2014. Valorisation of food waste to biofuel: current trends and technological challenges. *Sustainable Chemical Processes* 2(1): 1–4.

Kiran, E.U., Trzcinski, A.P., Ng, W.J., & Liu, Y. 2014. Bioconversion of food waste to energy: A review. *Fuel* 134: 389–399.

Kondusamy, D., & Kalamdhad, A.S. 2014. Pre-treatment and anaerobic digestion of food waste for high rate methane production-A review. *Journal of Environmental Chemical Engineering* 2(3): 1821–1830.

Kushwaha, J.P., Srivastava, V.C., & Mall, I.D. 2011. An overview of various technologies for the treatment of dairy wastewaters. *Critical Reviews in Food Science and Nutrition* 51(5), 442–452.

Labatut, R.A., Angenent, L.T., & Scott, N.R. 2014. Conventional mesophilic vs. thermophilic anaerobic digestion: a trade-off between performance and stability? *Water research* 53: 249–258.

Lewis, J.J., Hollingsworth, J.W., Chartier, R.T., Cooper, E.M., Foster, W.M., Gomes, G.L., Kussin, P.S., MacInnis, J.J., Padhi, B.K., Panigrahi, P., & Rodes, C.E. 2017. Biogas stoves reduce firewood use, household air pollution, and hospital visits in Odisha, India. *Environmental Science & Technology* 51(1): 560–569.

Lin, C.S.K., Pfaltzgraff, L.A., Herrero-Davila, L., Mubofu, E.B., Abderrahim, S., Clark, J.H., Koutinas, A.A., Kopsahelis, N., Stamatelatou, K., Dickson, F., & Thankappan, S. 2013. Food waste as a valuable resource for the production of chemicals, materials and fuels. Current situation and global perspective. *Energy & Environmental Science* 6(2): 426–464.

Lomascolo, A., Uzan-Boukhris, E., Sigoillot, J.C., & Fine, F. 2012. Rapeseed and sunflower meal: a review on biotechnology status and challenges. *Applied Microbiology and Biotechnology* 95: 1105–1114.

Mago, M., Yadav, A., Gupta, R., & Garg, V.K. 2021. Management of banana crop waste biomass using vermi-composting technology. *Bioresource Technology* 326: 124742.

Mikucka, W. and M. Zielinska. 2020. Distillery stillage: characteristics, treatment, and valorization. *Applied Biochemistry and Biotechnology*. 192: 770–793.

Mirmohamadsadeghi, S., Karimi, K., Tabatabaei, M., & Aghbashlo, M. 2019. Biogas production from food wastes: A review on recent developments and future perspectives. *Bioresource Technology Reports* 7: 100202.

Mittal, S., Ahlgren, E.O., & Shukla, P.R. 2018. Barriers to biogas dissemination in India: A review. *Energy Policy* 112: 361–370.

Mohanty, A., Rout, P.R., Dubey, B., Meena, S.S., Pal, P., & Goel, M. 2021. A critical review on biogas production from edible and non-edible oil cakes. *Biomass Conversion and Biorefinery* 1–18.

Moreno, J., López, M.J., Vargas-García, M.C., & Suárez-Estrella, F. 2011. Recent advances in microbial aspects of compost production and use. In International Symposium on Growing Media, Composting and Substrate Analysis. 17 October, 443–457.

Morken, J., Gjetmundsen, M., & Fjortoft, K. 2018. Determination of kinetic constants from the co-digestion of dairy cow slurry and municipal food waste at increasing organic loading rates. *Renewable Energy* 117: 46–51.

Munnoli, P.M., & Bhosle, S. 2009. Effect of soil and cow dung proportion on vermi-composting by deep burrower and surface feeder species. *Journal of Scientific and Industrial Research* 68: 57–60.

Osorio, L.L.D.R., Flórez-López, E., & Grande-Tovar, C.D. 2021. The potential of selected agri-food loss and waste to contribute to a circular economy: Applications in the food, cosmetic and pharmaceutical industries. *Molecules* 26(2): 515.

Othman, N., Irwan, J.M., & Roslan, M.A. 2012. Vermicomposting of food waste. *International Journal of Integrated Engineering* 4(2): 39–48.

Otles, S., & Kartal, C. 2018. Food waste valorization. In: Galanakis, C.M. (ed), Sustainable food systems from agriculture to industry 371–399. Academic Press.

Paini, J., Benedetti, V., Ail, S.S., Castaldi, M.J., Baratieri, M., & Patuzzi, F. 2021. Valorization of Wastes from the Food Production Industry: A Review Towards an Integrated Agri-Food Processing Biorefinery. *Waste and Biomass Valorization* 13: 31–50.

Palaniveloo, K., Amran, M.A., Norhashim, N.A., Mohamad-Fauzi, N., Peng-Hui, F., Hui-Wen, L., Kai-Lin, Y., Jiale, L., Chian-Yee, M.G., Jing-Yi, L., & Gunasekaran, B. 2020. Food waste composting and microbial community structure profiling. *Processes* 8(6):723.

Patel, A., Hrůzová, K., Rova, U., Christakopoulos, P., & Matsakas, L. 2019. Sustainable biorefinery concept for biofuel production through holistic volarization of food waste. *Bioresource Technology* 294: 122247.

Patwa, A., Parde, D., Dohare, D., Vijay, R., & Kumar, R. 2020. Solid waste characterization and treatment technologies in rural areas: An Indian and international review. *Environmental Technology & Innovation* 20:101066.

Paulo, F., & Santos, L. 2021. Deriving valorization of phenolic compounds from olive oil by-products for food applications through microencapsulation approaches: a comprehensive review. *Critical Reviews in Food Science and Nutrition* 61(6): 920–945.

Qasim, W., & Mane, A.V. 2013. Characterization and treatment of selected food industrial effluents by coagulation and adsorption techniques. *Water Resources and Industry* 4: 1–12.

Ramachandran, S., Singh, S.K., Larroche, C., Soccol, C.R., & Pandey, A. 2007. Oil cakes and their biotechnological applications–a review. *Bioresource Technology* 98: 2000–2009.

Raposo, F., Borja, R., Rincon, B., & Jimenez, A.M. 2008. Assessment of process control parameters in the biochemical methane potential of sunflower oil cake. *Biomass Bioenergy* 32: 1235–1244.

Raposo, F., De la Rubia, M.A., Fernández-Cegrí, V., & Borja, R. 2012. Anaerobic digestion of solid organic substrates in batch mode: an overview relating to methane yields and experimental procedures. *Renewable and Sustainable Energy Reviews* 16(1): 861–877.

Rose P.K. 2022. Bioconversion of agricultural residue into biofuel and high-value biochemicals: Recent advancement. In: Nandabalan Y.K., Garg V.K., Labhsetwar N.K., Singh A. (eds) Zero Waste Biorefinery. Energy, Environment, and Sustainability 233–268. Springer, Singapore.

Saer, A., Lansing, S., Davitt, N.H., & Graves, R.E. 2013. Life cycle assessment of a food waste composting system: environmental impact hotspots. *Journal of Cleaner Production* 52: 234–244.

Salmani, Y., Mohammadi-Nasrabadi, F., & Esfarjani, F. 2022. A mixed-method study of edible oil waste from farm to table in Iran: SWOT analysis. *Journal of Material Cycles and Waste Management* 24: 111–121.

Sánchez, Ó.J., Ospina, D.A., & Montoya, S. 2017. Compost supplementation with nutrients and microorganisms in composting process. *Waste Management* 69: 136–153.

Singh, G., Daultani, Y., & Sahu, R. 2021. Investigating the barriers to growth in the Indian food processing sector. *OPSEARCH* 1–19.

Sorathiya, L.M., Fulsoundar, A.B., Tyagi, K.K., Patel, M.D., & Singh, R.R. 2014. Eco-friendly and modern methods of livestock waste recycling for enhancing farm profitability. *International Journal of Recycling of Organic Waste in Agriculture* 3(1): 50.

Sounni, F., Elgnaoui, Y., El Bari, H., Merzouki, M., & Benlemlih, M. 2021. Effect of mixture ratio and organic loading rate during anaerobic co-digestion of olive mill wastewater and agro-industrial wastes. *Biomass Conversion and Biorefinery* 1–21. https://doi.org/10.1007/s13399-021-01463-4

TERIENVIS, 2022. The Energy and Resources Institute and ENVIS Resource Partner on Renewable Energy and Climate Change. Viewed on 28 January 2022 http://terienvis.nic.in/index3.aspx?sslid=3850&subsublinkid=1318&langid=1&mid=1

Truong, L., Morash, D., Liu, Y., & King, A. 2019. Food waste in animal feed with a focus on use for broilers. *International Journal of Recycling of Organic Waste in Agriculture* 8(4): 417–429.

UNEP, 2021. Food Waste Index Report 2021. United Nations Environment Programme, Nairobi.

Valta, K., P. Damala, V. Panaretou, E. Orli, K. Moustakas and M. Loizidou. 2016. Review and assessment of waste and wastewater treatment from fruits and vegetables processing industries in Greece. Waste and Biomass Valorization. doi:10.1007/s12649-016-9672-4.

Vijay, V.K. 2011. Biogas enrichment and bottling technology for vehicular use. *Biogas Forum* 1(1):12–15.

Xu, F., Li, Y., Ge, X., Yang, L., & Li, Y. 2018. Anaerobic digestion of food waste–Challenges and opportunities. *Bioresource Technology* 247:1047–1058.

Xue, C., Liu, F., Xu, M., Zhao, J., Chen, L., Ren, J., Bai, F., & Yang, S.T. 2016. A novel in situ gas stripping-pervaporation process integrated with acetone-butanol-ethanol fermentation for hyper n-butanol production. *Biotechnology and Bioengineering* 113(1): 120–129.

Yadav, A., & Garg, V.K. 2011. Industrial wastes and sludges management by vermicomposting. *Reviews in Environmental Science and Biotechnology* 10(3): 243–276.

Yang, G., Zhang, P., Zhang, G., Wang, Y., & Yang, A. 2015. Degradation properties of protein and carbohydrate during sludge anaerobic digestion. *Bioresource Technology* 192: 126–130.

Yang, Y., Zheng, S., Ai, Z., & Jafari, M.M.M. 2021. On the prediction of biogas production from vegetables, fruits, and food wastes by ANFIS- and LSSVM-based Models. *BioMed Research International* 1–8. https://doi.org/10.1155/2021/9202127

Yasin, N.H.M., Mumtaz, T., & Hassan, M.A. 2013. Food waste and food processing waste for biohydrogen production: a review. *Journal of Environmental Management* 130: 375–385.

Yu, D., Wang, X., Fam, X., Ren, H., Hu, S., Wang, L., Shi, Y., Liu, N., & Qiao, N. 2018. Refned soybean oil wastewater treatment and its utilization for lipid production by the oleaginous yeast *Trichosporon fermentans*. *Biotechnology for Biofuels* 11(1): 1–12.

Yu, H., Xie, B., Khan, R., & Shen, G. 2019. The changes in carbon, nitrogen components and humic substances during organic-inorganic aerobic co-composting. *Bioresource Technology* 271: 228–235.

Zamanzadeh, M., Hagen, L.H., Svensson, K., Linjordet, R., & Horn, S.J. 2017. Biogas production from food waste via co-digestion and digestion-effects on performance and microbial ecology. *Scientific Reports* 7(1): 1–12.

Zhang, L., Lee, Y.W., & Jahng, D. 2011. Anaerobic co-digestion of food waste and piggery wastewater: Focusing on the role of trace elements. *Bioresource Technology* 102: 5048–5059.

Zhang, Z., O'Hara, I.M., Mundree, S., Gao, B., Ball, A.S., Zhu, N., Bai, Z., & Jin, B. 2016. Biofuels from food processing wastes. *Current Opinion in Biotechnology* 38: 97–105.

Zhao, X., Chen, J., & Du, F. 2012. Potential use of peanut by-products in food processing: a review. 49(5):521–529.

Zhu, J., Zheng, Y., Xu, F. & Li, Y. 2014. Solid-state anaerobic co-digestion of hay and soybean processing waste for biogas production. *Bioresource Technology* 154: 240–247.

Index

For Product Safety Concerns and Information please contact our EU
representative GPSR@taylorandfrancis.com
Taylor & Francis Verlag GmbH, Kaufingerstraße 24, 80331 München, Germany